Doing Bayesian Data Analysis

Doing Bayesian Data Analysis
A Tutorial with R and BUGS

John K. Kruschke

Department of Psychological & Brain Sciences
Indiana University
Bloomington, IN

AMSTERDAM • BOSTON • HEIDELBERG • LONDON
NEW YORK • OXFORD • PARIS • SAN DIEGO
SAN FRANCISCO • SINGAPORE • SYDNEY • TOKYO
Academic Press is an imprint of Elsevier

Academic Press is an imprint of Elsevier
30 Corporate Drive, Suite 400, Burlington, MA 01803, USA
Elsevier, The Boulevard, Langford Lane, Kidlington, Oxford, OX5 1GB, UK

Copyright © 2011 Elsevier Inc. All rights reserved.

No part of this publication may be reproduced or transmitted in any form or by any means, electronic or
mechanical, including photocopying, recording, or any information storage and retrieval system, without
permission in writing from the publisher. Details on how to seek permission, further information about the
Publisher's permissions policies and our arrangements with organizations such as the Copyright Clearance
Center and the Copyright Licensing Agency, can be found at our website: www.elsevier.com/permissions.

This book and the individual contributions contained in it are protected under copyright by the Publisher
(other than as may be noted herein).

Notices
Knowledge and best practice in this field are constantly changing. As new research and experience broaden
our understanding, changes in research methods, professional practices, or medical treatment may become
necessary.

Practitioners and researchers must always rely on their own experience and knowledge in evaluating and
using any information, methods, compounds, or experiments described herein. In using such information or
methods they should be mindful of their own safety and the safety of others, including parties for whom they
have a professional responsibility.

To the fullest extent of the law, neither the Publisher nor the authors, contributors, or editors, assume any
liability for any injury and/or damage to persons or property as a matter of products liability, negligence or
otherwise, or from any use or operation of any methods, products, instructions, or ideas contained in the
material herein.

Library of Congress Cataloging-in-Publication Data
Application submitted

British Library Cataloguing-in-Publication Data
A catalogue record for this book is available from the British Library.

ISBN: 978-0-12-381485-2

For information on all Academic Press publications
visit our Web site at *www.elsevierdirect.com*

Printed in the United States of America
10 11 12 13   9 8 7 6 5 4 3 2 1

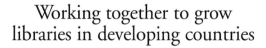

Working together to grow
libraries in developing countries

www.elsevier.com | www.bookaid.org | www.sabre.org

ELSEVIER   BOOK AID
International   Sabre Foundation

Dedicated to my mother, Marilyn A. Kruschke,
and to the memory of my father, Earl R. Kruschke,
who both brilliantly exemplified and taught sound reasoning.
And, in honor of my father,
who dedicated his first book to his children,
I also dedicate this book to mine:
Claire A. Kruschke and Loren D. Kruschke

Contents

Part 3   **Applied to the Generalized Linear Model**

CHAPTER 1

This Book's Organization

Read Me First!

CONTENTS

Oh honey I'm searching for love that is true,
But driving through fog is so dang hard to do.
Please paint me a line on the road to your heart,
I'll rev up my pick up and get a clean start.

1.1 REAL PEOPLE CAN READ THIS BOOK

This book explains how to actually *do* Bayesian data analysis, by real people (like you), for realistic data (like yours). The book starts at the basics, with notions of probability and programming, then progresses to advanced hierarchical models that are used in realistic data analysis. In other words, you do not need to already know statistics and programming. This book is speaking to a first-year graduate student or advanced undergraduate in the social or biological sciences: someone who grew up in Lake Wobegon,[1] but who is not the

---

[1]A popular weekly radio show on National Public Radio, called *A Prairie Home Companion*, features fictional anecdotes about a small town named Lake Wobegon. The stories, written and orated by Garrison Keillor, always end with the phrase, "And that's the news from Lake Wobegon, where all the women are strong, all the men are good looking, and all the children are above average." So, if you grew up there, ...

Doing Bayesian Data Analysis: A Tutorial with R and BUGS. DOI: 10.1016/B978-0-12-381485-2.00001-8
© 2011, Elsevier Inc. All rights reserved.

mythical being who has the previous training of a nuclear physicist and then decided to learn about Bayesian statistics.

This book provides broad coverage and ease of access. Section 1.3 describes the contents in a bit more detail, but here are some highlights. This book covers Bayesian analogues of all the traditional statistical tests that are presented in introductory statistics textbooks, including *t*-tests, analysis of variance (ANOVA), regression, chi-square tests, and so on. This book also covers crucial issues for *designing* research, such as statistical power and methods for determining the sample size needed to achieve a desired research goal. And you don't need to already know statistics to read this book, which starts at the beginning, including introductory chapters about concepts of probability and an entire chapter devoted to Bayes' rule. The important concept of hierarchical modeling is introduced with unique simple examples, and the crucial methods of Markov chain Monte Carlo sampling are explained at length, starting with simple examples that, again, are unique to this book. Computer programs are thoroughly explained throughout the book and are listed in their entirety, so you can use and adapt them to your own needs.

But wait, there's more. As you may have noticed from the beginning of this chapter, the chapters commence with a stanza of elegant and insightful verse composed by a famous poet. The quatrains[2] are formed of dactylic[3] tetrameter[4] or, colloquially speaking, "country waltz" meter. The poems regard conceptual themes of the chapter via allusion from immortal human motifs often expressed by country western song lyrics, all in waltz timing.

> If you do not find them to be all that funny,
> if they leave you wanting back all of your money,
> well honey some waltzing's a small price to pay,
> for all the good learning you'll get if you stay.

1.2 PREREQUISITES

There is no avoiding mathematics when doing statistics. On the other hand, this book is definitely not a mathematical statistics textbook in that it does not emphasize theorem proving, and any mathematical statistician would be totally bummed at the informality, dude. But I do expect that you are coming to this book with a dim knowledge of basic calculus. For example, if you understand expressions like $\int dx\, x = \frac{1}{2}x^2$, you're probably good to go. Notice

---

[2]*quatrain* [noun]: Four lines of verse. (Unless it's written "*qua* train," in which case it's a philosopher comparing something to a locomotive.)

[3]*dactylic* [adj.]: A metrical foot in poetry comprising one stressed and two unstressed syllables. (Not to be confused with a pterodactyl, which was a flying dinosaur and which probably sounded nothing like a dactyl unless it fell from the sky and bounced twice: THUMP-bump-bump.)

[4]*tetrameter* [noun]: A line of verse containing four metrical feet. (Not to be confused with a quadraped, which has four feet but is averse to lines.)

the previous sentence said "understand" the statement of the integral, not "generate" the statement on your own! When mathematical derivations are helpful for understanding, they will usually be presented with a thorough succession of intermediate steps, so you can actually come away feeling secure and familiar with the trip and destination rather than just feeling car sick after being thrown blindfolded into the trunk and driven around curves at high speed.

The beginnings of your journey will go more smoothly if you have had some basic experience programming a computer, but previous programming experience is not crucial. A computer program is just a list of commands that the computer can execute. For example, if you've ever typed an "=" in an Excel spreadsheet cell, you've written a programming command. If you've ever written a list of commands in Basic or Pascal or Java or any other language, then you're set. We will be using a language called *R*, which is *free*. More on R later.

1.3 THE ORGANIZATION OF THIS BOOK

This book has three major parts. The first part covers foundations: the basic ideas of probabilities, models, Bayesian reasoning, and programming in R.

The second main part covers all the crucial ideas of modern Bayesian data analysis while using the simplest possible type of data, namely, dichotomous data such as agree/disagree, remember/forget, male/female, and the like. Because the data are so simplistic, the focus can be on the Bayesian techniques. In particular, the modern techniques of "Markov chain Monte Carlo" (MCMC) are explained thoroughly and intuitively. And the ideas of hierarchical models are thoroughly explored. Because the models are kept simple in this part of the book, intuitions about the meaning of hierarchical dependencies can be developed in glorious graphic detail. This second part of the book also explores methods for planning how much data will need to be collected to achieve a desired degree of precision in the conclusions. This is called *sample size planning* or *power analysis*.

The third main part of the book applies the Bayesian methods to realistic data. The applications are organized around the type of data being analyzed, and the type of measurements that are used to explain or predict the data. For example, suppose you are trying to predict college grade point average (GPA) from high school Scholastic Aptitude Test (SAT) score. In this case the data to be predicted, the GPAs, are values on a *metric* scale, and the predictor, the SAT scores, are also values on a *metric* scale. Suppose, on the other hand, that you are trying to predict college GPA from gender. In this case the predictor is a *dichotomous* value, namely, male versus female. Different types of measurement scales require different types of mathematical models, but otherwise the underlying concepts are always the same. Table 14.1 (p. 385) shows various combinations of measurement scales and their corresponding models that are explored in detail in the third part of this book.

1.3.1 What Are the Essential Chapters?

The foundations established in the first part of the book, and the Bayesian ideas of the second part, are important to understand. The applications to particular types of data, in the third part, can be more selectively perused as needed. Within those parts, however, some chapters are essential:

- Chapter 4 explains Bayes' rule.
- Chapter 7 explains Markov chain Monte Carlo methods.
- Chapter 9 explains hierarchical models.
- Chapter 14 overviews the generalized linear model and various types of data analyses that can be conducted.

As an emphasis of the book is *doing* Bayesian data analysis, it is also essential to learn the programming languages R and BUGS:

- Section 2.3 introduces R.
- Section 7.4 introduces BUGS.

Finally, the ultimate purpose of data analysis is to convince other people that their beliefs should be altered by the data. The results need to be communicated to a skeptical audience, and therefore additional essential reading is as follows:

- Section 23.1 summarizes how to report a Bayesian data analysis.

Another important topic is the planning of research, as opposed to the analysis of data after they have been collected. Bayesian techniques are especially nicely suited for estimating the probability that specified research goals can be achieved as a function of the sample size for the research. Therefore, although it might not be essential on a first reading, it is essential eventually to read the following:

- Chapter 13 regarding power analysis.

Figure 1.1 puts these recommendations in a convenient reference box, rearranged to match the order presented in the book.

1.3.2 Where's the Equivalent of Traditional Test X in This Book?

Because many readers will be coming to this book after having already been exposed to traditional 20th-century statistics that emphasize null hypothesis significance testing (NHST), this book provides Bayesian approaches to the usual topics in NHST textbooks. Table 1.1 lists various tests covered by standard introductory statistics textbooks, along with their Bayesian analogues. If you have been previously contaminated by NHST but want to know how to do an analogous Bayesian analysis, Table 1.1 may be useful.

- Section 2.3 introduces R.
- Chapter 4 explains Bayes' rule.
- Chapter 7 explains Markov chain Monte Carlo methods.
- Section 7.4 introduces BUGS.
- Chapter 9 explains hierarchical models.
- Chapter 13 explains varieties of power analysis.
- Chapter 14 overviews the generalized linear model and various types of data analyses that can be conducted.
- Section 23.1 summarizes how to report a Bayesian data analysis.

FIGURE 1.1
Essential sections of the book.

Table 1.1 Bayesian Analogues of 20th-Century Null Hypothesis Significance Tests

Traditional Analysis Name	Bayesian Analogue
t-test for a single mean	Chapter 15
t-test for two independent groups	Chapter 18 (Section 18.3)
Simple linear regression	Chapter 16
Multiple linear regression	Chapter 17
Oneway ANOVA	Chapter 18
Multifactor ANOVA	Chapter 19
Logistic regression	Chapter 20
Ordinal regression	Chapter 21
Binomial test	Chapters 5 to 9, 20
Chi-square test (contingency table)	Chapter 22
Power analysis (sample size planning)	Chapter 13

A superficial conclusion from Table 1.1 might be, "Gee, the table shows that traditional statistical tests do something analogous to Bayesian analysis in every case: therefore, it's pointless to bother with Bayesian analysis." Such a conclusion would be wrong. First, traditional NHST has deep problems, some of which are discussed in Chapter 11. Second, Bayesian analysis yields richer and more informative inferences than NHST, as will be shown in numerous examples throughout the book.

Computer programs, solutions to exercises, and other supplementary materials can be obtained at the book's web sites: http://www.elsevierdirect.com/ISBN/9780123814852/Doing-BayesianDataAnalysis and http://www.indiana.edu/~kruschke/DoingBayesianDataAnalysis/.

1.4 GIMME FEEDBACK (BE POLITE)

I have worked thousands of hours on this book, and I want to make it better. If you have suggestions regarding any aspect of this book, please do e-mail me: JohnKruschke@gmail.com. Let me know if you've spotted egregious errors or innocuous infelicities, typos, or thoughtos. Let me know if you have a suggestion for how to clarify something. Especially let me know if you have a good example that would make things more interesting or relevant. I'm also interested in complete raw data from research that is interesting to a broad audience and that can be used with acknowledgment but without fee. Let me know also if you have more elegant programming code than what I've cobbled together. The outside margins of these pages are intentionally made wide so that you have room to scribble your ridicule and epithets before rephrasing them into kindly stated suggestions in your e-mail to me. Rhyming couplets are especially appreciated. If I don't respond to your e-mail in a timely manner, it is only because I can't keep up with the deluge of fan mail, not because I don't appreciate your input. Thank you in advance!

1.5 ACKNOWLEDGMENTS

This book has been six years in the making, and many colleagues and students have provided helpful comments. The most extensive comments have come from Drs. Luiz Pessoa, Mike Kalish, Jerry Busemeyer, and Adam Krawitz; thank you all! Particular sections were insightfully improved by helpful comments from Drs. Michael Erickson, Robert Nosofsky, Geoff Iverson and James L. (Jay) McClelland. Various parts of the book benefited indirectly from communications with Drs. Woojae Kim, Charles Liu, Eric-Jan Wagenmakers, and Jeffrey Rouder. Leads to data sets were offered by Drs. Teresa Treat and Michael Trosset, among others. Very welcome supportive feedback was provided by Dr. Michael Lee, and also by Dr. Adele Diederich. Many colleagues provided a Bayesian-supportive working environment, including Drs. Richard Shiffrin, Jerome Busemeyer, Peter Todd, James Townsend, Robert Nosofsky, and Luiz Pessoa. Other department colleagues have been supportive of integrating Bayesian statistics into the curriculum, including Drs. Linda Smith and Amy Holtzworth-Munroe. Various teaching assistants have provided helpful comments; in particular I especially thank Drs. Noah Silbert and Thomas Wisdom for their excellent assistance. As this book has evolved over the years, suggestions have been contributed by numerous students, including Aaron Albin, Thomas Smith, Sean Matthews, Thomas Parr, Kenji Yoshida, Bryan Bergert, and perhaps dozens of others who have contributed insightful questions or comments that helped me tune the presentation in the book. To all the people who have made suggestions but whom I have inadvertently forgotten to mention by name, I extend my apologies and appreciation.

The Basics
Parameters, Probability, Bayes' Rule, and R

CHAPTER 2

Introduction

Models We Believe In

I just want someone who I can believe in,
Someone at home who will not leave me grievin'.
Show me a sign that you'll always be true,
and I'll be your model of faith and virtue.

Inferential statistical methods help us decide what to believe in. With inferential statistics, we don't just introspect to find the truth. Instead, we rely on data from observations. Based on the data, what should we believe in? Should we believe that the tossed coin is fair if it comes up heads in 7 of 10 flips? Should we believe that we have cancer when the test comes back positive? Should we believe that she loves me when the daisy has 17 petals? Our beliefs can be

Doing Bayesian Data Analysis: A Tutorial with R and BUGS. DOI: 10.1016/B978-0-12-381485-2.00002-X
© 2011, Elsevier Inc. All rights reserved.

modified when we have data, and this book is about techniques for making inferences *from* data *to* uncertain beliefs.

There might be some beliefs that cannot be decided by data, but such beliefs are dogmas that lie (double entendre intended) beyond the reach of evidence. If you are wondering about a belief that has no specific implications for concrete facts in the observable world, then inferential statistics won't help.

Why do we need hefty tomes full of mathematics to help us make decisions based on data? After all, we make lots of decisions every day without math. If we're driving, we look at the signal light and effortlessly decide whether it's red or green. We don't (consciously) go through a laborious process of mathematical statistics and finally conclude that it is probably the case that the light is red. Two attributes of this situation make the decision easy. First, the data about the light are numerous. An unobstructed view of the light results in a whole lot of photons striking our eyes. Second, there are only a few possible beliefs about the light that make distinct predictions about the photons: If the light is red, the photons are rather different than if the light is green. Consequently, the decision is easy because there is little variance in the data and little uncertainty across possible beliefs.

The math is most helpful when there is lots of variance in the data and lots of uncertainty in our beliefs. Data from scientific experiments, especially those involving humans or animals, are unmitigated heaps of variability. Theories in science tend to be rife with parameters of uncertain magnitude, and competing theories are numerous. In these situations, the mathematics of statistical inference provide precise numerical bounds on our uncertainty. The math allows us to determine accurately what the data imply for different possible beliefs. The math can tell us exactly how likely or unlikely each possibility is, even when there is an infinite spectrum of possibilities. It is this power of precisely defining our uncertainty that makes inferential statistics such a useful tool, worth the effort of learning.

2.1 MODELS OF OBSERVATIONS AND MODELS OF BELIEFS

Suppose we flip a coin to decide which team kicks off. The teams agree to this procedure for deciding the kickoff because they believe that the coin is fair. But how do we determine whether the coin really is fair? Even if we could study the exact minting process of the coin and x-ray every nuance of the coin's interior, we would still need to test whether the coin really is fair when it's actually flipped. Ultimately, all we can do is flip the coin a few times and watch its behavior. From these observations, we can modify our beliefs about the fairness of the coin.

Suppose we have a coin from our friend the numistmatist.[1] We notice that on the obverse is embossed the head of Tanit (of ancient Carthage) and on the reverse side is embossed a horse. The coin is gold and shows the date 350 BCE. Do you believe that the coin is fair? Maybe you do, but maybe you're not very certain.[2] Let's flip it a few times. Suppose we flip it 10 times and we obtain two heads and eight tails. Now what do you think? Do you have a suspicion that maybe the coin is biased to come up tails more often than heads?

We've seen that the coin comes up horse tails a lot. Whoa! Let's dismount and have a heart-to-heart 'round the campfire. In that simple coin-flipping scenario we have made two sets of assumptions. First, we have assumed that the coin has some inherent fairness or bias that we can't directly observe. All we can actually observe is an inherently probabilistic effect of that bias, namely, whether the coin comes up heads or tails on any given flip. We've made lots of assumptions about exactly how the observable head or tail relates to the unobservable bias of the coin. For instance, we've assumed that the bias stays the same, flip after flip. We've assumed that the coin can't remember what way it came up last flip, so that its flip this time is uncorrupted by its previous landings. All these assumptions are about the process that converts the unobservable bias into a probabilistic observable event. This collection of assumptions about the coin-flipping process is our model of the head-tail observations.

The second set of assumptions is about our beliefs regarding the bias of the coin. We assume that we believe most strongly in the coin being fair, but we also allow for the possibility that the coin could be biased. Thus, we have a set of assumptions about how likely it is for the coin to be fair or to be biased to different amounts. This collection of assumptions is our model of our beliefs.

When we want to get specific about our model assumptions, then we have to use mathematical descriptions. A "formal" model uses mathematical formulas to precisely describe something. In this book, we'll almost always be using formal models, so *whenever the term "model" comes up, you can assume it means a mathematical description*. In the context of statistical models, the models are typically models of probabilities. Some models describe the probabilities of observable events (e.g., we can have a formula that describes the probability that a coin will come up heads). Other models describe the extent to which we believe in various underlying possibilities (e.g., we can have a formula that describes how much we believe in each possible bias of the coin).

---

[1] *Numistmatist* [noun]: A person who studies or collects coins.
[2] A tale about coins marked BCE is a well-known joke because any coin actually minted BCE could not have been marked BCE at the time it was minted. But even a coin marked with a bogus date might be a fair flipper.

Mathematical models are formulas with variables. Some variables have values supplied as data from the world. For example, the data variable of a coin flip can have the values head or tail. But other variables refer to underlying characteristics, such as the bias of the coin. Variables that refer to underlying characteristics are called *parameters*. A parameter can take on many possible values. For example, the bias parameter of a coin can take on a value anywhere on the continuum between zero and one.

2.1.1 Prior and Posterior Beliefs

We could believe that the coin is fair—that is, that the probability of coming up heads is 50%. We could instead have other beliefs about the coin, especially if it's dated 350 BCE, which no coin would be labeled if it were really minted BCE (because the people alive in 350 BCE didn't yet know they were BCE). Perhaps, therefore, we also think it's possible for the coin to be biased to come up heads 20% of the time, or 80% of the time. Before observing the coin flips, we might believe that each of these three dispositions is equally likely—that is, we believe that there is a one-in-three chance that the bias is 20%, a one-in-three chance that the bias is 50%, and a one-in-three chance that the bias is 80%.

After flipping the coin and observing 2 heads in 10 flips, we will want to modify our beliefs. It makes sense that we should now believe more strongly that the bias is 20%, because we observed 20% heads in the sample. This book is about determining *exactly* how much more strongly we should believe that the bias is 20%.

Before observing the flips of the coin, we had certain beliefs about the possible biases of the coin. These are called *prior* beliefs because they are our beliefs before taking into account some particular set of observations. After observing the flips of the coin, we modified our beliefs. These are called the *posterior* beliefs because they are computed after taking into account a particular set of observations. Bayesian inference gets us from prior to posterior beliefs.

There is an infelicity in the terms "prior" and "posterior," however. The terms connote the passage of time, as if the prior beliefs were held temporally before the posterior beliefs. But that is a misconception. There is no temporal ordering in the prior and posterior beliefs! Rather, the prior is simply the distribution of beliefs we hold by *excluding* a particular set of data, and the posterior is the distribution of beliefs we hold by *including* the set of data. Despite this misleading temporal connotation, the terms "prior" and "posterior" are firmly entrenched in the literature, so we'll use them too.

2.2 THREE GOALS FOR INFERENCE FROM DATA

When we make observations of the world, we typically have one of three goals in mind. Each of these goals can be illustrated with the coin-flipping scenario.

2.2.1 Estimation of Parameter Values

One goal we may have is deciding the extent to which we should believe in each of the possible values of an underlying parameter. In the case of the coin, we may use the observed data to determine the extent to which we should believe that the bias is 20%, 50%, or 80%. What we are determining is how much we believe in each of the available parameter values. In most real applications, we allow for a continuum of possible biases from zero to one, and the Bayesian mathematics reveal the credibility of every possible value on that continuum.

Because the flip of the coin is a random process, we cannot be certain of the underlying true probability of getting heads. So our posterior beliefs are an estimate. The posterior beliefs typically increase the magnitude of belief in some parameter values, while lessening the degree of belief in other parameter values. So this process of shifting our beliefs in various parameter values is called *estimation of parameter values*.

2.2.2 Prediction of Data Values

Another goal we may have is predicting other data values, given our current beliefs about the world. For example, given that we have just observed the ball leaving the pitcher's hand and we now believe it's a curve ball, where do we predict the ball will be when it gets near the plate?

Notice that "prediction" is another of those words that connotes temporal order but isn't always used that way in statistics. Prediction simply means inferring the values of some missing data based on some other included data, regardless of the actual temporal relationship of the included and missing data.

An ability to make specific predictions is one of the primary uses of mathematical models. Models help us predict the effectiveness of a flu vaccine when distributed to the general public. Models help us predict the paths of hurricanes. And models can help us predict whether the next coin flip will be heads or tails.

In Bayesian inference, to predict data values we typically take a weighted average of our beliefs. We let each belief make its individual prediction, and then we weigh each of those predictions according to how strongly we believe in them. For example, if we believe strongly that the coin has a bias of 20% heads and we only weakly believe in biases of 50% or 80%, then our prediction will be a mixture of the three beliefs weighted strongly toward 20%.

2.2.3 Model Comparison

A third goal of statistical inference is model selection, also known as model comparison. If we have two different models of how something might happen, then an observation of what really does happen can influence which model we believe in most. What Bayesian inference tells us is how to shift our magnitude of belief across the available models.

As a somewhat contrived example, suppose we have two different models of the coin. One model assumes what we've described before, that the coin could have biases of 20%, 50%, or 80% heads. The second model assumes that the coin is either a perfectly fair coin or else it's a trick coin with two heads or two tails. This model allows the coin to have biases of 0%, 50%, or 100% heads. Notice that the second model assumes different available parameter values than the first model.

After observing 10 flips that had 2 tails, which model do we believe in more? Let's think about the second model. Because our observations were not purely heads or purely tails, we know that the posterior beliefs for that model must load all belief on 50% heads, because we did not observe all heads or all tails. This model then is stuck asserting that the 10 observed flips with just 2 tails were generated by a fair coin, which is not likely. The first model, on the other hand, has the belief of 20% heads available to it, which can generate the observed data with high likelihood. Therefore, we should believe the first model more strongly than the second. The mathematics of Bayesian inference can tell *exactly* how much more to believe the first model than the second.

One of the nice qualities of Bayesian model comparison is that it intrinsically adjusts for model complexity. More complex models usually will fit data better than simple models, merely because the complex models have more flexibility. Unfortunately, more complex models will also fit random noise better than simpler models. We are interested in the model that best fits the real trends in the data, not just the model that best fits the noise. As we will see in later chapters, Bayesian methods naturally take into account the complexity of models.

2.3 THE R PROGRAMMING LANGUAGE

With this book you will learn how to actually *do* Bayesian statistics. For any but the simplest models, that means using a computer. Because the computer results are so central to doing real Bayesian statistics, examples of using the R computer programming language will be integrated into the simplest "toy" problems, so that R will not be an extra hurdle later.

The R language is great at doing Bayesian statistics for a number of reasons. First, it's free! You can get it via the web and easily install it on your computer. Second, it's already a popular language for doing Bayesian statistics, so there are lots of resources available. Third, it is a powerful and easy general-purpose computing language, so you can use it for many other applications too.

2.3.1 Getting and Installing R

It's easy to get and install R, but there are a lot of optional details in the process, and the hardest part of installation is figuring out which little details do *not* apply to you!

Basic installation is easy. Go to `http://cran.r-project.org`. At the top of that web page is a section headed "Download and Install R" followed by three links: Linux, MacOS, and Windows. These three links refer to the type of operating system used on your computer. Although R can be installed on any of those three types of operating systems, it turns out that another package we'll be using extensively, called *BUGS*, only works on Windows. Macintosh users report that if they first install the freeware WINE (WINE Is Not an Emulator) from `www.winehq.org`, and then install R and BUGS from within WINE, everything works seamlessly. The same is supposed to be true of Unix/Linux users. *Therefore, from here on, I will assume that you are using Windows or WINE.*

After clicking the Windows link, you will see a page with two links: base and contrib. Click "base." At the top of the resulting page is the link for downloading R for Windows. Click it and follow the installation instructions. (If you are using MacOS or Linux, download the Windows executable and install it from within WINE.) There may be a few details that you have to navigate on your own, but remember that centuries ago lots of people crossed the oceans in tiny wooden boats without any electronics, so you can navigate the small perils of R installation.[3]

2.3.2 Invoking R and Using the Command Line

Invoke R by double-clicking the R icon in Windows. A user interface should open. In particular, one of the windows will be a command line interface. This window is constantly attentive to your every whim (well, every whim you can express in R). All you have to do is type in your wish and R will execute it as a command. For example, if you type in `show(2 + 2)`, followed by pressing the Enter key, R will reply with 4. In fact, if you just type in `2 + 2`, without the `show` function, R will still reply with 4.

A *program* (a.k.a. *script*) is just a list of commands that R executes. For example, you could first type in x = 2 and then, as a second command, type in x + x, to which R will reply 4. This is because R assumes that when you type in x + x, you are really asking for the value of the sum of the value of x with the value of x, not an algebraic reformulation such as 2 ∗ x that some systems assume.[4]

---

[3] Of course, lots of people failed to cross the ocean, but that's different.

[4] You might ask, if R were a Bayesian reasoner, when you typed in 2+2, wouldn't it reply something like "Well, I believe most strongly in 4, but the answer might be a little higher or a little lower"? A Bayesian reasoner would reply that way only if uncertainty were introduced somewhere along the way. If the values to be added were uncertain or if summation itself were an uncertain process, then the sum would be uncertain too. R assumes that numerals and arithmetic have no uncertainty.

2.3.3 A Simple Example of R in Action

As a simple example of what R can do, let's plot a quadratic function: $y = x^2$. What looks like a smooth curve on a graph is actually a set of points connected by straight lines, but the lines are so small that the graph looks like a smooth curve. So what we have to do is tell R where all those densely packed points go.

Every point is specified by its x and y coordinates, so we have to provide R with a list of x values and a list of corresponding y values. Let's arbitrarily select x values from -2 to $+2$, separated by intervals of 0.1. We have R set up the list of x values by using the built-in *sequence* function: `x = seq( from = -2 , to = 2 , by = 0.1 )`. Inside R, the variable x now refers to a list of 31 values: -2.0, -1.9, $-1.8, \ldots, +2.0$. This sort of ordered list of numerical values is called a *vector* in R. In this textbook, programming commands are typeset in a distinctive font, `like this`, to distinguish them from English prose and to help demarcate the scope of the programming command when it is embedded in an English sentence.

Next we tell R to create the corresponding y values. We type in `y = x^2`. R interprets "^" to mean raising values to a power. Inside R, the variable y now refers to a vector of 31 values: 4.0, 3.61, $3.24, \ldots, 4.0$.

All that remains is telling R to make a plot of the x and y points, connected by lines. Conveniently, R has a built-in function called `plot`, which we call by entering `plot( x , y , type="l" )`. The segment of code, `type="l"`, tells R to plot lines only and not points. If we left that part of the command out, then R would plot only points by default, not the connecting lines. The resulting plot is shown in Figure 2.1, and the complete R code that generated the graph is shown here (SimpleGraph.R):

```
1  x = seq( from = -2 , to = 2 , by = 0.1 )    # Specify vector of x values.
2  y = x^2                                      # Specify corresponding y values.
3  plot( x , y , type = "l" )                   # Make a graph of the x,y points.
4  dev.copy2eps( file = "SimpleGraph.eps" )     # Save the plot to an EPS file.
```

This code listing has a few display features that will be standard throughout this book. First, the listing begins with the script's filename. This filename is relevant when you want to find the script on the website for this book. The filename is also in the index of this book. Second, the listing has line numbers in the margin. These line numbers are not part of the script, they are only part of the printed display. The line numbers are especially useful for the long programs encountered later in the book.

The last line of the program uses the function `dev.copy2eps` to save the graph to a file using a format called *encapsulated PostScript* (eps). The command might not work on all installations of R, because some systems have not installed the eps driver. If the command doesn't work on your system, don't worry! You can save the graph in other formats, such as pdf. But how? Type

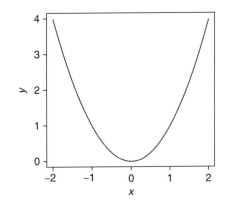

FIGURE 2.1
A simple graph drawn by R. The R code
that generated this graph is on p. 16.

help(`dev.copy2eps`). Although the resulting help text can be somewhat cryptic, it reveals alternative commands such as dev.copy2pdf, which save the graph in pdf format.

2.3.4 Getting Help in R

The plot function has many optional details that you can specify, such as the axis limits and labels, font sizes, and so on. You can learn more about those details by getting help from R. Type the command help(`plot`) and you can read all about it. In particular, the information directs you to another command, par, that controls all the plot parameters. To learn about it, type help(`par`). In general, it actually *is* helpful to use the help command. To get a list of all sorts of online documentation, much of it written in readable prose instead of telegraphic lists, type help.start(). Another useful way to find help with R is through web search. In your favorite web search engine, type in the R terms you want help with. When searching with the term "R" it helps to enclose it in square brackets, like this: [R].

A highly recommended resource is a summary of basic R commands that can be found on a compact list available at this URL: http://cran.r-project.org/doc/contrib/Short-refcard.pdf. Other versions of reference cards can be found by searching the web with the phrase "R reference card."

Much of the time you'll learn about features of R on an as-needed basis, and usually that means you'll look for examples of the sort of thing you want to do and then imitate the example. (Or the example might at least provoke you into realizing that there is a better way to do it than the method in the example!) Therefore, most of the examples in this book have their full R code included. The hope is that it will help you to study those examples as needed.

If you are already familiar with the programming languages Matlab or Python, you can find thesauruses of synonymous commands in R at this website: http://mathesaurus.sourceforge.net.

2.3.5 Programming in R

Instead of typing all the commands one at a time at the command line, you can type them into a text document and then have R execute the document. The document is called a *program*.

Some important points for newbie programmers:

- Be sure you save your program in a file where you can find it again, with a filename that is easy to recognize weeks later.
- Be sure to save the program every time you make a small *working* change.
- If you are about to make a big change, save the current working version and start the modified version with a new filename. This way, when your modified version doesn't work, you still have the old working version to fall back on.
- Put lots of explanatory comments in your code, so that you can understand what the heck you were doing when you come back to the program months later. To include a comment in a program, simply type a "#" character, and the R interpreter will ignore everything after that character, on the same line. You can see examples of comments in the code listings at the ends of the chapters.

2.3.5.1 Editing Programs in R

Programs such as the one on p. 16 can be typed in any text-based word processor, but it can help enormously to use an editor that is "R friendly." The basic editor built into R is okay for small programs, but larger programs become unwieldy. One free editor is Tinn-R, available at www.sciviews.org/Tinn-R. It has many useful features such as displaying comments in different colors and fonts than the command lines, highlighting matching parentheses, and the like.

If you are using Windows Vista or Windows7, and if Tinn-R will not communicate with R, try the following fix: Browse to the folder Program Files > R > R-2.*. Right-click the "etc" folder and open its Properties. On the security tab, change the permissions so that the folder contents can be written to by you. This permits the Rconsole.site file to be overwritten by Tinn-R. Now go to Tinn-R, and from its menu click R > Configure > Permanent (Rprofile.site).

2.3.5.2 Variable Names in R

You should name variables meaningfully, so that the programming commands are easy for a reader to understand. If you name your variables cryptically, you will curse your poor judgment when you return to the program weeks later and you have no idea what your program does.

You can use fairly long, descriptive names. If the names get too long, however, then the program becomes unwieldy to type and read. For example, suppose you want to name the crucial final output of a program. You could name it `tempfoo`, but that's not very meaningful, and might even lead you to think that the variable is unimportant. Instead, you could name it `the-CrucialFinalOutputThatWillChangeTheWorldForever`, but that would be burdensome to type and read as it is reused in the program. So you might best name it something like `finalOutput`, which is meaningful but not too long.

Computer programmers typically use a naming convention called *camelBack notation*. This is a way of connecting several words into a contiguous variable name without using spaces between words. For example, suppose you want to name a variable "final output." You are not allowed to name a variable with a space in it because computer compilers interpret spaces as separators of variables. One way to avoid using spaces is to connect the words with explicit connectors such as an underscore or a dot, like this: `final_output` or `final.output`. Many programmers do use those naming conventions. But the underscore notation can be difficult to read sometimes, and the dot notation is interpreted by some programming languages (other than R) as referring to subcomponents of structured variables, which confuses people who are familiar with that meaning of a dot. Therefore, the spaces are simply dropped, with successive words capitalized: `finalOutput`. The initial word is typically not capitalized, but some people have different uses for initial-capitalized variable names. R is case sensitive: the variable `myVar` is different than the variable `myvar`!

I will try to use camelBack notation in all the programs in this book. I may occasionally lapse from Bactrian beauty, instead slithering into snake-back notation (`finaloutput`) or gooseneck notation (`final_output`) or ant notation (`final.output`). If you see these lower forms, quietly shun them, knowing that when you create your own programs, you will use the more highly evolved dromedary design.

2.3.5.3 Running a Program
Running a program is easy, but exactly how to do it depends on how you are interacting with R.

If you are working in R's command console, first make sure that R has its working directory specified as the folder in which the program resides. Do this by selecting menu items *File* then *Change dir...* and browsing to the appropriate folder in the pop-up dialogue box. Then you can run the program by "`source`"-ing it. Do this by selecting menu items *File* then *Source R code...* and browsing to the program in the pop-up dialogue box. You can also type the `source("yourProgramName.R")` command directly at the command line.

You will more often be working interactively with a program that is open in an editing window. To open an R editing window, select menu items *File* then

New script or *Open script*. Once you have a program open in an editing window, you can run the program, or merely a few lines within it, by selecting menu items *Edit* then *Run line or selection* or *Run all*. If you run the program from an editing window, every command is echoed in the command window. If you run the program by sourceing it in the command window, then the program is executed without displaying the lines of code.

If you are working on a program in the editor Tinn-R, you will see menu buttons on the top tool bar that are equivalent to the R commands reviewed here. There is a button for setting the current working directory, there is another button for sourceing the program, and there is yet another button for running only the lines selected in the program being edited.

2.4 EXERCISES

Exercise 2.1. [Purpose: To get you to think about what beliefs can be altered by inference from data.] Suppose I believe that exactly 47 angels can dance on my head. (These angels cannot be seen or felt in any way.) Can you provide any evidence that would change my belief? Suppose I believe that exactly 47 anglers[5] can dance on the floor of the bait shop. Is there any evidence you could provide that would change my belief?

Exercise 2.2. [Purpose: To get you to actively manipulate mathematical models of probabilities. Notice, however, that these models have no parameters.] Suppose we have a four-sided die from a board game. (On a tetrahedral die, each face is an equilateral triangle. When you roll the die, it lands with one face down and the other three visible as the faces of a three-sided pyramid. To read the value of the roll, you pick up the die and see what landed face down.) One side has one dot, the second side has two dots, the third side has three dots, and the fourth side has four dots. Denote the value of the bottom face as x. Consider the following three mathematical descriptions of the probabilities of x. Model A: $p(x) = 1/4$. Model B: $p(x) = x/10$. Model C: $p(x) = 12/(25x)$. For each model, determine the value of $p(x)$ for each value of x. Describe in words what kind of bias (or lack of bias) is expressed by each model.

Exercise 2.3. [Purpose: To get you to think actively about how data cause beliefs to shift.] Suppose we have the tetrahedral die introduced in the previous exercise, along with the three candidate models of the die's probabilities. Suppose that initially we are not sure what to believe about the die. On the one hand, the die might be fair, with each face landing with the same probability. On the other hand, the die might be biased, with the faces that have more dots landing down more often (because the dots are created by embedding heavy jewels in

---

[5] *Angler* [noun]: A person who fishes with a hook and line.

the die, so that the sides with more dots are more likely to land on the bottom). On yet another hand, the die might be biased such that more dots on a face make it less likely to land down (because maybe the dots are bouncy rubber or protrude from the surface). So, initially, our beliefs about the three models can be described as $p(A) = p(B) = p(C) = 1/3$. Now we roll the die 100 times and find these results: #1's = 25, #2's = 25, #3's = 25, and #4's = 25. Do these data change our beliefs about the models? Which model now seems most likely? Suppose when we rolled the die 100 times, we found these results: #1's = 48, #2's = 24, #3's = 16, and #4's = 12. Now which model seems most likely?

Exercise 2.4. [Purpose: To actually do Bayesian statistics, eventually, and the next exercises, immediately.] Install R on your computer. (And if that's not exercise, I don't know what is.)

Exercise 2.5. [Purpose: To be able to record and communicate the results of your analyses.] Run the code listed on p. 16 (SimpleGraph.R). The last line of the code saves the graph to a file in a format called *encapsulated PostScript* (abbreviated as eps), which your favorite word processor might be able to import. If your favorite word processor does not import eps files, then read the R documentation and find some other format that your word processor likes better; try `help('dev.copy2eps')`. You may find that you can just copy and paste the displayed graph directly into your document, but it can be useful to save the graph as a stand-alone file for future reference. Include the code listing and the resulting graph in a document that you compose using a word processor of your choice.

Exercise 2.6. [Purpose: To gain experience with the details of the command syntax within R.] Adapt the code listed on p. 16 (SimpleGraph.R). so that it plots a cubic function ($y = x^3$) over the interval $x \in [-3, +3]$. Save the graph in a file format of your choice. Include a listing of your code, commented, and the resulting graph.

What Is This Stuff Called Probability?

Doing Bayesian Data Analysis: A Tutorial with R and BUGS. DOI: 10.1016/B978-0-12-381485-2.00003-1
© 2011, Elsevier Inc. All rights reserved.

Oh darlin' you change from one day to the next,
I'm feelin' deranged and just plain ol' perplexed.
I've learned to put up with your raves and your rants,
The mean I can handle but not variance.

Inferential statistical techniques provide precision to our uncertainty about possibilities. Uncertainty is measured in terms of *probability*, so we have to establish the basic properties of probability before we can make inferences about it. This chapter introduces the basic ideas of probability. If this chapter seems too abbreviated for you, an excellent beginner's introduction to the topics of this chapter was written by Albert and Rossman (2001, pp. 227–320).

3.1 THE SET OF ALL POSSIBLE EVENTS

Suppose I have a coin that I am going to flip. How likely is it to come up a head? How likely is it to come up a tail? How likely is it to come up a torso? Notice that when we contemplate the likelihood of each outcome, we have a space of all possible outcomes in mind. Torso is not one of the possible outcomes. Notice also that a single flip of a coin can result in only one outcome; it can't be both heads and tails in a single flip. The outcomes are mutually exclusive.

Whenever we ask how likely an event is, we always ask with a set of possible events in mind. This set exhausts all possible outcomes, and the outcomes are all mutually exclusive. This set is called the *sample space*.

In the previous chapter, we talked about the probability of a flipped coin coming up heads. The probability of coming up heads can be denoted with parameter label θ (Greek letter theta); for example, a coin is fair when $\theta = 0.5$ (spoken "theta equals point five"). We also have talked about the degree of belief that the coin is fair. The degree of belief about a parameter can be denoted $p(\theta)$. If the coin is minted by the federal government, we might have a strong belief that the coin is fair; for example, $p(\theta = 0.5) = 0.99$, spoken "the probability (or degree of belief) that theta equals 0.5 is 99 percent."

Both "probability" (of head or tail) and "degree of belief" (in fairness) refer to sample spaces. The sample space for flips of a coin consists of two possible events: head and tail. The sample space for coin bias consists of a continuum of possible values: $\theta = 0$, $\theta = 0.01$, $\theta = 0.02$, $\theta = 0.03$, and all values in between, up to $\theta = 1$. When we flip a given coin, we are sampling from the space of head or tail. When we grab a coin at random from a sack of coins, we are sampling from the space of possible biases.

3.1.1 Coin Flips: Why You Should Care

The fairness of a coin might be hugely consequential for high stakes games, but it isn't often in life that we flip coins. So why bother studying the statistics of coin flips?

Because coin flips are a surrogate for myriad other real-life events that we care about. For a given type of heart surgery, what is the probability that patients survive more than 1 year? For a given type of drug, what is the probability of headache as a side effect? For a particular training method, what is the probability of at least 10% improvement? For a survey question, what is the probability that people will agree or disagree? In a two-candidate election, what is the probability that a person will vote for each candidate?

Whenever we are discussing coin flips, keep in mind that we could be talking about some domain in which you are actually interested! The coins are merely a generic representative of a universe of analogous applications.

3.2 PROBABILITY: OUTSIDE OR INSIDE THE HEAD

Sometimes we talk about probabilities of events that are "out there" in the world. The face of a flipped coin is such an event: We can observe the outcome of a flip, and the probability of coming up heads can be estimated by observing a bunch of flips.

But sometimes we talk about probabilities of events that are not so clearly "out there" and instead are just possible beliefs "inside the head." Our belief about the fairness of a coin is an example of an event inside the head. (The coin may have an intrinsic bias, but right now I'm referring to our *belief* about the bias.) Our beliefs refer to a space of mutually exclusive and exhaustive possibilities, but it might be strange to say that we randomly sample from our beliefs, like we randomly sample from a sack of coins. Nevertheless, the mathematical properties, of probabilities outside the head and beliefs inside the head, are the same in their essentials, as we will see.

3.2.1 Outside the Head: Long-Run Relative Frequency

For events outside the head, it's intuitive to think of probability as being the long-run relative frequency of each possible event. For example, if I say that for a fair coin the probability of heads is 0.5, what I mean is that if we flipped the coin many times, about 50% of the flips would come up heads. In the long run, after flipping the coin many, many times, the relative frequency of heads would be nearly 0.5.

We can determine the long-run relative frequency two different ways. One way is to approximate it by actually sampling from the space many times and tallying the number of times each event happens. A second way is by deriving it mathematically. We now explore these two methods in turn.

3.2.1.1 Simulating a Long-Run Relative Frequency

Suppose we want to know the long-run relative frequency of getting heads from a fair coin. It might seem blatantly obvious that we should get about 50% heads in any long sequence of flips. But let's pretend that it's not so obvious. All we know is that there's some underlying process that generates an "H" or a "T" when we sample from it. The process has a parameter called θ, whose value is $\theta = 0.5$. If that's all we know, then we can approximate the long-run probability of getting an "H" by simply repeatedly sampling from the process. We sample from the process N times, tally the number of times an "H" appeared, and estimate the probability of H by the relative frequency: est. $\theta = \hat{\theta} = \#H/N$.

It gets tedious and time consuming to manually sample a process, such as flipping a coin. Instead, we can let the computer do the repeated sampling much faster (and we hope the computer feels less tedium than we do). Figure 3.1

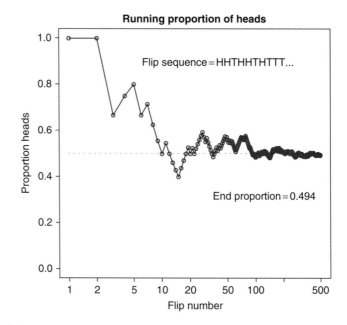

FIGURE 3.1

Running proportion of heads when flipping a coin. The x-axis is plotted on a logarithmic scale so that you can see the details of the first few flips but also the long-run trend after many flips. (The R code that generated this graph is in Section 3.5.1 (RunningProportion.R). When you run the code, your graph will look different than this one because you will generate a different random sequence of flips.)

shows the results of a computer simulating many flips of a fair coin. The R programming language has pseudo-random number generators built into it, which we will use often.[1] On the first flip, the computer randomly generates a head or a tail. It then computes the proportion of heads obtained so far. If the first flip was a head, then the proportion of heads is $1/1 = 1.0$. If the first flip was a tail, then the proportion of heads is $0/1 = 0.0$. Then the computer randomly generates a second head or tail and computes the proportion of heads obtained so far. If the sequence so far is HH, then the proportion of heads is $2/2 = 1.0$. If the sequence so far is HT or TH, then the proportion of heads is $1/2 = 0.5$. If the sequence so far is TT, then the proportion of heads is $0/2 = 0.0$. Then the computer generates a third head or tail, computes the proportion of heads so far, and so on for many flips. Figure 3.1 shows the running proportion of heads as the sequence continues.

Notice in Figure 3.1 that at the end of the long sequence, the proportion of heads is *near* 0.5 but not necessarily exactly equal to 0.5. This discrepancy reminds us that even this long run is still just a finite random sample, and there is no guarantee that the relative frequency of an event will match the true underlying probability of the event. That's why we say we are *approximating* the probability by the long-run relative frequency.

3.2.1.2 Deriving a Long-Run Relative Frequency

Sometimes, when the situation is simple enough mathematically, we can derive the exact long-run relative frequency. The case of the fair coin is one such simple situation.

The sample space of the coin consists of two possible outcomes: head and tail. By the assumption of fairness, we know that each outcome is equally likely. Therefore the long-run relative frequency of heads should be exactly one out of two (i.e., 1/2, and the long-run relative frequency of tails should also be exactly 1/2).

This technique is easily extended to other simple situations. Consider, for example, a standard six-sided die. It has six possible outcomes, namely, one dot, two dots, ..., six dots. If we assume that the die is fair, then the long-run relative frequency of each outcome should be exactly 1/6.

Suppose that we put different dots on the faces of the six-sided die. In particular, suppose that we put one dot on one face, two dots on two faces, and three dots on the remaining three faces. We still assume that each of the six faces is equally likely. Then the long-run relative frequency of one dot is exactly

---

[1] Pseudo-random number generators are not actually random; they are, in fact, deterministic. But the properties of the sequences they generate mimic the properties of random processes.

1/6, the long-run relative frequency of two dots is exactly 2/6, and the long-run relative frequency of three dots is exactly 3/6.

3.2.2 Inside the Head: Subjective Belief

How strongly do you believe that a coin minted by the U.S. government is fair? If you believe that the coin could be slightly different than exactly fair, then how strongly do you believe that $\theta = 0.51$? Or $\theta = 0.49$? If instead you are considering a coin that is ancient, asymmetric, and lopsided, do you believe that it inherently has $\theta = 0.50$? How about a coin purchased at a magic shop? We are not talking here about the true, inherent probability that the coin will come up heads, we are talking about our degree of belief in each possible probability.

To specify our subjective beliefs, we have to specify how likely we think each possible outcome is. It can be hard to pin down mushy intuitive beliefs. In the next section we explore one way to "calibrate" subjective beliefs, and in the subsequent section we discuss ways to mathematically describe degrees of belief.

3.2.2.1 Calibrating a Subjective Belief by Preferences

Consider a simple question that might affect travelers: How strongly do you believe that there will be a snowstorm that closes the interstate highways near Indianapolis next New Year's Day? Your job is to provide a number between 0 and 1 that accurately reflects your belief probability. One way to come up with such a number is to calibrate your beliefs relative to other events with clear probabilities.

As a comparison event, consider a marbles-in-sack experiment. In a sack we put 10 marbles, 5 red and 5 white. We shake the sack and then draw a marble at random. The probability of getting a red marble is, of course, $5/10 = 1/2 = 0.5$. We will use this sack of marbles as a comparison for considering snow in Indianapolis on New Year's Day.

Consider the following two gambles that you can choose from:

- *Gamble A*. You get $100 if there is a traffic-stopping snowstorm in Indianapolis next New Year's Day.
- *Gamble B*. You get $100 if you draw a red marble from a sack of marbles with 5 red and 5 white marbles.

Which gamble would you prefer? If you prefer gamble B, that means you think there is less than a 50-50 chance of a traffic-stopping snowstorm in Indy. So at least you now know that your subjective belief about the probability of a traffic-stopping snowstorm is less than 0.5.

We can narrow down the degree of belief by considering other comparison gambles. Consider these two gambles:

- *Gamble A.* You get \$100 if there is a traffic-stopping snowstorm in Indianapolis next New Year's Day.
- *Gamble C.* You get \$100 if you draw a red marble from a sack of marbles with 1 red and 9 white marbles.

Which gamble would you prefer? If you now prefer gamble A, that means you think there is more than a 10% chance of a traffic-stopping snowstorm in Indy on New Year's Day. Taken together, the two comparison gambles have told you that your subjective probability lies somewhere between 0.1 and 0.5. We could continue to consider preferences against other candidate gambles to calibrate your subjective belief more accurately.

3.2.2.2 Describing a Subjective Belief Mathematically

When there are several possible outcomes in a sample space, it might be too much effort to try to calibrate your subjective belief about every possible outcome. Instead, you can use a mathematical function to summarize your beliefs.

For example, you might believe that the average American woman is 5′4″ tall, but be open to the possibility that the average might be somewhat above or below that value. It is too tedious and maybe impossible to specify your degree of belief that the average height is 4′1″, 4′2″, 4′3″, and so on up through 6′1″, 6′2″, 6′3″, and so forth. So you might instead describe your degree of belief by a bell-shaped curve that is highest at 5′4″ and drops off symmetrically above and below that most likely height. You can change the width and center of the curve until it seems best to capture your subjective belief. Later in the book we will talk about exact mathematical formulas for functions like these, but the point now is merely to understand the idea that mathematical functions can define curves that can be used to describe degrees of belief.

3.2.3 Probabilities Assign Numbers to Possibilities

In general, a probability, whether it's outside the head or inside the head, is just a way of assigning numbers to a set of mutually exclusive possibilities. The numbers, called *probabilities*, merely need to satisfy three properties (Kolmogorov, 1956):

1. A probability value must be non-negative (i.e., zero or positive).
2. The sum of the probabilities across all events in the entire sample space must be 1.0 (i.e., one of the events in the space must happen, otherwise the space does not exhaust all possibilities).

3. For any two mutually exclusive events, the probability that one *or* the other occurs is the *sum* of their individual probabilities. For example, the probability that a fair six-sided die comes up three dots *or* four dots is $1/6 + 1/6 = 2/6$. As another example, if you believe there is a 60% chance of 0 to <3 inches of snow and a 20% chance of 3 to <6 inches of snow, then you should believe that there is an 80% (= 60% + 20%) chance of 0 to <6 inches of snow.

Any assignment of numbers to events that respects those three properties will also have all the properties of probabilities that we will discuss later. So whether a probability is thought of as a long-run relative frequency of outcomes in the world or as a magnitude of a subjective belief, it behaves the same way mathematically.

3.3 PROBABILITY DISTRIBUTIONS

A probability *distribution* is simply a list of all possible outcomes and their corresponding probabilities. For a coin, the probability distribution is trivial: We list two outcomes (head and tail) and their two corresponding probabilities (θ and $1 - \theta$). For other sets of outcomes, however, the distribution can be more complex. For example, consider the total number of calories consumed by a person in a day. There is some probability that the number of calories consumed in a day will be 2000, some probability that the number will be less, say 1898.3, some probability that the number will be more, say 2447.9, and so forth. When the outcomes are continuous, like calories, then the notion of probability takes on some subtleties, as we will see.

3.3.1 Discrete Distributions: Probability Mass

When the sample space consists of discrete outcomes, then we can talk about the probability of each distinct outcome. For example, the sample space of a flipped coin has two discrete outcomes, and we talk about the probability of head or tail. The sample space of a six-sided die has six discrete outcomes, and we talk about the probability of one dot, two dots, and so forth.

For continuous outcome spaces, we can *discretize* the space into a finite set of mutually exclusive and exhaustive "bins." For example, although calories consumed in a day is a continuous scale, we can divide up the scale into a finite number of intervals, such as <1500, 1500 to 2000, 2000 to 2500, 2500 to 3000, and >3000. Then we can talk about the probability of any one of those five intervals occurring. The probability of 2000 to 2500 is perhaps highest, with the probabilities of the other intervals dropping off from that high. Of course, the sum of the probabilities across the five intervals must sum to 1.

The probability of a discrete outcome is sometimes referred to as a probability *mass* to distinguish it from the probability density of an infinitesimal outcome, which will be defined next.

3.3.2 Continuous Distributions: Rendezvous with Density[2]

If you think carefully about a continuous outcome space, you realize that it becomes problematic to talk about the probability of a specific value on the continuum, as opposed to an interval on the continuum. For example, the probability that I eat exactly 2319.58372019... calories today is essentially nil, and that is true for *any* exact value you care to think of. We can, however, talk about the probability of intervals. The probability that I eat between 2000 and 2500 calories today is, say, 0.43. The problem with using intervals, however, is that their widths and edges are arbitrary, and very wide intervals are not precise. So what we will do is make the intervals infinitesimally narrow, and instead of talking about the infinitesimal probability of that infinitesimal interval, we will talk about the ratio of the probability to the interval width. That ratio is called the probability *density*. Examples and further explanation follow.

Consider a spinner of the kind often found with board games. It has an arrow mounted on a hub in the center, and a flick of the finger makes the arrow spin around. Friction causes the arrow to stop eventually, pointing in a random direction. Along the perimeter of points reached by the arrowhead, there is a numerical scale that reads continuously from 0 to 1, wrapping around the circumference of the circle so that the 1 touches the 0, as shown in the top-left section of Figure 3.2. We assume that the spinner is fair, so that any value from 0 to 1 is equally likely to be pointed at.

Let's divide the spinner into two equal sectors, one from 0 to 0.5 and the other from 0.5 to 1. When we spin the spinner, what is the probability that the outcome is in the first sector? Obviously the answer is 1/2. Suppose instead we divide the spinner into N equal sectors. What is the probability that the spinner stops in any one of the N sectors? Obviously, $1/N$. Notice that as we divide the spinner into more and more sectors, the probability of stopping in any one of them gets smaller and smaller. So if we want to know the probability of getting within a narrow range of a specific value, the probability is arbitrarily tiny.

But what if we instead consider the ratio of probability to sector width? The probability of stopping in a sector is $1/N$, and the width of a sector is $1/N$, so the ratio is $(1/N)/(1/N) = 1$. That ratio is called the probability density. It is called probability density by analogy with material density, which is defined as mass divided by volume. More loosely speaking, density is defined as the amount of stuff divided by the space it takes up. Applying that to the spinner, probability density in a sector is the amount of probability in that sector

---

[2] "There is a mysterious cycle in human events. To some generations much is given. Of other generations much is expected. This generation of Americans has a rendezvous with destiny." Franklin Delano Roosevelt, 1936.

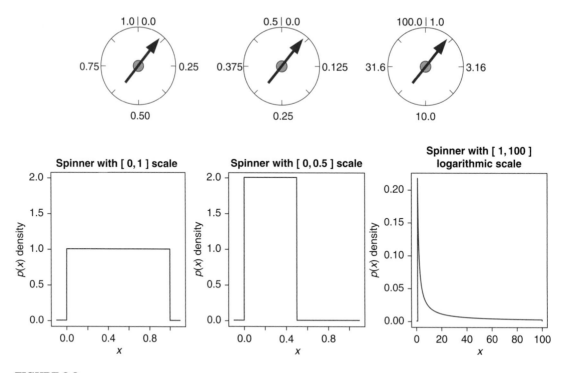

FIGURE 3.2
Spinners with different scales (*above*) and graphs of their probability density functions (*below*).

divided by the size of the sector. Again by analogy to material density, the amount of probability in the sector is called the probability *mass*. To reiterate, the probability density in an interval is the probability mass of that interval divided by the interval width.

Our goal, however, is to find the probability density *at a point*. To do that, we consider the probability density in the limit as the interval width around the point becomes infinitesimal. In the case of the fair spinner, this is easy: As N gets infinitely large, $1/N$ gets infinitesimally small. The density around a point in an interval is always $(1/N)/(1/N) = 1$, even as N grows infinitely large. In conclusion, for this case of a fair spinner ranging from 0 to 1, the probability density at every value is exactly 1.

Probability densities are not always exactly 1. For example, suppose that we relabel the spinner scale such that it starts at zero, but instead of uniformly going up to 1.0 in a 360° sweep, it goes only up to 0.5, as shown in the top-middle section of Figure 3.2. Consider the sector from 0 to 0.1. This sector covers one-fifth of the spinner, not one-tenth. Therefore, the probability mass within the sector is $1/5 = 0.2$, the sector width is 0.1, and the probability density within the sector is $0.2/0.1 = 2$. In general for this spinner, for a sector of width w, the probability mass is $2w$, and so the probability density in the sector is $2w/w = 2$. The probability density at a point is just the probability density of an infinitesimal interval around the point. Thus, for this case of a fair spinner ranging from 0 to 0.5, the probability density at every value is exactly 2. Notice that all we have changed from the previous example is the scale that goes around the circumference of the spinner. When the scale went from 0 to 1, the density was uniformly 1.0. When the scale went from 0 to 0.5, the density was uniformly 2.0.

Probability densities can be greater than 1, whereas probability masses cannot be greater than 1. Consider an analogy to a sponge. A 1-gram sponge has a mass of 1 gram regardless of how expanded or compressed it is. But the 1-gram sponge, when extremely tightly compressed into a tiny volume, can have a density as high as metal.

Probability densities are not always uniform. Continuing with the spinner example, suppose that the scale on the circumference of the spinner is logarithmic base 10, starting at 1 and wrapping around to 100 (like the x axis of Figure 3.1, but going only to 100), as shown in the top-right section of Figure 3.2. For this scale, the value of 10 appears 180° around the circle from 1, and the value 100 meets the value 1. The sector from 1 to 10 spans half the circle, so the probability mass of the interval is 0.5, and the (average) probability density of this interval is $0.5/(10 - 1) = 0.05556$. The sector from 10 to 100 spans the other half of the circle, and its (average) probability density is $0.5/(100 - 10) = 0.00556$, 10 times less than the first half. As we consider smaller intervals, you can see that the density differences will persist.

The lowly sponge can again educate our intuition. The 1-gram sponge can be squeezed at one end while the other end remains unsqueezed. The overall mass of the sponge remains 1 gram, but the density in the compressed end is much higher than the density in the uncompressed end.

3.3.2.1 Properties of Probability Density Functions
Consider again the basic spinner with a uniform scale from 0 to 1. What is the probability that the spinner lands somewhere between 0 and 1? The answer is, of course, 1, because the spinner must, by definition, have an outcome in the range 0 to 1. When we partition the spinner into N intervals, the probability of landing in any one interval is $1/N$. Notice, therefore, that the sum of the probabilities of the intervals is 1: $1/N + \cdots + 1/N = N \times 1/N = 1$.

In general, for any continuous value that is split up into intervals, the sum of the probabilities of the individual intervals must be 1, because some particular value must happen, by definition. We can write that as an equation, but we need to define some notation first. Let the continuous variable be denoted x. The width of an interval on x is denoted Δx (the symbol "Δ" is Greek letter capital delta). Let i be an index for the intervals, and let $[x_i, x_i+\Delta x]$ denote the interval between x_i and $x_i+\Delta x$. The probability mass of the ith interval is denoted $p([x_i, x_i+\Delta x])$. Then—and this is the point—the sum of those probability masses is 1:

$$\sum_i p([x_i, x_i+\Delta x]) = 1.0 \tag{3.1}$$

Recall now the definition of probability density: It is the ratio of probability mass over interval width. We can rewrite Equation 3.1 in terms of the density of each interval, by dividing and multiplying by Δx, as follows:

$$\sum_i \Delta x \frac{p([x_i, x_i+\Delta x])}{\Delta x} = 1.0 \tag{3.2}$$

In the limit, as the interval width becomes infinitesimal, we denote the width of the interval around x as dx instead of Δx, and we denote the probability *density* in the infinitesimal interval around x simply as $p(x)$ (not to be confused with $p([x_i, x_i+\Delta x])$, which was the probability mass in an interval). Then the summation in Equation 3.2 becomes an integral:

$$\underbrace{\sum_i}_{\int} \underbrace{\Delta x}_{dx} \underbrace{\frac{p([x_i, x_i+\Delta x])}{\Delta x}}_{p(x)} = 1.0 \quad \text{that is,} \quad \int dx\, p(x) = 1.0 \tag{3.3}$$

In this book, integrals are written with the dx term next to the integral sign, as in Equation 3.3, instead of at the far right end of the expression. Although this placement is not the most conventional notation, it is neither wrong nor unique to this book. The placement of dx next to the integral sign makes it easy to see what variable is being integrated over, without having to put subscripts on the integral sign. This usage is especially helpful later when we encounter integrals of functions that involve multiple variables.

The lower half of Figure 3.2 plots the probability density functions of the three spinner examples. The first example with the spinner had a scale that went from 0 to 1. As you may recall, we figured out that the density is 1.0 for all values of x on that scale. That density is plotted in the lower-left panel of Figure 3.2. Does Equation 3.3 work for this case? That is, does the probability density integrate to 1.0? Consider the density graph: It is a rectangle with height 1 and width 1, so its area is 1. The area under the density graph *is* the integral of the density function, so clearly Equation 3.3 is verified.

The second example with the spinner had a scale that went from 0 to 0.5, and we determined that the density was 2.0 for all values of x. This density is plotted in the lower-middle panel of Figure 3.2. Does Equation 3.3 work for this case? Does the probability density integrate to 1.0? The density function is a rectangle, with height 2.0 and width 0.5, so its area is 1.0.

The third example with the spinner involved a logarithmic scale that went from 1 to 100. We determined that its density on the low end of the scale was greater than its density on the high end of the scale. Although I won't derive it here, it turns out that the density at the point x is $1/(2\ln(10)x)$. (The mathematically inclined can find the method for deriving this result in Section 23.4.) This density is plotted in the lower-right panel of Figure 3.2. Notice that over the interval from 1 to 10, a typical density is around 0.05, which we determined earlier. For the interval from 10 to 100, a typical density is around 0.005, as we found before. This density function also integrates to 1.0, as it must.

To reiterate, in Equation 3.3, $p(x)$ is the probability density in the infinitesimal interval around x. Typically we let context tell us whether we are referring to a probability mass or a probability density, and we use the same notation, $p(x)$, for both. For example, if x is the value of the face of a six-sided die, then $p(x)$ is a probability mass. If x is the exact point value of number of calories consumed, then $p(x)$ is a probability density. There can be "slippage" in the usage, however. For example, if x refers to calories consumed but the scale is discretized into intervals, then $p(x)$ is really referring to the probability mass of the interval in which x falls. In the end, you'll have to be careful and tolerant of ambiguity.

3.3.2.2 The Normal Probability Density Function

Any function that has only non-negative values and integrates to 1 (i.e., satisfies Equation 3.3) can be construed as a probability density function. Perhaps the most famous probability density function is the *normal* distribution, also known as the Gaussian distribution. A graph of the normal curve is a bell shape; an example is shown in Figure 3.3.

The mathematical formula for the normal probability density has two parameters: μ (Greek mu) is called the *mean* of the distribution and σ (Greek sigma) is called the *standard deviation*. The exact definitions of these terms will be provided in the next section, but for now all you need to know is that the value of μ governs where the middle of the bell shape falls on the x-axis, and the value of σ governs how wide the bell is. The exact mathematical formula for the normal probability density is

$$p(x) = \frac{1}{\sigma\sqrt{2\pi}} \exp\left(-\frac{1}{2}\left[\frac{x-\mu}{\sigma}\right]^2\right) \tag{3.4}$$

Figure 3.3 shows an example of the normal distribution. The figure panel indicates the mean and standard deviation of the particular normal distribution

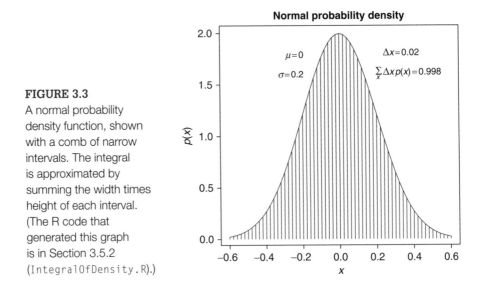

FIGURE 3.3
A normal probability
density function, shown
with a comb of narrow
intervals. The integral
is approximated by
summing the width times
height of each interval.
(The R code that
generated this graph
is in Section 3.5.2
(IntegralOfDensity.R).)

that is displayed. Notice that the peak probability density can be greater than 1.0 when the standard deviation, σ, is small. In other words, when the standard deviation is small, a lot of probability mass has to get squeezed into a small interval, and consequently the probability density in that interval is high.

Figure 3.3 also illustrates that the area under the normal curve is, in fact, 1. The x axis is divided into a dense comb of small intervals, with width denoted Δx. The integral of the normal density is approximated by summing the masses of all the tiny intervals. As the text within the graph shows, the sum of the interval areas is nearly 1.0. Only rounding error and the fact that the extreme tails of the distribution are not included in the sum prevent the sum from being exactly 1.

The normal probability density function can be used to describe the relative frequencies of x values generated by a random process. The normal probability density function can also be used to describe our degree of belief in different x values. Let's apply these two notions to the case of guessing the height of a person selected at random. Under the first notion, the idea is that the height of the person is influenced by myriad random factors, so the person's height will tend to be around μ but could be somewhat larger or smaller, with a spread indicated by σ. In this case, $p(x)$ refers to the probability density for observing a value x. Under the second notion, the idea is that we believe most strongly that the person's height is μ, but we also entertain the possibility that the person's height could be larger or smaller, with a spread indicated by σ. In this case, $p(x)$ refers to the density of our belief in value x. The mathematics of $p(x)$ are the same for either meaning, whether we think of $p(x)$ as referring to

(1) the relative frequency that a random process will generate various values or (2) degrees of belief in alternative possibilities.

3.3.3 Mean and Variance of a Distribution

When we have a numerical (not just categorical) value x that is generated with probability $p(x)$, we can wonder what would be its average value in the long run, if we repeatedly sampled values of x. For example, if we have a fair six-sided die, then each of its six values should come up 1/6th of the time in the long run, so the long-run average value of the die is $(1/6)1 + (1/6)2 + (1/6)3 + (1/6)4 + (1/6)5 + (1/6)6 = 3.5$. As another example, if we play a slot machine for which we win $100 with probability 0.001, we win $5 with probability 0.14, and otherwise we lose $1, then in the long run our payoff is $(0.001)(\$100) + (0.14)(\$5) + (0.859)(-\$1) = -\$0.059$ (i.e., in the long run we lose about 6 cents per pull of the bandit's arm). Notice what we did in those calculations. We weighted each possible outcome by the probability that it happens. This procedure defines the *mean* of a probability distribution, which is also called the *expected value* and which is denoted $E[x]$:

$$E[x] = \sum_x p(x)\,x \qquad (3.5)$$

Equation 3.5 applies when the values of x are discrete, so $p(x)$ denotes a probability mass. When the values of x are continuous, then $p(x)$ denotes a probability density and the sum becomes an integral over infinitesimal intervals:

$$E[x] = \int dx\, p(x)\,x \qquad (3.6)$$

The conceptual meaning is the same: the long-run average of the values.

The mean value of a distribution typically lies near the distribution's middle, intuitively speaking. For example, the mean of a normal distribution turns out to be the value of its parameter μ—that is, $E[x] = \mu$. A specific case appears in Figure 3.3, where the bulk of the distribution is centered over $x = \mu$; see the text in the figure for the exact value of μ.

Here's an example of computing the mean of a continuous distribution, using Equation 3.6. Consider the probability density function $p(x) = 6x(1 - x)$ defined over the interval $x \in [0, 1]$. That really is a probability density function: It's an upside-down parabola starting at $x = 0$, peaking over $x = 0.5$, and dropping down to baseline again at $x = 1$. Because it is a symmetric distribution, intuition tells us that the mean should be at its midpoint—that is, $x = 0.5$.

Let's check that it really is:

$$E[x] = \int_0^1 dx\, p(x)\, x$$

$$= \int_0^1 dx\, 6x(1-x)\, x$$

$$= 6\int_0^1 dx\, \left(x^2 - x^3\right)$$

$$= 6\left[\frac{1}{3}x^3 - \frac{1}{4}x^4\right]_0^1$$

$$= 6\left[\left(\frac{1}{3}1^3 - \frac{1}{4}1^4\right) - \left(\frac{1}{3}0^3 - \frac{1}{4}0^4\right)\right]$$

$$= 0.5 \tag{3.7}$$

The *variance* of a probability distribution is a number that represents the dispersion of the distribution away from its mean. There are many conceivable definitions of how far the values of x are dispersed from their mean, but the definition used for the specific term "variance" is based on the squared difference between x and the mean. The definition of variance is simply the mean squared deviation (MSD) of the x values from their mean:

$$\text{var}_x = \int dx\, p(x)\, \left(x - E[x]\right)^2 \tag{3.8}$$

Notice that Equation 3.8 is just like the formula for the mean (Equation 3.6) except that instead of integrating x weighted by x's probability, we're integrating $\left(x - E[x]\right)^2$ weighted by x's probability. In other words, the variance is just the average value of $\left(x - E[x]\right)^2$. For a discrete distribution, the integral in Equation 3.8 becomes a sum, analogous to the relationship between Equations 3.6 and 3.5. The square root of the variance, sometimes referred to as root mean squared deviation (RMSD), is called the *standard deviation* of the distribution.

The variance of the normal distribution turns out to be the value of its parameter σ squared (i.e., for the normal, $\text{var}_x = \sigma^2$). In other words, the standard deviation of the normal distribution is the value of the parameter σ. In a normal distribution, about 34% of the distribution lies between μ and $\mu + \sigma$

(see Exercise 3.4). Take a look at Figure 3.3 and visually identify where μ and $\mu + \sigma$ lie on the x axis (the values of μ and σ are indicated in the text within the figure) to get a visual impression of how far one standard deviation lies from the mean. Be careful, however, not to overgeneralize to distributions with other shapes: Non-normal distributions can have very different areas between their mean and first standard deviation.

3.3.3.1 Mean as Minimized Variance

An alternative conceptual emphasis starts with the definition of variance and derives a definition of mean, instead of starting with the mean and working to a definition of variance. Under this alternative conception, the goal is to define a value for the *central tendency* of a probability distribution. A value represents the central tendency of the distribution if the value is close to the highly probable values of the distribution. Therefore, we define the central tendency of a distribution as the value M that minimizes the long-run expected distance between it and all the other values of x. But how should we define "distance" between values? One way to define distance is as squared difference: The distance between x and M is $(x - M)^2$. One virtue of this definition is that the distance from x to M is the same as the distance from M to x, because $(x - M)^2 = (M - x)^2$. But the primary virtue of this definition is that it makes a lot of subsequent algebra tractable (which will not be rehearsed here). The central tendency is, therefore, the value M that minimizes the expected value of $(x - M)^2$. Thus, we want the value M that minimizes $\int dx\, p(x)\, (x - M)^2$. Does that look familiar? It's essentially the formula for the variance of the distribution, in Equation 3.8, but here thought of as a function of M. Here's the punch line: The value of M that minimizes $\int dx\, p(x)\, (x - M)^2$ is, it turns out, $E[x]$. In other words, the mean of the distribution is the value that minimizes the expected squared deviation. In this way, the mean is a central tendency of the distribution.

As an aside, if the distance between M and x is defined instead as $|x - M|$, then the value that minimizes the expected distance is called the *median* of the distribution. An analogous statement applies to the *modes* of a distribution, with distance defined as zero for any exact match, and one for any mismatch.

3.3.4 Variance as Uncertainty in Beliefs

If $p(\theta)$ represents degrees of belief in values of θ, instead of the probability of sampling θ, then the mean of $p(\theta)$ can be thought of as the value of θ that represents our typical or central belief. The variance of θ, which measures how spread out the distribution is, can be thought of as our uncertainty about possible beliefs. If the variance is small, then we believe strongly in values of θ near the mean. If the variance is large, then we are not certain about what value of

θ to believe in. This notion of variance (or standard deviation) as representing uncertainty will reappear often.

3.3.5 Highest Density Interval (HDI)

Another way of summarizing a belief distribution, which we will use often, is the highest density interval, abbreviated HDI.[3] The HDI indicates which points of a distribution we believe in most strongly and which cover most of the distribution. Thus, the HDI summarizes the distribution by specifying an interval that spans most of the distribution, say 95% of it, such that every point inside the interval has higher believability than any point outside the interval.

If you think of the probability distribution as the profile of an island rising out of the water, then the 95% HDI marks the waterline on the beach such that 95% of the island's mass is within the waterline. All the points within the waterline (above water) have higher believability than any point outside the waterline (below water). Figure 3.4 shows examples of HDIs; take a look now.

The formal definition of an HDI is just a mathematical expression of the waterline idea. What we want in the HDI is all those values of x for which we have a belief density at least as big as some value W (which is the depth of water), such that the integral over all those x values is 95% (or 99%, or whatever). Formally, the values of x in the 95% HDI are those such that $p(x) > W$, where W satisfies $\int_{x \text{ s.t. } p(x) > W} dx\, p(x) = 0.95$.

The width of the HDI is another way of measuring uncertainty of beliefs. If the HDI is wide, then beliefs are uncertain. If the HDI is narrow, then beliefs are fairly certain. Sometimes the goal of research is to obtain data that achieve a reasonably high degree of certainty about a particular value x. The desired degree of certainty can be measured as the width of the 95% HDI. For example, if x is a measure of how much a drug decreases blood pressure, the researcher may want to have an estimate with a 95% HDI no larger than five units on the blood pressure scale. As another example, if x is a measure of a population's preference for candidate A over candidate B, the researcher may want to have an estimate with a 95% HDI no larger than 10 percentage points.

---

[3] Some authors refer to the HDI as the HDR, which stands for highest density *region*, because a region can refer to multiple dimensions, but an interval refers to a single dimension. Because we will almost always consider the HDI of one parameter at a time, I will use "HDI" in an effort to reduce confusion. Some authors refer to the HDI as the HPD, to stand for highest probability density, but which I prefer not to use because it takes more space to write "HPD interval" than "HDI". Some authors refer to the HDI as the HPD, to stand for highest *posterior* density, but which I prefer not to use because *prior* distributions also have HDIs.

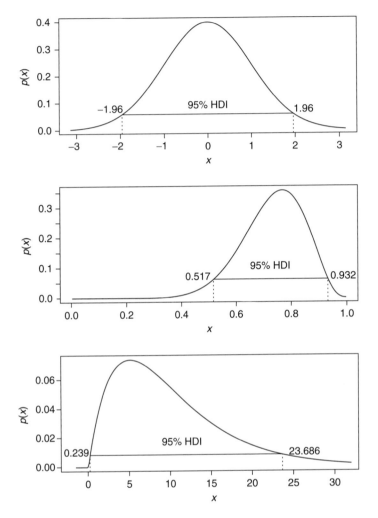

FIGURE 3.4
Examples of 95% highest density intervals (HDIs). For each example, all the x values inside the interval have higher density than any x value outside the interval and the total mass of the points inside the interval is 95%. The ends of the HDIs are marked with their x values. The top panel shows a normal distribution with mean zero and standard deviation one. The middle panel shows a beta distribution (defined in Chapter 5), and the lower panel shows a gamma distribution (defined in Chapter 9). Don't fret over the definitions of the density functions. The point is to get a visual intuition for HDIs in differently shaped distributions.

3.4 TWO-WAY DISTRIBUTIONS

In many situations we are interested in the conjunction of two outcomes. What is the probability of being dealt a card that is both a queen *and* a heart? What is the probability of meeting a person with both red hair *and* green eyes? When playing a board game involving a die and a spinner, we have degrees of belief about both the die *and* the spinner being fair.

As a specific example for developing these ideas, imagine tossing a coin three times in a row. The sequence of flips might be TTT, or TTH, and so on. Table 3.1 shows all possible sequences of three outcomes in its left column. As you can tell by counting the rows of the table, there are eight possible sequences. Because the coin is assumed to be fair, each row is equally likely, so each row has probability 1/8, which is indicated in the far right column.

For each sequence of three tosses, we can count the number of heads in the sequence and the number of times that the outcome switched between heads or tails. Table 3.1 also lists the number of heads and the number of switches in each sequence. We can list the combinations of head counts and switches in a *two-way* table, as shown in Table 3.2. This two-way table shows the probability of getting a particular combination of number of heads *and* number of switches. The probability of two things happening together is called their *conjoint* probability.

Table 3.1 Sample Space for Tossing a Coin Three Times

Outcome	Number of Heads	Number of Switches	Probability
TTT	0	0	1/8
TTH	1	1	1/8
THT	1	2	1/8
THH	2	1	1/8
HTT	1	1	1/8
HTH	2	2	1/8
HHT	2	1	1/8
HHH	3	0	1/8

Table 3.2 Conjoint Probabilities for Tossing a Coin Three Times, Compiled from Table 3.1

Number of Switches	Number of Heads			
	0	1	2	3
0	1/8	0	0	1/8
1	0	2/8	2/8	0
2	0	1/8	1/8	0

One of the interesting characteristics revealed by Table 3.2 is that not all combinations of events are equally likely, and some combinations don't happen at all. For example, the probability of getting a sequence with one head and one switch is 2/8 (i.e., this combination happens 25% of the time in the long run). On the other hand, it never happens that there is a sequence with one head and zero switches. We will use this table of conjoint probabilities to develop other concepts.

3.4.1 Marginal Probability

When we flip the coin three times, we can count the number of heads and the number of switches, but we might be interested only in one or the other type of outcome. We can ask, what's the probability of getting two heads in three flips, without worrying about the number of switches involved. One way to determine the probabilities of these different classes of outcomes is to sum across the conjoint probabilities we've already compiled. For example, if we want to determine the probability of getting zero heads, regardless of the number of switches in the sequence, we simply sum the probability of getting zero heads with zero switches and the probability of getting zero heads with one switch and the probability of getting zero heads with two switches. In other words, we simply sum across rows in Table 3.2. We can do this for every number of heads and write the sums in the lower margin of the table, as shown in Table 3.3. These summed probabilities are called the *marginal* probabilities. We can also compute the probability of each number of switches, regardless of the number of heads, by summing across the number of heads. Table 3.3 also shows these probabilities in the right-hand margin.

Let's now establish some general notation for the type of example we've been considering. We suppose that we have a sample space in which every outcome has two attributes, x and y. In the previous example, the outcomes in the sample space were sequences of three flips, and the two attributes were number of heads (x) and number of switches (y). The conjoint probability of a particular combination of x and y values is denoted $p(x, y)$. For example, the conjoint

Table 3.3 Marginal Probabilities When Tossing a Coin Three Times (this table extends Table 3.2)

Number of Switches	Number of Heads				Marginal (Number of Switches)
	0	1	2	3	
0	1/8	0	0	1/8	2/8
1	0	2/8	2/8	0	4/8
2	0	1/8	1/8	0	2/8
Marginal (Number of Heads):	1/8	3/8	3/8	1/8	

Table 3.4 Beliefs Regarding Two Coins

Dime	Nickel Two Tails	Fair	Two Heads	Marginal (Dime)
Two tails	0.1	0	0.1	0.2
Fair	0	0.6	0	0.6
Two heads	0.1	0	0.1	0.2
Marginal (Nickel):	0.2	0.6	0.2	

probability of two heads and one switch is $p(x=2, y=1) = 2/8$. Notice that conjoint probabilities are symmetric: $p(x, y) = p(y, x)$.

To compute the probability distribution for x by itself, we sum $p(x, y)$ across all values of y:

$$p(x) = \sum_y p(x, y) \tag{3.9}$$

When the x and y variables are continuous, then $p(x, y)$ is a probability density, and the summation becomes an integral:

$$p(x) = \int dy\, p(x, y) \tag{3.10}$$

where the resulting marginal distribution, $p(x)$, is also a density. This summation process is called *marginalizing over y* or *integrating out* the variable y. Of course, we can also determine the probability of y by marginalizing over x.

These notions of conjoint and marginal probabilities also apply to beliefs. Consider, for example, two coins: a nickel and a dime. Suppose that we believe that they might be fair, or that they are trick coins with heads on both sides or with tails on both sides. We believe most strongly that they are both fair, but that there is a small chance that they are trick coins. Moreover, we believe that if one is a trick coin, then the other is a trick coin too. These beliefs can be captured by a joint probability table, as shown in Table 3.4.

Table 3.4 indicates that our belief that both coins are fair is 60%, that we believe there is a 10% chance that both coins are two tailed, and that we believe it is impossible for one coin to be fair while the other is a trick coin. Table 3.4 also shows the marginal distributions of our beliefs. This example illustrates that we can talk about conjoint and marginal structure of belief distributions just as we do about data distributions.

3.4.2 Conditional Probability

We often want to know the probability of one event, given that we know another event is true. For example, what is the probability that it will rain in the next 24 hours given that there is a thunderstorm 400 miles due west of you?

What is the probability that you will pass the statistics class given that you scored 88/100 on the first assignment? What is the probability that a sequence of three coin flips has one switch given that it has one head?

Let's think about that last question in detail. Refer back to Table 3.3. We want to know the probability that a sequence of three flips has one switch, given that it has one head. This means that we are only considering sequences that we know to have one head, which means only the column in Table 3.3 labeled "1" under "Number of Heads." The question is, of the probability within that column, how much of it occurs for one switch? We can see that the conjoint probability of one switch and one head is 2/8, and the total probability of one-head column is $0 + 2/8 + 1/8 = 3/8$. Therefore, the probability of getting one switch, given that the sequence has one head, is $(2/8)/(3/8) = 2/3$.

Notice the conditional probability of getting one switch, given that there is one head, is different from the marginal probability of getting one switch. The marginal probability indicates the probability of getting one switch on average across all numbers of heads, whereas the conditional probability restricts consideration to a particular number of heads.

Conditional probabilities have their own notation. The probability of a value of y given a value of x is denoted $p(y\,|\,x)$. For the previous example, with number of heads denoted x and number of switches denoted y, we write $p(y=1\,|\,x=1) = 2/3$.

Now that we have some general notation, we can generalize our computations from the example. Recall that to compute the conditional probability $p(y=1\,|\,x=1)$, we divided the conjoint probability $p(y=1, x=1)$ by the sum of conjoint probabilities for the given value, $\sum_y p(y, x=1)$. Notice also that the sum of the conjoint probabilities is the marginal probability. So in general, the conditional probability is

$$p(y\,|\,x) = \frac{p(y, x)}{\sum_y p(y, x)} = \frac{p(y, x)}{p(x)} \tag{3.11}$$

(Notice that the equality of the denominators in Equation 3.11 was already discussed in Equation 3.9.) When y is continuous, the sum becomes an integral over the conjoint density:

$$p(y\,|\,x) = \frac{p(y, x)}{\int_y dy\, p(y, x)} = \frac{p(y, x)}{p(x)} \tag{3.12}$$

(Notice that the equality of the denominators in Equation 3.12 was already discussed in Equation 3.10.)

Of course, we can conditionalize on the other variable, instead. That is, we can consider $p(x\,|\,y)$ instead of $p(y\,|\,x)$. It is important to recognize that, in general, $p(x\,|\,y) \neq p(y\,|\,x)$. For example, the probability that the ground is wet, given that it's raining, is different than the probability that it's raining, given that the ground is wet. The next chapter provides an extended discussion of the relationship between $p(x \mid y)$ and $p(y \mid x)$. The relationship is called *Bayes' rule*.

It is also important to recognize that there is no temporal order in conditional probabilities. When we say "the probability of x given y," we do *not* mean that y has already happened and x has yet to happen. All we mean is that we are restricting our calculations of probability to a particular subset of possible events. A better gloss of $p(x \mid y)$ is to say, "among all events with value y, this proportion of them also have value x." So, for example, we can talk about the probability that it rained last night given that there are clouds this morning. This simply refers to the proportion of all cloudy mornings like this one that had rain the night before.

Finally, as I have repeatedly emphasized, the notions of conditional probability apply to belief distributions, just as they apply to data distributions. Refer back to Table 3.4 regarding beliefs about the fairness of a nickel and a dime. Consider the probability that the dime is fair given that the nickel is fair. Using Equation 3.11, we find that the conditional probability is 1. This simply means that of all our beliefs for which the nickel is fair, 100% of them have the dime being fair.

3.4.3 Independence of Attributes

Suppose I have a six-sided die and a spinner. Suppose they are fair. I flick the spinner and it points to 0.123. Given this result on the spinner, what is the probability that the rolled die will come up 3? In answering this question, you probably thought, "the spinner has no influence on the die, so the probability of the die coming up 3 is 1/6 regardless of what the spinner is pointing at." If that's what you thought, you were assuming that the spinner and the die are *independent*.

In general, when the value of y has no influence on the value of x, we know that $p(x \mid y) = p(x)$, for all values of x and y. Let's think a moment about what that implies. We know from the definition of conditional probability, in Equations 3.11 or 3.12, that $p(x \mid y) = p(x, y)/p(y)$. Combining those equations implies that $p(x) = p(x, y)/p(y)$ for all values of x and y. After multiplying both sides by $p(y)$, we get the implication that $p(x, y) = p(x)p(y)$ for all values of x and y. The implication goes the other way too: When $p(x, y) = p(x)p(y)$ for all values of x and y, then $p(x \mid y) = p(x)$ for all values of x and y. Therefore, either of these conditions is our mathematical definition of independence of attributes.

Consider the example back in Table 3.3 (page 43), regarding sequences of three flips of a coin. Are the attributes of number of heads and number of switches independent? You can quickly see that the answer is no. Consider, for example, the top-left cell, which contains the conjoint probability of zero heads and zero switches (namely, 1/8). Does it equal the product of the marginal probability of zero heads and the marginal probability of zero switches (namely, $1/8 \times 2/8$)? No, clearly not.

As a second case, consider the example in Table 3.4 (page 44), regarding beliefs about two coins. The beliefs in that scenario were explicitly that the coins were *not* independent: If one coin was fair, then so was the other one, but if one coin was a trick coin, so was the other one. The lack of independence can be verified mathematically in Table 3.4. Consider the top-left cell: Is our conjoint belief probability that both coins are two tailed (namely, 0.1) equal to the product of our marginal belief probabilities that the nickel is two tailed and the dime is two tailed (namely, 0.2×0.2)?

As a simple example of two attributes that *are* independent, consider the suit and value of cards in a standard deck. There are four suits, and 13 values of each suit, making 52 cards altogether. Consider a randomly dealt card. What is the probability that it is a heart? (Answer: $13/52 = 1/4$.) Suppose I look at the card without letting you see it, and I tell you that it is a queen. Now what is the probability that it is a heart? (Answer: 1/4.) Telling you the card's value does not change the probabilities of the suits, so value and suit are independent. We can verify this in terms of cross multiplying marginal probabilities too: Each combination of value and suit has a 1/52 chance of being dealt (in a fairly shuffled deck). Notice that 1/52 is exactly the marginal probability of any one suit (1/4) times the marginal probability of any one value (1/13).

Among other contexts, independence will come up again when we are constructing mathematical descriptions of our beliefs about more than one attribute. We will create a mathematical description of our beliefs about one attribute and another mathematical description of our beliefs about the other attribute. Then, to describe what we believe about combinations of attributes, we will often assume independence and simply multiply the separate beliefs to specify the conjoint beliefs.

3.5 R CODE

3.5.1 R Code for Figure 3.1

(RunningProportion.R)

```
1   # Goal: Toss a coin N times and compute the running proportion of heads.
2   N = 500          # Specify the total number of flips, denoted N.
3   # Generate a random sample of N flips for a fair coin (heads=1, tails=0);
4   # the function "sample" is part of R:
5   #set.seed(47405) # Uncomment to set the "seed" for the random number generator.
6   flipsequence = sample( x=c(0,1) , prob=c(.5,.5) , size=N , replace=TRUE )
7   # Compute the running proportion of heads:
```

```
8   r = cumsum( flipsequence ) # The function "cumsum" is built in to R.
9   n = 1:N                    # n is a vector.
10  runprop = r / n            # component by component division.
11  # Graph the running proportion:
12  # To learn about the parameters of the plot function,
13  # type help('par') at the R command prompt.
14  # Note that "c" is a function in R.
15  plot( n , runprop , type="o" , log="x" ,
16          xlim=c(1,N) , ylim=c(0.0,1.0) , cex.axis=1.5 ,
17          xlab="Flip Number" , ylab="Proportion Heads" , cex.lab=1.5 ,
18          main="Running Proportion of Heads" , cex.main=1.5 )
19  # Plot a dotted horizontal line at y=.5, just as a reference line:
20  lines( c(1,N) , c(.5,.5) , lty=3 )
21  # Display the beginning of the flip sequence. These string and character
22  # manipulations may seem mysterious, but you can de-mystify by unpacking
23  # the commands starting with the innermost parentheses or brackets and
24  # moving to the outermost.
25  flipletters = paste( c("T","H")[ flipsequence[ 1:10 ] + 1 ] , collapse="" )
26  displaystring = paste( "Flip Sequence = " , flipletters , "..." , sep="" )
27  text( 5 , .9 , displaystring , adj=c(0,1) , cex=1.3 )
28  # Display the relative frequency at the end of the sequence.
29  text( N , .3 , paste("End Proportion =",runprop[N]) , adj=c(1,0) , cex=1.3 )
30  # Save the plot to an EPS file.
31  dev.copy2eps( file = "RunningProportion.eps" )
```

3.5.2 R Code for Figure 3.3

(IntegralOfDensity.R)
```
1   # Graph of normal probability density function, with comb of intervals.
2   meanval = 0.0               # Specify mean of distribution.
3   sdval = 0.2                 # Specify standard deviation of distribution.
4   xlow  = meanval - 3*sdval   # Specify low end of x-axis.
5   xhigh = meanval + 3*sdval   # Specify high end of x-axis.
6   dx = 0.02                   # Specify interval width on x-axis
7   # Specify comb points along the x axis:
8   x = seq( from = xlow , to = xhigh , by = dx )
9   # Compute y values, i.e., probability density at each value of x:
10  y = ( 1/(sdval*sqrt(2*pi)) ) * exp( -.5 * ((x-meanval)/sdval)^2 )
11  # Plot the function. "plot" draws the intervals. "lines" draws the bell curve.
12  plot( x , y , type="h" , lwd=1 , cex.axis=1.5
13      , xlab="x" , ylab="p(x)" , cex.lab=1.5
14      , main="Normal Probability Density" , cex.main=1.5 )
15  lines( x , y )
16  # Approximate the integral as the sum of width * height for each interval.
17  area = sum( dx * y )
18  # Display info in the graph.
19  text( -sdval , .9*max(y) , bquote( paste(mu ," = " ,.(meanval)) )
20      , adj=c(1,.5) )
21  text( -sdval , .8*max(y) , bquote( paste(sigma ," = " ,.(sdval)) )
22      , adj=c(1,.5) )
```

```
23    text( sdval , .9*max(y) , bquote( paste(Delta , "x = " ,.(dx)) )
24           , adj=c(0,.5) )
25    text( sdval , .8*max(y) ,
26           bquote(
27              paste( sum(,x,) , " " , Delta , "x p(x) = " , .(signif(area,3)) )
28           ) , adj=c(0,.5) )
29    # Save the plot to an EPS f{i}le.
30    dev.copy2eps( f{i}le = "IntegralOfDensity.eps" )
```

3.6 EXERCISES

Exercise 3.1. [Purpose: To give you some experience with random number generation in R.] Modify the coin-flipping program in Section 3.5.1 (RunningPropor-tion.R) to simulate a biased coin that has $p(H) = 0.8$. Change the height of the reference line in the plot to match $p(H)$. Comment your code. Hint: Read the help for sample.

Exercise 3.2. [Purpose: To have you work through an example of the logic presented in Section 3.2.1.2.] Determine the exact probability of drawing a 10 from a shuffled pinochle deck. (A pinochle deck has 48 cards. There are six values: 9, 10, jack, queen, king, and ace. There are two copies of each value in each of the standard four suits: hearts, diamonds, clubs, and spades.)

(A)   What is the probability of getting a 10?
(B)   What is the probability of getting a 10 or jack?

Exercise 3.3. [Purpose: To give you hands-on experience with a simple probability density function, in R and in calculus, and to reemphasize that density functions can have values larger than 1.] Consider a spinner with a $[0,1]$ scale on its circumference. Suppose that the spinner is slanted or magnetized or bent in some way such that it is biased, and its probability density function is $p(x) = 6x(1 - x)$ over the interval $x \in [0, 1]$.

(A)   Adapt the code from Section 3.5.2 (IntegralOfDensity.R) to plot this density function and approximate its integral. Comment your code. Be careful to consider values of x only in the interval $[0, 1]$. Hint: You can omit the first couple lines regarding meanval and sdval, because those parameter values pertain only to the normal distribution. Then set xlow=0 and xhigh=1.
(B)   Derive the exact integral using calculus. Hint: See the example, Equation 3.7.
(C)   Does this function satisfy Equation 3.3?
(D)   From inspecting the graph, what is the maximal value of $p(x)$?

Exercise 3.4. [Purpose: To have you use a normal curve to describe beliefs. It's also handy to know the area under the normal curve between μ and σ.]

(A) Adapt the code from Section 3.5.2 (`IntegralOfDensity.R`) to determine (approximately) the probability mass under the normal curve from $x = \mu - \sigma$ to $x = \mu + \sigma$. Comment your code. Hint: Just change `xlow` and `xhigh` appropriately, and change the `text` location so that the `area` still appears within the plot.

(B) Now use the normal curve to describe the following belief. Suppose you believe that women's heights follow a bell-shaped distribution, centered at 162 cm with about two-thirds of all women having heights between 147 cm and 177 cm.

Exercise 3.5. [Purpose: To recognize and work with the fact that Equation 3.11 can be solved for the conjoint probability, which will be crucial for developing Bayes' theorem.] Schoolchildren were surveyed regarding their favorite foods. Of the total sample, 20% were 1st graders, 20% were 6th graders, and 60% were 11th graders. For each grade, the following table shows the proportion of respondents who chose each of three foods as their favorite:

	Ice Cream	Fruit	French Fries
1st graders	0.3	0.6	0.1
6th graders	0.6	0.3	0.1
11th graders	0.3	0.1	0.6

From that information, construct a table of conjoint probabilities of grade and favorite food. Also, say whether grade and favorite food are independent and how you ascertained the answer. Hint: You are given p(grade) and p(food | grade). You need to determine p(grade, food).

CHAPTER 4

Bayes' Rule

CONTENTS

I'll love you forever in every respect
(I'll marginalize all your glaring defects)
But if you could change some to be more like me
I'd love you today unconditionally.

If you see that there are clouds, what is the probability that soon there will be rain? If you know that it is raining, by hearing it patter on the roof, what is the probability that there are clouds? Notice that p(clouds | rain) is not equal

Doing Bayesian Data Analysis: A Tutorial with R and BUGS. DOI: 10.1016/B978-0-12-381485-2.00004-3
© 2011, Elsevier Inc. All rights reserved.

to p(rain | clouds). If someone smiles at you, what is the probability that they love you? If someone loves you, what is the probability that they will smile at you? Notice that p(smile | love) is not equal to p(love | smile).

Let's consider an example for which we can determine specific numbers. Suppose I have a standard deck of playing cards, which has 52 cards altogether. There are four suits: hearts, diamonds, clubs, and spades. Within each suit, there are 13 values: ace, two, three,..., ten, jack, queen, and king. I shuffle the cards and draw one at random without showing it to you. I look at the card, and tell you (truthfully) that it is a queen. Given that you know it is a queen, what is the probability that it is a heart? Think about it a moment: There are four queens in the deck, and only one of them is a heart. So the probability that the card is a heart is 1/4. We can write this as a conditional probability:

$$p(\heartsuit \,|\, Q) = \frac{1}{4}.$$

Now I put the card back into the deck and reshuffle. I draw another card from the deck, and this time I tell you that it is a heart. Given that you know it is a heart, what is the probability that it is a queen? Think about it a moment: There are 13 hearts in the deck, and only one of them is a queen. So the probability that the card is a queen is 1/13. We can write this as a conditional probability:

$$p(Q \,|\, \heartsuit) = \frac{1}{13}.$$

Notice that $p(\heartsuit \,|\, Q)$ does not equal $p(Q \,|\, \heartsuit)$. Despite the inequality, the reversed conditional probabilities must be related somehow, right? Answer: Yes! What Bayes' rule tells us is the relationship between the two conditional probabilities.

4.1 BAYES' RULE

Thomas Bayes (1702–1761) was a reputable mathematician and Presbyterian minister in England. His famous theorem was published posthumously in 1764. The simple rule that relates conditional probabilities has vast ramifications for statistical inference, and therefore as long as his name is attached to the rule, we'll continue to see his name in textbooks.

A crucial application of Bayes' rule is to determine the probability of a model when given a set of data. What the model itself provides is the probability of the data, given specific parameter values and the model structure. We use Bayes' rule to get from the probability of the data, given the model, to the probability of the model, given the data. This process will be explained during the course of this chapter and, indeed, during the rest of this book.

There is another branch of statistics, called *null hypothesis significance testing* (NHST), that relies on the probability of data given the model and does *not* use Bayes' rule. Chapter 11 describes NHST and its perils. This approach is often identified with another towering figure from England who lived about 200 years later than Bayes, named Ronald Fisher (1890–1962). His name, or at least the first letter of his last name, is immortalized in the most common statistic used in NHST, the F-ratio.[1] It is curious and reassuring that the overwhelmingly dominant approach of the 20th century (i.e., NHST) is giving way in the 21st century to a Bayesian approach that had its genesis in the 18th century.

4.1.1 Derived from Definitions of Conditional Probability

Recall from the definition of conditional probability, back in Equations 3.11 and 3.12 on p. 45, that $p(y \mid x) = p(y, x)/p(x)$. In words, the definition simply says that the probability of y given x is the probability that they happen together relative to the probability that x happens at all. We used this definition quite naturally when computing the conditional probabilities for the example, presented earlier, regarding hearts and queens in a deck of cards.

Now we just do some very simple algebraic manipulations. First, multiply both sides of $p(y \mid x) = p(y, x)/p(x)$ by $p(x)$ to get $p(y \mid x)p(x) = p(y, x)$. Second, notice that we can do the analogous manipulation starting with $p(x \mid y) = p(y, x)/p(y)$ to get $p(x \mid y)p(y) = p(y, x)$. Now we have two different expressions equal to $p(y, x)$, so we know those expressions equal each other: $p(y \mid x)p(x) = p(x \mid y)p(y)$. Divide both sides of that last expression by $p(x)$ to arrive at

$$p(y \mid x) = \frac{p(x \mid y)p(y)}{p(x)} \tag{4.1}$$

But we are not done yet, because we can rewrite the denominator in terms of $p(x \mid y)$ also. Toward that goal, recall that $p(x) = \sum_y p(x, y)$. That was Equation 3.9, on p. 44, if you're keeping score. We also know that $p(x, y) = p(x \mid y)p(y)$. Combining those equations yields $p(x) = \sum_y p(x, y) = \sum_y p(x \mid y)p(y)$. Substitute that into the denominator of Equation 4.1 to get

$$p(y \mid x) = \frac{p(x \mid y)p(y)}{\sum_y p(x \mid y)p(y)} \tag{4.2}$$

In Equation 4.2, the y in the numerator is a specific fixed value, whereas the y in the denominator is a variable that takes on all possible values of y over the summation. Equations 4.1 and 4.2 are called *Bayes' rule*. This simple relationship lies at the core of Bayesian inference.

---

[1] But Fisher did not advocate the type of NHST ritual that contemporary social science performs; see Gigerenzer, Krauss, & Vitouch (2004).

4.1.2 Intuited from a Two-Way Discrete Table

It's easy to derive Bayes' rule (we just did!), but let's now get an intuition for what it means and how it works. First, let's confirm that it works for the simple case of the queen of hearts. Earlier we figured out that $p(Q|\heartsuit) = \dfrac{1}{13}$ and $p(\heartsuit|Q) = \dfrac{1}{4}$. Do those conditional probabilities satisfy Bayes' rule? Let's find out: $p(\heartsuit|Q)p(Q)/p(\heartsuit) = \dfrac{1}{4}\dfrac{4}{52}\Big/\dfrac{13}{52} = \dfrac{1}{13} = p(Q|\heartsuit)$. It works!

The suit and value on playing cards are independent. (The idea of independent attributes was discussed in Section 3.4.3.) Let's now confirm Bayes' rule for two attributes that are not independent. Recall the case of tossing a coin three times and counting the number of heads and the number of switches between heads and tails, as tabulated back in Table 3.3 (p. 43), and repeated here for convenience:

Number of Switches	Number of Heads				Marginal (Number of Switches)
	0	**1**	**2**	**3**	
0	1/8	0	0	1/8	2/8
1	0	2/8	2/8	0	4/8
2	0	1/8	1/8	0	2/8
Marginal (Number of Heads)	1/8	3/8	3/8	1/8	

Consider the probability of getting one switch given that there is one head—that is, $p(1S|1H)$—versus the probability of getting one head given that there is one switch, that is, $p(1H|1S)$. From the table, we can determine that $p(1S|1H) = p(1S, 1H)/p(1H) = (2/8)/(3/8) = 2/3$, and $p(1H|1S) = p(1H, 1S)/p(1S) = (2/8)/(4/8) = 1/2$. Notice that $p(1S|1H)$ does not equal $p(1H|1S)$. Then we can verify Bayes' rule: $p(1H|1S)p(1S)/p(1H) = (1/2)(4/8)/(3/8) = 2/3 = p(1S|1H)$. It works! In going through that arithmetic, essentially what we did was go through the motions of deriving Bayes' rule, using specific values instead of variables.

A valuable intuition, for understanding conditional probabilities and Bayes' rule, comes from restricting our spatial attention to a single row or column of the conjoint probability table. Suppose someone tosses a coin three times and tells us that the sequence contains one switch. Given that knowledge, we can restrict our attention to the row of the table corresponding to one switch. We know that one of the conjoint events *within that row* must have happened, but we don't know which one. The relative probabilities of events within that row have not changed, but we know that the total probability within that row must now sum to 1.0. To achieve that transformation mathematically, we simply divide the cell probabilities in the one-switch row by its original row total. This preserves the relative probabilities within the row but makes the total probability equal to 1.0. Dividing a set of values by their sum is called *normalizing* the values. When we normalize the cell probabilities in the ith row, we get the conditional probabilities of the columns, given the row value. In particular, when we normalize the one-switch row, we get the conditional probabilities for number of heads: $p(0H \mid 1S) = 0/(4/8) = 0$, $p(1H \mid 1S) = (2/8)/(4/8) = 0.5$, $p(2H \mid 1S) = (2/8)/(4/8) = 0.5$, and $p(3H \mid 1S) = 0/(4/8) = 0$.

The idea of restricting attention to a single column or row of the conjoint probability table yields a way of intuiting Bayes' rule in general. The key to Bayes' rule is to notice, from the definition of conditional probability (Equations 3.11 and 3.12 on p. 45), that the conjoint probability of the ith row (R_i) and the jth column (C_j) can be reexpressed either as $p(R_i \mid C_j)p(C_j)$ or as $p(C_j \mid R_i)p(R_i)$. These alternative expressions of the conjoint probability $p(R_i, C_j)$ have been entered into the i, j^{th} cell of Table 4.1.

Suppose we know that event R_i has happened, but we don't know the column value. In this case, the remaining possibilities are the cells in row R_i, and therefore we can restrict our attention to only the ith row of Table 4.1. Because we know that R_i is true, our universe of remaining possibilities has collapsed to that row, and therefore we know that the sum of the probabilities in the row must be 1, instead of $p(R_i)$. This promotion of $p(R_i)$ to 1.0 is mathematically like dividing everything in the ith row by $p(R_i)$. As mentioned before, this operation is called normalizing the probabilities in the ith row so they sum to 1.0. When we normalize, the equation in the i, j^{th} cell becomes $p(R_i, C_j)/p(R_i) = p(R_i \mid C_j)p(C_j)/p(R_i) = p(C_j \mid R_i)$. This is Bayes' rule.

In summary, the key idea is that conditionalizing on a known row value is like restricting attention to only the row for which that known value is true and then normalizing the probabilities in that row by dividing by the row's total probability. This act of spatial attention, when expressed in algebra, yields Bayes' rule.

Table 4.1 Making Bayes' Rule Not Merely Special but Spatial

Row		Column			Marginal
	...	j		...	
i	...	$p(R_i, C_j)$ $= p(R_i \mid C_j)p(C_j)$ $= p(C_j \mid R_i)p(R_i)$		...	$p(R_i)$
Marginal		$p(C_j)$			

Of course, the same relationship applies to columns instead of rows. It is arbitrary which attribute to place down the rows and which attribute to place across the columns. Thus, the analogous spatial relationship applies to columns: If we know the column value, then we restrict attention to that column and normalize the cell probabilities to yield Bayes' rule again.

4.1.3 The Denominator as an Integral over Continuous Values

Up to this point, Bayes' rule has been presented only in the context of discrete-valued variables. It also applies to continuous variables, but probability masses become probability densities and sums become integrals. For continuous variables, Bayes' rule (Equation 4.2) becomes

$$p(y \mid x) = \frac{p(x \mid y)p(y)}{\int dy \, p(x \mid y)p(y)} \tag{4.3}$$

In Equation 4.3, the y in the numerator is a specific fixed value, whereas the y in the denominator is a variable that takes on all possible values of y over the integral. It is this continuous-variable version of Bayes' rule that we will deal with most often.

4.2 APPLIED TO MODELS AND DATA

One of the key applications that makes Bayes' rule so useful is when the row and column variables are data values and model parameter values, respectively. A model specifies the probability of particular data values given the model's structure and particular parameter values. For example, our usual model of coin flips says that $p(\text{datum}=H \mid \theta) = \theta$ and $p(\text{datum}=T \mid \theta) = 1 - \theta$. More generally, a model specifies

$$p(\text{data values} \mid \text{parameters values and model structure})$$

We use Bayes' rule to convert that to what we really want to know, which is how strongly we should believe in the model, given the data:

$$p(\text{parameters values and model structure} \mid \text{data values})$$

When we have observed some data, we use Bayes' rule to determine our beliefs across competing parameter values in a model, and to determine our beliefs across competing models.

It helps to think about the application of Bayes' rule to data and models in terms of a two-way table, shown in Table 4.2. The columns of Table 4.2 correspond to specific values of the model parameter, and the rows of Table 4.2 correspond to specific values of the data. Each cell of the table holds the conjoint probability of the specific combination of parameter value θ and data value D.

Table 4.2 Applying Bayes' Rule to Data and Model Parameter

Data	...	Model Parameter θ value	...	Marginal
$\vdots$		$\vdots$		
D value	...	$p(D, \theta)$ $= p(D \mid \theta)p(\theta)$ $= p(\theta \mid D)p(D)$	...	$p(D)$
$\vdots$		$\vdots$		
Marginal		$p(\theta)$		

That is, $p(D, \theta)$ is the probability of getting that particular combination of data value and parameter value, across all possible combinations of data values and parameter values.

The prior probability of the parameter values is the marginal distribution, $p(\theta)$, which appears in the lower margin of Table 4.2. This is simply the probability of each possible value of θ, collapsed across all possible values of data.

When we observe a particular data value, D, so we know it is true, we are restricting our attention to one specific row of Table 4.2, namely, the row corresponding to the observed value, D. The posterior distribution on θ is obtained by dividing the conjoint probabilities in that row by the row marginal, $p(D)$. Thus, the posterior probability of θ is just the conjoint probabilities in that row, normalized by $p(D)$ to sum to 1.

We need to define some notation and terms at this point. The factors of Bayes' rule have names as indicated here:

$$\underbrace{p(\theta \mid D)}_{\text{posterior}} = \underbrace{p(D \mid \theta)}_{\text{likelihood}} \; \underbrace{p(\theta)}_{\text{prior}} / \underbrace{p(D)}_{\text{evidence}} \qquad (4.4)$$

where the evidence is (from the denominator of Equation 4.3)

$$p(D) = \int d\theta \, p(D \mid \theta)p(\theta) \qquad (4.5)$$

The "prior," $p(\theta)$, is the strength of our belief in θ without the data D. The "posterior," $p(\theta \mid D)$, is the strength of our belief in θ when the data D have been taken into account. The "likelihood," $p(D \mid \theta)$, is the probability that the data could be generated by the model with parameter values θ. The "evidence," $p(D)$, is the probability of the data according to the model, determined by summing across all possible parameter values weighted by the strength of belief in those parameter values.

We talk about parameter values θ only in the context of a particular model; it's the model that gives meaning to the parameter. In some applications, it can help to make the model explicit in Bayes' rule. Let's call the model M. Then, because all the probabilities are defined given that model, we can rewrite Equation 4.4 as

$$\underbrace{p(\theta \mid D, M)}_{\text{posterior}} = \underbrace{p(D \mid \theta, M)}_{\text{likelihood}} \, \underbrace{p(\theta \mid M)}_{\text{prior}} / \underbrace{p(D \mid M)}_{\text{evidence}} \qquad (4.6)$$

where the evidence is

$$p(D \mid M) = \int d\theta \, p(D \mid \theta, M) p(\theta \mid M) \qquad (4.7)$$

It's especially handy to have the model explicitly annotated as in Equation 4.6 when you have more than one model in mind and you're using the data to help determine the strength of belief in each model. Suppose we have two models, creatively named M1 and M2. Then, by Bayes' rule, $p(M1 \mid D) = p(D \mid M1)p(M1)/p(D)$ and $p(M2 \mid D) = p(D \mid M2)p(M2)/p(D)$, where $p(D) = \sum_i p(D \mid M_i)p(M_i)$. Taking the ratio of those equations, we get

$$\frac{p(M1 \mid D)}{p(M2 \mid D)} = \underbrace{\frac{p(D \mid M1)}{p(D \mid M2)}}_{\text{Bayes factor}} \frac{p(M1)}{p(M2)} \qquad (4.8)$$

Equation 4.8 says that the ratio of the posterior beliefs is the ratio of the evidences (as defined in Equation 4.7) times the ratio of the prior beliefs. The ratio of the evidences is called the *Bayes factor*. Examples of all these abstract terms will be provided soon.

Terminological aside: The quantity $p(D \mid M)$, which is called the *evidence* in this book, is sometimes instead called the *marginal likelihood* or *prior predictive* by other authors. The term "evidence" is common in the machine learning literature (e.g., Bishop, 2006, MacKay, 2003). Whenever I refer to the "evidence" for a model, I am referring to $p(D \mid M)$ as defined in Equation 4.7. This usage might be a little confusing in the context of model comparison when considering the equation $p(M1 \mid D) = p(D \mid M1)p(M1)/p(D)$, where $p(D \mid M1)$ plays the *role* of the likelihood, not the evidence. This apparent confusion is cleared up when abbreviated terminology is expanded to its full specificity. The factor $p(D \mid M)$ is not merely "the evidence," it is "the evidence for model M." On the other hand, the factor $p(D)$, in the context of the equation $p(M1 \mid D) =$

$p(D|M1)p(M1)/p(D)$, is not the evidence for *a* model but is the evidence for the entire *set* of models under consideration: $p(D) = \sum_i p(D|M_i)p(M_i)$. The term "likelihood" also deserves expansion. In Equation 4.6, the likelihood is more fully stated as "the likelihood of parameter value θ in model M for data D." That is, the likelihood is referring to the parameter θ. On the other hand, in the context of model comparison, the factor $p(D|M1)$, in the equation $p(M1|D) = p(D|M1)p(M1)/p(D)$, is the "likelihood of the *model* M1 for the data D." To reiterate, the term "evidence" is merely a word to refer to $p(D|M)$. As we will see, its value does not have much meaning by itself. Instead, $p(D|M)$ can only be interpreted in the context of other models.

4.2.1 Data Order Invariance

One more nuance about Bayesian updating of beliefs. Bayes' rule in Equation 4.4 gets us from a prior belief, $p(\theta)$, to a posterior belief, $p(\theta|D)$, when we take into account some data. Now suppose we observe some *more* data, which we'll denote D'. We can then update our beliefs again, from $p(\theta|D)$ to $p(\theta|D',D)$. Here's the question: Does our final belief depend on whether we update with D first and D' second, or update with D' first and D second?

The answer is, it depends. In particular, it depends on the model function that defines the likelihood, $p(D|\theta)$. In many models, $p(D|\theta)$ does not depend in any way on other data. That is, the conjoint probability $p(D, D'|\theta)$ equals $p(D|\theta)p(D'|\theta)$. The data probabilities are independent, according to this type of model. Moreover, in many models the probability function does not change in time or depend on how many data values have been generated. The probability function is stationary. Under these conditions, when $p(D|\theta)$ and $p(D'|\theta)$ are *independent and identically distributed* (commonly referred to as "i.i.d."), then the order of updating has no effect on the final posterior.

This invariance to ordering of the data makes sense intuitively: If the likelihood function has no dependence on time or data ordering, then the posterior shouldn't have any dependence on time or data ordering either. But it's easy to prove mathematically too. First, we'll unpack $p(\theta|D',D)$ by applying Bayes' rule on D':

$$p(\theta|D',D) = \frac{p(D'|\theta,D)\,p(\theta|D)}{\int d\theta\,p(D'|\theta,D)\,p(\theta|D)}$$

Now, notice that $p(D'|\theta,D) = p(D'|\theta)$, because the model asserts that the probability of a data value depends only on the value of θ and not on anything else, such as other data. Therefore, the preceding equation can be rewritten as

$$p(\theta|D',D) = \frac{p(D'|\theta)\,p(\theta|D)}{\int d\theta\,p(D'|\theta)\,p(\theta|D)}$$

Now we use Bayes' rule again, this time for $p(\theta \mid D)$, which converts the equation into

$$p(\theta \mid D', D) = \frac{p(D' \mid \theta) \, p(D \mid \theta) \, p(\theta)/p(D)}{\int d\theta \, p(D' \mid \theta) \, p(D \mid \theta) \, p(\theta)/p(D)}$$

Notice that $p(D)$ in that equation is a constant and cancels out. This last equation, presented earlier, involves the product of $p(D' \mid \theta)$ and $p(D \mid \theta)$. Because multiplication can be done in either order (i.e., it is "commutative" in technical terminology), we arrive at the same formula if we start with the data in the opposite order: $p(\theta \mid D, D')$.

In all of the examples in this book, the likelihood functions generate i.i.d. data. One way of thinking about this assumption is as follows: We assume that every datum is equally representative of the underlying process, regardless of when the datum was observed. Older observations are just as valid and representative as more recent observations, and the underlying process that generates the data has not changed during the course of making the observations.

4.2.2 An Example with Coin Flipping

With all the emphasis on coin flipping, by now you must be imagining flipping coins over pasture fences as you try to fall asleep. Nevertheless, imagine flipping coins once again, and try not to fall asleep. We will start with some prior beliefs about the possible bias of the coin, then flip the coin a few times, and then update our beliefs based on the observed flips.

First, we specify our prior beliefs. We denote the bias as $\theta = p(H)$, the probability of the coin coming up heads. To keep the example straightforward, suppose that we believe there are only three possible values for the coin's bias. Either the coin is fair, with $\theta = 0.50$, or the coin is biased, with $\theta = 0.25$ or $\theta = 0.75$. We believe that the coin is probably fair, but there's some smaller chance it could be biased high or low. This prior probability is graphed in the top panel of Figure 4.1. It shows three "spikes," one over each value of θ that we think could be possible. The spike over $\theta = 0.5$ is tallest, indicating that we believe it to be most likely. Note that the heights of the spikes are probability masses, not densities, because each spike indicates the probability of its specific, discrete value of θ.

Next, we flip the coin to get some data, D, and determine the likelihood, $p(D \mid \theta)$. Suppose we flip the coin 12 times and it comes up heads 3 times. According to our model of the coin, the probability of coming up heads is θ, and the probability of coming up tails is $1 - \theta$. Moreover, the flips are independent of each other, and therefore we can multiply the probabilities of the individual flips to get the probability of the combination of flips. Consequently, the probability of a specific sequence of three heads and nine tails is $p(D \mid \theta) = \theta^3 (1 - \theta)^9$. The resulting likelihood for each value of θ is plotted in the middle panel of Figure 4.1. Notice that the likelihood is highest for

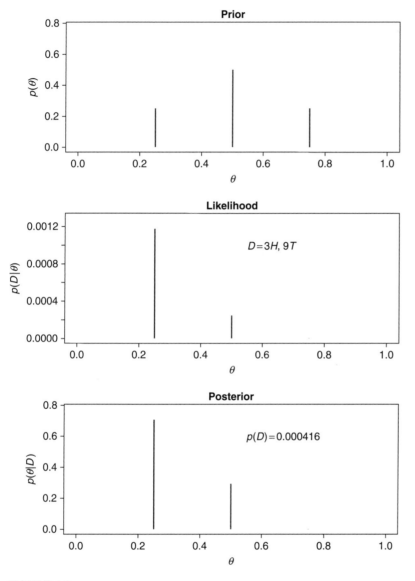

FIGURE 4.1
Bayesian updating of beliefs about the bias of a coin. The prior and posterior
distributions indicate probability masses at discrete candidate values of θ. (The R code
that generated this graph is in Section 4.4.1 (`BayesUpdate.R`).)

$\theta = 0.25$ and lowest for $\theta = 0.75$. This peak at $\theta = 0.25$ makes sense, because
the data have 25% heads, so they are more likely if $\theta = 0.25$ than if $\theta = 0.50$ or
$\theta = 0.75$. The value of θ that maximizes the likelihood is called the *maximum
likelihood estimate* of θ.

The lower panel of Figure 4.1 includes the value of $p(D|M)$, the evidence for the model. Recall from Equation 4.7 that the evidence is the overall probability of the data, averaging across the available parameter values weighted by the degree to which we believe in them: $p(D|M) = \sum_\theta p(D|\theta, M)p(\theta, M)$. This is the normalizer for the posterior distribution, hence it is displayed in the plot of the posterior distribution. The value is displayed as $p(D)$ instead of as $p(D|M)$ because there is only one model in this context, and therefore the M notation is suppressed. When you see the value of $p(D)$ in Figure 4.1, you might think that $p(D)$ is terribly small, until you remember that we are talking about the conjoint probability of several things happening together (i.e., exactly the 12 flips we observed). The probability of 1 head is θ. The probability of 2 heads is θ^2, which is smaller than θ. The probability of 3 heads is θ^3, which is smaller yet. As the set of data D gets bigger, in terms of containing more observations, $p(D)$ gets smaller, regardless of how closely the model θ matches the true bias in the coin.

The bottom panel of Figure 4.1 displays the posterior beliefs for each value of θ. According to Bayes' rule, the posterior is proportional to the product of the prior and the likelihood. So the shape of the posterior is influenced by both the prior and the likelihood. You can see this dual influence in Figure 4.1 by inspecting the relative heights of the left and middle spikes. In the prior, the middle spike is much taller than the left spike. In the likelihood, the middle spike is much shorter than the left spike. In the posterior, there is a compromise between the prior and the likelihood: The middle spike is shorter than the left spike, but not so short as in the likelihood because it (the middle spike) is buoyed up by the prior. Notice how our beliefs have changed from prior to posterior. Initially we believed most strongly in a fair coin. After accounting for the data, we believed most strongly in a biased coin. The Bayesian mathematics let us compute exactly how much our beliefs changed.

4.2.2.1 $p(D|\theta)$ *Is Not* θ

In the examples involving coin flips, it is easy to lose sight of the important fact that $p(D|\theta)$ is different from θ, even though they both are values between 0 and 1 for our current examples. The likelihood $p(D|\theta)$ is a mathematical function of θ. The value of the likelihood function is always a probability (a probability mass if θ has a finite number of values, and a probability density otherwise). The value of the parameter, however, could be on any scale, depending on the meaning of the parameter. In our examples so far, the meaning of the parameter is itself a probability, so it is easy to confuse the parameter value with the likelihood value. Adding to the confusability is the fact that, in our examples so far, the function that maps θ to $p(D=H|\theta)$ has been the identity function:

$$p(D=H|\theta) = \theta \qquad (4.9)$$

and, of course, $p(D=T|\theta) = 1.0 - p(D=H|\theta) = 1.0 - \theta$. It is easy to confuse $p(D|\theta)$ with θ in our examples because the function that relates them is the identity. Later in the book, we will see many examples for which the likelihood function is not the identity function.

The point of this subsection has been to remind you that θ is a parameter that has a scale and meaning in the context of a model. The value $p(D|\theta)$, on the other hand, is a probability, and is a function of the parameter θ. Thus, $p(D|\theta)$ and θ are distinct entities, despite the fact that in simple models of coin flipping, $p(D=H|\theta) = \theta$.

4.3 THE THREE GOALS OF INFERENCE

Back in Section 2.2 (p. 12), I introduced three goals of inference: estimation of parameter values, prediction of data values, and model comparison. Each of these goals are now given precise mathematical expressions.

4.3.1 Estimation of Parameter Values

Estimation of parameter values means determining the extent to which we believe in each possible parameter value. This is precisely what Equation 4.6 tells us. The posterior distribution over the parameter values θ *is* our estimate of those values.

The posterior distribution can be narrow, with most of the probability piled heavily over a small range of θ. In this case, we are fairly certain about the possible values of θ. On the other hand, the posterior probability distribution could be wide, spread over a large range of θ. In this case, we have high uncertainty about the possible values of θ.

4.3.2 Prediction of Data Values

Using our current beliefs, we may want to predict the probability of future data values. To avoid notational conflicts later, I'll denote a data value as y. The predicted probability of data value y is determined by averaging the predicted data probabilities across all possible parameter values, weighted by the belief in the parameter values:

$$p(y) = \int d\theta\, p(y|\theta) p(\theta)$$

Notice that this is exactly the evidence, discussed after Equation 4.4, except that the evidence refers to a specific observed value of y, whereas here we are computing the probability of any possible value of y.

As an example, consider the prior beliefs in the top panel of Figure 4.1. For those beliefs, the predicted probability of getting a head is

$$p(y=H) = \sum_\theta p(y=H|\theta)p(\theta)$$

$$= p(y=H|\theta=0.25)p(\theta=0.25)$$
$$+ p(y=H|\theta=0.50)p(\theta=0.50)$$
$$+ p(y=H|\theta=0.75)p(\theta=0.75)$$
$$= 0.25 \times 0.25 + 0.50 \times 0.50 + 0.75 \times 0.25$$
$$= 0.5$$

and the probability of getting a tail is computed analogously to be $p(y=T) = 0.5$. Notice that the predictions are probabilities of each possible data value, given the current model beliefs.

If we want to predict a particular point value for the next datum, instead of a distribution across all possible data values, it is typical to use the mean (i.e., expected value) of the predicted data distribution. Thus, the predicted value is $\bar{y} = \int dy \, y \, p(y)$. This integral only makes sense if y is on a continuum. If y is nominal, like the result of a coin flip, then the most probable value can be used as "the" predicted value. The decision to use the mean of the predicted values as our single best prediction, instead of, say, the mode or median, relies implicitly on the costs of being wrong and the benefits of being correct. These costs and benefits, called the *utilities*, are considered in more advanced treatments of Bayesian decision theory. For our purposes, we will default to the mean, purely for convenience.

4.3.3 Model Comparison

You may recall from earlier discussion (p. 58) that Bayes' rule is also useful for comparing models. Equation 4.8 indicated that the posterior beliefs in the models involve the evidences of the models. Notice that in this third goal (i.e., model comparison), the evidence term appears again, just as it appeared for the goals of parameter estimation and data prediction.

One of the nice features of Bayesian model comparison is that there is an automatic accounting for model complexity when assessing the degree to which we should believe in the model. This might be best explained with an example. Recall the coin-flipping example discussed earlier, illustrated in Figure 4.1 and reproduced in the left side of Figure 4.2. In that example, we supposed that the bias θ could take on only three possible values. This restriction made the model rather simple. We could instead entertain a more complex model that allows for many more possible values of θ. One such model is illustrated in

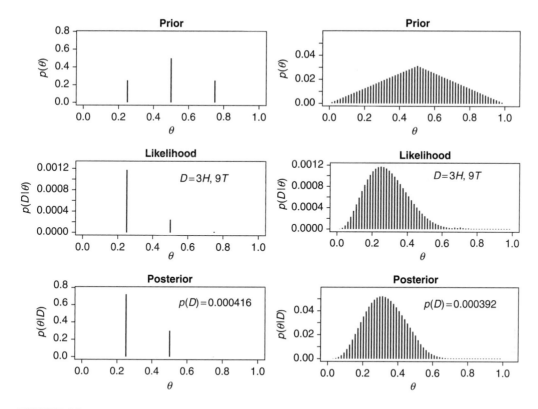

FIGURE 4.2
A simple model in the left column and a complex model in the right column. The prior and posterior distributions indicate probability masses at discrete candidate values of θ. The same data are addressed by both models. The evidence $p(D \mid M_{simple})$ for the simple model is displayed as $p(D)$ in the lower-left panel, and the evidence $p(D \mid M_{complex})$ for the complex model is displayed as $p(D)$ in the lower-right panel. In this case, the data are such that the simple model is favored. The R code that generated these graphs is in Section 4.4.1 (BayesUpdate.R).

the right side of Figure 4.2. This model has 63 possible values of θ instead of only 3. The shape of the prior beliefs in the complex model follows the same triangular shape as in the simple model; there is highest belief in $\theta = 0.50$, with lesser belief in more extreme values.

The complex model has many more available values for θ, and so it has much more opportunity to fit arbitrary data sets. For example, if a sequence of coin flips has 37% heads, the simple model does not have a θ value very close to that outcome, but the complex model does. On the other hand, for θ values that are in both the simple and complex models, the prior probability on those values in the simple model is much higher than in the complex model. Because there are so many possibilities in the complex model, the prior beliefs have to get spread out, very shallowly, over a larger range of possibilities. This can be seen in Figure 4.2 by inspecting the numerical scales on the vertical axes of the prior beliefs. The scale on the simple model is much larger than the scale on the complex model.

Therefore, if the actual data we observe happens to be well accommodated by a θ value in the simple model, we will believe in the simple model more than the complex model, because the prior on that θ value in the simple model is so high. Figure 4.2 shows a case in which this happens. The data have 25% heads, so the evidence in the simple model is larger than the evidence in the complex model. The complex model has its prior spread too thin for us to believe in it as much as we believe in the simple model.

The complex model can be the winner if the data are not adequately fit by the simple model. For example, consider a case in which the observed data have just 1 head and 11 tails. None of the θ values in the simple model is close to this outcome. But the complex model does have some θ values near the observed proportion, even though there is not a strong belief in those values. Figure 4.3 shows that the simple model has less evidence in this situation, and we have stronger belief in the complex model.

The evidence for a model, $p(D|M)$, is not particularly meaningful as an absolute magnitude for a single model. The evidence is most meaningful only in the context of the Bayes factor, $p(D|M1)/p(D|M2)$, which is the *relative* evidence for two models, when considering an observed data set D.[2] Regardless of which model wins, the winning model does not need to be a good model of the data. The model comparison process merely tells us about the *relative* evidence for each model. The winning model is better than the other models in the competition, but the winning model might merely be less bad than the horrible competitors. In later chapters we will explore ways to assess whether the winning model is actually a viable model of the data.

We will see in Chapter 10 that Bayesian model comparison is "really" just a case of Bayesian parameter estimation, in which a parameter that indexes the models is estimated. The individual model parameters depend on the indexical parameter, and thus the scheme involves a hierarchy of dependencies. Hierarchial models are introduced in Chapter 9. The fact that model comparison is a case of parameter estimation is mentioned here only to fend off any mistaken impression that parameter estimation and model comparison are fundamentally different.

---

[2]The Bayes factor, $p(D|M1)/p(D|M2)$, is quite different than considering evidences of a single model for different candidate data sets. Specifically, $p(D1|M)/p(D2|M)$ is *not* a Bayes factor and is not further discussed.

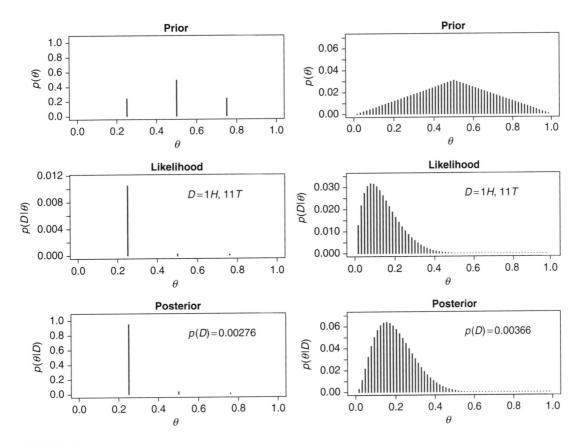

FIGURE 4.3
A simple model in the left column and a complex model in the right column. The prior and posterior distributions indicate probability masses at discrete candidate values of θ. The same data are addressed by both models. The evidence $p(D\,|\,M_{simple})$ for the simple model is displayed as $p(D)$ in the lower-left panel, and the evidence $p(D\,|\,M_{complex})$ for the complex model is displayed as $p(D)$ in the lower-right panel. In this case, the data are such that the complex model is favored. The R code that generated these graphs is in Section 4.4.1 (BayesUpdate.R).

4.3.4 Why Bayesian Inference Can Be Difficult

All three goals involve the denominator of Bayes' formula (i.e., the evidence), which usually means computing a difficult integral. There are a few ways out of this difficulty. The traditional way is to use likelihood functions with "conjugate" prior functions. A prior function that is conjugate to the likelihood function simply makes the posterior function come out with the same functional form as the prior. That is, the math works out nicely. If this method doesn't work, an alternative is to approximate the actual functions with other functions that are easier to work with, and then show that the approximation is reasonably good under typical conditions. But this method is still pure, analytical mathematics. Yet another method is to numerically approximate the integral. When the parameter space is small, then it can be covered with a comb or grid of points and the integral can be computed by exhaustively summing across that grid. But when the parameter space gets even moderately large, there are too many grid points, and therefore other methods must be used. A large class of random sampling methods have been developed, which can be referred to as Markov chain Monte Carlo (MCMC) methods, that can numerically approximate probability distributions even for large spaces. It is the development of these MCMC methods that has allowed Bayesian statistical methods to gain practical use. The next major part of this book explains these various methods in some detail. For applications to complex situations, we will ultimately focus on MCMC methods.

Another potential difficulty of Bayesian inference is determining a reasonable prior. What distribution of beliefs should we start with, over all possible parameter values or over competing models? This question may seem daunting, but in practice it is typically addressed in a straightforward manner. As we will discuss more in Chapter 11, it is actually advantageous and rational to start with an explicit prior. Prior beliefs *should* influence rational inference from data, because the role of new data is to modify our beliefs from whatever they were without the new data. Prior beliefs are *not* capricious and idiosyncratic and unknowable, but instead they are based on publicly agreed facts and theories. Prior beliefs used in data analysis must be admissible by a skeptical scientific audience. When scientists disagree about prior beliefs, the analysis can be conducted with various priors to assess the robustness of the posterior against changes in the prior. Or the priors can be mixed together into a joint prior, with the posterior thereby incorporating the uncertainty in the prior. In summary, for most applications, specification of the prior turns out to be technically *un*problematic, although it is conceptually very important to understand the consequences of one's assumptions about the prior. Thus, the main reason that Bayesian analysis can be difficult is the computation of the evidence, and that computation is tractable in many situations via MCMC methods.

4.3.5 Bayesian Reasoning in Everyday Life
4.3.5.1 Holmesian Deduction
Despite the difficulty of exact Bayesian inference in complex mathematical models, the essence of Bayesian reasoning is frequently used in everyday life. One example has been immortalized in the words of Sherlock Holmes to his friend Dr. Watson: "How often have I said to you that when you have eliminated the impossible, whatever remains, however improbable, must be the truth?" (Arthur Conan Doyle, *The Sign of Four*, 1890, Chapter 6). This reasoning is actually a consequence of Bayesian belief updating, as expressed in Equation 4.4. Let me restate it this way: "How often have I said to you

that when $p(D|\theta_i) = 0$ for all $i \neq j$, then, no matter how small the prior $p(\theta_j) > 0$ is, the posterior $p(\theta_j|D)$ must equal one." Somehow it sounds better the way Holmes said it. The intuition behind Holmes's deduction is clear, though: When we reduce belief in some possibilities, we necessarily increase our belief in the remaining possibilities (*if* our set of possibilities exhausts all conceivable options). Thus, according to Holmesian deduction, when the data make some options less believable, we increase belief in the other options.

4.3.5.2 Judicial Exoneration

The reverse of Holmes's logic is also commonplace. For example, when an object d'art is found fallen from its shelf, our prior beliefs may indict the house cat, but when the visiting toddler is seen dancing next to the shelf, then the cat is exonerated. This downgrading of a hypothesis is sometimes called *explaining away* of a possibility by verifying a different one. This sort of exoneration also follows from Bayesian belief updating: When $p(D|\theta_j)$ is higher, then, even if $p(D|\theta_i)$ is unchanged for all $i \neq j$, $p(\theta_i|D)$ is lower. This logic of exoneration is based on competition of mutually exclusive possibilities: If the culprit is suspect A, then suspect B is exonerated.

Holmesian deduction and judicial exoneration are both expressions of the essence of Bayesian reasoning: We have a space of beliefs that are mutually exclusive and exhaust all possibilities. Therefore, if the data cause us to decrease belief in some possibilities, we must increase belief in other possibilities (as said Holmes), or, if the data cause us to increase belief in some possibilities, we must decrease belief in other possibilities (as in exoneration). What Bayes' rule tells us is exactly how much to shift our beliefs across the available possibilities.

4.4 R CODE

4.4.1 R Code for Figure 4.1

Several new commands are used in this program. When you encounter a puzzling command in an R program, it usually helps to try the help command. For example, when perusing this code, you'll come across the command matrix. To find out about the syntax and usage of this command, do this: At the R command line, type help("matrix") and you'll get some clues about how it works. Then experiment with the command at the interactive command line until you're confident about what its various arguments do. For example, try typing at the command line:

```
matrix( 1:6 , nrow=2 , ncol=3 , byrow=TRUE )
```

Then try

```
matrix( 1:6 , nrow=2 , ncol=3 , byrow=FALSE )
```

The listing that follows includes line numbers in the margins, to facilitate track-
ing the code across page splits and to facilitate referring to specific lines of the
code when you have enthusiastic conversations about it at parties.

Mac users: If you are running R under MacOS instead of in a Windows emulator such as WINE, you will need to change all the windows() commands to
quartz(). Later in the book, when we use BUGS, there is no Mac equivalent
and you must run the programs under WINE or windows.

(BayesUpdate.R)

```
1    # Theta is the vector of candidate values for the parameter theta.
2    # nThetaVals is the number of candidate theta values.
3    # To produce the examples in the book, set nThetaVals to either 3 or 63.
4    nThetaVals = 3
5    # Now make the vector of theta values:
6    Theta = seq( from = 1/(nThetaVals+1) , to = nThetaVals/(nThetaVals+1) ,
7                 by = 1/(nThetaVals+1) )
8    }
9    # pTheta is the vector of prior probabilities on the theta values.
10   pTheta = pmin( Theta , 1-Theta ) # Makes a triangular belief distribution.
11   pTheta = pTheta / sum( pTheta )  # Makes sure that beliefs sum to 1.
12
13   # Specify the data. To produce the examples in the book, use either
14   # Data = c(1,1,1,0,0,0,0,0,0,0,0,0,0) or Data = c(1,0,0,0,0,0,0,0,0,0,0,0,0).
15   Data = c(1,1,1,0,0,0,0,0,0,0,0,0,0)
16   nHeads = sum( Data == 1 )
17   nTails = sum( Data == 0 )
18
19   # Compute the likelihood of the data for each value of theta:
20   pDataGivenTheta = Theta^nHeads * (1-Theta)^nTails
21
22   # Compute the posterior:
23   pData = sum( pDataGivenTheta * pTheta )
24   pThetaGivenData = pDataGivenTheta * pTheta / pData   # This is Bayes' rule!
25
26   # Plot the results.
27   windows(7,10) # create window of specified width,height inches.
28   layout( matrix( c( 1,2,3 ) ,nrow=3 ,ncol=1 ,byrow=FALSE ) ) # 3x1 panels
29   par(mar=c(3,3,1,0))          # number of margin lines: bottom,left,top,right
30   par(mgp=c(2,1,0))            # which margin lines to use for labels
31   par(mai=c(0.5,0.5,0.3,0.1)) # margin size in inches: bottom,left,top,right
32
33   # Plot the prior:
34   plot( Theta , pTheta , type="h" , lwd=3 , main="Prior" ,
35         xlim=c(0,1) , xlab=bquote(theta) ,
36         ylim=c(0,1.1*max(pThetaGivenData)) , ylab=bquote(p(theta)) ,
37         cex.axis=1.2 , cex.lab=1.5 , cex.main=1.5 )
38
39   # Plot the likelihood:
40   plot( Theta , pDataGivenTheta , type="h" , lwd=3 , main="Likelihood" ,
41         xlim=c(0,1) , xlab=bquote(theta) ,
```

```
42        ylim=c(0,1.1*max(pDataGivenTheta)) , ylab=bquote(paste("p(D|",theta,")")),
43        cex.axis=1.2 , cex.lab=1.5 , cex.main=1.5 )
44   text( .55 , .85*max(pDataGivenTheta) , cex=2.0 ,
45        bquote( "D=" * .(nHeads) * "H," * .(nTails) * "T" ) , adj=c(0,.5) )
46
47   # Plot the posterior:
48   plot( Theta , pThetaGivenData , type="h" , lwd=3 , main="Posterior" ,
49        xlim=c(0,1) , xlab=bquote(theta) ,
50        ylim=c(0,1.1*max(pThetaGivenData)) , ylab=bquote(paste("p(",theta,"|D)")),
51        cex.axis=1.2 , cex.lab=1.5 , cex.main=1.5 )
52   text( .55 , .85*max(pThetaGivenData) , cex=2.0 ,
53        bquote( "p(D)=" * .(signif(pData,3)) ) , adj=c(0,.5) )
```

4.5 EXERCISES

Exercise 4.1. [Purpose: Application of Bayes' rule to disease diagnosis, to see the important role of prior probabilities.] Suppose that in the general population, the probability of having a particular rare disease is 1 in a 1000. We denote the true presence or absence of the disease as the value of a parameter, θ, that can have the value $\theta = \frown$ if the disease is present, or the value $\theta = \smile$ if the disease is absent. The base rate of the disease is therefore denoted $p(\theta = \frown) = 0.001$. This is our prior belief that a person selected at random has the disease.

Suppose that there is a test for the disease that has a 99% hit rate, which means that if a person has the disease, then the test result is positive 99% of the time. We denote a positive test result as $D = +$ and a negative test result as $D = -$. The observed test result is a bit of data that we will use to modify our belief about the value of the underlying disease parameter. The hit rate is expressed as $p(D = + | \theta = \frown) = 0.99$. The test also has a false alarm rate of 5%. This means that 5% of the time when the disease is not present, the test falsely indicates that the disease is present. We denote the false alarm rate as $p(D = + | \theta = \smile) = 0.05$.

Suppose we sample a person at random from the population, administer the test, and it comes up positive. What is the posterior probability that the person has the disease? Mathematically expressed, we are asking, what is $p(\theta = \frown | D = +)$? Before determining the answer from Bayes' rule, generate an intuitive answer and see if your intuition matches the Bayesian answer. Most people have an intuition that the probability of having the disease is near the hit rate of the test (which in this case is 0.99).

Hint: The following table of conjoint probabilities might help you understand the possible combinations of events. (The following table is a specific case of Table 4.2, p. 57.) The prior probabilities of the disease are on the bottom marginal. When we know that the test result is positive, we restrict our attention to the row marked $D = +$.

	$\theta = \overset{\frown}{\smile}$	$\theta = \smile$	
D = +	$p(D=+,\theta=\overset{\frown}{\smile})$ $= p(D=+\mid\theta=\overset{\frown}{\smile})p(\theta=\overset{\frown}{\smile})$	$p(D=+,\theta=\smile)$ $= p(D=+\mid\theta=\smile)p(\theta=\smile)$	$p(D=+)$
D = −	$p(D=-,\theta=\overset{\frown}{\smile})$ $= p(D=-\mid\theta=\overset{\frown}{\smile})p(\theta=\overset{\frown}{\smile})$	$p(D=-,\theta=\smile)$ $= p(D=-\mid\theta=\smile)p(\theta=\smile)$	$p(D=-)$
	$p(\theta=\overset{\frown}{\smile})$	$p(\theta=\smile)$	

Caveat regarding interpreting the results: Remember that here we have assumed that the person was selected at random from the population; there were no other symptoms that motivated getting the test.

Exercise 4.2. [Purpose: Iterative application of Bayes' rule, to see how posterior probabilities change with inclusion of more data.] Continuing from the previous exercise, suppose that the same randomly selected person as in the previous exercise is retested after the first test comes back positive, and on the retest the result is negative. Now what is the probability that the person has the disease? Hint: *For the prior probability of the retest, use the posterior computed from the previous exercise.* Also notice that $p(D=-\mid\theta=\overset{\frown}{\smile}) = 1 - p(D=+\mid\theta=\overset{\frown}{\smile})$ and $p(D=-\mid\theta=\smile) = 1 - p(D=+\mid\theta=\smile)$.

Exercise 4.3. [Purpose: To get an intuition for the previous results by using "natural frequency" and "Markov" representations.]

(A) Suppose that the population consists of 100,000 people. Compute how many people should fall into each cell of the table in the hint shown in Exercise 4.1. To compute the expected frequency of people in a cell, just multiply the cell probability by the size of the population. To get you started, a few of the cells of the frequency table are filled in here:

	$\theta = \overset{\frown}{\smile}$	$\theta = \smile$	
D = +	$\mathrm{freq}(D=+,\theta=\overset{\frown}{\smile})$ $= p(D=+,\theta=\overset{\frown}{\smile})N$ $= p(D=+\mid\theta=\overset{\frown}{\smile})p(\theta=\overset{\frown}{\smile})N$ $= 99$	$\mathrm{freq}(D=+,\theta=\smile)$ $= p(D=+,\theta=\smile)N$ $= p(D=+\mid\theta=\smile)p(\theta=\smile)N$ $=$	$\mathrm{freq}(D=+)$ $= p(D=+)N$ $=$
D = −	$\mathrm{freq}(D=-,\theta=\overset{\frown}{\smile})$ $= p(D=-,\theta=\overset{\frown}{\smile})N$ $= p(D=-\mid\theta=\overset{\frown}{\smile})p(\theta=\overset{\frown}{\smile})N$ $= 1$	$\mathrm{freq}(D=-,\theta=\smile)$ $= p(D=-,\theta=\smile)N$ $= p(D=-\mid\theta=\smile)p(\theta=\smile)N$ $=$	$\mathrm{freq}(D=-)$ $= p(D=-)N$ $=$
	$\mathrm{freq}(\theta=\overset{\frown}{\smile})$ $= p(\theta=\overset{\frown}{\smile})N$ $= 100$	$\mathrm{freq}(\theta=\smile)$ $= p(\theta=\smile)N$ $= 99{,}900$	N $= 100{,}000$

Notice the frequencies on the lower margin of the table. They indicate that out of 100,000 people, only 100 have the disease, whereas 99,900 do not have the disease. These marginal frequencies instantiate the prior probability that $p(\theta = \frown) = 0.001$. Notice also the cell frequencies in the column $\theta = \frown$, which indicate that of 100 people with the disease, 99 have a positive test result and 1 has a negative test result. These cell frequencies instantiate the hit rate of 0.99. Your job for this part of the exercise is to fill in the frequencies of the remaining cells of the table.

(B)   Take a good look at the frequencies in the table you just computed for the previous part. These are the so-called *natural frequencies* of the events, as opposed to the somewhat unintuitive expression in terms of conditional probabilities (Gigerenzer & Hoffrage, 1995). From the cell frequencies alone, determine the proportion of people who have the disease, given that their test result is positive. Before computing the exact answer arithmetically, first give a rough intuitive answer merely by looking at the relative frequencies in the row $D = +$. Does your intuitive answer match the intuitive answer you provided back in Exercise 4.1? Probably not. Your intuitive answer here is probably much closer to the correct answer. Now compute the exact answer arithmetically. It should match the result from applying Bayes' rule in Exercise 4.1.

(C)   Now we'll consider a related representation of the probabilities in terms of natural frequencies, which is especially useful when we accumulate more data. Krauss, Martignon, & Hoffrage (1999) called this type of representation a *Markov* representation. Suppose now we start with a population of $N = 10,000,000$ people. We expect 99.9% of them (i.e., 9,990,000) not to have the disease, and just 0.1% (i.e., 10,000) to have the disease. Now consider how many people we expect to test positive. Of the 10,000 people who have the disease, 99% (i.e., 9900), will be expected to test positive. Of the 9,990,000 people who do not have the disease, 5%, (i.e., 499,500) will be expected to test positive. Now consider retesting everyone who has tested positive on the first test. How many of them are expected to show a negative result on the retest? Use this diagram to compute your answer:

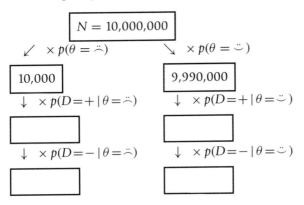

When computing the frequencies for the empty boxes, be careful to use the proper conditional probabilities.

(D)   Use the diagram in the previous part to answer this question: What proportion of people who test positive at first and then negative on retest actually have the disease? In other words, of the total number of people at the bottom of the diagram in the previous part (those are the people who tested positive then negative), what proportion of them are in the left branch of the tree? *How does the result compare with your answer to Exercise 4.2?*

Exercise 4.4. [Purpose: To see a hands-on example of data-order invariance.] Consider again the disease and diagnostic test of the previous two exercises. Suppose that a person selected at random from the population gets the test and it comes back negative. Compute the probability that the person has the disease. The person then is retested, and on the the second test the result is positive. Compute the probability that the person has the disease. *How does the result compare with your answer to Exercise 4.2?*

Exercise 4.5. [Purpose: An application of Bayes' rule to neuroscience, to infer cognitive function from brain activation.] Cognitive neuroscientists investigate which areas of the brain are active during particular mental tasks. In many situations, researchers observe that a certain region of the brain is active and infer that a particular cognitive function is therefore being carried out. Poldrack (2006) cautioned that such inferences are not necessarily firm and need to be made with Bayes' rule in mind. Poldrack (2006) reported the following frequency table of previous studies that involved any language-related task (specifically phonological and semantic processing) and whether or not a particular region of interest (ROI) in the brain was activated:

	Language Study	Not Language Study
Activated	166	199
Not activated	703	2154

Suppose that a new study is conducted and finds that the ROI is activated. If the prior probability that the task involves language processing is 0.5, what is the posterior probability, given that the ROI is activated? (Hint: Poldrack (2006) reports that it is 0.69. Your job is to derive this number.)

Exercise 4.6. [Purpose: To make sure you really understand what is being shown in Figure 4.1.] Derive the posterior distribution in Figure 4.1 by hand. The prior has $p(\theta=0.25) = 0.25$, $p(\theta=0.50) = 0.50$, and $p(\theta=0.75) = 0.25$. The data consist of a specific sequence of flips with three heads and nine tails, so $p(D|\theta) = \theta^3 (1 - \theta)^9$. Hint: Check that your posterior probabilities sum to 1.

Exercise 4.7. [Purpose: For you to see, hands on, that $p(D)$ lives in the denominator of Bayes' rule.] Compute $p(D)$ in Figure 4.1 by hand. Hint: Did you notice that you already computed $p(D)$ in the previous exercise?

2 PART

All the Fundamentals Applied to Inferring a Binomial Proportion

CHAPTER 5

Inferring a Binomial Proportion via Exact Mathematical Analysis

I built up my courage to ask her to dance
By drinking too much before taking the chance.
I fell on my butt when she said see ya later;
Less priors might make my posterior beta.

This part of the book addresses a simple scenario: estimating the underlying probability that a coin comes up heads. The methods don't require that we are referring to a coin, of course. All we require in this scenario is that the space of possibilities for each datum has just two possible values that are mutually exclusive. These two values have no ordinal or metric relationship with each other, they are just nominal values. Because there are two nominal values, I refer to this sort of data as "binomial," or sometimes as "dichotomous."

Doing Bayesian Data Analysis: A Tutorial with R and BUGS. DOI: 10.1016/B978-0-12-381485-2.00005-5
© 2011, Elsevier Inc. All rights reserved.

We also assume that each datum is independent of the others and that the underlying probability is stationary through time. Coin flipping is the standard example of this situation: There are two possible outcomes (head or tail), the flips are (we assume) independent of each other, and the probability of getting a head is stationary through time (again, by assumption). Other examples include the proportion of free throws hit by a player in basketball, the proportion of babies born that are girls, the proportion of heart surgery patients who survive more than a year after surgery, the proportion of people who agree with a statement on a survey, the proportion of widgets on an assembly line that are faulty, and so on. While we talk about heads and tails for coins, keep in mind that the methods could be applied to many other interesting real-world situations.

In a Bayesian analysis, we begin with some prior beliefs over possible probabilities of the coin coming up heads. Then we observe some data that consist of a set of results from flipping the coin. Then we infer the posterior distribution of our beliefs using Bayes' rule. Bayes' rule requires us to specify the likelihood function, and that is the topic of the next section.

5.1 THE LIKELIHOOD FUNCTION: BERNOULLI DISTRIBUTION

When we flip a coin, the result can be a head or a tail. We will denote the result by y, with $y = 1$ for head and $y = 0$ for tail. Giving the head or tail a numerical value (i.e., 1 or 0) is helpful for mathematically expressing the probabilities. But do not be lulled into thinking that somehow a head is "greater than" a tail because $1 > 0$ or that the "distance" between a head and a tail is 1 because $|1 - 0| = 1$. We will use $y = 1$ for head and $y = 0$ for tail only for convenience, but we must remember that the data are truly nominal (i.e., categorical) values without any metric or ordinal properties.

As discussed in Section 4.2.2.1, p. 62, the probability of the coin coming up heads is a function of an underlying parameter: $p(y = 1|\theta) = f(\theta)$. We assume a particularly simple function, namely, the identity: $p(y = 1|\theta) = \theta$. Consequently, the probability of tails is the complement—that is, $p(y = 0|\theta) = 1 - \theta$. These two equations can be combined into a single expression as follows:

$$p(y|\theta) = \theta^y (1 - \theta)^{(1-y)} \tag{5.1}$$

for y in the set $\{1, 0\}$ and θ in the interval $[0, 1]$. Notice that when $y = 1$, the right-hand side of Equation 5.1 reduces to θ, and when $y = 0$, the right-hand side of Equation 5.1 reduces to $1 - \theta$.

The formula in Equation 5.1 expresses the *Bernoulli distribution*. The Bernoulli distribution is a probability distribution over the two discrete values of y, for any fixed value of θ. In particular, the sum of the probabilities is 1, as must be true of a probability distribution: $\sum_y p(y|\theta) = p(y = 1|\theta) + p(y = 0|\theta) = \theta + (1 - \theta) = 1$.

Another perspective on Equation 5.1 is to think of the data value y as fixed by an observation and the value of θ as variable. Equation 5.1 then specifies the probability of the fixed y value if θ has some particular value. Different values of θ yield different probabilities of the datum y. When thought of in this way, Equation 5.1 is the *likelihood function* of θ.

Notice that the likelihood function is a function of a continuous value θ, whereas the Bernoulli distribution is a discrete distribution over the two values of y. The likelihood function, though it specifies a probability at each value of θ, is *not* a probability distribution. In particular, it does not integrate to 1. For example, suppose that $y = 1$. Then $\int_0^1 d\theta\, \theta^y\, (1 - \theta)^{(1-y)} = \int_0^1 d\theta\, \theta = \frac{1}{2} \neq 1$.

In Bayesian inference, the function $p(y|\theta)$ is usually thought of with the data, y, known and fixed, and the parameter, θ, uncertain and variable. Therefore, $p(y|\theta)$ is usually called the likelihood function for θ, and Equation 5.1 is called the *Bernoulli likelihood function*. Don't forget, however, that the same function is also the probability of the datum, y.

When we flip the coin N times, we have a set of data, $D = \{y_1, \ldots, y_N\}$, where each y_i is 0 or 1. By assumption, each flip is independent of the others. (Recall the definition of independence from Section 3.4.3, p. 46.) Therefore, the probability of getting the set of N flips $D = \{y_1, \ldots, y_N\}$ is the product of the individual outcome probabilities:

$$p(\{y_1, \ldots, y_N\}|\theta) = \prod_i p(y_i|\theta)$$

$$= \prod_i \theta^{y_i}\, (1 - \theta)^{(1-y_i)} \tag{5.2}$$

If the number of heads in the set of flips is denoted $z = \sum_i^N y_i$, then Equation 5.2 can be written as

$$p(z, N|\theta) = \theta^z\, (1 - \theta)^{(N-z)} \tag{5.3}$$

I often lapse terminologically sloppy and refer to Equation 5.3 as the Bernoulli likelihood function for a set of flips, but please remember that the Bernoulli distribution is really Equation 5.1 and refers to a single flip.[1]

---

[1] Some readers might be familiar with the binomial distribution, $p(z|N, \theta) = \binom{N}{z}\theta^z\, (1 - \theta)^{(N-z)}$, and wonder why it is not used here. The reason is that here we are considering each flip of the coin to be a distinct event, whereby each observation has just two possible values, $y \in \{0, 1\}$. The probability of the *set* of

5.2 A DESCRIPTION OF BELIEFS: THE BETA DISTRIBUTION

In this chapter, we use purely mathematical analysis, with no numerical approximation, to derive the mathematical form of the posterior distribution of beliefs. To do this, we need a mathematical description of our prior beliefs. That is, we need a mathematical formula that describes the prior belief probability for each value of the bias θ in the interval $[0, 1]$.

In principle, we could use any probability density function supported on the interval $[0, 1]$. When we intend to apply Bayes' rule (Equation 4.4), however, there are two desiderata for mathematical tractability. First, it would be convenient if the product of $p(y|\theta)$ and $p(\theta)$, which is in the numerator of Bayes' rule, results in a function of the same form as $p(\theta)$. When this is the case, the prior and posterior beliefs are described using the same form of function. This quality allows us to include subsequent additional data and derive another posterior distribution, again of the same form as the prior. Therefore, no matter how much data we include, we always get a posterior of the same functional form. Second, we desire the denominator of Bayes' rule, namely, $\int d\theta\, p(y|\theta)p(\theta)$, to be solvable analytically. This quality also depends on how the form of the function $p(\theta)$ relates to the form of the function $p(y|\theta)$. When the forms of $p(y|\theta)$ and $p(\theta)$ combine so that the posterior distribution has the same form as the prior distribution, then $p(\theta)$ is called a *conjugate prior* for $p(y|\theta)$. Notice that the prior is conjugate only with respect to a particular likelihood function.

In the present situation, we are seeking a functional form for a prior density over θ that is conjugate to the Bernoulli likelihood function in Equation 5.1. If you think about it a minute, you'll notice that if the prior is of the form $\theta^a(1 - \theta)^b$, then when you multiply the Bernoulli likelihood with the prior, you'll again get a function of the same form, namely, $\theta^{(y+a)}(1 - \theta)^{(1-y+b)}$. So, to express the prior beliefs over θ, we seek a probability density function involving $\theta^a(1 - \theta)^b$.

A probability density of that form is called a *beta distribution*. Formally, a beta distribution has two parameters, called a and b, and the density itself is defined as

$$p(\theta|a, b) = \text{beta}(\theta|a, b)$$
$$= \theta^{(a-1)}(1 - \theta)^{(b-1)}/B(a, b) \tag{5.4}$$

---

events is then the product of the individual event probabilities, as in Equation 5.2. If we instead considered a single "event" to be the flipping of N coins, then an observation of a *single* event could have $N + 1$ possible values, $z \in \{0, 1, \ldots, N\}$, and the probability of those values would be given by the binomial distribution. The binomial distribution is explained in Section 11.1.1, p. 267.

where $B(a, b)$ is simply a normalizing constant that ensures that the area under the beta density integrates to 1.0, as all probability density functions must. In other words, the normalizer for the beta distribution is $B(a, b) = \int_0^1 d\theta\, \theta^{(a-1)} (1 - \theta)^{(b-1)}$.

Remember that the beta distribution is only defined for values of θ in the interval [0, 1]. The values of a and b must be positive; zero and negative values don't work. Notice that in the definition of the beta distribution (Equation 5.4), the value of θ is raised to the power $a - 1$, not the power a, and the value of $(1 - \theta)$ is raised to the power $b - 1$, not the power b. Be careful to distinguish the beta *function*, $B(a, b)$, from the beta *distribution*, beta$(\theta|a, b)$. The beta function is not a function of θ because θ has been "integrated out." In the programming language R, beta$(\theta|a, b)$ is `dbeta(`θ`,a,b)`, and $B(a, b)$ is `beta(a,b)`.[2]

Examples of the beta distribution are shown in Figure 5.1. Each panel of Figure 5.1 shows the beta distribution for particular values of a and b, as indicated inside each panel. Notice that as a gets bigger, the bulk of the distribution moves rightward over higher values of θ, but as b gets bigger, the bulk of the distribution moves leftward over lower values of θ. Notice that as a and b get bigger together, the beta distribution gets narrower.

5.2.1 Specifying a Beta Prior

We would like to specify a beta distribution that describes our prior beliefs. For this goal it is useful to know the mean and variance (recall Equations 3.6, p. 37, and 3.8, p. 38) of the beta distribution, so we can get a sense of which a and b values correspond to reasonable descriptions of our prior beliefs about θ. It turns out that the mean of the beta$(\theta|a, b)$ distribution is $\bar\theta = a/(a + b)$. Thus, when $a = b$, the mean is 0.5, and the bigger a is relative to b, the bigger the mean is. The standard deviation of the beta distribution is $\sqrt{\bar\theta(1 - \bar\theta)/(a + b + 1)}$. Notice that the standard deviation gets smaller when $a + b$ gets larger.

You can think of a and b in the prior as if they were previously observed data, in which there were a heads and b tails in a total of $a + b$ flips. For example, if we have no prior knowledge other than the knowledge that the coin has a head side and a tail side, that's tantamount to having previously observed one head and one tail, which corresponds to $a = 1$ and $b = 1$. You can see in Figure 5.1 that when $a = 1$ and $b = 1$ the beta distribution is uniform: All values of θ are equally probable. As another example, if we think that the coin is probably fair

---

[2] Whereas it is true that $B(a, b) = \int_0^1 d\theta\, \theta^{(a-1)} (1 - \theta)^{(b-1)}$, the beta function can also be expressed as $B(a, b) = \Gamma(a)\Gamma(b)/\Gamma(a + b)$, where Γ is the *gamma function*: $\Gamma(a) = \int_0^\infty dt\, t^{(a-1)} \exp(-t)$. The gamma function is a generalization of the factorial function, because, for integer valued a, $\Gamma(a) = (a - 1)!$. In R, $\Gamma(a)$ is `gamma(a)`. Many sources define the beta function this way, in terms of the gamma function.

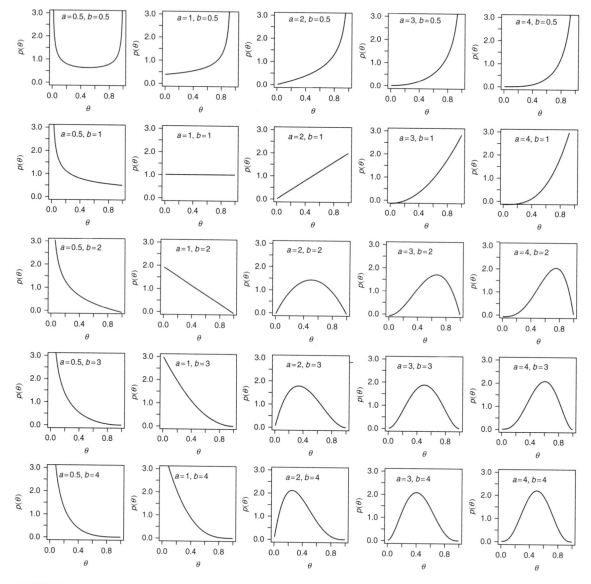

FIGURE 5.1
Examples of beta distributions.

but we're not very sure, then we can imagine that the previously observed data had, say, $a = 4$ heads and $b = 4$ tails. You can see in Figure 5.1 that when $a = 4$ and $b = 4$ the beta distribution is peaked at $\theta = 0.5$, but higher or lower values of θ are moderately probable too.

Instead of thinking in terms of a heads and b tails in the prior data, it's easier to think in terms of the mean proportion of heads in the prior data and its sample size. The mean proportion of heads is $m = a/(a+b)$ and the sample size is $n = a + b$. Solving those two equations for a and b yields

$$a = mn \quad \text{and} \quad b = (1 - m)n \tag{5.5}$$

where m is our guess for the prior mean value of the proportion θ, and n is our guess for the number of observations girding our prior belief. The value we choose for the prior n can be thought of as the number of new flips of the coin that we would need to make us teeter between the new data and the prior belief about m. If we would only need a few new flips to sway our beliefs, then our prior beliefs should be represented by a small n. If we would need a large number of new flips to sway us away from our prior beliefs about m, then our prior beliefs are worth a very large n. For example, suppose that I think the coin is fair, so $m = 0.5$, but I'm not highly confident about it, so maybe I imagine I've seen only $n = 8$ previous flips. Then $a = mn = 4$ and $b = (1 - m)n = 4$, which, as we saw before, is a beta distribution peaked at $\theta = 0.5$ and with higher or lower values less probable.

Another way of establishing the shape parameters is by starting with the mean and standard deviation of the desired beta distribution. You must be careful with this approach, because the standard deviation must make sense in the context of a beta density. In particular, the standard deviation should typically be less than 0.289, which is the standard deviation of a uniform density. For a beta density with mean m and standard deviation s, the shape parameters are

$$a = m \left(\frac{m(1-m)}{s^2} - 1 \right) \quad \text{and} \quad b = (1 - m) \left(\frac{m(1-m)}{s^2} - 1 \right) \tag{5.6}$$

For example, if $m = 0.5$ and $s = 0.28867$, Equation 5.6 implies that $a = 1$ and $b = 1$. As another example, if $m = 0.5$ and $s = 0.1$, then $a = 12$ and $b = 12$. In other words, a beta$(\theta|12,12)$ density has a standard deviation of 0.1.

In most applications, we will deal with beta distributions for which $a \geq 1$ and $b \geq 1$ (i.e., $n \geq 2$), which reflects prior knowledge that the coin has a head side and a tail side. There are some situations, however, in which it may be convenient to use beta distributions in which $a < 1$ and/or $b < 1$. For example, we might believe that the coin is a trick coin that nearly always comes up heads or nearly always comes up tails, but we don't know which. In this case, the bimodal beta$(\theta; 0.5, 0.5)$ prior might be a useful description of our prior belief, as shown in the top-left panel of Figure 5.1. Exercise 5.4 has you explore this a bit more.

5.2.2 The Posterior Beta

Now that we have determined a convenient prior for the Bernoulli likelihood function, let's figure out exactly what the posterior distribution is when we apply Bayes' rule (Equation 4.4, p. 57). Suppose we have a set of data comprising N flips with z heads. Substituting the Bernoulli likelihood (Equation 5.3) and the beta prior distribution (Equation 5.4) into Bayes' rule yields

$$p(\theta|z, N) = p(z, N|\theta)p(\theta)/p(z, N)$$

$$= \theta^z (1-\theta)^{(N-z)} \; \theta^{(a-1)} (1-\theta)^{(b-1)} / [B(a,b)p(z,N)]$$

$$= \theta^{((z+a)-1)} (1-\theta)^{((N-z+b)-1)} \Bigg/ \underbrace{\frac{[B(a,b)\,p(z,N)]}{B(z+a, N-z+b)}} \tag{5.7}$$

In that sequence of equations, you probably followed the collection of powers of θ and of $(1-\theta)$, but you may have balked at the transition, underbraced in the denominator, from $B(a,b)p(z,N)$ to $B(z+a, N-z+b)$. This transition was not made via some elaborate analysis of integrals. Instead, the transition was made by simply thinking about what the normalizing factor for the numerator must be. The numerator is $\theta^{((z+a)-1)} (1-\theta)^{((N-z+b)-1)}$, which is the numerator of a beta$(\theta|z+a, N-z+b)$ distribution. For the function in Equation 5.7 to be a probability distribution, as it must be, the denominator must be the normalizing factor for the corresponding beta distribution.

In other words, Equation 5.7 says this: If the prior distribution is beta$(\theta|a, b)$ and the data have z heads in N flips, then the posterior distribution is beta$(\theta|z+a, N-z+b)$. The simplicity of that updating rule is one of the beauties of the mathematical approach to Bayesian inference.

It is also revealing to think about the relationship between the prior and posterior means. The prior mean of θ is $a/(a+b)$. The posterior mean is $(z+a)/[(z+a)+(N-z+b)] = (z+a)/(N+a+b)$. The posterior mean can be algebraically rearranged into a weighted average of the prior mean, $a/(a+b)$, and the data proportion, z/N:

$$\underbrace{\frac{z+a}{N+a+b}}_{\text{posterior}} = \underbrace{\frac{z}{N}}_{\text{data}} \underbrace{\frac{N}{N+a+b}}_{\text{weight}} + \underbrace{\frac{a}{a+b}}_{\text{prior}} \underbrace{\frac{a+b}{N+a+b}}_{\text{weight}} \tag{5.8}$$

Equation 5.8 indicates that the posterior mean is always somewhere between the prior mean and the proportion in the data. The mixing weight on the prior mean has N in its denominator, so it decreases as N increases. The mixing weight on the data proportion increases as N increases. So the more data we have, the less is the influence of the prior, and the posterior mean gets closer to the proportion in the data. In particular, when $N = a + b$, the mixing weights are 0.5, which indicates that the prior mean and the data proportion have equal influence in the posterior. This result echoes what was said earlier (Equation 5.5) regarding how to set a and b to represent our prior beliefs: The choice of prior n should represent the size of the new data set that would sway us away from our prior toward the data proportion.

5.3 THREE INFERENTIAL GOALS

5.3.1 Estimating the Binomial Proportion

The posterior distribution over θ tells us exactly how much we believe in each possible value of θ. When the posterior is a beta distribution, we can make a graph of the distribution and see in glorious detail what our new beliefs look like. We can extract numerical details of the distribution by using handy functions in R.

Figure 5.2 shows examples of posterior beta distributions. Each column of graphs show a prior beta distribution, a likelihood graph, and the resulting posterior distribution. Both columns use the same data and therefore have the same likelihood graphs. The columns have different priors, however, hence different posteriors. The prior in the left column is uniform, which represents a prior of tremendous uncertainty wherein any bias is equally believable. The prior in the right column loads most belief over a bias of $\theta = 0.5$, indicating a moderately strong prior belief that the coin is fair. As indicated in the graphs of the likelihood, the coin is flipped 14 times and comes up heads 11 times. The posterior beta distributions are graphed in the bottom row. You can see in the left column that when the prior is uniform, then the posterior exactly mirrors the likelihood. When the prior is loaded over $\theta = 0.5$, however, the posterior is only slightly shifted away from the prior toward the proportion of heads in the data. This small shift is a graphic depiction of the relationship expressed in Equation 5.8.

The posterior distribution indicates which values of θ are relatively more credible than others. One way of summarizing where the bulk of the posterior resides is with the highest density interval (HDI), which was introduced in Section 3.3.5. The 95% HDI is an interval that spans 95% of the distribution, such that every point inside the interval has higher believability than any point outside the interval. Figure 5.2 shows the 95% HDI in the two posteriors. You can see that the 95% HDI is fairly wide when the prior is uncertain, but it is narrower when the prior is more certain. Thus, in this case, the posterior inherits the relative uncertainty of the prior, and the width of the 95% HDI is one measure of uncertainty.

The 95% HDI is also one way for declaring which values of the parameter are deemed "credible." Suppose we want to decide whether or not a value of

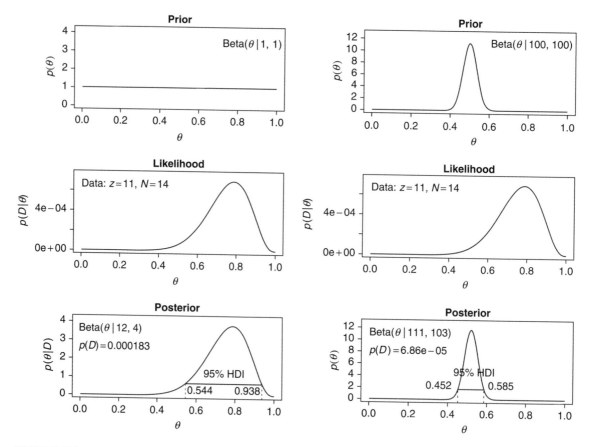

FIGURE 5.2
Two different beta priors, updated using the same data. (The R code that generated these graphs is in Section 5.5.1 (BernBeta.R).)

interest for θ is credible, given the data. Consider, for example, $\theta = 0.5$, which indicates that the coin is fair. We first establish a *region of practical equivalence* (ROPE) around the value of interest, which means a small interval such that any value within the ROPE is equivalent to the value of interest for all practical purposes. Suppose that we declare the ROPE for $\theta = 0.5$ to be 0.48 to 0.52. We want to know if any values within the ROPE are reasonably credible, given the data. How should we define "reasonably credible"? One way is by saying that any points within the 95% HDI are reasonably credible. Hence, we use the following heuristic decision rule: A value of interest, such as $\theta = 0.5$, is declared to be incredible if no point in its ROPE falls within the 95% HDI of the posterior.

It is important to distinguish the two roles for the HDI just mentioned. One role for the HDI is acting as a summary of the distribution. A different role for the HDI is using it for deciding whether a value of interest is or is not credible. This latter process, of converting a rich posterior distribution to a discrete yes/no decision about credibility, involves many extra assumptions that have nothing to do with the HDI. The HDI can be a useful summary apart from whether or not it is used to decide the credibility of a point. These issues are explored at length in Chapter 12.

5.3.2 Predicting Data

As introduced back in Section 4.3.2 (p. 63), the predicted probability of a datum value y is determined by averaging that value's probability across all possible parameter values, weighted by the belief in the parameter values: $p(y) = \int d\theta\, p(y|\theta)p(\theta)$. The belief in the parameter values, $p(\theta)$, is the current posterior belief, including the data observed so far, which we can indicate explicitly as $p(\theta|z, N)$.

In the present application, the predicted probability of heads is particularly simple, because $p(y = 1|\theta) = \theta$, and therefore

$$p(y = 1) = \int d\theta\, p(y = 1|\theta)\, p(\theta|z, N)$$

$$= \int d\theta\, \theta\, p(\theta|z, N)$$

$$= \bar{\theta}|z, N$$

$$= (z + a)/(N + a + b) \tag{5.9}$$

In other words, the predicted probability of heads is just the mean of the posterior distribution over θ. Recall from Equation 5.8 that the posterior mean is a weighted mixture of the prior mean and the data proportion. So the predicted probability of getting a head on the next flip is somewhere between the prior mean and the proportion of heads in the flips observed so far.

Let's make that concrete by considering a particular prior and sequence of flips. Suppose that we start with a uniform prior, beta($\theta|1, 1$). We flip the coin once and get a head. The posterior is then beta($\theta|2, 1$), which has a mean of 2/3. Thus, after the first flip comes up heads, the predicted probability of heads on the next flip is 2/3. Suppose we flip the coin a second time and again get a head. The posterior is then beta($\theta|3, 1$), and the predicted probability of heads on the next flip is 3/4. Notice that even though we have flipped the coin twice and observed heads both times, we do not predict that there is 100% chance of coming up heads on the next flip, because the uncertainty of the prior beliefs is mixed with the observed data.

Consider a variation of that example in which we start with a prior of beta(θ|50, 50), which expresses a fairly strong prior belief that the coin is fair (about 95% of the probability mass lies between $\theta = 0.40$ and $\theta = 0.60$). Suppose we flip the coin twice and get heads both times. The posterior is beta(θ|52, 50), and hence the predicted probability of getting a head on the next flip is $52/102 \approx 51\%$ (which is much different than what we predicted when starting with a uniform prior). Because the prior $a + b$ was so large, it will take a large N to overpower the prior belief.

5.3.3 Model Comparison

You may recall from the previous chapter (particularly Equation 4.8 on page 58) that Bayes' rule can be used to compare models. To do this, we compute the evidence, $p(D|M)$, for each model. The evidence is the weighted total probability of the newly observed data across all possible parameter values, where the total is computed by weighting each parameter value by its prior probability. That is, $p(D|M) = \int d\theta\, p(D|\theta, M)\, p(\theta|M)$.

In the present scenario, the data D are expressed by the values z and N. When using a Bernoulli likelihood and a beta prior, then the evidence $p(D|M)$ is $p(z, N)$ and it is especially easy to compute. In Equation 5.7, the denominator (i.e., the part with the underbrace) showed that

$$B(a, b)\, p(z, N) = B(z + a, N - z + b)$$

Solving for $p(z, N)$ reveals that

$$p(z, N) = B(z + a, N - z + b) \,/\, B(a, b) \tag{5.10}$$

Thus, we can determine $p(z, N)$ using well-established beta functions, and we do not need to do any difficult integral calculus.

The lower panels of Figure 5.2 show the values of $p(z, N)$ for two different priors, for a fixed data set $z = 11, N = 14$. One prior is uniform, while the other prior is strongly peaked over $\theta = 0.50$. The data have a proportion of 1's that is not very close to 0.5, and therefore the prior that is peaked over 0.5 does not capture the data very well. The peaked prior has very low belief in values of θ near the data proportion, which means that $p(z, N)$ for the peaked-prior model is relatively small. The uniform prior, on the other hand, has relatively more belief in values of θ near the data proportion, so its $p(D|M)$ is relatively high.

Consider a data set in which half the flips were heads (e.g., $z = 7$ and $N = 14$). Then which prior would produce the larger $p(z, N)$? Figure 5.3 shows the answer: The prior peaked over $\theta = 0.5$ is now preferred.

When we are evaluating the veracity of a model, the prior distribution for its parameters must be considered along with the likelihood function. In this chapter we are using a Bernoulli likelihood function and a beta prior. We can-not say whether the Bernoulli likelihood is a "good" model of the coin flips without also specifying the values of θ that we believe in. Some values of θ may match the data well, but other values may not. If the values of θ that we believe in are not the values that match the data, then the model is not very good.

Because the prior distributions are part of the model, we can think of differ-ent prior distributions as constituting different models. We already took this approach, back in Figures 4.2 (p. 65) and 4.3 (p. 67). The simple model had a prior with nonzero belief on just a few values of θ, whereas the complex model had a prior with nonzero belief on many values of θ. We compared the models

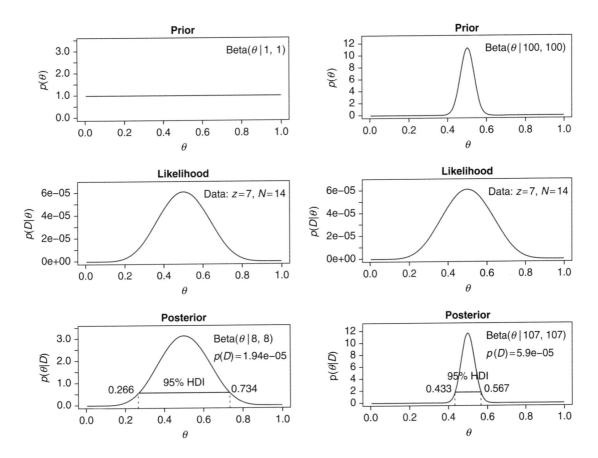

FIGURE 5.3
Two different beta priors, updated using the same data ($z = 7, N = 14$). The evidences are denoted by $p(D)$ (with M suppressed) in the lower panels of each column, and they favor the prior that is peaked over $\theta = 0.5$. Contrast with Figure 5.2, for which the evidences favored the uniform prior.

by considering the relative magnitudes of $p(D|M_{simple})$ and $p(D|M_{complex})$. We can do the same thing for the two priors in Figure 5.2. The peaked prior constitutes one model, and the uniform prior constitutes another model. We evaluate their relative veracity by comparing their values of $p(D|M_{peaked})$ and $p(D|M_{uniform})$.

We prefer the model with the higher value of $p(D|M)$, but the preference is not absolute. A tiny advantage in $p(D|M)$ should not be translated into a strong preference for one model over another. After all, the data themselves are just a random sample from the world, and they could have been somewhat different. It is only when the relative magnitudes of $p(D|M)$ are very different that we can feel confident in a preference of one model over another.

Moreover, we need to take into account our prior beliefs in the models, $p(M1)$ and $p(M2)$, as indicated in Equation 4.8 (p. 58). Typically we will go into a model comparison with equal prior probabilities on the models, and so the ratio of the posterior probabilities is just the ratio of evidences (also known as the Bayes factor). But if the prior probabilities of the models are not equal, then they must be factored into the posteriors. In particular, if we have a strong prior belief in a model, it takes more evidence for the other model to overcome our prior belief. For further applications of model comparison, see Exercises 5.6 and 5.7

5.3.3.1 Is the Best Model a Good Model?

Suppose we have two models, we collect some data, and find that the evidence for one model is much larger than the evidence for the other model. If our prior beliefs were equal, then the posterior beliefs strongly favor the winning model.

But is the winning model actually a good model of the data? The model comparison process has merely told us the models' *relative* believabilities, not their *absolute* believabilities. The winning model might merely be a less bad model than the horrible losing model.

Exercise 5.8 explains one way to assess whether the winning model can actually account for the data. The method used there is called a *posterior predictive check*.

5.4 SUMMARY: HOW TO DO BAYESIAN INFERENCE

In this chapter, we've covered a lot of important conceptual points that are crucial for understanding what you are doing when you do Bayesian inference. But I don't want the volume of concepts to overwhelm the underlying simplicity of what you actually do. Here are the steps:

1. For the methods of this chapter to apply to your situation, the data must have two nominal values, like heads and tails. The two values must come up randomly, independently across observations, and with a single and

fixed (i.e., stationary through time) probability. Denote the underlying probability of "heads" by the value θ.

2. Establish a description of your prior beliefs regarding values of θ by using a beta($\theta|a, b$) distribution. Decide what you think is the most probable value for θ; call it m. Decide how strongly you believe in m by considering how many new data points (e.g., flips of the coin) it would take to sway you away from your prior belief. Call that number n. (In real research, the prior is established by considering previously published results and the audience to whom the analysis is addressed.) Then convert the m, n values to a, b values for the beta distribution by using Equation 5.5. You should check that the resulting beta distribution really captures your beliefs by look-ing at its graph! This is easily done by using the program in Section 5.5.1 (BernBeta.R). If you cannot express your prior beliefs as a beta distribu-tion, then the methods of this chapter do not apply.

3. Observe some data. Enter them as a vector of 1's and 0's into the program of Section 5.5.1 (BernBeta.R). For N data points (e.g., flips of a coin), the vector should have N elements. "Heads" are coded by the value 1, and "tails" are coded by the value 0. The total number of 1's (i.e., heads) is denoted z.

4. Determine the posterior distribution of beliefs regarding values of θ. When the prior is a beta distribution, the posterior is a beta distribution too. The program of Section 5.5.1 (BernBeta.R) displays it graphically, along with a credible interval.

5. Make inferences from the posterior, depending on your goal. If your goal is to estimate θ, use the posterior distribution, perhaps as summarized by the 95% HDI. If your goal is to predict new data, then your predicted proba-bility of "heads" on the next flip is the mean of the posterior, which is $(z + a)/(N + a + b)$. If your goal is to compare models (i.e., priors), then use $p(D)$ to decide which model's prior better accounts for the data. Use a posterior predictive check to get a sense of whether the better model actually mimics the data well.

5.5 R CODE

5.5.1 R Code for Figure 5.2

This program defines a *function* in R instead of a script. A function takes input values, called "arguments," and does something with them. In general, a func-tion in R in defined by code of the form

```
function_name = function( arguments ) { commands }
```

The commands inside the braces can extend over many lines of code. When the function is called, it takes the values of the arguments in parentheses and uses them in the commands in the braces. You invoke the function by commanding

R thus:

```
function_name( argument_values )
```

As a simple example, consider this definition:

```
asqplusb = function( a , b ) { a^2 + b }
```

We can then type

```
asqplusb( a = 2 , b = 1 )
```

and R returns 5. We can get the same result by typing

```
asqplusb( 2 , 1 )
```

because unlabeled arguments are assumed to be provided in the order they were listed in the function definition. If you want to use some other ordering of arguments when calling the function, that is okay as long as the arguments are explicitly labeled. So, for example, we can get the same result by typing

```
asqplusb( b = 1 , a = 2 )
```

By default, a function in R returns the last value it computed in the list of commands. If you want to be sure that the function returns the intended value, you should include an explicit `return` command at the end of your function; for example:

```
asqplusb = function( a , b ) {
    c = a^2 + b
    return( c )
}
```

A useful feature of function definitions in R is that arguments can be given default values. For example, in the following function definition, the argument b is given a default value of 1:

```
asqplusb = function( a , b =1 ) { a^2 + b }
```

The function can then be called *without specifying a value for* b, and the value b = 1 will be assumed by default. For example, typing `asqplusb(a = 2)` would return 5. A default value can be overridden by explicitly specifying its value in the function call. For example, `asqplusb(a = 2,b = 2)` would return 6.

To use a function, R must know that the function exists. To point out a function to R and make it ready to call, do the following. First, for this function `BernBeta.R`, save the full text of the function in a file, named "BernBeta.R," in a folder of your choice. Second, you must `source` that file, which essentially just runs the code. Running the function file simply defines the function for R, it does not call the function with arguments. To `source` the file, you can type the command `source("<pathtofile>/BernBeta.R")`, or, in the R command window (not an editing or graph window), click the File menu item, then the Source menu item, and then browse to the file and select it.

See the comments at the top of the code listing for details of how to use this function. When you examine the code, you may notice that the Bayesian computations constitute just a few lines; the bulk of the code is devoted to all the details of displaying the distributions. On line 57, a large window is opened. Line 58 specifies the layout of the subplots. The `layout` command takes a matrix as an argument; the integers in the matrix indicate which subpanels in the layout should be used for which plot. Thus, the first subplot is put into the matrix cells that have a 1, the second subplot is put into the matrix cells that have a 2, and so forth. Line 59 adjusts how the axis information is displayed. The `par` command has arguments such as `mar` and `mgp`, which adjust the margins of the plot; type `help(par)` in R for details, and try varying the numbers in the `mar` and `mgp` vectors to explore their effects on the plot. Mathematical text in the plots uses the `expression` and `bquote` functions, which interpret their arguments as specifications for mathematical characters. For help with how to plot math characters in R, at the command line type `demo(plotmath)` and `help(plotmath)`.

The function that computes the HDI uses some advanced techniques, and its explanation is deferred to Section 23.3.3. Nevertheless, the program that computes the HDI, called `HDIofICDF.R`, must be available for use by the program listed here, called `BernBeta.R`. Therefore, be sure that the two programs are in the same folder.

(BernBeta.R)

```
1   BernBeta = function( priorShape , dataVec , credMass=0.95 , saveGraph=F ) {
2   # Bayesian updating for Bernoulli likelihood and beta prior.
3   # Input arguments:
4   #   priorShape
5   #      vector of parameter values for the prior beta distribution.
6   #   dataVec
7   #      vector of 1's and 0's.
8   #   credMass
9   #      the probability mass of the HDI.
10  # Output:
11  #   postShape
12  #      vector of parameter values for the posterior beta distribution.
13  # Graphics:
14  #   Creates a three-panel graph of prior, likelihood, and posterior
15  #   with highest posterior density interval.
16  # Example of use:
17  # > postShape = BernBeta( priorShape=c(1,1) , dataVec=c(1,0,0,1,1) )
18  # You will need to "source" this function before using it, so R knows
19  # that the function exists and how it is defined.
20
21  # Check for errors in input arguments:
22  if ( length(priorShape) != 2 ) {
23    stop("priorShape must have two components.") }
24  if ( any( priorShape <= 0 ) ) {
```

```
25       stop("priorShape components must be positive.") }
26   if ( any( dataVec != 1 & dataVec != 0 ) ) {
27       stop("dataVec must be a vector of 1s and 0s.") }
28   if ( credMass <= 0 | credMass >= 1.0 ) {
29       stop("credMass must be between 0 and 1.") }
30
31   # Rename the prior shape parameters, for convenience:
32   a = priorShape[1]
33   b = priorShape[2]
34   # Create summary values of the data:
35   z = sum( dataVec == 1 ) # number of 1's in dataVec
36   N = length( dataVec )   # number of flips in dataVec
37   # Compute the posterior shape parameters:
38   postShape = c( a+z , b+N-z )
39   # Compute the evidence, p(D):
40   pData = beta( z+a , N-z+b ) / beta( a , b )
41   # Determine the limits of the highest density interval.
42   # This uses a home-grown function called HDIofICDF.
43   source( "HDIofICDF.R" )
44   hpdLim = HDIofICDF( qbeta , shape1=postShape[1] , shape2=postShape[2] )
45
46   # Now plot everything:
47   # Construct grid of theta values, used for graphing.
48   binwidth = 0.005 # Arbitrary small value for comb on Theta.
49   Theta = seq( from = binwidth/2 , to = 1-(binwidth/2) , by = binwidth )
50   # Compute the prior at each value of theta.
51   pTheta = dbeta( Theta , a , b )
52   # Compute the likelihood of the data at each value of theta.
53   pDataGivenTheta = Theta^z * (1-Theta)^(N-z)
54   # Compute the posterior at each value of theta.
55   pThetaGivenData = dbeta( Theta , a+z , b+N-z )
56   # Open a window with three panels.
57   windows(7,10)
58   layout( matrix( c( 1,2,3 ) ,nrow=3 ,ncol=1 ,byrow=FALSE ) ) # 3x1 panels
59   par( mar=c(3,3,1,0) , mgp=c(2,1,0) , mai=c(0.5,0.5,0.3,0.1) ) # margin specs
60   maxY = max( c(pTheta,pThetaGivenData) ) # max y for plotting
61   # Plot the prior.
62   plot( Theta , pTheta , type="l" , lwd=3 ,
63           xlim=c(0,1) , ylim=c(0,maxY) , cex.axis=1.2 ,
64           xlab=bquote(theta) , ylab=bquote(p(theta)) , cex.lab=1.5 ,
65           main="Prior" , cex.main=1.5 )
66   if ( a > b ) { textx = 0 ; textadj = c(0,1) }
67   else { textx = 1 ; textadj = c(1,1) }
68   text( textx , 1.0*max(pThetaGivenData) ,
69           bquote( "beta(" * theta * "|" * .(a) * "," * .(b) * ")" ) ,
70           cex=2.0 ,adj=textadj )
71   # Plot the likelihood: p(data|theta)
72   plot( Theta , pDataGivenTheta , type="l" , lwd=3 ,
73           xlim=c(0,1) , cex.axis=1.2 , xlab=bquote(theta) ,
74           ylim=c(0,1.1*max(pDataGivenTheta)) ,
75           ylab=bquote( "p(D|" * theta * ")" ) ,
76           cex.lab=1.5 , main="Likelihood" , cex.main=1.5 )
77   if ( z > .5*N ) { textx = 0 ; textadj = c(0,1) }
78   else { textx = 1 ; textadj = c(1,1) }
```

```
79     text( textx , 1.0*max(pDataGivenTheta) , cex=2.0 ,
80              bquote( "Data: z=" * .(z) * ",N=" * .(N) ) ,adj=textadj )
81     # Plot the posterior.
82     plot( Theta , pThetaGivenData ,type="l" , lwd=3 ,
83              xlim=c(0,1) , ylim=c(0,maxY) , cex.axis=1.2 ,
84              xlab=bquote(theta) , ylab=bquote( "p(" * theta * "|D)" ) ,
85              cex.lab=1.5 , main="Posterior" , cex.main=1.5 )
86     if ( a+z > b+N-z ) { textx = 0 ; textadj = c(0,1) }
87     else { textx = 1 ; textadj = c(1,1) }
88     text( textx , 1.00*max(pThetaGivenData) , cex=2.0 ,
89              bquote( "beta(" * theta * "|" * .(a+z) * "," * .(b+N-z) * ")" ) ,
90              adj=textadj )
91     text( textx , 0.75*max(pThetaGivenData) , cex=2.0 ,
92              bquote( "p(D)=" * .(signif(pData,3)) ) , adj=textadj )
93     # Mark the HDI in the posterior.
94     hpdHt = mean( c( dbeta(hpdLim[1],a+z,b+N-z) , dbeta(hpdLim[2],a+z,b+N-z) ) )
95     lines( c(hpdLim[1],hpdLim[1]) , c(-0.5,hpdHt) , type="l" , lty=2 , lwd=1.5 )
96     lines( c(hpdLim[2],hpdLim[2]) , c(-0.5,hpdHt) , type="l" , lty=2 , lwd=1.5 )
97     lines( hpdLim , c(hpdHt,hpdHt) , type="l" , lwd=2 )
98     text( mean(hpdLim) , hpdHt , bquote( .(100*credMass) * "% HDI" ) ,
99            adj=c(0.5,-1.0) , cex=2.0 )
100    text( hpdLim[1] , hpdHt , bquote(.(round(hpdLim[1],3))) ,
101            adj=c(1.1,-0.1) , cex=1.2 )
102    text( hpdLim[2] , hpdHt , bquote(.(round(hpdLim[2],3))) ,
103            adj=c(-0.1,-0.1) , cex=1.2 )
104    # Construct file name for saved graph, and save the graph.
105    if ( saveGraph ) {
106      filename = paste( "BernBeta_",a,"_",b,"_",z,"_",N,".eps" ,sep="")
107      dev.copy2eps( file = filename )
108    }
109    return( postShape )
110    } # end of function
```

5.6 EXERCISES

Exercise 5.1. [Purpose: To see the influence of the prior in each successive flip, and to see another demonstration that the posterior is invariant under reorderings of the data.] For this exercise, use the R function of Section 5.5.1 (`BernBeta.R`). (Read the comments at the top of the code for an example of how to use it, and don't forget to `source` the function before calling it.) Notice that the function returns the posterior beta values each time it is called, so you can use the returned values as the prior values for the next function call.

(A) Start with a prior distribution that expresses some uncertainty that a coin is fair: $\text{beta}(\theta|4, 4)$. Flip the coin once; suppose we get a head. What is the posterior distribution?

(B) Use the posterior from the previous flip as the prior for the next flip. Suppose we flip again and get a head. Now what is the new posterior? (Hint: If you type `post = BernBeta( c(4,4) , c(1) )` for the first part, then you can type `post = BernBeta( post , c(1) )` for the next part.)

(C)   Using that posterior as the prior for the next flip, flip a third time and get T. Now what is the new posterior? (Hint: Type `post = BernBeta( post , c(0) )`.)

(D)   Do the same three updates but in the order T, H, H instead of H, H, T. Is the final posterior distribution the same for both orderings of the flip results?

Exercise 5.2. [Purpose: To connect HDIs to the real world, with iterative data collection.] Suppose an election is approaching, and you are interested in knowing whether the general population prefers candidate A or candidate B. A just-published poll in the newspaper states that of 100 randomly sampled people, 58 preferred candidate A and the remainder preferred candidate B.

(A)   Suppose that before the newspaper poll, your prior belief was a uniform distribution. What is the 95% HDI on your beliefs after learning of the newspaper poll results?

(B)   Based in the newspaper poll, is it credible to believe that the population is equally divided in its preferences among candidates?

(C)   You want to conduct a follow-up poll to narrow down your estimate of the population's preference. In your follow-up poll, you randomly sample 100 people and find that 57 prefer candidate A and the remainder prefer candidate B. Assuming that peoples' opinions have not changed between polls, what is the 95% HDI on the posterior?

(D)   Based on your follow-up poll, is it credible to believe that the population is equally divided in its preferences among candidates?

Exercise 5.3. [Purpose: To apply the Bayesian method to real data analysis. These data are representative of real data (Kruschke, 2009).] Suppose you train people in a simple learning experiment, as follows. When people see the two words "radio" and "ocean," on the computer screen, they should press the F key on the computer keyboard. They see several repetitions and learn the response well. Then you introduce another correspondence for them to learn: Whenever the words "radio" and "mountain" appear, they should press the J key on the computer keyboard. You keep training them until they know both correspondences well. Now you probe what they've learned by asking them about two novel test items. For the first test, you show them the word "radio" by itself and instruct them to make the best response (F or J) based on what they learned before. For the second test, you show them the two words "ocean" and "mountain" and ask them to make the best response. You do this procedure with 50 people. Your data show that for "radio" by itself, 40 people chose F and 10 chose J. For the word combination "ocean" and "mountain," 15 chose F and 35 chose

J. Are people biased toward F or toward J for either of the two probe types? To answer this question, assume a uniform prior, and use a 95% HDI to decide which biases can be declared to be credible.

Exercise 5.4. [Purpose: To explore an unusual prior and learn about the beta distribution in the process.] Suppose we have a coin that we know comes from a magic-trick store, and therefore we believe that the coin is strongly biased either usually to come up heads or usually to come up tails, but we don't know which. Express this belief as a beta prior. (Hint: See Figure 5.1, upper-left panel.) Now we flip the coin five times and it comes up heads in four of the five flips. What is the posterior distribution? (Use the R function of Section 5.5.1 (BernBeta.R) to see graphs of the prior and posterior.)

Exercise 5.5. [Purpose: To get hands-on experience with the goal of predicting the next datum, and to see how the prior influences that prediction.]

(A) Suppose you have a coin that you know is minted by the federal government and has not been tampered with. Therefore, you have a strong prior belief that the coin is fair. You flip the coin 10 times and get 9 heads. What is your predicted probability of heads for the 11th flip? Explain your answer carefully; justify your choice of prior.

(B) Now you have a different coin, this one made of some strange material and marked (in fine print) "Patent Pending, International Magic, Inc." You flip the coin 10 times and get 9 heads. What is your predicted probability of heads for the 11th flip? Explain your answer carefully; justify your choice of prior. Hint: Use the prior from Exercise 5.4.

Exercise 5.6. [Purpose: To get hands-on experience with the goal of model comparison.] Suppose we have a coin, but we're not sure whether it's a fair coin or a trick coin. We flip it 20 times and get 15 heads. Is it more likely to be fair or trick? To answer this question, consider the value of the Bayes factor (i.e., the ratio of the evidences of the two models). When answering this question, justify your choice of priors to express the two hypotheses. Use the R function of Section 5.5.1 (BernBeta.R) to graph the priors and check that they reflect your beliefs; the R function will also determine the evidences from Equation 5.10.

Exercise 5.7. [Purpose: To see how very small data sets can give strong leverage in model comparison when the model predictions are very different.] Suppose we have a coin that we strongly believe is a trick coin, so it almost always comes up heads or it almost always comes up tails; we just don't know if the coin is the head-biased type or the tail-biased type. Thus, one model is a beta prior heavily biased toward tails, beta($\theta|1, 100$), and the other model is a beta prior heavily biased toward heads, beta($\theta|100, 1$). We flip the coin once and it comes up heads. Based on that single flip, what is the value of the Bayes factor (i.e., the ratio of the evidences of the two models)? Use the R function of Section 5.5.1 (BernBeta.R) to determine the evidences from Equation 5.10.

Exercise 5.8. [Purpose: Hands-on learning about the method of posterior predictive checking.] Following the scenario of the previous exercise, suppose we flip the coin a total of $N = 12$ times and it comes up heads in $z = 8$ of those flips. Suppose we let a beta$(\theta|100, 1)$ distribution describe the head-biased trick coin, and we let a beta$(\theta|1, 100)$ distribution describe the tail-biased trick coin.

(A)   What are the evidences for the two models, and what is the value of the Bayes factor?

Now for the new part, a *posterior predictive check*. Is the winning model actually a good model of the data? In other words, one model can be whoppingly better than the other, but that does not necessarily mean that the winning model is a good model; it might mean merely that the winning model is less bad than the losing model. One way to examine the veracity of the winning model is to simulate data sampled from the winning model and see if the simulated data "look like" the actual data. To simulate data generated by the winning model, we do the following: First, we will randomly generate a value of θ from the posterior distribution of the winning model. Second, using that value of θ, we will generate a sample of coin flips. Third, we will count the number of heads in the sample, as a summary of the sample. Finally, we determine whether the number of heads in a typical *simulated* sample is close to the number of heads in our *actual* sample. The following program carries out these steps. Study it, run it, and answer the questions that follow.

The program uses a for loop to repeat an action. For example, if you tell R for (i in 1:5) { show(i) }, it replies with 1 2 3 4 5. What for really does is execute the commands within the braces for every element in the specified vector. For example, if you tell R for (i in c(7,-.2)) { show(i) } it replies with 7 -.2.

(BetaPosteriorPredictions.R)
```
1   # Specify known values of prior and actual data.
2   priorA = 100
3   priorB = 1
4   actualDataZ  = 8
5   actualDataN  = 12
6   # Compute posterior parameter values.
7   postA = priorA + actualDataZ
8   postB = priorB + actualDataN - actualDataZ
9   # Number of flips in a simulated sample should match the actual sample size:
10  simSampleSize = actualDataN
11  # Designate an arbitrarily large number of simulated samples.
12  nSimSamples = 10000
```

```
13    # Set aside a vector in which to store the simulation results.
14    simSampleZrecord = vector( length=nSimSamples )
15    # Now generate samples from the posterior.
16    for ( sampleIdx in 1:nSimSamples ) {
17            # Generate a theta value for the new sample from the posterior.
18            sampleTheta = rbeta( 1 , postA , postB )
19            # Generate a sample, using sampleTheta.
20            sampleData = sample( x=c(0,1) , prob=c( 1-sampleTheta , sampleTheta ) ,
21                            size=simSampleSize , replace=TRUE )
22            # Store the number of heads in sampleData.
23            simSampleZrecord[ sampleIdx ] = sum( sampleData )
24    }
25    # Make a histogram of the number of heads in the samples.
26    hist( simSampleZrecord )
```

(B) How many samples (each of size N) were simulated?

(C) Was the same value of θ used for every simulated sample, or were different values of θ used in different samples? *Why?*

(D) Based on the simulation results, does the winning model seem to be a good model? Why or why not?

CHAPTER 6

Inferring a Binomial Proportion via Grid Approximation

I'm kinda coarse while the lady's refined,
I kinda stumble while she holds the line. But
both of us side-step and guess what the answer is;
Both might feel better with (psycho-)analysis.

The previous chapter considered how to make inferences about a binomial proportion when the prior could be specified as a beta distribution. Using the beta distribution was convenient because it made the integrals work out easily by direct formal analysis. But what if no beta distribution adequately expresses our prior beliefs? For example, our beliefs could be trimodal: The coin might be heavily biased toward tails, or be approximately fair, or be heavily biased toward heads. No beta distribution has three "humps" like that.

In this chapter we explore one technique for numerically approximating the posterior distribution by defining the prior distribution over a fine grid of θ

Doing Bayesian Data Analysis: A Tutorial with R and BUGS. DOI: 10.1016/B978-0-12-381485-2.00006-7
© 2011, Elsevier Inc. All rights reserved.

values. In this situation, we do not need a mathematical function of the prior over θ; we can specify any prior probability values we desire at each of the θ values. Moreover, we do not need to do any analytical (i.e., formulas only) integration. The denominator of Bayes' rule becomes a sum over many discrete θ values instead of an integral.

6.1 BAYES' RULE FOR DISCRETE VALUES OF θ

As in the previous chapter, the parameter θ denotes the value of a binomial proportion, such as the underlying propensity for a coin to come up heads. Previously we assumed that θ was continuous over the interval [0, 1]. We assumed that θ could have any value in that continuous domain. The prior probability on θ was, therefore, a probability *density* at each value of θ, such as a beta distribution.

Instead, we could assume that there are only a finite number of θ values in which we have any nonzero belief. For example, we might believe that θ can only have the values 0.25, 0.50, or 0.75. We already saw an example like this back in Figure 4.1 (p. 61). When there are a finite number of θ values, then our prior distribution expresses the probability *mass* at each value of θ. In this situation, Bayes' rule is expressed as

$$p(\theta \mid D) = \frac{p(D \mid \theta)\, p(\theta)}{\sum_{\theta} p(D \mid \theta)\, p(\theta)} \qquad (6.1)$$

where the sum in the denominator is over the finite number of discrete values of θ that we are considering, and $p(\theta)$ denotes the probability *mass* at θ.

There are two niceties of dealing with the discrete version of Bayes' rule in Equation 6.1. One attraction is that some prior beliefs are easier to express with discrete values than with continuous density functions. Another felicity is that some mathematical functions that are difficult to integrate analytically can be approximated by evaluating the function on a fine grid of discrete values.

6.2 DISCRETIZING A CONTINUOUS PRIOR DENSITY

If we could approximate a continuous prior density with a grid of discrete prior masses, then we could use the discrete form of Bayes' rule (in Equation 6.1) instead of the continuous form, which requires mathematically evaluating an integral. Fortunately, in some situations we can, in fact, make such an approximation. Figure 6.1 illustrates how a continuous prior density can be partitioned into a set of narrow rectangles that approximate the continuous prior. This process of discretizing the prior is straightforward: Divide the domain into a large number of narrow intervals. Draw a rectangle over each

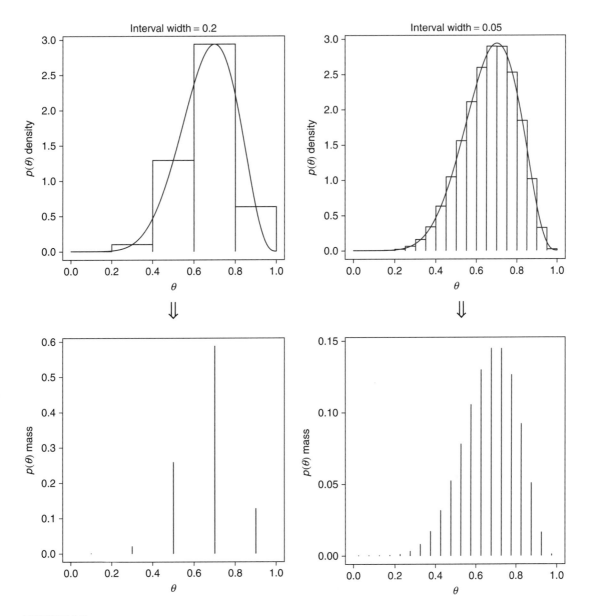

FIGURE 6.1
Approximation of a continuous density by a grid of discrete masses. Upper panels show a continuous density and its partitioning into rectangles. Lower panels plot the area of each rectangle (i.e., the masses). Left side shows approximation with a coarse grid; right side shows approximation with a finer grid.

narrow interval, with height equal to the value of the density at the middle of the narrow interval. Approximate the area under the continuous density in each narrow interval by the area in the corresponding rectangle. This much of the process is illustrated in the top panels of Figure 6.1. The approximation gets better as the rectangles get narrower.

To discretize the density, we consider only the discrete θ values at the middles of each interval and set the probability mass at that value to be the area of the corresponding rectangle. To make sure that the resulting discrete probabilities sum to exactly 1.0, we set each discrete probability to the corresponding interval area and then divide by the sum of those probabilities. This discrete representation is shown in the lower panels of Figure 6.1. Notice that the scale on the y axis has changed in going from upper to lower panels. In the upper panels, $p(\theta)$ refers to probability density at continuous values. In the lower panels, $p(\theta)$ refers to probability mass at discrete values.

When the prior density is discretized into a grid of masses, we can apply the discrete version of Bayes' rule (Equation 6.1). It is only an approximation to the true integral form, but if the grid is dense enough, the approximation can be accurate.

6.2.1 Examples Using Discretized Priors

Figure 6.2 shows a uniform prior discretized. Notice that because the prior is only defined at discrete points, the likelihood and posterior are only defined at those same points. The prior distribution is not represented by some mathematical function such as beta($\theta|1, 1$); it is merely a list of probability masses at each discrete value of θ. Likewise, the shape of the posterior distribution is not stored as a mathematical function; it too is merely a list of probability masses at each discrete value of θ. The computations for the posterior did not involve any finesse with functions that described shapes of distributions; instead, the computations were just brute-force application of the discrete version of Bayes' rule (Equation 6.1).

The left and right sides of Figure 6.2 show the results for a coarse discretization and a finer discretization, respectively. Compare the approximation in Figure 6.2 with the exact beta-distribution updating in the left side of Figure 5.2, p. 86. You can see that the fine-grid approximation is accurate. It turns out in this case that even the coarse discretization does a good job at estimating $p(D)$; (i.e., the denominator of Bayes' rule). The highest density interval (HDI), on the other hand, does differ noticeably between the two approximations. This is because the credible interval can only be specified to as fine a resolution as the subintervals in the discretization.

Figure 6.3 shows an arbitrarily shaped prior, discretized. Here we see the advantaged gained by discretized priors, because no beta function could possibly

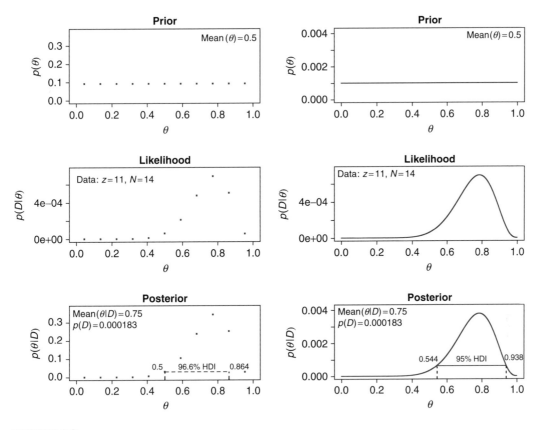

FIGURE 6.2
Grid approximations to Bayesian updating. The left side uses θ values at intervals of 1/11. The right side uses θ values at intervals of 1/1001. Compare this with the left side of Figure 5.2, p. 86. (R code for this graph is in Section 6.7.1 (BernGrid.R).)

imitate this prior accurately. Nevertheless, application of Bayes' rule yields the posterior distribution shown in the figure, which is as accurate as the discretized prior allows. If we desired, we could represent the prior beliefs on a finer comb of values over the parameter θ.

The joy of grid approximation is freedom from the siren song of beta priors. Eloquent as they are, beta priors can only express a limited range of beliefs. No beta function approximates the prior used in Figure 6.3, for example. You might say that these are cases for which beta ain't better. Even if you could find some complex mathematical function of θ that expressed the contours of your beliefs, that function would probably be difficult to integrate in the denominator of Bayes' rule, so you still couldn't determine an exact mathematical solution. Instead, you can express those beliefs approximately by specifying your degree of belief for each value of θ on a fine grid.

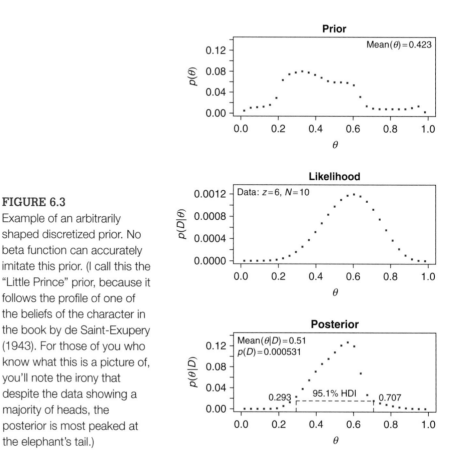

FIGURE 6.3

Example of an arbitrarily shaped discretized prior. No beta function can accurately imitate this prior. (I call this the "Little Prince" prior, because it follows the profile of one of the beliefs of the character in the book by de Saint-Exupery (1943). For those of you who know what this is a picture of, you'll note the irony that despite the data showing a majority of heads, the posterior is most peaked at the elephant's tail.)

6.3 ESTIMATION

The full list of posterior probability masses provides a complete estimate of the parameter values. Those masses can be summarized in whatever manner is convenient and meaningful. Figures 6.2 and 6.3 provided two summary descriptors of the posterior, namely, the mean value of θ and the 95% HDI.

The mean of θ is just the sum of the available parameter values weighted by the probability that they occur. Formally, that is expressed by

$$\bar{\theta}\,|\,D = \sum_{\theta} \theta\, p(\theta\,|\,D) \tag{6.2}$$

where the sum is over discrete values of θ at its grid points, and $p(\theta\,|\,D)$ is the probability mass at each grid point. The mean value is computed explicitly as this sum by the program in Section 6.7.1 (BernGrid.R), and the value is displayed in the plots it produces, as in Figures 6.2 and 6.3.

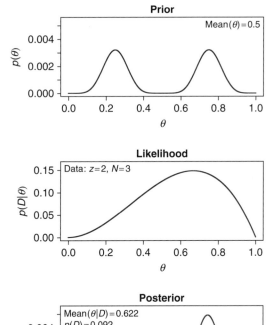

FIGURE 6.4
The 95% HDI of the posterior is *split* across more two distinct subintervals. Only the extreme left and right ends of the HDIs are marked in the plot, but the unmarked internal divisions *are* also end points of the split HDI region.

Recall that the HDI is defined such that the probability of any point inside the HDI is greater than the probability of any point outside the HDI, and the total probability of points in the 95% HDI is 95%. Because we are dealing with discrete masses, the sum of the masses in an interval usually will not be exactly 95%, and therefore we define the 95% HDI so it has total mass as small as possible but greater than or equal to 95%. That is why the left sides of Figure 6.2 and Figure 6.3 show HDI masses slightly larger than 95%.

Figure 6.4 illustrates the 95% HDI for a bimodal posterior. The HDI is split into two segments. This split makes good sense: We want the HDI to represent those values of the parameter that are believable. For a bimodal posterior, we should have a region of believability in two pieces. One of the attractions of grid approximation is that multimodal HDIs are easily determined.

6.4 PREDICTION OF SUBSEQUENT DATA

A second typical goal of Bayesian inference is predicting subsequent data after incorporating an observed set of data. As has been our habit, let's denote the observed set of data as D and the posterior distribution over parameter θ as $p(\theta \,|\, D)$. Our predicted probability for the next value of y is just the probability of that value happening for each value of θ, weighted by the posterior believability of each θ:

$$p(y \,|\, D) = \int d\theta \, p(y \,|\, \theta) \, p(\theta \,|\, D)$$

$$\approx \sum_{\theta} p(y \,|\, \theta) \, p(\theta \,|\, D) \tag{6.3}$$

where $p(\theta \,|\, D)$ in the first line is a probability density, and $p(\theta \,|\, D)$ in the second line is a probability mass at discrete values of θ. In particular, for $y = 1$, Equation 6.3 becomes

$$p(y \!=\! 1 \,|\, D) \approx \sum_{\theta} p(y \!=\! 1 \,|\, \theta) \, p(\theta \,|\, D)$$

$$= \sum_{\theta} \theta \, p(\theta \,|\, D) \tag{6.4}$$

which is just the average value of θ in the posterior distribution of θ. This fact is not news; we've seen this idea before in Equation 5.9 (p. 87). What's new is that we are not relying on the posterior having the form of a beta distribution. For an example, see Exercise 6.6.

6.5 MODEL COMPARISON

A third typical goal of Bayesian inference is model comparison. Suppose we have two models, denoted $M1$ and $M2$, with prior beliefs $p(M1)$ and $p(M2)$. We want to determine the posterior beliefs, $p(M1 \,|\, D)$ and $p(M2 \,|\, D)$. Recall from Equation 4.8, p. 58, that

$$\frac{p(M1 \,|\, D)}{p(M2 \,|\, D)} = \frac{p(D \,|\, M1) \, p(M1)}{p(D \,|\, M2) \, p(M2)} \tag{6.5}$$

where (as in Equation 4.7)

$$p(D \,|\, M) = \int d\theta \, p(D \,|\, \theta, M) \, p(\theta \,|\, M)$$

is the "evidence" for model M. The integral for the evidence becomes a sum when we have a discrete grid approximation,

$$p(D \,|\, M) \approx \sum_{\theta} p(D \,|\, \theta, M) \, p(\theta \,|\, M) \tag{6.6}$$

where the sum is over discrete values of θ and $p(\theta \mid M)$ is a probability mass at each value of θ.

In our current application, we are assuming a Bernoulli likelihood function for all models. In other words, we can drop the "M" from $p(D \mid \theta, M)$ in Equation 6.6. The only differentiation among models is the specification of the prior beliefs over θ. For example, one model might posit that the coin is head biased, whereas the other model posits that the coin is tail biased. Therefore, combining Equations 6.5 and 6.6 yields

$$\frac{p(M1 \mid D)}{p(M2 \mid D)} = \frac{\sum_\theta p(D \mid \theta)\, p(\theta \mid M1)}{\sum_\theta p(D \mid \theta)\, p(\theta \mid M2)} \frac{p(M1)}{p(M2)} \tag{6.7}$$

The expression in Equation 6.7 is useful when the priors are not beta distributions. (If the priors are beta distributions, then exact mathematical analysis yields the results described in Section 5.3.3, p. 88.) Conveniently, the R function provided in Section 6.7.1 (BernGrid.R) displays $p(D \mid M)$ in its graphical output, computed by Equation 6.6. For examples, see Exercises 6.7 and 6.8.

6.6 SUMMARY

This chapter showed that Bayesian inference, regarding a continuous binomial proportion θ, can be achieved by approximating the continuous θ with a dense grid of discrete values. A disadvantage of this approach is that the approximation is only as good as the density of the grid. But there are several advantages of the approach. One advantage is that we have great freedom in the type of prior distributions we can specify; we are not restricted to beta distributions. Another advantage is that we can use the discrete approximation to find approximate HDI regions. We can also do posterior prediction and model comparison with arbitrary priors. The program presented in Section 6.7.1 (BernGrid.R) is provided to help you conduct these analyses.

6.7 R CODE

6.7.1 R Code for Figure 6.2 and the Like

For this function you need to set up a comb (i.e., grid) of θ values and a prior on those values. The extensive comments at the beginning of the function provide an example for how to do this. See also Exercise 6.1 for a caveat regarding how *not* to do this.

The Bayesian computations form only a few lines of this function (namely, lines 30 to 37). The bulk of the program is devoted to all the details of plotting the information. Section 23.3.1 (HDIofGrid.R) explains how the HDI is approximated.

(BernGrid.R)

```
1   BernGrid = function( Theta , pTheta , Data ,
2                        credib=.95 , nToPlot=length(Theta) ) {
3   # Bayesian updating for Bernoulli likelihood and prior specified on a grid.
4   # Input arguments:
5   #   Theta is a vector of theta values, all between 0 and 1.
6   #   pTheta is a vector of corresponding probability _masses_.
7   #   Data is a vector of 1's and 0's, where 1 corresponds to a and 0 to b.
8   #   credib is the probability mass of the credible interval, default is 0.95.
9   #   nToPlot is the number of grid points to plot; defaults to all of them.
10  # Output:
11  #   pThetaGivenData is a vector of posterior probability masses over Theta.
12  #   Also creates a three-panel graph of prior, likelihood, and posterior
13  #   probability masses with credible interval.
14  # Example of use:
15  #   # Create vector of theta values.
16  #   > binwidth = 1/1000
17  #   > thetagrid = seq( from=binwidth/2 , to=1-binwidth/2 , by=binwidth )
18  #   # Specify probability mass at each theta value.
19  #   > relprob = pmin(thetagrid,1-thetagrid) # relative prob at each theta
20  #   > prior = relprob / sum(relprob) # probability mass at each theta
21  #   # Specify the data vector.
22  #   > datavec = c( rep(1,3) , rep(0,1) ) # 3 heads, 1 tail
23  #   # Call the function.
24  #   > posterior = BernGrid( Theta=thetagrid , pTheta=prior , Data=datavec )
25  # Hints:
26  #   You will need to "source" this function before calling it.
27  #   You may want to define a tall narrow window before using it; e.g.,
28  #   > windows(7,10)
29
30  # Create summary values of Data
31  z = sum( Data==1 ) # number of 1's in Data
32  N = length( Data ) # number of flips in Data
33  # Compute the likelihood of the Data for each value of Theta.
34  pDataGivenTheta = Theta^z * (1-Theta)^(N-z)
35  # Compute the evidence and the posterior.
36  pData = sum( pDataGivenTheta * pTheta )
37  pThetaGivenData = pDataGivenTheta * pTheta / pData
38
39  # Plot the results.
40  layout( matrix( c( 1,2,3 ) ,nrow=3 ,ncol=1 ,byrow=FALSE ) ) # 3x1 panels
41  par( mar=c(3,3,1,0) , mgp=c(2,1,0) , mai=c(0.5,0.5,0.3,0.1) ) # margin settings
42  dotsize = 4 # how big to make the plotted dots
43  # If the comb has a zillion teeth, it's too many to plot, so plot only a
44  # thinned out subset of the teeth.
45  nteeth = length(Theta)
46  if ( nteeth > nToPlot ) {
47    thinIdx = seq( 1, nteeth , round( nteeth / nToPlot ) )
48    if ( length(thinIdx) < length(Theta) ) {
49      thinIdx = c( thinIdx , nteeth ) # makes sure last tooth is included
50    }
51  } else { thinIdx = 1:nteeth }
52  # Plot the prior.
```

```
53   meanTheta = sum( Theta * pTheta ) # mean of prior, for plotting
54   plot( Theta[thinIdx] , pTheta[thinIdx] , type="p" , pch="." , cex=dotsize ,
55        xlim=c(0,1) , ylim=c(0,1.1*max(pThetaGivenData)) , cex.axis=1.2 ,
56        xlab=bquote(theta) , ylab=bquote(p(theta)) , cex.lab=1.5 ,
57        main="Prior" , cex.main=1.5 )
58   if ( meanTheta > .5 ) {
59      textx = 0 ; textadj = c(0,1)
60   } else {
61      textx = 1 ; textadj = c(1,1)
62   }
63   text( textx , 1.0*max(pThetaGivenData) ,
64        bquote( "mean(" * theta * ")=" * .(signif(meanTheta,3)) ) ,
65        cex=2.0 , adj=textadj )
66   # Plot the likelihood: p(Data|Theta)
67   plot(Theta[thinIdx] ,pDataGivenTheta[thinIdx] ,type="p" ,pch="." ,cex=dotsize
68           ,xlim=c(0,1) ,cex.axis=1.2 ,xlab=bquote(theta)
69           ,ylim=c(0,1.1*max(pDataGivenTheta))
70           ,ylab=bquote( "p(D|" * theta * ")" )
71           ,cex.lab=1.5 ,main="Likelihood" ,cex.main=1.5 )
72   if ( z > .5*N ) { textx = 0 ; textadj = c(0,1) }
73   else { textx = 1 ; textadj = c(1,1) }
74   text( textx ,1.0*max(pDataGivenTheta) ,cex=2.0
75           ,bquote( "Data: z=" * .(z) * ",N=" * .(N) ) ,adj=textadj )
76   # Plot the posterior.
77   meanThetaGivenData = sum( Theta * pThetaGivenData )
78   plot(Theta[thinIdx] ,pThetaGivenData[thinIdx] ,type="p" ,pch="." ,cex=dotsize
79           ,xlim=c(0,1) ,ylim=c(0,1.1*max(pThetaGivenData)) ,cex.axis=1.2
80           ,xlab=bquote(theta) ,ylab=bquote( "p(" * theta * "|D)" )
81           ,cex.lab=1.5 ,main="Posterior" ,cex.main=1.5 )
82   if ( meanThetaGivenData > .5 ) { textx = 0 ; textadj = c(0,1) }
83   else { textx = 1 ; textadj = c(1,1) }
84   text(textx ,1.00*max(pThetaGivenData) ,cex=2.0
85           ,bquote( "mean(" * theta * "|D)=" * .(signif(meanThetaGivenData,3)) )
86           ,adj=textadj )
87   text(textx ,0.75*max(pThetaGivenData) ,cex=2.0
88           ,bquote( "p(D)=" * .(signif(pData,3)) ) ,adj=textadj )
89   # Mark the highest density interval. HDI points are not thinned in the plot.
90   source("HDIofGrid.R")
91   HDIinfo = HDIofGrid( pThetaGivenData  )
92   points( Theta[ HDIinfo$indices ] ,
93           rep( HDIinfo$height , length( HDIinfo$indices ) ) , pch="-" , cex=1.0 )
94   text( mean( Theta[ HDIinfo$indices ] ) , HDIinfo$height ,
95           bquote( .(100*signif(HDIinfo$mass,3)) * "% HDI" ) ,
96           adj=c(0.5,-1.5) , cex=1.5 )
97   # Mark the left and right ends of the waterline. This does not mark
98   # internal divisions of an HDI waterline for multi-modal distributions.
99   lowLim = Theta[ min( HDIinfo$indices ) ]
100  highLim = Theta[ max( HDIinfo$indices ) ]
101  lines( c(lowLim,lowLim) , c(-0.5,HDIinfo$height) , type="l" , lty=2 , lwd=1.5)
102  lines( c(highLim,highLim) , c(-0.5,HDIinfo$height) , type="l" , lty=2 , lwd=1.5)
103  text( lowLim , HDIinfo$height , bquote(.(round(lowLim,3))) ,
104           adj=c(1.1,-0.1) , cex=1.2 )
105  text( highLim , HDIinfo$height , bquote(.(round(highLim,3))) ,
```

```
106          adj=c(-0.1,-0.1) , cex=1.2 )
107
108   return( pThetaGivenData )
109   } # end of function
```

6.8 EXERCISES

Exercise 6.1. [Purpose: To understand the discretization used for the priors in the R functions of Section 6.7.1 (BernGrid.R) and throughout this chapter.] Consider this R code for discretizing a beta(θ, 8, 4) distribution:

```
nIntervals = 10
width = 1 / nIntervals
Theta = seq( from = width/2 , to = 1-width/2 , by = width )
approxMass = dbeta( Theta , 8 , 4 ) * width
pTheta = approxMass / sum( approxMass )
```

(A) What is the value of sum(approxMass)? Why is it not exactly 1?

(B) Suppose we use instead the following code to define the grid of points:

```
Theta = seq( from = 0 , to = 1 , by = width )
```

Why is this not appropriate? (Hint: Consider exactly what intervals are represented by the first and last values in Theta. Do those first and last intervals have the same widths as the other intervals? If they do, do they fall entirely within the domain of the beta distribution?)

Exercise 6.2. [Purpose: To practice specifying a nonbeta prior.] Suppose we have a coin that has a head on one side and a tail on the other. We think it might be fair, or it might be a trick coin that is heavily biased toward heads or tails. We want to express this prior belief with a single prior over θ. Therefore, the prior needs to have three peaks: one near zero, one around 0.5, and one near 1.0. But these peaks are not just isolated spikes, because we have uncertainty about the actual value of θ.

(A) Express your prior belief as a list of probability masses over a fairly dense grid of θ values. Remember to set a gradual decline around the three peaks. Briefly justify your choice. Hint: You can specify the peaks however you want, but one simple way is something like this:

```
pTheta = c( 50:1 , rep(1,50) , 1:50 , 50:1,...
pTheta = pTheta / sum( pTheta )
width = 1 / length(pTheta)
Theta = seq( from = width/2 , to = 1-width/2 , by = width )
```

(B) Suppose you flip the coin 20 times and get 15 heads. Use the R function of Section 6.7.1 (BernGrid.R) to display the posterior beliefs. Include the R code you used to specify the prior values.

Exercise 6.3. [Purpose: To use the function of Section 6.7.1 (`BernGrid.R`) for sequential updating (i.e., use output of one function call as the prior for the next function call). Observe that data ordering does not matter]

(A) Using the same prior that you used for the previous exercise, suppose you flip the coin just 4 times and get 3 heads. Use the R function of Section 6.7.1 (`BernGrid.R`) to display the posterior.

(B) Suppose we flip the coin an additional 16 times and get 12 heads. Now what is the posterior distribution? To answer this question, use the posterior distribution that is output by the function in the previous part as the prior for this part. Show the R commands you used to call the function. (Hint: The final posterior should match the posterior of Exercise 6.2, except that the graph of the prior should look like the posterior from the previous part. Figure 6.5 shows an example.)

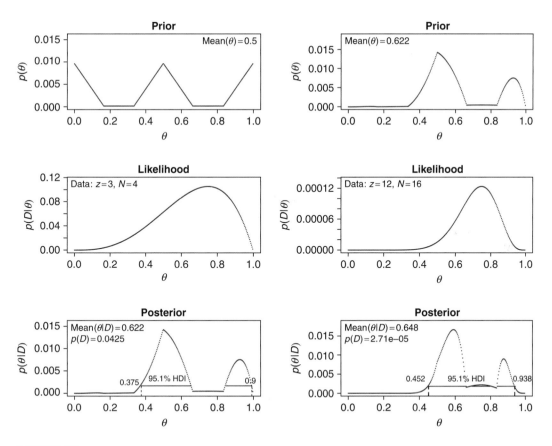

FIGURE 6.5

For Exercise 6.3. The posterior from the first 4 flips, in the left column, is used as the prior for the next 16 flips, in the right column. Your original prior (top-left panel) may look different from the original prior used here.

Exercise 6.4. [Purpose: To connect HDIs to the real world, with iterative data collection.] Suppose an election is approaching, and you are interested in knowing whether the general population prefers candidate A or candidate B. A just-published poll in the newspaper states that of 100 randomly sampled people, 58 preferred candidate A and the remainder preferred candidate B.

(A) Suppose that before the newspaper poll, your prior belief was a uniform distribution. What is the 95% HDI on your beliefs after learning of the newspaper poll results? Use the function of Section 6.7.1 (`BernGrid.R`) to determine your answer.

(B) Based in the newspaper poll, is it credible to believe that the population is equally divided in its preferences among candidates?

(C) You want to conduct a follow-up poll to narrow down your estimate of the population's preference. In your follow-up poll, you randomly sample 100 people and find that 57 prefer candidate A and the remainder prefer candidate B. Assuming that peoples' opinions have not changed between polls, what is the 95% HDI on the posterior?

(D) Based on your follow-up poll, is it credible to believe that the population is equally divided in its preferences among candidates? (Hint: Compare your answer here to your answer for Exercise 5.2.)

Exercise 6.5. [Purpose: To explore HDIs in the (almost) real world.] Suppose that the newly hired quality control manager at the Acme Widget factory is trying to convince the CEO that the proportion of defective widgets coming off the assembly line is less than 10%. No previous data are available regarding the defect rate at the factory. The manager randomly samples 500 widgets, and she finds that 28 of them are defective. What do you conclude about the defect rate? Justify your choice of prior. Include graphs to explain/support your conclusion.

Exercise 6.6. [Purpose: To use grid approximation for prediction of subsequent data.] Suppose we believe that a coin is biased to come up heads, and we describe our prior belief as quadratically increasing: $p(\theta) \propto \theta^2$. Suppose we flip the coin four times and observe two heads and two tails. Based on the posterior distribution, what is the predicted probability that the next flip will yield a head? To answer this question, use the function of Section 6.7.1 (`BernGrid.R`). Define `thetagrid` as in the example in the comments at the beginning of the function. Then define `relprob = thetagrid ^ 2`, and normalize it to specify the prior. The function returns a vector of discrete posterior masses, which you might call `posterior`. Apply Equation 6.4 by computing `sum( thetagrid * posterior )`. (Bonus hint: The answer is also displayed in the output graphics.)

Exercise 6.7. [Purpose: To use grid approximation to compare models.] Suppose we have competing beliefs about the bias of a coin: One person believes the coin is head biased, and the second person believes the coin is tail biased. To make this specific, suppose the head-biased prior is $p(\theta \mid M1) \propto \theta^2$ and the tail-biased prior is $p(\theta \mid M2) \propto (1 - \theta)^2$. Suppose that we are equally willing to entertain the two models, so $p(M1) = p(M2) = 0.5$. We flip the coin $N = 8$ times and observe $z = 6$ heads. What is the ratio of posterior beliefs? To answer this question, read the coding suggestion in Exercise 6.6 and look at $p(D)$ in the graphical output.

Exercise 6.8. [Purpose: To model comparison in the (almost) real world.] A pharmaceutical company claims that its new drug increases the probability that couples who take the drug will conceive a boy. The company has published no studies regarding this claim, so there is no public knowledge regarding the efficacy of the drug. Suppose you conduct a study in which 50 couples, sampled at random from the general population, take the drug during a period of time while trying to conceive a baby. Suppose that eventually all couples conceive; there are 30 boys and 20 girls (no multiple births).

(A) You want to estimate the probability of conceiving a boy for couples who take the drug. What is an appropriate prior belief distribution? It cannot be the general population probability, because that is a highly peaked distribution near 0.5 that refers to nondrugged couples. Instead, the prior needs to reflect our preexperiment uncertainty in the effect of the drug. Discuss your choice of prior with this in mind.

(B) Using your prior from the previous part, show a graph of the posterior and decide whether it is credible that couples who take the drug have a 50% chance of conceiving a boy.

(C) Suppose that the drug manufacturers make a strong claim that their drug sets the probability of conceiving a boy to nearly 60%, with high certainty. Suppose you represent that claim by a beta(60,40) prior. Compare that claim against the skeptic who says there is no effect of the drug, and the probability of conceiving a boy is represented by a beta(50,50) prior. What is the value of $p(D)$ for each prior? What is the posterior belief in each claim? (Hint: Be careful when computing the posterior belief in each model, because you need to take into account the prior belief in each model. Is the prior belief in the manufacturer's claim as strong as the prior belief in the skeptical claim?

CHAPTER 7

Inferring a Binomial Proportion via the Metropolis Algorithm

Doing Bayesian Data Analysis: A Tutorial with R and BUGS. DOI: 10.1016/B978-0-12-381485-2.00007-9
© 2011, Elsevier Inc. All rights reserved.

You furtive posterior: coy distribution.
Alluring, curvaceous, evading solution.
Although I can see what you hint at is ample,
I'll settle for one representative sample.

In this chapter we continue with the goal of inferring the underlying probability θ that a coin comes up heads, given an observed set of flips. In Chapter 5, we considered the scenario when the prior distribution is specified by a function that is conjugate to the likelihood function and thus yields an analytically solvable posterior distribution. In Chapter 6, we considered the scenario when the prior is specified on a dense grid of points spanning the range of θ values, and thus the posterior is numerically generated by summing across the discrete values.

But there are situations in which neither of those methods will work. We already recognized the possibility that our prior beliefs about θ could not be adequately represented by a beta distribution, or by any function that yields an analytically solvable posterior function. The chapter on grid approximation was one approach to addressing such situations. When we have just one parameter with a finite range, then approximation by a grid is a useful procedure. But what if we have several parameters? Although we have only been dealing with models involving a single parameter, it is much more typical to have models involving several parameters, as we will see in later chapters. In these situations, the parameter space cannot be spanned by a grid with a reasonable number of points. Consider, for example, a model with, say, six parameters. If we set up a grid on each parameter that has 1000 values, then the six-dimensional parameter space has $1000^6 = 1,000,000,000,000,000,000$ combinations of parameter values, which is too many for any computer to evaluate. In anticipation of those situations when grid approximation will not work, we explore a new method in the simple context of estimating a single parameter. In real research you would probably not want to apply this method for such simple one-parameter models, instead going with mathematical analysis or grid approximation. But it is useful to *learn* about this new method in the one-parameter context.

The method described in this chapter assumes that the prior distribution is specified by a function that is easily evaluated. This simply means that if you specify a value for θ, then the value of $p(\theta)$ is easily determined, especially by a computer. The method also assumes that the value of the likelihood function, $p(D | \theta)$, can be computed for any specified values of D and θ. Actually, all that the method really demands is that the product of the prior and likelihood be easily computed for any given value of θ, and then only up to a multiplicative constant. What the method produces for us is an approximation of the posterior distribution, $p(\theta | D)$, in the form of a large number of θ values sampled from that distribution. This heap of representative θ values can be used to estimate the mean and median of the posterior, its credible region, and so on. The posterior distribution is estimated by randomly generating a lot of values from it, and therefore, by analogy to the random events at games in a casino, this approach is called a *Monte Carlo method*.

7.1 A SIMPLE CASE OF THE METROPOLIS ALGORITHM

Our goal in Bayesian inference is to get a good handle on the posterior distribution over the parameters. One way to do that is to sample a large number of representative points from the posterior and then, from those points, compute descriptive statistics about the distribution. For example, consider a beta($\theta \mid a, b$) distribution. We learned earlier that the mean and standard deviation of the distribution can be analytically derived, and it can be expressed exactly in terms of a and b. We also learned earlier that the cumulative probability is easily computed and is implemented by a command (qbeta) in R, so we can determine equal tailed credible intervals.

But suppose we did not know the analytical formulas for the mean and standard deviation, and we did not have a direct way to calculate cumulative probabilities. Suppose, however, that we do have a way to generate representative values from the distribution. The random values could be generated from a spinner. The spinner is marked on its circumference with values from zero to one; these are the possible θ values. Imagine that the spinner is biased to point at θ values exactly according to a beta($\theta \mid a, b$) distribution. We spin the spinner a few thousand times and record the values that it points to. These values are representative of the underlying beta($\theta \mid a, b$) distribution (i.e., the population from which the sample came). In particular, the mean of the sampled values should be a close approximation to the true mean of the underlying distribution. The standard deviation of the sampled values should be a close approximation to the true standard deviation of the distribution. And the percentiles in the sampled values should be a close approximation to the true percentiles in the population. In other words, if we can get a large sample of representative values from the distribution, then we can approximate all sorts of useful characteristics of the distribution.

The question then becomes, how can we sample a large number of representative values from a distribution? For an answer, let's ask a politician.

7.1.1 A Politician Stumbles on the Metropolis Algorithm

Suppose an elected politician lives on a long chain of islands. He is constantly traveling from island to island, wanting to stay in the public eye. At the end of a grueling day of photo opportunities and fundraising,[1] he has to decide whether to (1) stay on the current island, (2) move to the adjacent island to the west, or (3) move to the adjacent island to the east. His goal is to visit all the islands proportionally to their relative population, so that he spends the most time on the most populated islands and proportionally less time on the less populated islands. Unfortunately, he holds his office despite having no idea what the total population of the island chain is. He doesn't even know exactly how many islands there are! His entourage of advisers is capable of some minimal information-gathering abilities, however. When they are not busy fundraising, they can ask the mayor of the island they are on how many people are on the island. And, when the politician proposes to visit an adjacent island, they can ask the mayor of that adjacent island how many people are on that island.

The politician has a simple heuristic for deciding whether to travel to the proposed island: First, he flips a (fair) coin to decide whether to propose the adjacent island to the east or the adjacent island to the west. If the proposed island has a larger population than the current island, then he definitely goes to the proposed island. On the other hand, if the proposed island has a smaller population than the current island, then he goes to the proposed island only probabilistically, to the extent that the proposed island has a population as big as the current island. In more detail, denote the population of the proposed island as $P_{proposed}$, and the population of the current island as $P_{current}$. Then he moves to the less populated island with probability $p_{move} = P_{proposed}/P_{current}$. The politician does this by spinning a fair spinner marked on its circumference with uniform values from zero to one. If the pointed-to value is between zero and p_{move}, then he moves. What's amazing about this heuristic is that it works: In the long run, the probability that the politician is on any one of the islands in the chain exactly matches the relative population of the island!

7.1.2 A Random Walk

Let's consider the island-hopping heuristic in a bit more detail. Suppose that there are seven islands in the chain, with relative populations as shown in the bottom panel of Figure 7.1. The islands are indexed by the value θ, whereby the leftmost, western island is $\theta = 1$ and the rightmost, eastern island is $\theta = 7$. The relative populations of the islands increase linearly such that $P(\theta) = \theta$.

The middle panel of Figure 7.1 shows one possible trajectory taken by the politician. Each day corresponds to one time increment, indicated on the vertical axis. The plot of the trajectory shows that on the first day ($t = 1$) the politician happens to be on the middle island in the chain (i.e., $\theta_{current} = 4$).

---

[1] Maybe I shouldn't blithely make cynical jokes about politicians, because I believe that most elected representatives really do try to do some good for their constituencies. But saying so isn't as entertaining as the cheap joke.

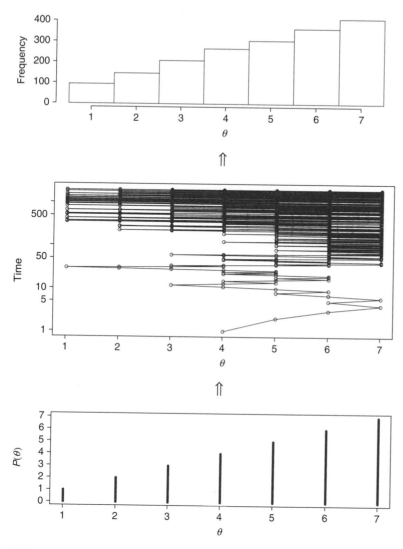

FIGURE 7.1

A simple Metropolis algorithm. The bottom panel shows the values of the target distribution. The middle panel shows one random walk, at each time step proposing to move either one unit right or one unit left, and accepting the proposed move according the heuristic described in the main text. The top panel shows the frequency distribution of the positions in the walk.

To decide where to go on the second day, he flips a coin to propose moving either one position left or one position right. In this case the coin proposed moving right (i.e., $\theta_{proposed} = 5$). Because the relative population at the proposed position is greater than the relative population at the current position, the proposed move is accepted. The trajectory shows this move, because when $t = 2$, then $\theta = 5$.

Continue considering the middle panel. Count along the trajectory to the fourth position, when $t = 4$ and $\theta = 7$. At the end of this time step, the coin flip proposes moving to the left. The probability of accepting this proposal is $p_{move} = P(\theta_{proposed})/P(\theta_{current}) = 6/7$. The politician then spins a fair spinner, which happens to come up with a value between 0 and $6/7$, and therefore the politician makes the move. Hence, the trajectory shows that $\theta = 6$ when $t = 5$.

The upper panel of Figure 7.1 shows a histogram of the frequencies with which each position is visited during this junket. Notice that the sampled relative frequencies closely mimic the actual relative populations in the bottom panel! In fact, a sequence generated this way will converge, as the sequence gets longer, to an arbitrarily close approximation of the actual relative probabilities.

7.1.3 General Properties of a Random Walk

The trajectory shown in Figure 7.1 is just one possible sequence of positions when the movement heuristic is applied. At each time step, the direction of the proposed move is random, and, if the relative probability of the proposed position is less than that of the current position, then acceptance of the proposed move is also random. Because of the randomness, if the process were started over again, then the specific trajectory would almost certainly be different.

Figure 7.2 shows the probability of being in each position as a function of time. At time $t = 1$, the politician starts at $\theta = 4$. This starting position is indicated in the upper-left panel of Figure 7.2, labeled $t = 1$, by the fact that there is 100% probability of being at $\theta = 4$.

We want to figure out the probability of ending up in each position at the next time step. To determine the probabilities of positions for time $t = 2$, consider the possibilities from the movement process. The process starts with the flip of a fair coin to decide which direction to propose moving. There is a 50% probability of proposing to move right (i.e., to $\theta = 5$). By inspecting the target distribution of relative probabilities in the lower-right panel of Figure 7.2, you can see that $P(\theta = 5) > P(\theta = 4)$, and therefore a rightward move is always accepted whenever it is proposed. Thus, at time $t = 2$, there is a 0.5 (i.e., 50%) probability of ending up at $\theta = 5$. The panel labeled $t = 2$ in Figure 7.2 plots this probability as a bar of height 0.5 at $\theta = 5$. The other 50% of the time, the proposed move is to the left (i.e., $\theta = 3$). By inspecting the target distribution of relative probabilities in the lower-right panel of Figure 7.2, you can see that $P(\theta = 3) = 3$, whereas $P(\theta = 4) = 4$, and therefore a leftward move is accepted only 3/4 of the times it is proposed. Therefore, at time $t = 2$, the

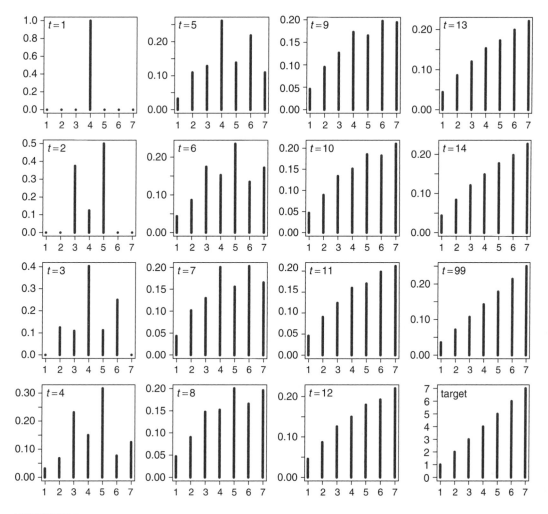

FIGURE 7.2
The probability of being at position θ, as a function of time t, when a simple Metropolis algorithm is applied to the target distribution in the lower-right panel. The time in each panel corresponds to the step in random walk, an example of which is shown in Figure 7.1. The target distribution is shown in the lower-right panel.

probability of ending up at $\theta = 3$ is $50\% \times 3/4 = 0.375$. The panel labeled $t = 2$ in Figure 7.2 shows this as a bar of height 0.375 at $\theta = 3$. Finally, if a leftward move is proposed but not accepted, we just stay at $\theta = 5$. The probability of this happening is only 0.125 (i.e., $50\% \times (1 - 3/4)$).

This process repeats for the next time step. I won't go through the arithmetic details for each value of θ. But it is important to notice that after two proposed moves (i.e., when $t = 3$, the politician could be at any of the positions $\theta = 2$

through $\theta = 6$, because he could be at most two positions away from where he started).

The probabilities continue to be computed the same way at every time step. You can see that in the early time steps, the probability distribution is not a straight incline like the target distribution. Instead, the probability distribution has a bulge over the starting position. This influence of the arbitrary starting point is not desired, so we exclude the early steps from the positions that we will treat as representative of the target distribution. This early period that we exclude is the "burn-in" period. As Figure 7.2 shows, by time $t = 99$, the position probability is virtually indistinguishable from the target distribution, at least for this simple distribution. More complex distributions require a longer burn-in period.

The graphs of Figure 7.2 show the *probability* that the moving politician is at each value of θ. But remember, at any given time step, the politician is at just one particular position, as shown in Figure 7.1. To approximate the target distribution, we let the politician meander around for many time steps while we keep track of where he has been. The record of where he has been is our approximation to the distribution. Moreover, we have to be careful not to use the portion of the meandering that still has the influence of the initial position. That is, we have to exclude the burn-in period. But once we have a long record of where the traveler has been, after the burn-in period, then we can approximate the target probability at each value of θ by simply counting the relative number times that the traveler visited that value.

Here is a summary of our algorithm for moving from one position to another. We are currently at position $\theta_{current}$. We then propose to move one position right or one position left. The specific proposal is determined by flipping a coin, which can result in 50% heads (move right) or 50% tails (move left). The range of possible proposed moves, and the probability of proposing each, is called the *proposal distribution*. In the present algorithm, the proposal distribution is simple: It has only two values with 50-50 probabilities.

Having proposed a move, we then decide whether or not to accept it. The acceptance decision is based on the value of the target distribution at the proposed position, relative to the value of the target distribution at our current position. Specifically, if the target distribution is greater at the proposed position than at our current position, then we definitely accept the proposed move: We always move higher if we can. On the other hand, if the target position is less at the proposed position than at our current position, we accept the move probabilistically: We move to the proposed position with probability $p_{move} = P(\theta_{proposed})/P(\theta_{current})$, where $P(\theta)$ is the value of the target distribution at θ. We can combine these two possibilities, of the target distribution being higher or lower at the proposed position than at our current position, into a single expression for the probability of moving to the proposed position:

$$p_{move} = \min\left(\frac{P(\theta_{proposed})}{P(\theta_{current})}, 1\right) \quad (7.1)$$

Notice that Equation 7.1 says that when $P(\theta_{proposed}) > P(\theta_{current})$, then $p_{move} = 1$. Notice also that the target distribution, $P(\theta)$, does not need to be normalized (i.e., it does not need to sum to 1 as a probability distribution must). This is because what matters for our choice is the *ratio*, $P(\theta_{proposed})/P(\theta_{current})$, not the absolute magnitude of $P(\theta)$. This property was used in the example of the island-hopping politician: The target distribution was the population of each island, not a normalized probability.

Having proposed a move by sampling from the proposal distribution and having then determined the *probability* of accepting the move according to Equation 7.1, we then actually accept or reject the proposed move by sampling a value from a uniform distribution over the interval $[0, 1]$. If the sampled value is between 0 and p_{move}, then we actually make the move. Otherwise, we reject the move, and stay at our current position. The whole process repeats at the next time step.

7.1.4 Why We Care

Notice what we must be able to do in the random-walk process:

- We must be able to generate a random value from the proposal distribution (to create $\theta_{proposed}$).
- We must be able to evaluate the target distribution at any proposed position (to compute $P(\theta_{proposed})/P(\theta_{current})$).
- We must be able to generate a random value from a uniform distribution (to make a move according to p_{move}).

By being able to do those three things, we are able to do *indirectly* something we could not necessarily do directly: We can generate random samples from the target distribution. Moreover, we can generate those random samples from the target distribution even when the target distribution is not normalized.

This technique is profoundly useful when the target distribution $P(\theta)$ is a posterior proportional to $p(D|\theta)p(\theta)$. Merely by evaluating $p(D|\theta)p(\theta)$, we can generate random representative values from the posterior distribution. This result is wonderful because the method obviates direct computation of the evidence $p(D)$, which, as you'll recall, is one of the most difficult aspects of Bayesian inference. By using techniques such as these, we can do Bayesian inference in rich and complex models. It has only been with the development of these techniques that Bayesian inference is applicable to complex data analysis, and it has only been with the production of fast and cheap computers that Bayesian inference is accessible to a wide audience.

7.1.5 Why It Works

In this section I'll explain a bit of the mathematics behind why the algorithm works. My goal in providing the mathematics for the simple case is to justify an intuitive leap, in the next section, to the more general algorithm. If you get a good running start in this section, it can be easier to make the leap in the next section.

To get an intuition for why this algorithm works, consider two adjacent positions and the probabilities of moving from one to the other. We'll see that the relative transition probabilities, between adjacent positions, exactly match the relative values of the target distribution. Extrapolate that result across all the positions, and you can see that, in the long run, each position will be visited proportionally to its target value. Now the details. Suppose we are at position θ. The probability of moving to $\theta + 1$, denoted $p(\theta \to \theta + 1)$, is the probability of proposing that move times the probability of accepting it if proposed, which is $p(\theta \to \theta + 1) = 0.5 \times \min(P(\theta + 1)/P(\theta), 1)$. On the other hand, if we are presently at position $\theta + 1$, the probability of moving to θ is the probability of proposing that move times the probability of accepting it if proposed, which is $p(\theta + 1 \to \theta) = 0.5 \times \min(P(\theta)/P(\theta + 1), 1)$. The ratio of the transition probabilities is

$$
\begin{aligned}
\frac{p(\theta \to \theta + 1)}{p(\theta + 1 \to \theta)} &= \frac{0.5 \min(P(\theta + 1)/P(\theta), 1)}{0.5 \min(P(\theta)/P(\theta + 1), 1)} \\[2mm]
&= \begin{cases} \dfrac{1}{P(\theta)/P(\theta + 1)} & \text{if } P(\theta + 1) > P(\theta) \\[3mm] \dfrac{P(\theta + 1)/P(\theta)}{1} & \text{if } P(\theta + 1) < P(\theta) \end{cases} \\[2mm]
&= \frac{P(\theta + 1)}{P(\theta)}
\end{aligned}
\tag{7.2}
$$

Equation 7.2 tells us that during transitions back and forth between adjacent positions, the relative probability of the transitions exactly matches the relative values of the target distribution. That might be enough for you to get the intuition that, in the long run, adjacent positions will be visited proportionally to their relative values in the target distribution. If that's true for adjacent positions, then, by extrapolating from one position to the next, it must be true for the whole range of positions.

To make that intuition more defensible, we have to fill in some more details. To do this, I'll use matrix arithmetic. This is the only place in the book where matrix arithmetic appears, so if the details here are unappealing, feel free to skip ahead to Section 7.2, p. 129. What you'll miss is an explanation of the mathematics underlying Figure 7.2, which depicts the key idea that the target distribution is *stable*: If the current probability of being in a position matches the target probabilities, then the Metropolis algorithm keeps it that way.

Consider the probability of transitioning from position θ to some other position. The proposal distribution, in the present simple scenario, considers only positions $\theta + 1$ and $\theta - 1$. If the proposed position is not accepted, we stay at the current position θ. The probability of moving to position $\theta - 1$ is the probability of proposing that position times the probability of accepting the move if it is proposed: $0.5 \min (P(\theta - 1)/P(\theta), 1)$. The probability of moving to position $\theta + 1$ is the probability of proposing that position times the probability of accepting the move if it is proposed: $0.5 \min (P(\theta + 1)/P(\theta), 1)$. The probability of staying at position θ is simply the complement of those two move-away probabilities: $0.5 [1 - \min (P(\theta - 1)/P(\theta), 1)] + 0.5 [1 - \min (P(\theta + 1)/P(\theta), 1)]$.

We can put those transition probabilities into a matrix. Each row corresponds to the current position, and each column corresponds to the candidate moved-to position. A *submatrix* from the full transition matrix T follows, showing rows $\theta - 2$ to $\theta + 2$ and columns $\theta - 2$ to $\theta + 2$:

$$
\begin{bmatrix}
\ddots & p(\theta - 2 \to \theta - 1) & 0 & 0 & 0 \\
\ddots & p(\theta - 1 \to \theta - 1) & p(\theta - 1 \to \theta) & 0 & 0 \\
0 & p(\theta \to \theta - 1) & p(\theta \to \theta) & p(\theta \to \theta + 1) & 0 \\
0 & 0 & p(\theta + 1 \to \theta) & p(\theta + 1 \to \theta + 1) & \ddots \\
0 & 0 & 0 & p(\theta + 2 \to \theta + 1) & \ddots
\end{bmatrix}
$$

which equals

$$
\begin{bmatrix}
\ddots & 0.5\min\left(\frac{P(\theta-1)}{P(\theta-2)}, 1\right) & 0 & 0 & 0 \\
\ddots & \begin{array}{l} 0.5\left[1 - \min\left(\frac{P(\theta-2)}{P(\theta-1)}, 1\right)\right] \\ + 0.5\left[1 - \min\left(\frac{P(\theta)}{P(\theta-1)}, 1\right)\right] \end{array} & 0.5\min\left(\frac{P(\theta)}{P(\theta-1)}, 1\right) & 0 & 0 \\
0 & 0.5\min\left(\frac{P(\theta-1)}{P(\theta)}, 1\right) & \begin{array}{l} 0.5\left[1 - \min\left(\frac{P(\theta-1)}{P(\theta)}, 1\right)\right] \\ + 0.5\left[1 - \min\left(\frac{P(\theta+1)}{P(\theta)}, 1\right)\right] \end{array} & 0.5\min\left(\frac{P(\theta+1)}{P(\theta)}, 1\right) & 0 \\
0 & 0 & 0.5\min\left(\frac{P(\theta)}{P(\theta+1)}, 1\right) & \begin{array}{l} 0.5\left[1 - \min\left(\frac{P(\theta)}{P(\theta+1)}, 1\right)\right] \\ + 0.5\left[1 - \min\left(\frac{P(\theta+2)}{P(\theta+1)}, 1\right)\right] \end{array} & \ddots \\
0 & 0 & 0 & 0.5\min\left(\frac{P(\theta+1)}{P(\theta+2)}, 1\right) & \ddots
\end{bmatrix}
$$

$$(7.3)$$

The usefulness of putting the transition probabilities into a matrix is that we can then use matrix multiplication to get from any current location to the probability of the next locations. Here's a reminder of how matrix multipli-cation operates. Consider a matrix T. The value in its rth row and c^{th} column is denoted T_{rc}. We can multiply the matrix on its *left* side by a *row* vector w, which yields another row vector. The cth component of the product wT is $\sum_r w_r T_{rc}$. In other words, to compute the cth component of the result, take the row vector w and multiply its components by the corresponding components in the cth *column* of W, and sum up those component products.[2]

To use the transition matrix in Equation 7.3, we put the *current* location prob-abilities into a row vector, which I will denote w because it indicates where we are. For example, if at the current time we are definitely in location θ, then w has 1.0 in its θ component, and zeros everywhere else: $w = [\ldots, 0, 1, 0, \ldots]$. To determine the probability of the locations at the next time step, we simply mul-tiply w by T. Here's a key example to think through: When $w = [\ldots, 0, 1, 0, \ldots]$ with a 1 only in the θ position, then wT is simply the row of T corresponding to θ, because the cth component of wT is $\sum_r w_r T_{rc} = T_{\theta c}$, where I'm using the subscript θ to stand for the index that corresponds to value θ.

Matrix multiplication is a useful procedure for keeping track of position proba-bilities. At every time step, we multiply the current position probability vector w by the transition probability matrix T to get the position probabilities for the next time step. We keep multiplying by T over and over again to derive the long-run position probabilities. This process is exactly what generated the graphs in Figure 7.2.

Here's the climactic implication: When the vector of position probabilities is the target distribution, it stays that way on the next time step! In other words, the position probabilities are stable at the target distribution. We can actually prove this result without much trouble. Suppose the current position probabilities are the target probabilities—that is, $w = [\ldots, P(\theta - 1), P(\theta), P(\theta + 1), \ldots]/Z$, where $Z = \sum_\theta P(\theta)$ is the normalizer for the target distribution. Consider the θ component of wT. We will demonstrate that the θ component of wT is the same as the θ component of w, for any component θ. The θ component of wT is $\sum_r w_r T_{r\theta}$. Look back at the transition matrix in Equation 7.3, and you can see then that the θ component of wT is

$$\sum_r w_r T_{r\theta} = P(\theta - 1)/Z \times 0.5\min\left(\frac{P(\theta)}{P(\theta - 1)}, 1\right)$$

$$+ P(\theta)/Z \times \left(0.5\left[1 - \min\left(\frac{P(\theta - 1)}{P(\theta)}, 1\right)\right] + 0.5\left[1 - \min\left(\frac{P(\theta + 1)}{P(\theta)}, 1\right)\right]\right)$$

$$+ P(\theta + 1)/Z \times 0.5\min\left(\frac{P(\theta)}{P(\theta + 1)}, 1\right) \tag{7.4}$$

---

[2] Although we don't do it here, we can also multiply a matrix on its *right* side by a *column* vector, which yields another column vector. For a column vector v, the rth component of Tv is $\sum_c T_{rc} v_c$.

To simplify that equation, we can consider separately the four cases: case 1: $P(\theta) > P(\theta - 1)$ and $P(\theta) > P(\theta + 1)$; case 2: $P(\theta) > P(\theta - 1)$ and $P(\theta) < P(\theta + 1)$; case 3: $P(\theta) < P(\theta - 1)$ and $P(\theta) > P(\theta + 1)$; case 4: $P(\theta) < P(\theta - 1)$ and $P(\theta) < P(\theta + 1)$. In each case, Equation 7.4 simplifies to $P(\theta)/Z$. For example, consider case 1, when $P(\theta) > P(\theta - 1)$ and $P(\theta) > P(\theta + 1)$. Equation 7.4 becomes

$$
\begin{aligned}
\sum_r w_r T_{r\theta} &= P(\theta - 1)/Z \times 0.5 \\
&\quad + P(\theta)/Z \times \left(0.5 \left[1 - \left(\frac{P(\theta - 1)}{P(\theta)} \right) \right] + 0.5 \left[1 - \left(\frac{P(\theta + 1)}{P(\theta)} \right) \right] \right) \\
&\quad + P(\theta + 1)/Z \times 0.5 \\
&= 0.5\, P(\theta - 1)/Z \\
&\quad + 0.5 P(\theta)/Z - 0.5 P(\theta)/Z \frac{P(\theta - 1)}{P(\theta)} + 0.5 P(\theta)/Z \\
&\quad - 0.5 P(\theta)/Z \frac{P(\theta + 1)}{P(\theta)} + 0.5\, P(\theta + 1)/Z \\
&= P(\theta)/Z
\end{aligned}
$$

If you work through the other cases, you'll find that it always reduces to $P(\theta)/Z$. In conclusion, when the θ component starts at $P(\theta)/Z$, it stays at $P(\theta)/Z$.

We have shown that the target distribution is stable under the Metropolis algorithm, for our special case of island hopping. To prove that the Metropolis algorithm realizes the target distribution, we would need to show that the process actually gets us *to* the target distribution regardless of where we start. For this I'll settle for intuition. You can see that no matter where you start, the distribution will naturally diffuse and explore other positions. Examples of this were shown in Figures 7.1 and 7.2. It's reasonable to think that the diffusion will settle into some stable state, and we've just shown that the target distribution is a stable state. To really make the argument complete, we'd have to show that there are no other stable states and that the target distribution is actually an attractor into which other states flow, rather than a state that is stable if it is ever obtained but impossible to actually attain. This complete argument is far beyond what would be useful for the purposes of this book, but if you're interested you could take a look at the book by Robert & Casella (2004).

7.2 THE METROPOLIS ALGORITHM MORE GENERALLY

The procedure described in the previous section was just a special case of a more general procedure known as the Metropolis algorithm, named after the first author of a famous article (Metropolis et al., 1953). In the previous section, we considered the simple case of (1) discrete positions (2) on one dimension (3) with moves that proposed just one position left or right. That simple situation made it relatively easy (believe it or not) to understand the procedure and how it works. The general algorithm applies to (1) continuous values (2) on any number of dimensions (3) with more general proposal distributions.

The essentials of the general method are the same as for the simple case. First, we have some target distribution, $P(\theta)$, over a multidimensional continuous parameter space from which we would like to generate representative sample values. We must be able to compute the value of $P(\theta)$ for any candidate value of θ. The distribution, $P(\theta)$, does not have to be normalized, however. It merely needs to be non-negative. In typical applications, $P(\theta)$ is the unnormalized posterior distribution on θ, which is to say, it is the product of the likelihood and the prior.

Sample values from the target distribution are generated by taking a random walk through the parameter space. The walk starts at some arbitrary point, specified by the user. The starting point should be someplace where $P(\theta)$ is nonzero. The random walk progresses at each time step by proposing a move to a new position in parameter space and then deciding whether or not to accept the proposed move. Proposal distributions can take on many different forms, with the goal being to use a proposal distribution that efficiently explores the regions of the parameter space where $P(\theta)$ has most of its mass. Of course, we must use a proposal distribution for which we have a quick way to generate random values! For our purposes, we will consider the generic case in which the proposal distribution is normal, centered at the current position. (Recall the discussion of the normal distribution back in Section 3.3.2.2, p. 35.) The idea behind using a normal distribution is that the proposed move will typically be near the current position, with the probability of proposing a more distant position dropping off according to the normal curve. Computer languages such as R have built-in functions for generating pseudo-random values from a normal distribution. For example, if we want to generate a proposed jump from a normal distribution that has a mean of zero and a standard deviation of 0.2, we could command R as follows: `proposedJump = rnorm( 1 , mean=0 , sd=0.2 )`, where the first argument, 1, indicates that we want a single random value, not a vector of many random values.

Having generated a proposed new position, the algorithm then decides whether or not to accept the proposal. The decision rule is exactly what was already specified in Equation 7.1. In detail, this is accomplished by computing the ratio $p(\theta_{move}) = P(\theta_{proposed})/P(\theta_{current})$. Then a random number from the uniform interval $[0, 1]$ is generated; in R, this can be accomplished with the command `runif(1)`. If the random number is between 0 and $p(\theta_{move})$, then the move is accepted. The process repeats and, in the long run, the positions visited by the random walk will closely approximate the target distribution.

7.2.1 "Burn-in," Efficiency, and Convergence

If the target distribution is very spread out, but the proposal distribution is very narrow, then it will take a long time for the random walk to cover the distribution with representative steps. This is like trying to make a geographical survey of an entire continent (many thousands of kilometers wide) by repeatedly tossing a stone up in the air (a few meters at best) and occasionally moving to where the stone lands. Thus, when the proposal distribution is too narrow, the Metropolis algorithm is not very efficient: It takes way too many steps to accumulate a representative sample.

This problem can be especially evident if the initial position of the random walk happens to be in a region of the target distribution that is flat and low, in which case the random walk moves only slowly away from the starting position into regions where the target distribution is denser. Even if the proposal distribution is not too narrow, an unrepresentative initial position can leave its mark in the random walk for a long time. To alleviate this problem, the early steps of the random walk are excluded from those steps considered to be representative of the target distribution. These excluded initial steps are referred to as the burn-in period.

Even after the random walk has meandered around for a long time, we cannot be sure that it is really exploring the main regions of the target distribution, especially if the target distribution is complex over many dimensions. There are various methods for trying to assess the convergence of the random walk, but we will not need to explore their nuances in this chapter because the application is so simple. In future chapters, however, we will discuss the idea of starting random walks from several different positions in the parameter space and checking that they converge to similar distributions.

The previous considerations, pointing out the problems of a proposal distribution that is too narrow, may cause you to think that it is best to make the proposal distribution very wide instead. But excessive width can also produce problems. Remember that the probability of accepting a move is $P(\theta_{proposed})/P(\theta_{current})$. Suppose that the current position has a relatively high $P(\theta_{current})$. In this case, when the proposed positions are far away, they will often fall outside of the main mass of the target distribution, and then $P(\theta_{proposed})$ will usually be less than $P(\theta_{current})$, and therefore the proposed move will rarely be accepted. Therefore, the random walk rejects many proposals before it accepts one, and the process again becomes inefficient.

All in all, selecting a proposal distribution with a well-tuned variance, excluding the right amount of burn-in steps, and making sure the random walk has converged appropriately can be a bit of a challenge. In the present chapter, however, the target distributions are well behaved enough that these issues will not be a big problem. Nevertheless, in Exercise 7.1, you will get to observe some of these infelicities of the Metropolis algorithm.

7.2.2 Terminology: Markov Chain Monte Carlo

Assessing the properties of a target distribution by generating representative random values is a case of a Monte Carlo simulation. Any simulation that samples a lot of random values from a distribution is called a *Monte Carlo simulation*, named after the dice and spinners and shufflings of the famous casino locale. This appellation is attributed to the mathematician von Neumann or sometimes to Metropolis (Gill, 2002, p. 239).

The Metropolis algorithm is a specific type of Monte Carlo process. It generates a random walk such that each step in the walk is completely independent of the steps before the current position. The proposed next step has no dependence on where the walk has been before, and the decision to reject or accept the proposed step has no dependence on where the walk has been before. Any process in which each step has no memory for states before the current one is called a (first-order) Markov process, and a succession of such steps is a Markov chain. The Metropolis algorithm is an example of a *Markov chain Monte Carlo (MCMC)* process.

7.3 FROM THE SAMPLED POSTERIOR TO THE THREE GOALS

Let's regain perspective on the forest of Bayesian inference after focusing on the trees of MCMC. In Bayesian inference, we need a good description of the posterior distribution. If we cannot achieve that description through formal analysis, nor through dense-grid approximation, then we can generate a lot of representative values from the posterior distribution and use those values to approximate the posterior. So far in this chapter, we have explored one process for generating representative values from a distribution, namely, the Metropolis algorithm. We have yet to consider the details of applying the Metropolis algorithm to inference about a binomial proportion.

As you will fondly recall, in this part of the book we are restricting our attention to the simple case of estimating the probability that a coin comes up heads. In other words, we are trying to estimate our posterior beliefs regarding the underlying probability θ. We start with a prior belief distribution, $p(\theta)$. In the present scenario, $p(\theta)$ is specified by a mathematical function of θ; it is not merely a list of probability masses at discrete values of θ. The value of the mathematical function $p(\theta)$ must be easily computable at any value of θ. One example of such a mathematical function is the beta distribution; R easily computes the value of the beta density with its dbeta function.

We also have a mathematical likelihood function that specifies $p(D|\theta)$, where the datum D for any one flip is $y = 1$ for heads and $y = 0$ for tails. For a single flip, the likelihood function is the Bernoulli distribution, $p(y|\theta) = \theta^y \, (1-\theta)^{(1-y)}$. When there are several independent flips, the likelihood is the product of the probabilities of the individual flips. When there are N flips with z heads, the likelihood (back in Equation 5.3, p. 79) is $p(z, N|\theta) = \theta^z \, (1-\theta)^{(N-z)}$. Notice that the value of the likelihood function is easily computable for any given value of θ.

The posterior distribution $p(\theta|D)$ is, by Bayes' rule, proportional to $p(D|\theta)p(\theta)$. We use that product as the target distribution in a Metropolis algorithm. The Metropolis algorithm only needs the relative posterior probabilities in the target distribution, not the absolute posterior probabilities, so we could use an unnormalized prior or unnormalized posterior when generating sample values of θ. (Later, when we consider the goal of model comparison, we will want to estimate $p(D)$, and for that we will want the actual, normalized values of the posterior.)

We start the random walk of the Metropolis algorithm at some candidate value of θ, such as $\theta = 0.5$; then we propose a jump to a new position. The proposal distribution could be a normal distribution, with some reasonable standard deviation such as $\sigma = 0.2$. Why is that choice of σ reasonable? One consideration is that the range of θ in this application is limited to $[0, 1]$, so certainly we would like the proposal distribution to be narrower than the range of θ. Another consideration is that the proposal distribution should be "tuned" to the width of the posterior, not too wide and not too narrow. When the sample size is small, the posterior is typically not very narrow, and so $\sigma = 0.2$ can work. But remember, the proposal function and its characteristics are chosen by you, the analyst. Choose wisely! Exercise 7.1 has you explore the consequences of different proposal distributions. Notice that when a normal distribution is used as the proposal distribution, there can be proposed values less than zero or greater than one, which are invalid values of θ in the present situation. This is okay as long as the prior and likelihood functions return values of zero when the proposed θ is inappropriate.

Section 7.6.1 (BernMetropolisTemplate.R), p. 146, has R code that provides a detailed template for the Metropolis algorithm. By inspecting the loquaciously commented code, you can see that the code begins by defining three functions: the likelihood function, the prior probability function, and the target distribution function. Typically the target distribution function is simply the product of the likelihood and prior, but the target distribution is coded

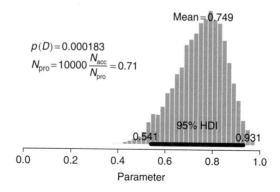

FIGURE 7.3

Graphical output from the Metropolis algorithm executed by the program of Section 7.6.1. (`BernMetropolisTemplate.R`) Compare this with the posterior distributions shown in Figures 5.2 (p. 86) and 6.2 (p. 105).

as a separate function because the target function for a Metropolis algorithm could be anything, it doesn't have to be a posterior from Bayes' rule. After defining the target distribution, the next chunk of code generates the random walk. After meandering around the parameter space a while, the code simply excludes an arbitrary burn-in portion and saves the remainder of the walk as its heap of representative values from the posterior distribution.

An example of the results of the Metropolis algorithm is shown in Figure 7.3. The θ values were generated from using a uniform prior, Bernoulli likelihood, and data in which $z = 11$ and $N = 14$. This prior and data match examples used in previous chapters. Compare Figure 7.3 with the posterior distributions shown in Figure 5.2 (p. 86), which was determined from mathematical analysis, and Figure 6.2 (p. 105), which was computed by grid approximation. You'll notice that they are very similar.

7.3.1 Estimation

From the heap of representative values of $p(\theta \,|\, D)$, we can estimate aspects of the actual $p(\theta \,|\, D)$ distribution. For example, to summarize the central tendency of the representative values, it is easy to compute their mean or their median. The mean of the distribution is displayed in Figure 7.3.

7.3.1.1 Highest Density Intervals from Random Samples

Highest density intervals (HDIs) can also be estimated from MCMC samples. One way to do it relies on computing the (relative) posterior probability at each point in the sample (Hyndman, 1996). Essentially, this method finds the water level such that 5% of the points are under water. The remaining points above water represent the 95% HDI region. In more detail, first, we compute the relative height at each point—that is, the relative posterior density $p(\theta_i \mid D) \propto p(D \mid \theta_i)p(\theta_i)$ at each θ_i in the sample. Let $p_{0.05}$ denote the 5th percentile of all those posterior heights. This is the height of the waterline at the edge of the HDI. The set of all points above the waterline represents the HDI. Think of it this way: Certainly that set of θ's is 95% of the sampled values, because the only ones excluded are in the 5th percentile or less. And certainly all the θ's in the set have a posterior density greater than all the ones excluded, by definition of membership in the set. Figure 7.3 shows the HDI interval estimated this way. It is nearly the same as the grid-based estimate in Figure 6.2 (p. 105) and the function-based estimate in Figure 5.2 (p. 86).

If computing the relative posterior probability at each point is onerous or inconvenient (it's not in the present application but will be in more complex situations), the HDI region can be estimated by another method that uses only the parameter values themselves. The method is limited, however, to treating one parameter at a time and assumes that the distribution is unimodal. Fortunately, these conditions are satisfied in most applications. The method is based on the fact that the width of the HDI turns out to be the *narrowest* of all credible intervals of the same mass. Consider, for example, the 95% credible interval (CI) that goes from the 1st percentile of the θ values to the 96th percentile. That CI has a certain width. Consider instead the 95% CI that goes from the 2.5th percentile to the 97.5th percentile. That CI has some other width. Of all possible 95% CIs, the narrowest one matches the 95% HDI. The program listed in Section 23.3.2 (HDIofMCMC.R) implements this method.

7.3.1.2 Using a Sample to Estimate an Integral

Suppose we have a large number of representative values from a distribution. What is a good estimate of the mean of the distribution? The intuitive answer is the mean of the sample, because the sample has a lot of values that represent the distribution. In other words, we are approximating the true mean, which is an integral (recall Equation 3.6, p. 37), by a summation over discrete values sampled from the distribution.

We can express that approximation formally. Let $p(\theta)$ be a distribution over parameter θ. Let θ_i (notice the subscript i) be representative values sampled from the distribution $p(\theta)$. We write $\theta_i \sim p(\theta)$ to indicate that the θ_i values are sampled according to the probability distribution $p(\theta)$. Then the true mean, which is an integral, is approximated by the mean of the sample:

$$\int d\theta \; \theta \, p(\theta) \approx \frac{1}{N} \sum_{\substack{i \\ \theta_i \sim p(\theta)}}^{N} \theta_i \qquad (7.5)$$

In Equation 7.5, the summation is over values θ_i sampled from $p(\theta)$, and N is the number of sampled values. The approximation gets better as N gets larger. My use of limits in the summation notation in Equation 7.5 is unconventional, because the usual notation would indicate $\sum_i^N$ instead of $\sum_{\theta_i \sim p(\theta)}^N$. The unconventional notation is extremely helpful, in my opinion, because it explicitly indicates the distribution of the discrete θ_i values, and shows how the distribution $p(\theta)$ on the left side of the equation has an influence on the right side of the equation.

Equation 7.5 is a special case of a general principle. For *any* function $f(\theta)$, the integral of that function, weighted by the probability distribution $p(\theta)$, is approximated by the average of the function values at the sampled points. In math,

$$\int d\theta\, f(\theta)\, p(\theta) \approx \frac{1}{N} \sum_{\theta_i \sim p(\theta)}^N f(\theta_i) \tag{7.6}$$

The approximation of the mean in Equation 7.5 merely has $f(\theta) = \theta$. Equation 7.6 is a workhorse for the remainder of the chapter. You should think about it long enough to convince yourself that it is correct, so I don't have to figure out a way to explain it better.

Well, okay, I'll give it a try. Consider discretizing the integral in the left side of Equation 7.6, so that it is approximated as a sum over many small intervals: $\int d\theta\, f(\theta)\, p(\theta) \approx \sum_j [\Delta\theta\, p(\theta_j)] f(\theta_j)$, where θ_j is a representative value of θ in the jth interval. The term in brackets, $\Delta\theta\, p(\theta_j)$, is the probability mass of the small interval around θ_j. That probability mass is approximated by the relative number of times we happen to get a θ value from that interval when sampling from $p(\theta)$. Denote the number of times we get a θ value from the jth interval as n_j, and the total number of sampled values as N. With a large sample, notice that $n_j/N \approx \Delta\theta\, p(\theta_j)$. Then $\int d\theta\, f(\theta)\, p(\theta) \approx \sum_j [\Delta\theta\, p(\theta_j)] f(\theta_j) \approx \sum_j [n_j/N] f(\theta_j) = \frac{1}{N} \sum_j n_j f(\theta_j)$. In other words, every time we sample a θ value from the jth interval, we add into the summation another iteration of the interval's representative value, $f(\theta_j)$. But there is no need to use the interval's representative value; just use the value of $f(\theta)$ at the sampled value of θ, because the sampled θ already is in the jth interval. So the approximation becomes $\int d\theta\, f(\theta)\, p(\theta) \approx \frac{1}{N} \sum_j n_j f(\theta_j) \approx \frac{1}{N} \sum_{\theta_i \sim p(\theta)}^N f(\theta_i)$, which is Equation 7.6.

7.3.2 Prediction

The second typical goal for Bayesian inference is predicting subsequent data values. For a given value $y \in \{0, 1\}$, the predicted probability of y is $p(y \mid D) = \int d\theta\, p(y \mid \theta)\, p(\theta \mid D)$. Notice that this has the form of the left-hand side of Equation 7.6. Therefore, applying that equation to the predicted probability that the next y equals 1, we have

$$p(y = 1 \,|\, D) = \int d\theta \; p(y = 1 \,|\, \theta) \, p(\theta \,|\, D)$$

$$= \int d\theta \; \theta \, p(\theta \,|\, D)$$

$$\approx \frac{1}{N} \sum_{\substack{i \\ \theta_i \sim p(\theta \,|\, D)}}^{N} \theta_i \qquad\qquad (7.7)$$

Notice that this happens to be the mean of the sample of posterior θ values (only because $p(y = 1 \,|\, \theta) = \theta$ in this case); the mean is displayed in Figure 7.3.

7.3.3 Model Comparison: Estimation of $p(D)$

This section is heavy on the math and may be of less immediate value to readers who wish to focus on applications in the latter part of the book. This section may be skipped on a first reading, without loss of dignity.

For the goal of model comparison, we want to compute $p(D) = \int d\theta \; p(D \,|\, \theta) \, p(\theta)$, where $p(\theta)$ is the prior. In principle, we could just apply Equation 7.6 directly:

$$p(D) = \int d\theta \; p(D \,|\, \theta) \, p(\theta)$$

$$\approx \frac{1}{N} \sum_{\substack{i \\ \theta_i \sim p(\theta)}}^{N} p(D \,|\, \theta_i)$$

This means that we are getting samples from the prior, $p(\theta)$, perhaps by using a Metropolis algorithm. But in practice, the prior is diffuse, and for most of its sampled values, $p(D \,|\, \theta)$ is tiny. Values of $p(D \,|\, \theta)$ will contribute notably to the sum only in a relatively small region of θ. So we would need a *ginormous* number of samples for the approximation to converge to a stable value.

Instead of sampling from the prior, we will use our sample from the posterior distribution, in a clever way. First, consider Bayes' rule:

$$p(\theta \,|\, D) = \frac{p(D \,|\, \theta) p(\theta)}{p(D)}$$

We can rearrange it to get

$$\frac{1}{p(D)} = \frac{p(\theta \,|\, D)}{p(D \,|\, \theta) p(\theta)}$$

Now a trick (from Gelfand & Dey, 1994; summarized by Carlin & Louis, 2000): For any probability density function $h(\theta)$, it is the case that $\int d\theta\, h(\theta) = 1$. We will multiply the rearranged Bayes' rule by 1:

$$\frac{1}{p(D)} = \frac{p(\theta \mid D)}{p(D \mid \theta)p(\theta)}$$

$$= \frac{p(\theta \mid D)}{p(D \mid \theta)p(\theta)} \int d\theta\, h(\theta)$$

$$= \int d\theta\, \frac{h(\theta)}{p(D \mid \theta)p(\theta)}\, p(\theta \mid D)$$

$$\approx \frac{1}{N} \sum_{\theta_i \sim p(\theta \mid D)}^{N} \frac{h(\theta_i)}{p(D \mid \theta_i)p(\theta_i)} \tag{7.8}$$

In this derivation, the transition from first to second lines was the trick of multiplying by 1. The transition from second to third lines was algebraic rearrangement.[3] Finally, the transition from third to last lines was application of Equation 7.6.

The only thing yet to do is to specify our choice for the function $h(\theta)$. It would be good for $h(\theta)$ to be similar to $p(D \mid \theta)p(\theta)$, so that their ratio, which appears in Equation 7.8, will not get extremely large or extremely small for different values of θ. If their ratio did get too big or small, that would upset the convergence of the sum as N grows.

When the likelihood function is Bernoulli, it is reasonable that $h(\theta)$ should be a beta distribution with mean and standard deviation corresponding to the mean and standard deviation of the samples from the posterior. The idea is that the posterior will tend to be beta-ish, especially as N gets larger, regardless of the shape of the prior, because the Bernoulli likelihood will overwhelm the prior as N gets large. Therefore, I want $h(\theta)$ to be a beta distribution with mean and standard deviation equal to the mean M and standard deviation S of the θ values sampled from the posterior. Equation 5.6, p. 83, provides the corresponding shape parameters for the beta distribution. To summarize, we approximate $p(D)$ by using Equation 7.8 with $h(\theta)$ being a beta distribution with its a and b values set as just described. Note that Equation 7.8 yields the reciprocal of $p(D)$, so we have to invert the result to get $p(D)$ itself. The R script

---

[3] Actually, there is a subtlety in the transition from second to third lines of Equation 7.8. The θ in $h(\theta)$ varies over the range of the integral, but the θ in $p(D \mid \theta)$, $p(\theta)$, and $p(\theta \mid D)$ is a specific value. Therefore, it might seem that we cannot "just rearrange." However, the ratio $\dfrac{p(\theta \mid D)}{p(D \mid \theta)p(\theta)}$ is the same value for *any* value of θ, because the ratio always equals the constant $1/p(D)$; therefore we can let the value of θ in $p(\theta)$ and so on *equal* the value of θ in $h(\theta)$. In other words, although the θ in $p(\theta)$ and so on began as a single value, it transitioned into being a value that varies along with the θ in $h(\theta)$. This transition preserves the equality of the expressions because the ratio $\dfrac{p(\theta \mid D)}{p(D \mid \theta)p(\theta)}$ is the same for any value of θ.

in Section 7.6.1 (`BernMetropolisTemplate.R`) implements this procedure, and the result is shown in Figure 7.3. Exercise 7.4 lets you explore other choices for $h(\theta)$.

In general, there might not be strong theoretical motivations to select a particular $h(\theta)$ density. No matter. All that's needed is any density that reasonably mimics the posterior. In many cases, this can be achieved by first generating a representative sample of the posterior and then finding an "off-the-shelf" density that describes it reasonably well. For example, if the parameter is limited to the range [0, 1], we might be able to mimic its posterior with a beta density that has the same mean and standard deviation as the sampled posterior, even if we have no reason to believe that the posterior really is exactly a beta distribution. If the parameter is limited to the range $[0, +\infty)$, then we might be able to mimic its posterior with a gamma density (see Figure 9.8, p. 209) that has the same mean and standard deviation as the sampled posterior, even if we have no reason to believe that the posterior really is exactly a gamma distribution.

7.4 MCMC IN BUGS

Back in Section 7.2.1, we worried over the fact that the Metropolis algorithm needs to be artfully tuned: The proposal distribution needs to be well matched to the posterior, the initial samples during the burn-in period need to be excluded, and the sampling chain needs to be run long enough to sample the whole distribution thoroughly. Wouldn't it be nice if some of those worries could be addressed automatically, or at least efficiently, in a software package? Fortunately such a package of software has been developed and is available, free, as a set of function calls in R. The original version of the software was called BUGS, for Bayesian inference Using Gibbs Sampling (Gilks, Thomas, & Spiegelhalter, 1994). A later chapter will define and explore Gibbs sampling; for now all you have to know is that it is a type of Metropolis algorithm. We will use the OpenBUGS (Thomas et al., 2006) version of BUGS, accessed from R via a package called BRugs (Thomas, 2004). OpenBUGS is a system for specifying Bayesian models and generating MCMC posterior samples.

What you need to keep in mind is that OpenBUGS is separate and distinct from R. OpenBUGS has its own language that is a lot like R, but is not R. BRugs is a set of R functions that lets R shuttle information to and from OpenBUGS. Thus, we will "live" in R and use BRugs to send information to OpenBUGS and get information back from it.

Installing BRugs in R is extraordinarily simple. Just invoke R and type the following at the command line:

```
install.packages("BRugs")
```

You must be connected to the Internet for the BRugs package to be retrieved from an Internet archive. Notice that the letters are uppercase "BR" and lowercase "ugs." (If you are using MacOS or Linux, you must be running the Windows version of R within WINE.) If you get an error message that the package is not available, then click the R console menu Packages > Select Repositories, and, in the resulting pop-up window, make sure to select both CRAN and CRAN(extras). You have to install BRugs on your computer only once; after that, BRugs should be locally accessible by R on your computer.

Once BRugs is installed, type the following to load BRugs into R's active list of libraries: `library(BRugs)`. You'll need to load BRugs into R's active library every time you restart R, but once you've loaded it during a session, R knows all the functions in BRugs for the rest of the session.

One way to get help for BRugs is by typing `help(BRugs)`. It's important to understand that BRugs is a library of functions that call OpenBUGS from R. In the BRugs help package, you can see the long list of BRugs functions. But to get help about OpenBUGS itself, see the OpenBUGS manual at

```
http://mathstat.helsinki.fi/openbugs/Manuals/Contents.html
```

Open the OpenBUGS User Manual and look at the Contents. Within the Contents, in the Model Specification section, the last subsection is "Appendix II: Function and Functionals," where you'll find information about the various mathematical functions built into OpenBUGS. For example, there you'll find the function for raising x to the power y. In R, we would do this by typing `x ^ y`. But OpenBUGS is not R. In OpenBUGS, the way to ask it to raise x to the power y is by typing `pow(x,y)`.

Henceforth, I'll often say BUGS as an abbreviation for OpenBUGS. It is understood that we'll be using BRugs to access OpenBUGS from R, but I'll just keep references short by using the term "BUGS."

7.4.1 Parameter Estimation with BUGS

MCMC sampling of a posterior distribution is simple in BUGS. We merely need to specify the prior, the likelihood, and the observed data. Sometimes we also specify the starting point for the random walk. We do not need to specify anything about the posterior distribution, nor do we need to specify anything about how to do the sampling. BUGS uses its own built-in sampling algorithms to generate a chain of random values from the posterior distribution.

As an example, consider the situation in Figures 5.2 (p. 86) and 6.2 (p. 105), which involved a beta(1,1) prior distribution, a Bernoulli likelihood function, and data consisting of 14 flips with 11 heads. The prior distribution and likelihood function are specified in the BUGS `model` statement, like this (BernBetaBugsFull.R):

```
9    model {
10       # Likelihood:
11       for ( i in 1:nFlips ) {
12           y[i] ~ dbern( theta )
13       }
14       # Prior distribution:
15       theta ~ dbeta( priorA , priorB )
16       priorA <- 1
17       priorB <- 1
18   }
```

The code in the `for` loop says that every flip of the coin comes from a Bernoulli distribution with parameter value `theta`, and the code at the end says that the value of `theta` comes from a prior distribution that is beta with shape parameters `priorA` and `priorB`.

The BUGS code may look like R language, but it is not R. It is merely a description of the likelihood and prior that BUGS interprets using its own compiler, distinct from R. Notice that the model statement begins with the word `model`, followed by a specification inside curly braces. The tilde (˜) notation means that a variable "is distributed as" the indicated distribution function. The `dbern` distribution is only known to BUGS, not to R. And the function `dbeta` in BUGS is different than the function `dbeta` in R; the BUGS version has two arguments but the R version has three arguments. The assignment operator, "`<-`", in BUGS works just like the "`=`" operator in R. When we type "`x <- 2`" it means "assign the value 2 to the variable named x." The equal sign is just a short-hand way of doing that in R. *R understands either* `<-` *or* =, *but BUGS understands only* `<-`. Finally, the statements in the model specification are processed by BUGS as a batch and then checked for overall consistency. Therefore, we could specify the prior *before* the likelihood if we wanted to.

What BRugs does for us is send the model specification to BUGS and command BUGS to check the model specification for syntactic consistency. We do this using some arcane and convoluted commands, but the commands never change across applications, so they are easy to get used to and then forget about. First, we express the model specification as a string in R. Then, we store the string in a file. Finally, we use BRugs to send the file to BUGS. The whole sequence looks like this (BernBetaBugsFull.R):

```
6    # Specify the model in BUGS language, but save it as a string in R:
7    modelString = "
8    # BUGS model specification begins ...
9    model {
10       # Likelihood:
11       for ( i in 1:nFlips ) {
12           y[i] ~ dbern( theta )
```

```
13          }
14          # Prior distribution:
15          theta ~ dbeta( priorA , priorB )
16          priorA <- 1
17          priorB <- 1
18      }
19      # ... BUGS model specification ends.
20      " # close quote to end modelString
21
22      # Write the modelString to a file, using R commands:
23      writeLines(modelString,con="model.txt")
24      # Use BRugs to send the model.txt file to BUGS, which checks the model syntax:
25      modelCheck( "model.txt" )
```

Again, the R+BRugs "wrapper" around the model specification never changes, so all you need to focus on is the model specification itself.

All the variables in the model specification need to have values specified either as constants or as values generated by MCMC sampling after their initial values are specified. In our example, the data determine the constant values of y[i] and nFlips, and our prior beliefs specify the constant values of priorA and priorB. We inform BUGS of these constants via the data section of the program (BernBetaBugsFull.R):

```
31      dataList = list(
32          nFlips = 14 ,
33          y = c( 1,1,1,1,1,1,1,1,1,1,1,0,0,0 )
34      )
35
36      # Use BRugs commands to put the data into a file and ship the file to BUGS:
37      modelData( bugsData( dataList ) )
```

As the comments in the code indicate, the values are first specified in R and then shipped off to BUGS by BRugs commands.

The only variables remaining to be specified are the initial values of the MCMC chains. Fortunately, we can ask BUGS to try to randomly generate some initial values according to the prior distribution. For BUGS to do this, it needs to first interpret the model and encode the prior parameter values. BRugs tells BUGS to do the interpretation with the modelCompile command (BernBetaBugsFull.R):

```
42      modelCompile() # BRugs command tells BUGS to compile the model.
```

Then BRugs tells BUGS to automatically generate some random initial values for the chains from the prior, as follows (BernBetaBugsFull.R):

```
43      modelGenInits() # BRugs command tells BUGS to randomly initialize a chain.
```

Later we will define our own functions for specifying the initial values of the chains, but automatic random initialization works fine for the present application.

BUGS now has everything it needs to generate MCMC chains. We merely have to tell it how long of a chain to generate and what variables to keep track of. By default, BUGS records nothing, because complex models will involve numerous parameters, and recording all their values for many thousands of steps could generate huge data files. The code for these commands is as follows (BernBetaBugsFull.R):

```
48    # BRugs tells BUGS to keep a record of the sampled "theta" values:
49    samplesSet( "theta" )
50    # R command defines a new variable that specifies an arbitrary chain length:
51    chainLength = 10000
52    # BRugs tells BUGS to generate a MCMC chain:
53    modelUpdate( chainLength )
```

In those commands, we did not specify any burn-in period to exclude; future models will do so.

BUGS will generate its MCMC chains while R "hangs" for an answer from BUGS. When BUGS reports back to R that it is done, nothing overt has happened. We still need to extract the chains from BUGS. We can get the actually sampled values, or a summary about the values. Here's how (BernBetaBugsFull.R):

```
58    thetaSample = samplesSample( "theta" ) # BRugs asks BUGS for the sample values.
59    thetaSummary = samplesStats( "theta" ) # BRugs asks BUGS for summary statistics.
```

The first line puts the sampled values of theta into a vector in R called thetaSample. We can graph those values in R however we want—for example, as shown in Figure 7.4. Notice that BUGS does not determine HDI intervals for us, but we could use the R code previously described (i.e., HDIofMCMC.R) for that purpose.

What BUGS has done is relieve us from figuring out the details of the sampling algorithm. Unlike the home-grown Metropolis algorithm, we did not need to specify how the MCMC chain is generated. In fact, BUGS is clever and doesn't even use a Metropolis algorithm for this case, opting instead for a more efficient algorithm we'll learn about in the next chapter.

7.4.2 BUGS for Prediction

The goal for prediction is determining the probability of subsequent data values. As described in Section 7.3.2 (p. 136), when the likelihood function is Bernoulli, then mathematical derivation tells us that $p(y = 1 | D)$ is the mean of the posterior distribution of θ. We can forgo the mathematical analysis when we use MCMC sampling, however. Instead, we can directly generate simulated values of y from the posterior sampled values of θ and then examine the distribution of y.

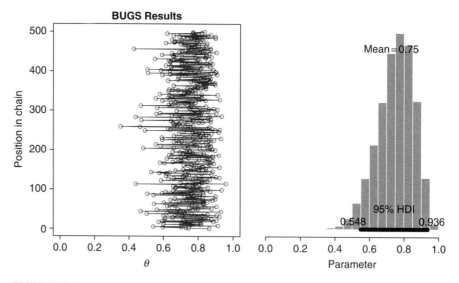

FIGURE 7.4
Values of theta sampled from the posterior by BUGS, using the program
BernBetaBugsFull.R. Compare the results with the home-grown Metropolis sampler
shown in Figure 7.3, p. 134 (they should appear to be similar).

To generate random *y* values based on the posterior sampled values of θ, we
add a few lines of R code at the end of the program. For each step in the
MCMC chain, we flip a coin using the θ value from that step in the chain.
In R, the command that flips a coin is sample(). The arguments of sample()
are x=c(0,1), which indicates the values to be sampled from, prob=c(1-
pHead,pHead), which indicates the probability of each value, and size=1,
which indicates to flip the coin just once. The sample() command lies inside
a loop that steps through the MCMC chain (BernBetaBugsFull.R):

```
72   # For each step in the chain, use posterior theta to flip a coin:
73   chainLength = length( thetaSample )
74   yPred = rep( NULL , chainLength )   # define placeholder for flip results
75   for ( stepIdx in 1:chainLength ) {
76     pHead = thetaSample[stepIdx]
77     yPred[stepIdx] = sample( x=c(0,1), prob=c(1-pHead,pHead), size=1 )
78   }
```

Figure 7.5 shows the results of the posterior predicted values of *y*. The actual
values of *y* are 1 or 0, but they have been jittered in the graph to make them
more visible. Each point plotted in the graph corresponds to a step in the
MCMC walk. We know from Equation 7.7, p. 137, that the mean value of *y*
should approximately equal the mean value of theta. The graph shows that
for this sample the approximation is very good.

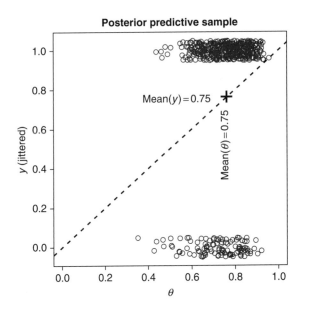

FIGURE 7.5

Posterior prediction. For each step in the MCMC chain, the value of θ is used as the bias to flip a simulated coin. The resulting y values are jittered in the graph to make them more visible. The "+" marks the mean θ value and the mean y value; they are equal in the long run, as suggested by their falling on the diagonal line.

7.4.3 BUGS for Model Comparison

BUGS provides no short cut for estimating $p(D)$ relative to the home-grown Metropolis algorithm. If we use the method of Section 7.3.3 to estimate $p(D)$, then we still have to invent a posterior-mimicking distribution and compute the likelihood times prior at every point of the MCMC chain. The only advantage of using BUGS in this case is that the sampling algorithm for generating the chain is done for us.

Later, in Section 10.2, we will explore a different method for using BUGS to compare models. In that method (called transdimensional MCMC), we will not need to compute $p(D)$ for each model. Instead, the Markov chain will jump from one model to another proportionally to how believable each model is. In the long run, the relative frequency of visiting each model is the relative believability of each model, given the data.

7.5 CONCLUSION

This chapter explained the essence of the Metropolis algorithm. Perhaps unique to this textbook, we applied the Metropolis algorithm to the simple case of estimating the posterior distribution of a single parameter. (Many other textbooks wait to introduce the Metropolis algorithm until multiparameter models are encountered.) I introduced the Metropolis algorithm with a single parameter model for pedagogical reasons. It is easier to understand the theory behind the Metropolis algorithm when its proposal distribution has just two options: move one step up or move one step down. In particular, it is easy to show the transition matrix concretely (Equation 7.3), and it is easy to show the exact evolution of position probabilities (Figure 7.2) through time. The payoff is that a good intuition is established for the simple case, so that more complex versions are easier to understand.

Another benefit of introducing the Metropolis algorithm here is that its results can be directly compared with the results of other analyses, such as exact formal analysis and grid approximation. For an example, see Exercise 7.3. If the Metropolis algorithm isn't introduced until it is needed for complex models that are intractable by other methods, then the other techniques cannot be compared!

The cost of introducing the Metropolis algorithm in this simple one-parameter scenario is that you may come away with the false belief that people would actually do this in real data analysis. They would not. For a single parameter with a limited range, grid approximation can be more accurate than MCMC, and grid approximation allows you to estimate $p(D)$ easily. Exercise 7.3 shows you an example of this.

This chapter also introduced the BUGS system for MCMC computations. BUGS is extremely useful, especially for the complicated models we will explore later. But it is not unbounded in its applicability; in particular, the prior distributions must be assembled from density functions that BUGS knows how to sample from. Exercise 7.3 provides an example in which BUGS is not easily applicable.

7.6 R CODE

7.6.1 R Code for a Home-Grown Metropolis Algorithm

The following code gives a complete example and template for applying the Metropolis algorithm to estimating a binomial proportion.

The code begins by defining three functions: the likelihood function, the prior probability function, and the target distribution for the Metropolis algorithm. These functions come first in the script because they are called by later commands.

The prior beliefs over θ are specified by the `prior` function. You can enter any function here that you want. The code supplies two possibilities as examples. The prior function does not need to be a proper density for the Metropolis algorithm to work, but the prior does need to be a proper density for the later computation of $p(D)$ to be meaningful.

The target distribution is defined in the `targetRelProb` function, which computes the (relative) probability in the target distribution. Typically this is just the unnormalized posterior probability (i.e., the numerator from Bayes' rule).

After those three functions are defined, the Metropolis algorithm itself is processed. The algorithm takes only a few lines of code, embedded in many lines of comments. *It is worth studying the code for the random walk to be sure you understand the process.*

The coding of the Metropolis algorithm, including the three initial functions and the generation of the random walk, takes up relatively little of the script. The remaining bulk of the script is devoted to extracting information from the random walk and displaying it meaningfully. The last section of the script computes $p(D)$ by implementing Equation 7.8.

(BernMetropolisTemplate.R)

```
1   # Use this program as a template for experimenting with the Metropolis
2   # algorithm applied to a single parameter called theta, defined on the
3   # interval [0,1].
4
5   # Specify the data, to be used in the likelihood function.
6   # This is a vector with one component per flip,
7   # in which 1 means a "head" and 0 means a "tail".
8   myData = c( 1, 1, 1, 1, 1, 1, 1, 1, 1, 1, 1, 0, 0, 0 )
9
10  # Define the Bernoulli likelihood function, p(D|theta).
11  # The argument theta could be a vector, not just a scalar.
12  likelihood = function( theta , data ) {
13      z = sum( data == 1 )
14      N = length( data )
15      pDataGivenTheta = theta^z * (1-theta)^(N-z)
16      # The theta values passed into this function are generated at random,
17      # and therefore might be inadvertently greater than 1 or less than 0.
18      # The likelihood for theta > 1 or for theta < 0 is zero:
19      pDataGivenTheta[ theta > 1 | theta < 0 ] = 0
20      return( pDataGivenTheta )
21  }
22
23  # Define the prior density function. For purposes of computing p(D),
24  # at the end of this program, we want this prior to be a proper density.
25  # The argument theta could be a vector, not just a scalar.
26  prior = function( theta ) {
27      prior = rep( 1 , length(theta) ) # uniform density over [0,1]
28      # For kicks, here's a bimodal prior. To try it, uncomment the next line.
29      #prior = dbeta( pmin(2*theta,2*(1-theta)) ,2,2 )
30      # The theta values passed into this function are generated at random,
31      # and therefore might be inadvertently greater than 1 or less than 0.
32      # The prior for theta > 1 or for theta < 0 is zero:
```

```
33      prior[ theta > 1 | theta < 0 ] = 0
34      return( prior )
35    }
36
37    # Define the relative probability of the target distribution,
38    # as a function of vector theta. For our application, this
39    # target distribution is the unnormalized posterior distribution.
40    targetRelProb = function( theta , data ) {
41      targetRelProb =  likelihood( theta , data ) * prior( theta )
42      return( targetRelProb )
43    }
44
45    # Specify the length of the trajectory, i.e., the number of jumps to try:
46    trajLength = 11112 # arbitrary large number
47    # Initialize the vector that will store the results:
48    trajectory = rep( 0 , trajLength )
49    # Specify where to start the trajectory:
50    trajectory[1] = 0.50 # arbitrary value
51    # Specify the burn-in period:
52    burnIn = ceiling( .1 * trajLength ) # arbitrary number, less than trajLength
53    # Initialize accepted, rejected counters, just to monitor performance:
54    nAccepted = 0
55    nRejected = 0
56    # Specify seed to reproduce same random walk:
57    set.seed(47405)
58
59    # Now generate the random walk. The 't' index is time or trial in the walk.
60    for ( t in 1:(trajLength-1) ) {
61        currentPosition = trajectory[t]
62        # Use the proposal distribution to generate a proposed jump.
63        # The shape and variance of the proposal distribution can be changed
64        # to whatever you think is appropriate for the target distribution.
65        proposedJump = rnorm( 1 , mean = 0 , sd = 0.1 )
66        # Compute the probability of accepting the proposed jump.
67        probAccept = min( 1,
68            targetRelProb( currentPosition + proposedJump , myData )
69            / targetRelProb( currentPosition , myData ) )
70        # Generate a random uniform value from the interval [0,1] to
71        # decide whether or not to accept the proposed jump.
72        if ( runif(1) < probAccept ) {
73            # accept the proposed jump
74            trajectory[ t+1 ] = currentPosition + proposedJump
75            # increment the accepted counter, just to monitor performance
76            if ( t > burnIn ) { nAccepted = nAccepted + 1 }
77        } else {
78            # reject the proposed jump, stay at current position
79            trajectory[ t+1 ] = currentPosition
80            # increment the rejected counter, just to monitor performance
81            if ( t > burnIn ) { nRejected = nRejected + 1 }
82        }
83    }
84
85    # Extract the post-burnIn portion of the trajectory.
```

```
86   acceptedTraj = trajectory[ (burnIn+1) : length(trajectory) ]
87
88   # End of Metropolis algorithm.
89
90   #------------------------------------------------------------------
91   # Display the posterior.
92
93   source("plotPost.R")
94   histInfo = plotPost( acceptedTraj , xlim=c(0,1) , breaks=30 )
95
96   # Display rejected/accepted ratio in the plot.
97   # Get the highest point and mean of the plot for subsequent text positioning.
98   densMax = max( histInfo$density )
99   meanTraj = mean( acceptedTraj )
100  sdTraj = sd( acceptedTraj )
101  if ( meanTraj > .5 ) {
102    xpos = 0.0 ; xadj = 0.0
103  } else {
104    xpos = 1.0 ; xadj = 1.0
105  }
106  text( xpos , 0.75*densMax ,
107      bquote( N[pro] * "=" * .(length(acceptedTraj)) * "  " *
108      frac(N[acc],N[pro]) * "=" * .(signif( nAccepted/length(acceptedTraj) , 3 ))
109      ) , adj=c(xadj,0)  )
110
111  #------------------------------------------------------------------
112  # Evidence for model, p(D).
113
114  # Compute a,b parameters for beta distribution that has the same mean
115  # and stdev as the sample from the posterior. This is a useful choice
116  # when the likelihood function is Bernoulli.
117  a =   meanTraj   * ( (meanTraj*(1-meanTraj)/sdTraj^2) - 1 )
118  b = (1-meanTraj) * ( (meanTraj*(1-meanTraj)/sdTraj^2) - 1 )
119
120  # For every theta value in the posterior sample, compute
121  # dbeta(theta,a,b) / likelihood(theta)*prior(theta)
122  # This computation assumes that likelihood and prior are proper densities,
123  # i.e., not just relative probabilities. This computation also assumes that
124  # the likelihood and prior functions were defined to accept a vector argument,
125  # not just a single-component scalar argument.
126  wtdEvid = dbeta( acceptedTraj , a , b ) / (
127          likelihood( acceptedTraj , myData ) * prior( acceptedTraj ) )
128  pData = 1 / mean( wtdEvid )
129
130  # Display p(D) in the graph
131  if ( meanTraj > .5 ) { xpos = 0.0 ; xadj = 0.0
132  } else { xpos = 1.0 ; xadj = 1.0 }
133  text( xpos , 0.9*densMax , bquote( p(D)==.( signif(pData,3) ) ) ,
134      adj=c(xadj,0) , cex=1.5 )
135
136  # Uncomment next line if you want to save the graph.
137  #dev.copy2eps(file="BernMetropolisTemplate.eps")
```

7.7 EXERCISES

Exercise 7.1. [Purpose: To see what happens in the Metropolis algorithm with different proposal distributions, and to get a sense how the proposal distribution must be "tuned" to the target distribution.] Use the home-grown Metropolis algorithm in the R script of Section 7.6.1 (BernMetropolisTemplate.R) for this exercise. See Figure 7.6 for examples of what your output might look like.

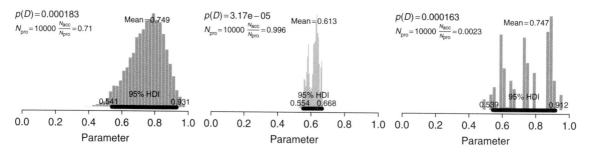

FIGURE 7.6

Results from the Metropolis algorithm when different proposal distributions are used. See Exercise 7.1. *Left panel*: sd = 0.1. *Middle panel*: sd = 0.001. *Right panel*: sd = 100.0.

(A) The proposal distribution generates candidate jumps that are normally distributed with mean zero. Set the standard deviation of the proposal distribution to 0.1 (if it isn't already) and run the script. Save/print the graph and annotate it with SD = 0.1.

(B) Set the standard deviation of the proposal distribution to 0.001 and run the script. Save/print the graph and annotate it with SD = 0.001.

(C) Set the standard deviation of the proposal distribution to 100.0 and run the script. Save/print the graph and annotate it with SD = 100.0.

(D) Which proposal distribution gave the most accurate representation of the posterior? Which proposal distribution had the fewest rejected proposals? Which proposal distribution had the most rejected proposals?

(E) If we didn't know from other techniques what the true posterior looked like, how would we know which proposal distribution generated the most accurate representation of the posterior? (This does not have a quick answer; it's meant mostly as a question for pondering and motivating techniques introduced in later chapters.)

Exercise 7.2. [Purpose: To understand the influence of the starting point of the random walk, and why the walk doesn't necessarily go back to that region.] Edit the home-grown Metropolis algorithm of Section 7.6.1 (BernMetropolisTemplate.R) for this exercise. It is best to save it as a differently named script so you don't mess up the original version. Set `trajlength = 100` and set `burnin = ceil-ing( 0.01 *trajlength)`. Finally, set `trajectory[1] = 0.001`. Now run the script and save the resulting histogram.

(A) How many jumps are proposed? How many steps are excluded as part of the burn-in portion? At what value of θ does the random walk start?

(B) Why does the histogram have so *many* points below $\theta = 0.5$? That is, why does the chain stay below $\theta = 0.5$ as long as it does?

(C) Why does the histogram have so *few* points below $\theta = 0.5$? That is, why does the chain not go back below 0.5?

Exercise 7.3. [Purpose: To get some hands-on experience with applying the Metropolis algorithm, and to compare its results with the other methods we've learned about.] Suppose you have a coin that you believe is either fair, or biased to come up heads, or biased to come up tails. As an expression of your prior belief, you define your prior on θ (the probability of heads) to be proportional to $[\cos(4\pi\theta) + 1]^2$. In other words, $p(\theta) = [\cos(4\pi\theta) + 1]^2/Z$, where Z is the appropriate normalizing constant. We flip the coin 12 times and we get 8 heads. See Figure 7.7 to see the prior, likelihood, and posterior.

(A) Determine the formula for the posterior distribution exactly, using formal integration in Bayes' rule. Just kidding. Instead, do the following: Explain the initial setup as if you wanted to try to determine the exact formula for the posterior. Show the explicit formulas involving the likelihood and prior in Bayes' rule. Do you think that the prior and likelihood are conjugate? That is, would the formula for the posterior have the "same form" as the formula for the prior?

(B) Use a fine grid over θ and approximate the posterior. Use the R function of Section 6.7.1 (`BernGrid.R`), p. 109. (The R function also plots the prior distribution, so you can see that it really is trimodal.)

(C) Use a Metropolis algorithm to approximate the posterior. Use the R script of Section 7.6.1, (`BernMetropolisTemplate.R`) adapted appropriately for the prior function. You'll need to alter the definitions of the likelihood and prior functions in the R script; include that portion of the code with what you hand in (but don't include the rest of the code). Must you normalize the prior to generate a sample from the posterior? Is the value of $p(D)$ displayed in the graph correct?

(D) Could you apply BUGS to this situation? In particular, can you think of a way to specify the prior density in terms of distributions that BUGS knows about?

Exercise 7.4. [Purpose: To approximate $p(D)$, explore other choices for $h(\theta)$ in Equation 7.8, and note that the one used in the R script of Section 7.6.1 (`BernMetropolisTemplate.R`) is a good one.] Edit the R script of Section 7.6.1

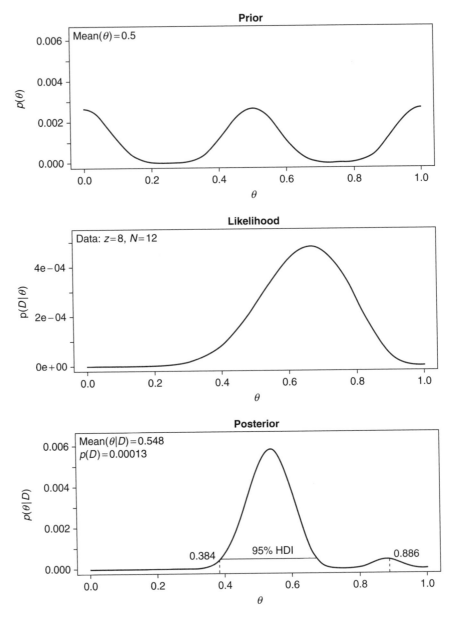

FIGURE 7.7
Example with trimodal prior, for Exercise 7.3.

(BernMetropolisTemplate.R) for this exercise. It is best to save it as a differently named script so you don't mess up the original version. At the end of the script, add this line: windows() ; plot(wtdEvid,type="l").

(A) Select (i.e., highlight with the cursor) that line in the R editor and run it. Save the plot. Explain what the plot is plotting. That is, what is wtdEvid (on the y axis) and what is on the x axis?

(B) Consider a different choice for the $h(\theta)$ in Equation 7.8. To do this, we'll leave it as a beta function, but change the choice of its a and b values. Find where a and b are specified in the R program (near the end, just before wtdEvid is defined), and type in a=1 and b=1 instead. Now select (i.e., highlight with the cursor) the portion of the program from the new a and b definitions, through the computation of wtdEvid, and the new plot command. Run the selection, and save the resulting plot.

(C) Repeat, but this time with a=10 and b=10.

(D) For which values of a and b are the values of wtdEvid most stable across the random walk? Which values of a and b would produce the most stable estimate of $p(D)$?

Exercise 7.5. [Purpose: To explore the use of BUGS and consider model comparison.] Suppose there are three people with different beliefs about a coin. One person (M1) believes that the coin is biased toward tails; we'll model this person's beliefs as a uniform distribution over θ values between 0 and 0.4. The second person (M2) believes that the coin is approximately fair; we'll model this person's beliefs as a uniform distribution between 0.4 and 0.6. The third person (M3) believes that the coin is biased toward heads; we'll model this person's beliefs as a uniform distribution over θ values between 0.6 and 1.0. We won't favor any person a priori, and therefore we start by assuming that $p(M1) = p(M2) = p(M3) = 1/3$. We now flip the coin 14 times and observe 11 heads. Use BUGS to determine the evidences for the three models. Hints: For each person, compute $p(D)$ by adapting the program BernBetaBugsFull.R of Section 7.4.1 in two steps. First, modify the model specification so that the prior is uniform over the limited range, instead of beta. Appendix I of the OpenBUGS User Manual (see Section 7.4) explains how to specify uniform distributions in BUGS. Second, include a new section at the end of the BUGS program that will compute $p(D)$. Do this by copying the last section of the program BernMetropolisTemplate.R that computes $p(D)$, pasting it onto the end of your BUGS program, and making additional necessary changes so that the output of BUGS can be processed by the newly added code. In particular, before the newly added code, you'll have to include these lines:

```
acceptedTraj = thetaSample
meanTraj = mean( thetaSample )
sdTraj = sd( thetaSample )
```

CHAPTER 8

Inferring Two Binomial Proportions via Gibbs Sampling

I'm hurtin' from all these rejected proposals;
My feelings, like peelings, down garbage disposals.
S'pose you should go your way and I should go mine,
We'll both be accepted somewhere down the line.

Doing Bayesian Data Analysis: A Tutorial with R and BUGS. DOI: 10.1016/B978-0-12-381485-2.00008-0
© 2011, Elsevier Inc. All rights reserved.

This chapter addresses the question of how to make inferences regarding two independent binomial proportions. For example, suppose we have two coins and we want to decide how different their biases are. This issue is about the relation between two proportions, not about the individual proportions only.

This topic appears at this location in the sequence of chapters for specific reasons. First, and obviously, because the topic involves inferences about proportions, it follows the preceding chapters that discussed inferences about single proportions. Second, and less obviously, the topic acts as an easy stepping stone to the next chapter, which addresses hierarchical models. Hierarchical models typically involve many parameters. Multiparameter models involve extra "administrative overhead" merely to annotate, graph, handle mathematically, and program in R and BUGS. These administrivia are introduced in this chapter in the context of a relatively simple two-parameter model. Subsequent chapters, involving multiparameter hierarchical models, can therefore focus on the concepts of hierarchical structure without getting bogged down in the logistics of dealing with multiple parameters.

Importantly, in this chapter we also get our first exposure to Gibbs sampling, previously only mentioned as the words behind the acronym BUGS. In Gibbs sampling, unlike the more general Metropolis algorithm, a proposed jump affects only one parameter at a time, and the proposed jump is never rejected. Gibbs sampling is terrific when it can be done; unfortunately it cannot always be done, as we will see.

Meanwhile, the issue of estimating the difference between two proportions is interesting in its own right. All the methods of the previous chapters are brought to bear on this issue, so this point in the sequence of topics is also a great opportunity for review and and consolidation.

In many real-world situations we observe two proportions, which differ by some amount in our specific random sample, but we want to infer what underlying difference is credible for the broader population from which the sample came. After all, the sample is just a noisy hint about the actual underlying proportions. For example, we may have a sample of 100 people suffering from a disease. We give a random half of them a promising drug and the other half a placebo. After one week, 12 people have gotten better in the drug treated group, and 5 people have gotten better in the placebo group. Did the drug actually do better than the placebo, and by how much? In other words, based on the observed difference in proportions, 12/50 versus 5/50, what underlying difference is actually credible? As another example, suppose you want to find out if mood affects cognitive performance. You manipulate mood by having 80 people sit through mood-inducing movies. A random half (i.e., 40) of your participants is shown a bittersweet film about lovers separated by circumstances of war but who never forget each other. The other random half of your participants is shown a light comedy about high school pranks. Immediately after seeing the film, all participants are given some cognitive tasks, including an arithmetic problem involving long division. Of the 40 people who saw the war movie, 32 correctly solved the long division problem. Of the 40 people who saw the comedy, 27 correctly solved the long division problem. Did the induced mood actually affect cognitive performance? In other words, based on the observed difference in proportions, 32/40 versus 27/40, what underlying difference is actually credible?

To talk about the problem more generally and with mathematical precision, we need to define some notation. We'll use the same sort of notation that we've used for previous chapters, but we'll add subscripts to indicate which of the two groups is being referred to. Thus, the hypothesized proportion of "heads" in group j (where $j = 1$ or $j = 2$) is denoted θ_j, and the actual number of heads observed in a sample of N_j "flips" is z_j, and the ith individual flip in group j is denoted y_{ji}.

Throughout this chapter we assume that the data from the two groups are *independent*. The performance of one group has no influence on the performance of the other. Typically we design research to make sure that the assumption of independence holds. In the previous examples, we assumed independence in the disease treatment scenario because we assumed that social interaction among the patients was minimal. We assumed independence in the mood-induction experiment because the experiment was designed to enforce zero social interaction between participants after the movie. The assumption of independence is crucial for all the mathematical analyses we will perform. If you have a situation in which two proportions are not independent, the methods of the this chapter do not directly apply. In situations where there are dependencies in the data, the model can try to formally express the dependency, but we will not be pursuing those situations.

8.1 PRIOR, LIKELIHOOD, AND POSTERIOR FOR TWO PROPORTIONS

We are considering situations in which there are *two* underlying proportions, namely θ_1 and θ_2 for the two groups. We are trying to determine what we should believe about these proportions after we have observed some data from the two groups.

In a Bayesian framework, we must first specify what we believe about the proportions without the data. Our prior beliefs are about *combinations* of parameter values. To specify a prior belief, we must specify our belief probability, $p(\theta_1, \theta_2)$, for every combination θ_1, θ_2. If we were to make a graph of $p(\theta_1, \theta_2)$, it would be three dimensional, with θ_1 and θ_2 on the two horizontal axes, and $p(\theta_1, \theta_2)$ on the vertical axis. Because the beliefs form a probability density function, the integral of the beliefs across the parameter space must be 1 (i.e., $\iint d\theta_1 d\theta_2 p(\theta_1, \theta_2) = 1$).

In some of the applications in this chapter, we will assume that our beliefs about θ_1 are independent of our beliefs about θ_2. This means that our belief about θ_2 is uninfluenced by our belief about θ_1. For example, if I have a coin from Canada and a coin from India, my belief about bias in the coin from Canada could be completely separate from my belief about bias in the coin from India. Independence of attributes was discussed in Section 3.4.3. Mathematically, independence means that $p(\theta_1, \theta_2) = p(\theta_1)p(\theta_2)$ for every value of θ_1 and θ_2, where $p(\theta_1)$ and $p(\theta_2)$ are the marginal belief distributions. When beliefs about two parameters are independent, the mathematical manipulations can be greatly simplified. Beliefs about the two parameters do not need to be independent, however. For example, perhaps I believe that coins are minted in similar ways across countries, and so if a Canadian coin is biased, then an Indian coin should be similarly biased. At the extreme, if I believe that θ_1 always exactly equals θ_2, then the two parameter values are completely dependent on each other. Dependence does not imply direct causal relationship, it merely implies that knowing the value of one parameter narrows beliefs about the value of the other parameter.

Along with the prior beliefs, we have some observed data. We assume that the flips within a group are independent of each other, and that the flips across groups are independent of each other. The veracity of this assumption depends on exactly how the observations are actually made, but, in properly designed experiments, we have reasons to trust this assumption. Notice that we will always assume independence of *data* within and across groups, regardless of whether we assume independence in our *beliefs* about the biases in the groups. Formally, the assumption of independence in the data means the following. Denote the result of a flip of coin 1 as y_1, where the result can be $y_1 = 1$ (heads) or $y_1 = 0$ (tails). Similarly, the result of a flip of coin 2 is denoted y_2. Independence of the data from the two coins means that $p(y_1 | \theta_1, \theta_2) = p(y_1 | \theta_1)$ and $p(y_2 | \theta_1, \theta_2) = p(y_2 | \theta_2)$.

From one group we observe the data D_1 consisting of z_1 heads out of N_1 flips, and from the other group we observe the data D_2 consisting of z_2 heads out of N_2 flips. In other words, $z_1 = \sum_{i=1}^{N_1} y_{1i}$ where y_{1i} denotes the ith flip of the first coin. Notice that $z_1 \in \{0, \ldots, N_1\}$ and $z_2 \in \{0, \ldots, N_2\}$. We denote the whole set of data as $D = \{z_1, N_1, z_2, N_2\}$. Because of independence of sampled flips, the probability of D is just the product of the Bernoulli distribution functions for the individual flips:

$$
\begin{aligned}
p(D \mid \theta_1, \theta_2) &= \prod_{y_{1i} \in D_1} p(y_{1i} \mid \theta_1, \theta_2) \prod_{y_j \in D_2} p(y_j \mid \theta_1, \theta_2) \\
&= \prod_{y_{1i} \in D_1} \theta_1^{y_{1i}} (1 - \theta_1)^{(1-y_{1i})} \prod_{y_{2j} \in D_2} \theta_2^{y_{2j}} (1 - \theta_2)^{(1-y_{2j})} \\
&= \theta_1^{z_1} (1 - \theta_1)^{(N_1 - z_1)} \theta_2^{z_2} (1 - \theta_2)^{(N_2 - z_2)} \quad\quad (8.1)
\end{aligned}
$$

The posterior distribution of our beliefs about the underlying proportions is derived in the usual way by applying Bayes' rule, but now the functions involve two parameters:

$$
\begin{aligned}
p(\theta_1, \theta_2 \mid D) &= p(D \mid \theta_1, \theta_2) p(\theta_1, \theta_2) / p(D) \\
&= p(D \mid \theta_1, \theta_2) p(\theta_1, \theta_2) \Big/ \iint d\theta_1 d\theta_2 \, p(D \mid \theta_1, \theta_2) p(\theta_1, \theta_2) \quad (8.2)
\end{aligned}
$$

Remember, as always in the expression of Bayes' rule, the θ_i's in left side of the equation and in the numerator of the right side are referring to specific values of θ_i, but the θ_i's in the integral in the denominator range over all possible values of θ_i.

What has just been described in the previous few paragraphs is the general Bayesian framework for making inferences about two proportions, when the likelihood function consists of independent Bernoulli distributions. What we have to do now is determine the details for specific prior distributions. After briefly considering formal analysis and grid approximation, the chapter is most devoted to Markov chain Monte Carlo (MCMC) approximation. In particular, we'll get our first look at Gibbs sampling. The final section of the chapter explores the issue of estimating the difference between two proportions.

8.2 THE POSTERIOR VIA EXACT FORMAL ANALYSIS

Suppose we want to pursue a solution to Bayes' rule (Equation 8.2) using formal analysis. What sort of prior probability function would make the analysis especially tractable? Perhaps you can guess the answer by recalling the discussion of Chapter 5. We learned there that the beta distribution is conjugate to the Bernoulli likelihood function for single proportions. This suggests that a product of beta distributions would be conjugate to a product of Bernoulli functions.

This suggestion turns out to be true. We pursue the same logic as was used for Equation 5.7 (p. 84). First, recall that a beta distribution has the form

$\mathrm{beta}(\theta \mid a, b) = \theta^{(a-1)}(1 - \theta)^{(b-1)}/B(a, b)$, where $B(a, b)$ is the beta normalizing function, which by definition is $B(a, b) = \int_0^1 d\theta\, \theta^{(a-1)}(1 - \theta)^{(b-1)}$. We assume a $\mathrm{beta}(\theta_1 \mid a_1, b_1)$ prior on θ_1, and an independent $\mathrm{beta}(\theta_2 \mid a_2, b_2)$ prior on θ_2. Then

$$p(\theta_1, \theta_2 \mid D) = p(D \mid \theta_1, \theta_2)p(\theta_1, \theta_2)/p(D)$$

$$= \frac{\theta_1^{(z_1+a_1-1)}(1 - \theta_1)^{(N_1-z_1+b_1-1)}\theta_2^{(z_2+a_2-1)}(1 - \theta_2)^{(N_2-z_2+b_2-1)}}{p(D)B(a_1, b_1)B(a_2, b_2)}$$

$$(8.3)$$

We know that the left side of Equation 8.3 must be a probability density function, and we see that the numerator of the right side has the form of a product of beta distributions, namely, $\mathrm{beta}(\theta_1 \mid z_1 + a_1, N_1 - z_1 + b_1)$ times $\mathrm{beta}(\theta_2 \mid z_2 + a_2, N_2 - z_2 + b_2)$. Therefore, the denominator of Equation 8.3 must be the corresponding normalizer for the product of beta distributions:

$$p(D)B(a_1, b_1)B(a_2, b_2) = B(z_1 + a_1, N_1 - z_1 + b_1)\, B(z_2 + a_2, N_2 - z_2 + b_2)$$

$$(8.4)$$

By rearranging terms, a convenient consequence of Equation 8.4 is that

$$p(D) = \frac{B(z_1 + a_1, N_1 - z_1 + b_1)\, B(z_2 + a_2, N_2 - z_2 + b_2)}{B(a_1, b_1)B(a_2, b_2)} \qquad (8.5)$$

This is exactly analogous to the result we found previously for one parameter, in Equation 5.10 (p. 88).

To recapitulate, when the prior is a product of independent beta distributions, the posterior is also a product of independent beta distributions, with each beta obeying the update rule we derived in Chapter 5. Explicitly, if the prior is $\mathrm{beta}(\theta_1 \mid a_1, b_1) \times \mathrm{beta}(\theta_2 \mid a_2, b_2)$, and the data are z_1, N_1, z_2, N_2, then the posterior is $\mathrm{beta}(\theta_1 \mid z_1 + a_1, N_1 - z_1 + b_1) \times \mathrm{beta}(\theta_2 \mid z_2 + a_2, N_2 - z_2 + b_2)$.

One way of understanding the posterior distribution is to visualize it. We want to plot the probability densities as a function of *two* parameters, θ_1 and θ_2. One way to do this is by placing the two parameters, θ_1 and θ_2, on two horizontal axes, and placing the probability density, $p(\theta_1, \theta_2)$, on a vertical axis. This three-dimensional surface can then be displayed in a picture as if it were a landscape viewed in perspective. This sort of plot is called a "perspective" plot in R. Alternatively, we can place the two parameter axes flat on the drawing paper and indicate the probability density with level contours, such that any one contour marks points that have the same specific level. An example of these plots is described next.

Figure 8.1 shows graphs for updating a product of beta distributions. In this example, the prior begins with moderate beliefs that each proportion is about

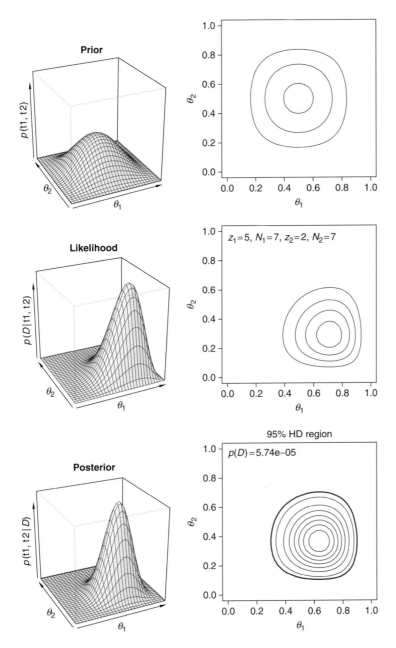

FIGURE 8.1
Bayesian updating of independent beta(θ | 3, 3) priors. Left panels show perspective surface plots; right panels show contour plots of the same distribution. The posterior contour plot (lower right) includes the value of $p(D)$ and shows the 95% highest density region as a darker contour line. The R code that generated this graph is in Section 8.8.1 (BernTwoGrid.R).

0.50, using a beta($\theta \mid 3, 3$) distribution for both proportions. The upper panels show a perspective plot and a contour plot for the prior distribution. Notice that it is gently peaked at the center of the parameter space, which reflects the prior belief that the two proportions are probably around 0.5, but without much certainty. The perspective plot shows vertical slices of the prior density parallel to θ_1 and θ_2. Consider slices parallel to the θ_1 axis, with different slices at different values of θ_2. The profile of the density on every slice has the same shape, with merely different heights. In particular, at every level of θ_2, the profile of the slice along θ_1 is a beta($\theta_1 \mid 3, 3$) shape, with merely an overall height that depends on the level of θ_2. When the shape of the profile on the slices does not change, as exemplified here, then the joint distribution is constructed from the product of independent marginal distributions.

The contour plot in the upper right panel of Figure 8.1 shows the same distribution as the upper-left panel. Instead of showing vertical slices through the distribution, the contour plot shows horizontal slices. Each contour corresponds to a slice at a particular height. Contour plots can be challenging to interpret because it is not immediately obvious whether two adjacent contours belong to different heights. Contours can be labeled with numerical values to indicate their heights, but then the plot can become very cluttered. Therefore, if the goal is a quick intuition about the general layout of the distribution, then a perspective plot is preferred over a contour plot. If the goal is instead a more detailed visual determination of the parameter values for a particular peak in the distribution, then a contour plot may be preferred.

The middle row of Figure 8.1 shows the likelihood functions for the specific data displayed in the panels. The plots show the likelihood for each possible combination of θ_1 and θ_2. Notice that the likelihood is maximized at θ values that match the proportions of heads in the data.

The resulting posterior distribution is shown in the lowest row of Figure 8.1. At each point in the θ_1, θ_2 parameter space, the posterior is the product of the prior and likelihood values at that point, divided by the normalizer, $p(D)$. As has been emphasized before, the posterior is a compromise between the prior and the likelihood. This can be seen in Figure 8.1 by looking at the peaks of the functions: The peak of the posterior is located between the peaks of the prior and likelihood in the θ_1, θ_2 parameter space.

The lower-right panel of Figure 8.1 also shows the 95% highest density region. This region encloses 95% of the posterior mass, such that all points within the region have higher believability than points outside the region. The highest density region can be difficult to determine from analysis of formulas alone. The highest density region in Figure 8.1 was instead estimated by a grid approximation, using the method described in Section 23.3.1 (HDIofGrid.R), p. 626.

In summary, the main point of this section was to graphically display the meaning of a prior, likelihood, and posterior on a two-parameter space. Figure 8.1 showed an example. In this section we emphasized the use of mathematical forms, with priors that are conjugate to the likelihood. The particular mathematical form, involving beta distributions, will be used again in a subsequent section that introduces Gibbs sampling, and that destination is another motivation for including the mathematical formulation of this section. Before getting to Gibbs sampling, however, we first consider grid approximation, to further bolster intuitions about probability distributions over two dimensions.

8.3 THE POSTERIOR VIA GRID APPROXIMATION

When the parameter space is small enough, we can approximate the integral in the denominator of Bayes' rule by a sum over densely placed points in the parameter space. The continuous form of Bayes' rule in Equation 8.2 becomes

$$p(\theta_1, \theta_2 \mid D) = p(D \mid \theta_1, \theta_2)p(\theta_1, \theta_2) \bigg/ \iint d\theta_1 d\theta_2 p(D \mid \theta_1, \theta_2)p(\theta_1, \theta_2)$$

$$\approx p(D \mid \theta_1, \theta_2)p(\theta_1, \theta_2) \bigg/ \sum_{\theta_1} \sum_{\theta_2} p(D \mid \theta_1, \theta_2)p(\theta_1, \theta_2) \qquad (8.6)$$

where $p(\theta_1, \theta_2)$ in the first row is a probability density, but $p(\theta_1, \theta_2)$ in the second row is a probability mass for the small area around the discrete point θ_1, θ_2. As always for these expressions of Bayes' rule, the θ_1 and θ_2 on the left side and in the numerator of the right side are specific values, but the θ_1 and θ_2 in the summation of the denominator take on a range of values over the interval $[0, 1]$. For convenience, we will choose discrete values placed on a regular grid over the parameter space. The discrete probability masses must sum to 1.0 when added across the entire parameter space.

One of the advantages of using a grid approximation is that we do not rely on formal analysis to derive a specification of the posterior distribution. Therefore, we can specify any prior distribution we like and still come up with an approximation of the posterior distribution. Figure 8.2 shows one such prior distribution that would be difficult to analyze formally. The particular prior in that figure would probably never be used in real research, but the purpose is to demonstrate that virtually any prior can be used.

Another advantage of grid approximation is that a highest density region can be approximated for any posterior distribution. As mentioned in the previous section, finding a multidimensional highest density region by formal analysis can be challenging, to say the least, but approximating one from a grid approximation is easy. Figure 8.2 shows the 95% highest density region for

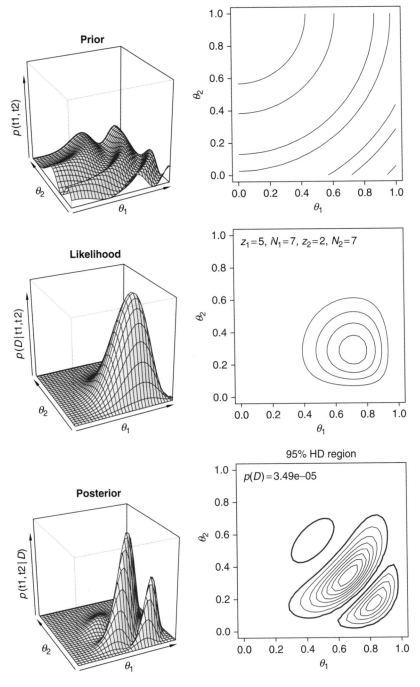

FIGURE 8.2

With grid approximation, even unusual prior distributions can be used. The R code for this graph is in Section 8.8.1 (`BernTwoGrid.R`).

a multimodal posterior. The program for approximating the highest density region is described in Section 23.3.1 (HDIofGrid.R), p. 626.

In summary, the main point of this section has been to illustrate Bayesian inference on two parameters, using a grid approximation. Grid approximation can be useful when the prior distribution cannot be expressed by a mathematical function that is conjugate to the likelihood function, or when the prior is not expressed by a mathematical function at all. Another benefit of grid approximation is that it makes all the computations explicit and relatively easy to understand as point-by-point multiplication across the parameter space, normalized by the simple sum. The main goal has been to enhance your intuition for how Bayesian updating works on a two-dimensional parameter space, especially via the example of the rippled prior in Figure 8.2.

8.4 THE POSTERIOR VIA MARKOV CHAIN MONTE CARLO

Imagine that we have a likelihood function and a prior distribution function that cannot be handled adequately by formal analysis, so we cannot determine the posterior distribution via formal mathematics alone. Imagine also that the parameter space is too big to be adequately spanned by a dense grid of points, so we cannot determine the posterior distribution via dense grid approximation. As described in the introduction to Chapter 7, when there are more than a few parameters, each represented by a dense comb of discrete values, then the number of grid points in the multidimensional space can grow to be huge and intractable. An alternative approach to approximating the posterior distribution, as introduced in Chapter 7, is to generate a large number of representative values from the posterior distribution and estimate the posterior from those representative values. Even though our present application, involving only two parameters on a limited domain, can be addressed using grid approximation, it is highly instructive to apply the Metropolis algorithm also. This will be our first application of the Metropolis algorithm to a two-dimensional parameter space.

8.4.1 Metropolis Algorithm

Recall from Chapter 7 the essential process of the Metropolis algorithm. The random walk starts at some arbitrary point in the parameter space, ideally somewhere not too far from the main bulk of the posterior distribution. We propose a jump to a new point in parameter space. The proposed jump is randomly generated from a *proposal distribution*, from which we assume it is easy to generate values. For our present purposes, the proposal distribution is a bivariate normal. You can visualize a bivariate normal distribution by imagining a one-dimensional normal (as in Figure 3.3, p. 36), sticking a pin down vertically through its peak, and spinning it around the pin to form a bell-shaped distribution. The use of a bivariate normal proposal distribution implies that the proposed jumps will usually be near the current position, but more distant jumps could be proposed with lesser probability.[1]

The proposed jump is definitely accepted if the posterior is more dense at the proposed position than at the current position, and the proposed jump is probabilistically accepted if the posterior is less dense at the proposed position than at the current position. The exact probability of accepting the proposed jump (i.e., the probability of moving) is

$$p_{move} = \min\left(\frac{P(\theta_{proposed})}{P(\theta_{current})}, 1\right) \tag{8.7}$$

where $P(\theta)$ is the (unnormalized) posterior probability at θ. (Equation 8.7 repeats Equation 7.1 from p. 125.) This probability of moving is turned into an actual discrete action to move or stay by generating a random number from the uniform density on [0, 1]: If the generated number lies between 0 and p_{move}, then the move is made; otherwise the current position is retained.

The Metropolis algorithm is useful because it generates representative values from an arbitrary posterior distribution with only a few simple tools. It needs to generate a random value from a proposal distribution such as a bivariate normal, which is easy in R. It needs to generate a random value from a uniform distribution, which again is easy in R. It needs to compute the likelihood and the prior at any given values of the parameters. Conveniently, the algorithm only needs the likelihood and prior up to a constant multiple. In other words, the prior does not need to be normalized and the likelihood can have "nuisance" multipliers dropped, if desired. (On the contrary, if $p(D)$ is being estimated, then the likelihood and prior may need to be computed exactly.)

Figure 8.3 shows the Metropolis algorithm applied to a case in which the prior distribution is a product of beta distributions on each proportion. The specifics of the prior distribution and the observed data match the ones used in Figure 8.1 (p. 161) so that you can directly compare the results of the Metropolis algorithm with the results of formal analysis and grid approximation. The sampled points in Figure 8.3 are connected by lines so that you can get a sense of the trajectory taken by the random walk. The ultimate estimates regarding

---

[1] The proposal distribution does not have to be a rotationally symmetric bivariate normal. For example, it could be a bivariate normal with nonzero covariances, so that proposals are more likely to be made in some diagonal directions more than others. The proposal distribution could even be non-normal. It is only for the present illustrative purposes that we assume a simple symmetric proposal distribution.

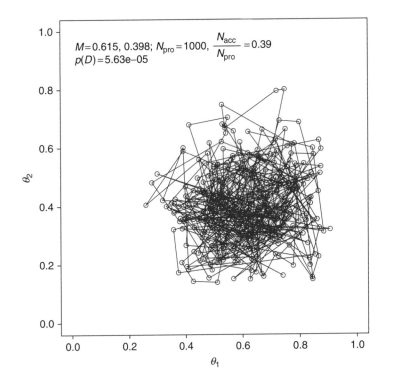

FIGURE 8.3

Metropolis algorithm applied to the prior and likelihood shown in Figure 8.1, p. 161. Compare this scatter plot with the lower right contour plot of Figure 8.1. N_{pro} is the number of proposed jumps (after burn-in), and N_{acc} is the number of accepted proposals. R code is in Section 8.8.2 (`BernTwoMetropolis.R`).

the posterior distribution do not care about the sequence in which the sampled points appeared, and so the trajectory is irrelevant to anything but your understanding of the Metropolis algorithm.

The trajectory shown in Figure 8.3 excludes an initial "burn-in" period, which may have been unduly affected by the arbitrary starting position. The trajectory also depended on the width of proposal distribution. In this case, the proposal distribution was a symmetric bivariate normal with standard deviations of 0.2. Exercise 8.4 has you explore the consequences of using a proposal distribution that is too narrow or too wide.

8.4.2 Gibbs Sampling

The Metropolis algorithm is very general and broadly applicable. One problem with it, however, is that the proposal distribution must be properly tuned to the posterior distribution if the algorithm is to work well. If the proposal distribution is too narrow or too broad, a large proportion of proposed jumps will be rejected or the trajectory will get bogged down in a localized region of the parameter space. It would be nice, therefore, if we had another method of sample generation that did not demand artful tuning of a proposal distribution. Gibbs sampling (first proposed by Geman & Geman, 1984) is one such method, and it will be described in this section.

Whereas Gibbs sampling obviates a separate proposal distribution, this different sampling method makes other demands: We must be able to generate samples from the posterior distribution conditioned on each individual parameter. In other words, Gibbs sampling will allow us to generate a sample from the *joint* posterior, $p(\theta_1, \theta_2, \theta_3 \mid D)$, if we are able to generate samples from all of the *conditional* posterior distributions, $p(\theta_1 \mid \theta_2, \theta_3, D)$, $p(\theta_2 \mid \theta_1, \theta_3, D)$, and $p(\theta_3 \mid \theta_1, \theta_2, D)$. Doing the formal analysis to determine the conditional posterior distributions can be difficult or impossible. Even if we can derive the conditional probabilities, we might not be able to directly generate samples from them. Therefore, Gibbs sampling has limited applicability. When it is applicable, however, Gibbs sampling can be more efficient and reliable than the Metropolis algorithm.

The procedure for Gibbs sampling is a type of random walk through parameter space, like the Metropolis algorithm. The walk starts at some arbitrary point, and at each point in the walk, the next step depends only on the current position and on no previous positions. Therefore, Gibbs sampling is another example of a Markov chain Monte Carlo process. What is different about Gibbs sampling, relative to the Metropolis algorithm, is how each step is taken. At each point in the walk, one of the component parameters is selected. The component parameter can be selected at random, but typically the parameters are cycled through, in order: $\theta_1, \theta_2, \theta_3, \ldots, \theta_1, \theta_2, \theta_3, \ldots$. (The reason that parameters are cycled rather than selected randomly is that for complex models with many dozens or hundreds of parameters, it would take too many steps to visit every parameter by random chance alone, even though they would be visited about equally often in the long run.) Suppose that parameter θ_i has been selected. Gibbs sampling then chooses a new value for that parameter by generating a random value directly from the conditional probability $p(\theta_i \mid \{\theta_{j \neq i}\}, D)$. The new value for θ_i, combined with the unchanged values of $\theta_{j \neq i}$, constitutes the new position in the random walk. The process then repeats: Select a component parameter and select a new value for that parameter from its conditional posterior distribution. Figure 8.4 illustrates two steps in this process.

Gibbs sampling can be thought of as just a special case of the Metropolis algorithm, in which the proposal distribution depends on the location in parameter space and the component parameter selected. At any point, a component parameter is selected, and then the proposal distribution for that

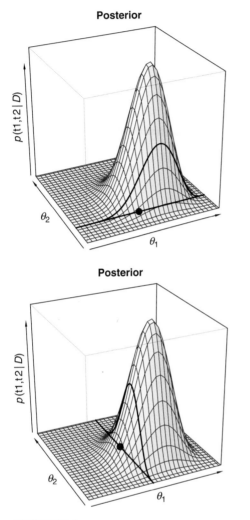

FIGURE 8.4

Two steps in a Gibbs sampling. The top panel shows a random generation of a value for θ_1, conditional on a value for θ_2. The heavy lines show a slice through the posterior at the conditional value of θ_2, and the large dot shows a random value of θ_1 sampled from the conditional density. The bottom panel shows a random generation of a value for θ_2, conditional on the value for θ_1 determined by the previous step. The heavy lines show a slice through the posterior at the conditional value of θ_1, and the large dot shows a random value of θ_2 sampled from the conditional density.

parameter's next value *is* the conditional posterior probability of that parameter. *Because the proposal distribution exactly mirrors the posterior probability for that parameter, the proposed move is always accepted.* A rigorous proof of this idea requires development of a generalized form of the Metropolis algorithm, called the Metropolis-Hastings algorithm. Details of the proof are described by Gelman et al. (2004, p. 293).

Gibbs sampling is especially useful when the complete joint posterior, $p(\theta_i \,|\, D)$, cannot be analytically determined and cannot be directly sampled, but all the conditional distributions, $p(\theta_i \,|\, \{\theta_{j \neq i}\}, D)$, can be determined and directly sampled. We will not encounter an example of such a situation until later in the book, but the process of Gibbs sampling can be illustrated now for a simpler situation.

Continuing the same scenario as started this chapter, suppose that we have two proportions, θ_1 and θ_2, and the prior belief distribution is a product of beta distributions. The posterior distribution is again a product of beta distributions, as was derived in Equation 8.3 (p. 160). This joint posterior is easily dealt with directly, and so there is no real need to apply Gibbs sampling, but we will do Gibbs sampling of this posterior distribution for purposes of illustration and comparison with other methods.

To accomplish Gibbs sampling, we must determine the conditional posterior distribution for each parameter. By definition of conditional probability,

$$p(\theta_1 \,|\, \theta_2, D) = p(\theta_1, \theta_2 \,|\, D)/p(\theta_2 \,|\, D)$$

$$= p(\theta_1, \theta_2 \,|\, D) \bigg/ \int d\theta_1 \, p(\theta_1, \theta_2 \,|\, D).$$

For our current application, the joint posterior is a product of beta distributions as in Equation 8.3. Therefore,

$$p(\theta_1 | \theta_2, D) = p(\theta_1, \theta_2 | D) \bigg/ \int d\theta_1 \, p(\theta_1, \theta_2 | D)$$

$$= \frac{\mathrm{beta}(\theta_1 | z_1 + a_1, N_1 - z_1 + b_1) \, \mathrm{beta}(\theta_2 | z_2 + a_2, N_2 - z_2 + b_2)}{\int d\theta_1 \, \mathrm{beta}(\theta_1 | z_1 + a_1, N_1 - z_1 + b_1) \, \mathrm{beta}(\theta_2 | z_2 + a_2, N_2 - z_2 + b_2)}$$

$$= \frac{\mathrm{beta}(\theta_1 | z_1 + a_1, N_1 - z_1 + b_1) \, \mathrm{beta}(\theta_2 | z_2 + a_2, N_2 - z_2 + b_2)}{\mathrm{beta}(\theta_2 | z_2 + a_2, N_2 - z_2 + b_2)}$$

$$= \mathrm{beta}(\theta_1 | z_1 + a_1, N_1 - z_1 + b_1) \tag{8.8}$$

We have just derived what may have already been intuitively clear: Because the posterior is a product of independent beta distributions, it makes sense that $p(\theta_1 \,|\, \theta_2, D) = p(\theta_1 \,|\, D)$. Nevertheless, the derivation illustrates the sort of analytical procedure needed in general. From these considerations, you can also see that the other conditional posterior probability is $p(\theta_2 \,|\, \theta_1, D) = \mathrm{beta}(\theta_2 \,|\, z_2 + a_2, N_2 - z_2 + b_2)$.

Having successfully determined the conditional posterior probability distributions, we now figure out whether we can directly sample from them. In this

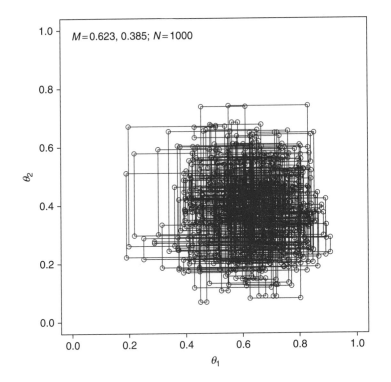

FIGURE 8.5
Gibbs sampling applied to the posterior shown in Figure 8.1, p. 161. Compare with the results of the Metropolis algorithm in Figure 8.3.

case the answer is yes, we can, because the conditional probabilities are beta distributions and our computer software comes prepackaged with generators of random values from beta distributions.

Figure 8.5 shows the result of applying Gibbs sampling to this scenario. Notice that each step in the random walk is parallel to a parameter axis, because each step changes only one component parameter. You can also see that at each point, the walk direction changes to the other parameter, rather than doubling back on itself or continuing in the same direction. This is because the walk cycled through the component parameters, $\theta_1, \theta_2, \theta_1, \theta_2, \theta_1, \theta_2, \ldots$, rather than selecting them at random at each point.

If you compare the results of Gibbs sampling in Figure 8.5 with the results of the Metropolis algorithm in Figure 8.3, (p. 167), you will see that the estimates of the mean and $p(D)$ are nearly equal. However, what Figure 8.3 does not show is that most of the points in the Metropolis-generated sample are multiply represented because the proposed moves away from them were rejected. Thus, in a trajectory with T proposed jumps, the Gibbs sample will have

T distinct points, but the Metropolis sample will typically have far fewer than T distinct points. In this sense, the Metropolis sample is "clumpier" than the Gibbs sample and might be a less effective estimator.

By helping us visualize *how* Gibbs sampling works, Figure 8.5 also helps us understand better *why* it works. Imagine that instead of changing the component parameter at every step, we linger on a component. Suppose, for example, that we have a fixed value of θ_1, and we keep generating new values of θ_2 for many steps. In terms of Figure 8.5, this amounts to lingering on a vertical slice of the parameter space, lined up over the fixed value of θ_1. As we continue sampling within that slice, we build up a good representation of the posterior distribution for that value of θ_1. Now we jump to a different value of θ_1, and again linger at the new value, filling in a new vertical slice of the posterior. If we do that enough, we will have many vertical slices, each representing the posterior distribution along that slice. We can use those vertical slices to represent the posterior, *if* we have also lingered in each slice proportionally to the posterior probability of being in that slice! Not all values of θ_1 are equally likely in the posterior, so we visit vertical slices according to the conditional probability of θ_1. Gibbs sampling carries out this process, but it lingers for only one step before jumping to a new slice.

8.4.2.1 Disadvantages of Gibbs Sampling

So far, I have emphasized the advantage of Gibbs sampling over the Metropolis algorithm, namely, that there is no need to tune a proposal distribution and no inefficiency of rejected proposals. I also mentioned a restriction: We must be able to derive the conditional probabilities of each parameter on the other and be able to generate samples from those conditional probabilities.

But there is one other disadvantage of Gibbs sampling. Because it only changes one parameter value at a time, its progress can be stalled by highly correlated parameters. We will encounter applications later in which credible parameters can be strongly correlated; see, for example, the right panel in Figure 16.4, p. 424. Here I hope merely to plant the seeds of an intuition for later development. Imagine a posterior distribution over two parameters. Its shape is a narrow ridge along the *diagonal* of the parameter space, and you are *inside, within*, this narrow ridge. Now imagine doing Gibbs sampling from this posterior. You are in the ridge somewhere, and you are contemplating a step along a parameter axis. Because the ridge is narrow and diagonal, a step along a parameter axis quickly encounters the wall of the ridge, and so your step size must be small. This is true no matter which parameter axis you face along. Therefore, you can take only small steps and only gradually explore the length of the diagonal ridge. On the other hand, if you were stepping according to a Metropolis sampler, whereby your proposal distribution included changes of both parameters at once, then you could jump in the diagonal direction and quickly explore the length of the ridge.

8.5 DOING IT WITH BUGS

The BUGS system for generating MCMC samples from a posterior distribution was introduced in Section 7.4 (p. 139). BUGS can easily be used for the present two-parameter situation. The model section of the code, shown here in lines 8 to 15, specifies that each flip of coin is distributed as a Bernoulli distribution, and the prior distribution for the bias of each coin is beta($\theta \mid 3, 3$) (BernTwoBugs.R):

```
8    model {
9        # Likelihood. Each flip is Bernoulli.
10       for ( i in 1 : N1 ) { y1[i] ~ dbern( theta1 ) }
11       for ( i in 1 : N2 ) { y2[i] ~ dbern( theta2 ) }
12       # Prior. Independent beta distributions.
13       theta1 ~ dbeta( 3 , 3 )
14       theta2 ~ dbeta( 3 , 3 )
15   }
```

The remainder of the code is similar to the code already explained in detail back in Section 7.4. For example, the data specification consists of this code (BernTwoBugs.R):

```
27   datalist = list(
28       N1 = 7 ,
29       y1 = c( 1,1,1,1,1,0,0 ) ,
30       N2 = 7 ,
31       y2 = c( 1,1,0,0,0,0,0 )
32   )
33   # Get the data into BRugs:
34   modelData( bugsData( datalist ) )
```

The complete program is listed in Section 8.8.3 (BernTwoBugs.R).

Figure 8.6 shows the results, which can be seen to be similar to the samples generated by the Metropolis algorithm in Figure 8.3 and by Gibbs sampling in Figure 8.5. BUGS actually has several different sampling algorithms at its disposal. Even when it uses Gibbs sampling, a graph of its chain will not look like the square-cornered trajectory in Figure 8.5, because BUGS does not consider a step to be complete until every parameter is sampled. This updating of all parameters, before declaring a "step" to have been taken, is important for models that have many parameters. Consider, for instance, a model with 200 parameters (as we will encounter later). If Gibbs sampling was used on every parameter, and a step was considered to be taken whenever a single parameter was changed, then every parameter value would remain unchanged for 199 steps, and the chain would have to be immensely long for the sample to be relatively unclumpy.

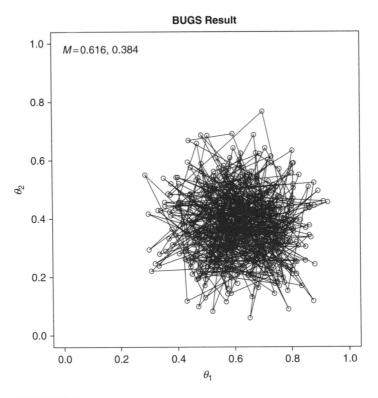

FIGURE 8.6

BUGS applied to the posterior shown in Figure 8.1, p. 161. Compare with the results of the Metropolis algorithm in Figure 8.3 and Gibbs sampling in Figure 8.5.

8.5.1 Sampling the Prior in BUGS

Up to now, we have used BUGS to generate a sample from the posterior distribution. We have merely assumed that BUGS has properly generated the prior distribution, and we have merely assumed that the prior we specified in BUGS language is the prior that we intended in our minds. But we know from everyday conversation that sometimes what we say is not what we mean. Unfortunately, this disconnection between intention and expression also applies to everyday programming. Therefore, it can be useful to have BUGS generate a representative sample from the prior, so we can check that the programmed prior is actually what we intended.

It is easy to sample the prior in BUGS simply by "detaching" the data from the model. The model specification remains the same; the data specification omits mention of the data (including only the values of nondata constants if there are any). For example, in the program described in the previous

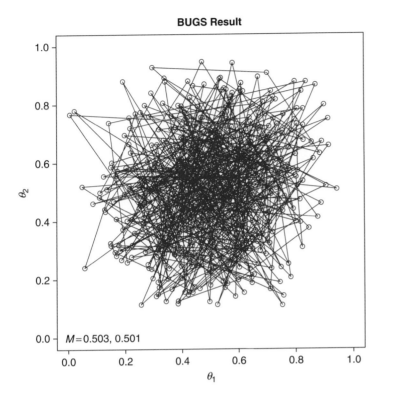

FIGURE 8.7
The prior as sampled by BUGS, when the data are detached from the model. Compare with the contour plot of the prior in Figure 8.1, p. 161.

section, all we have to do is comment out the two lines in the model specification that refer to the data (and also comment out the unneeded comma) (BernTwoBugsPriorOnly.R):

```
27   datalist = list(
28       N1 = 7 ,
29   #     y1 = c( 1,1,1,1,1,0,0 ) ,
30       N2 = 7 #,
31   #     y2 = c( 1,1,0,0,0,0,0 )
32   )
```

The result is shown in Figure 8.7. The sampled points are distributed as they were intended, like the prior shown in Figure 8.1.

8.6 HOW DIFFERENT ARE THE UNDERLYING BIASES?

In real applications, when we estimate the underlying biases for two coins, we are usually interested in determining how different the two biases are. One natural way to phrase this question is, *given the observed data, how believable is each possible difference in the underlying biases?* This question is easily answered by examining the chain of credible θ_1 and θ_2 values sampled from the posterior distribution. At each step in the chain, we compute the difference of the two theta values at that step. We then examine the distribution of the differences, collapsed across the steps of the chain.

As an example, let's continue with the case in which $N_1 = 7$, $z_1 = 5$, $N_2 = 7$, and $z_2 = 2$, with a prior of two independent beta$(\theta \mid 3, 3)$ distributions. The data suggest that coin 1 is more biased toward heads than coin 2, but what exactly are the credible differences in underlying biases? The posterior distribution of θ_1 and θ_2 is represented by the MCMC chain plotted in Figure 8.6 (p. 174). The plot indicates that believable values of θ_1 tend to be larger than believable values of θ_2, but we would like a precise picture of the differences. To get a representative sample of differences between θ_1 and θ_2, we compute the difference between the sampled θ_1 and θ_2 values at each step in the chain. This is coded at the end of the program (BernTwoBugs.R):

```
76  thetaDiff = theta1Sample - theta2Sample
77  source("plotPost.R")
78  windows(7,4)
79  plotPost( thetaDiff , xlab=expression(theta[1]-theta[2]) , compVal=0.0 ,
80           breaks=30 )
```

The variable `theta1Sample` contains the chain of sampled values for θ_1, and the variable `theta2Sample` contains the chain of sampled values for θ_2. Therefore, the variable `thetaDiff` contains the difference, $\theta_1 - \theta_2$, at each step in the chain. The differences are plotted as a histogram by the function `plotPost`, which is merely an enhanced histogram plot that I created, described in Section 8.8.4 (`plotPost.R`).

Figure 8.8 shows the resulting histogram of credible differences in the underlying biases. You can see that although the mean difference is 0.232, there is considerable uncertainty in that estimate because the 95% HDI goes from -0.128 to 0.576. In particular, a difference of zero is among the credible

FIGURE 8.8

Histogram of theta differences, derived from posterior sample in Figure 8.6. R code for the graphics is in Section 8.8.4 (`plotPost.R`).

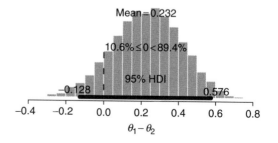

differences, with 10.6% of the credible differences falling below zero. (A different approach to deciding whether or not a "null" value is credible, which involves model comparison, will be discussed in Chapter 12.) The histogram shown in Figure 8.8 is especially informative because it reveals intuitively both the most credible differences and the nature of the uncertainty in those differences.

8.7 SUMMARY

This chapter has applied all the concepts of the previous chapters to a situation with two parameters. One of the main goals was to familiarize you with probability distributions over two parameters, and to provide graphical examples to develop your intuition for multidimensional distributions.

A second major goal was to introduce the concepts of Gibbs sampling. It is important to understand that Gibbs sampling can often (but not always) be more efficient than generic Metropolis sampling, because Gibbs sampling uses the exact conditional posterior for each individual parameter, thereby never suffering rejected proposals (and thereby inspiring the quatrain at the beginning of the chapter).

In our applications, we will not program samplers from scratch, because we will program BUGS to do the sampling for us. Nevertheless, to properly interpret the output of BUGS, we need to understand how sampling works. In this chapter we saw how to get BUGS to display a sample from the prior by disconnecting the data from the model. Displaying the prior can be helpful to check that BUGS is using the prior we intended, especially when the prior is complicated.

This chapter also provided our first look at the issue of asking the following question: Is an underlying difference nonzero? We answered the question by examining the posterior distribution of the credible differences. The posterior reveals the most credible differences and the uncertainty in those differences. If a decision is required regarding the credibility of a specific value (such as zero), then one approach is to declare the value to be incredible if it (or its ROPE) falls outside the 95% HDI. The topic of deciding whether a value is incredible will be revisited and expanded in Chapter 12.

This chapter has served as a place to review and consolidate the concepts of the previous chapters. Importantly, it has also been an introduction to two-parameter models, without introducing any new complexity to the model functions themselves. With the basic concepts of multiparameter models now established, we can move on to *hierarchical* models in the next chapter.

8.8 R CODE

8.8.1 R Code for Grid Approximation (Figures 8.1 and 8.2)

This program is a script, not a function, so you can run it as is rather than having to `source` it and then call it as a function with arguments that you have to specify.

The program uses some R commands that we have not previously used. One of the new functions is `outer(vec1,vec2,f)`, which creates the matrix outer product of vector `vec1` and vector `vec2`, applying the function `f` to each component combination. Suppose you've already defined a function `f`. Then you can apply that function to the pairwise components of two vectors by using the `outer` command. The result is the matrix shown here:

```
outer( c(r1,r2,r3,r4) ,  c(c1,c2,c3) ,f )
```

$$= \begin{bmatrix} f(r1,c1) & f(r1,c2) & f(r1,c3) \\ f(r2,c1) & f(r2,c2) & f(r2,c3) \\ f(r3,c1) & f(r3,c2) & f(r3,c3) \\ f(r4,c1) & f(r4,c2) & f(r4,c3) \end{bmatrix}$$

You can leave the function `f` unspecified, in which case it defaults to multiplication.

Also used for the first time are the plotting functions `persp` and `contour`. These functions have many optional arguments that you can read about by typing `help('persp')` and `help('contour')`. I will point out one caveat: The `persp` function uses an argument called `theta`, which specifies the degree of rotation for the 3D plot. Do not confuse that `theta` with the θ we have been using to specify the proportion of heads!

As usual, the Bayesian computations are accomplished in only a few lines of code, but the plotting consumes many lines.

(BernTwoGrid.R)

```
1   # Specify the grid on theta1,theta2 parameter space.
2   nInt = 501 # arbitrary number of intervals for grid on theta.
3   theta1 = seq( from=((1/nInt)/2) ,to=(1-((1/nInt)/2)) ,by=(1/nInt) )
4   theta2 = theta1
5
6   # Specify the prior probability _masses_ on the grid.
7   priorName = c("Beta","Ripples","Null","Alt")[1] # or define your own.
8   if ( priorName == "Beta" ) {
9       a1 = 3 ; b1 = 3 ; a2 = 3 ; b2 = 3
10      prior1 = dbeta( theta1 , a1 , b1 )
11      prior2 = dbeta( theta2 , a2 , b2 )
12      prior = outer( prior1 , prior2 ) # density
13      prior = prior / sum( prior ) # convert to normalized mass
14  }
```

```
15  if ( priorName == "Ripples" ) {
16         rippleAtPoint = function( theta1 , theta2 ) {
17               m1 = 0 ; m2 = 1 ; k = 0.75*pi
18               sin( (k*(theta1-m1))^2 + (k*(theta2-m2))^2 )^2 }
19         prior = outer( theta1 , theta2 , rippleAtPoint )
20         prior = prior / sum( prior ) # convert to normalized mass
21  }
22  if ( priorName == "Null" ) {
23      # 1's at theta1=theta2, 0's everywhere else:
24      prior = diag( 1 , nrow=length(theta1) , ncol=length(theta1) )
25      prior = prior / sum( prior ) # convert to normalized mass
26  }
27  if ( priorName == "Alt" ) {
28      # Uniform:
29      prior = matrix( 1 , nrow=length(theta1) , ncol=length(theta2) )
30      prior = prior / sum( prior ) # convert to normalized mass
31  }
32
33  # Specify likelihood
34  z1 = 5 ; N1 = 7 ; z2 = 2 ; N2 = 7 # data are specified here
35  likeAtPoint = function( t1 , t2 ) {
36      p = t1^z1 * (1-t1)^(N1-z1) * t2^z2 * (1-t2)^(N2-z2)
37      return( p )
38  }
39  likelihood = outer( theta1 , theta2 , likeAtPoint )
40
41  # Compute posterior from point-by-point multiplication and normalizing:
42  pData = sum( prior * likelihood )
43  posterior = ( prior * likelihood ) / pData
44
45  # ------------------------------------------------------------------------
46  # Display plots.
47
48  # Specify the complete filename for saving the plot
49  plotFileName = paste("BernTwoGrid",priorName,".eps" ,sep="")
50
51  # Specify the probability mass for the HDI region
52  credib = .95
53
54  # Specify aspects of perspective and contour plots.
55  rotate = (-25)
56  tilt = 25
57  parallelness = 5.0
58  shadeval = 0.05
59  perspcex = 0.7
60  ncontours = 9
61  zmax = max( c( max(posterior) , max(prior) ) )
62
63  # Specify the indices to be used for plotting. The full arrays would be too
64  # dense for perspective plots, so we plot only a thinned-out set of them.
65  nteeth1 = length( theta1 )
66  thindex1 = seq( 1, nteeth1 , by = round( nteeth1 / 30 ) )
67  thindex1 = c( thindex1 , nteeth1 ) # makes sure last index is included
68  thindex2 = thindex1
```

```
69
70    windows(7,10)
71    layout( matrix( c( 1,2,3,4,5,6 ) ,nrow=3 ,ncol=2 ,byrow=TRUE ) )
72    par(mar=c(3,3,1,0))              # number of margin lines: bottom,left,top,right
73    par(mgp=c(2,1,0))                # which margin lines to use for labels
74    par(mai=c(0.4,0.4,0.2,0.05))     # margin size in inches: bottom,left,top,right
75    par(pty="s")                     # makes contour plots in square axes.
76
77    # prior
78    persp( theta1[thindex1] , theta2[thindex2] , prior[thindex1,thindex2] ,
79          xlab="theta1" , ylab="theta2" , main="Prior" , cex=perspcex , lwd=0.1 ,
80          xlim=c(0,1) , ylim=c(0,1) , zlim=c(0,zmax) , zlab="p(t1,t2)" ,
81              theta=rotate , phi=tilt , d=parallelness , shade=shadeval )
82    contour( theta1[thindex1] , theta2[thindex2] , prior[thindex1,thindex2] ,
83            main=bquote(" ") , levels=signif(seq(0,zmax,length=ncontours),3) ,
84            drawlabels=FALSE , xlab=bquote(theta[1]) , ylab=bquote(theta[2]) )
85
86    # likelihood
87    persp( theta1[thindex1] , theta2[thindex2] , likelihood[thindex1,thindex2] ,
88          xlab="theta1" , ylab="theta2" , main="Likelihood" , lwd=0.1 ,
89              xlim=c(0,1) , ylim=c(0,1) , zlab="p(D|t1,t2)" , cex=perspcex ,
90              theta=rotate , phi=tilt , d=parallelness , shade=shadeval )
91    contour( theta1[thindex1] , theta2[thindex2] , likelihood[thindex1,thindex2] ,
92            main=bquote(" ") , nlevels=(ncontours-1) ,
93                xlab=bquote(theta[1]) , ylab=bquote(theta[2]) , drawlabels=FALSE )
94    # Include text for data
95    maxlike = which( likelihood==max(likelihood) , arr.ind=TRUE )
96    if ( theta1[maxlike[1]] > 0.5 ) { textxpos = 0 ; xadj = 0
97    } else { textxpos = 1 ; xadj = 1 }
98    if ( theta2[maxlike[2]] > 0.5 ) { textypos = 0 ; yadj = 0
99    } else { textypos = 1 ; yadj = 1 }
100   text( textxpos , textypos , cex=1.5 ,
101           bquote( "z1="* .(z1) *",N1="* .(N1) *",z2="* .(z2) *",N2="* .(N2) ) ,
102           adj=c(xadj,yadj) )
103
104   # posterior
105   persp( theta1[thindex1] , theta2[thindex2] , posterior[thindex1,thindex2] ,
106           xlab="theta1" , ylab="theta2" , main="Posterior" , cex=perspcex ,
107           lwd=0.1           , xlim=c(0,1) , ylim=c(0,1) , zlim=c(0,zmax) ,
108           zlab="p(t1,t2|D)" , theta=rotate , phi=tilt , d=parallelness ,
109           shade=shadeval )
110   contour( theta1[thindex1] , theta2[thindex2] , posterior[thindex1,thindex2] ,
111           main=bquote(" ") , levels=signif(seq(0,zmax,length=ncontours),3) ,
112           drawlabels=FALSE , xlab=bquote(theta[1]) , ylab=bquote(theta[2]) )
113   # Include text for p(D)
114   maxpost = which( posterior==max(posterior) , arr.ind=TRUE )
115   if ( theta1[maxpost[1]] > 0.5 ) { textxpos = 0 ; xadj = 0
116   } else { textxpos = 1 ; xadj = 1 }
117   if ( theta2[maxpost[2]] > 0.5 ) { textypos = 0 ; yadj = 0
118   } else { textypos = 1 ; yadj = 1 }
119   text( textxpos , textypos , cex=1.5 ,
120       bquote( "p(D)=" * .(signif(pData,3)) ) , adj=c(xadj,yadj) )
121
122   # Mark the highest posterior density region
```

```
123    source("HDIofGrid.R")
124    HDIheight = HDIofGrid( posterior )$height
125    par(new=TRUE) # don't erase previous contour
126    contour( theta1[thindex1] , theta2[thindex2] , posterior[thindex1,thindex2] ,
127           main=bquote(.(100*credib)*"% HD region") ,
128           levels=signif(HDIheight,3) , lwd=3 , drawlabels=FALSE ,
129           xlab=bquote(theta[1]) , ylab=bquote(theta[2]) )
130
131    ## Change next line if you want to save the graph.
132    wantSavedGraph = T # TRUE or FALSE
133    if ( wantSavedGraph ) { dev.copy2eps(file=plotFileName) }
```

8.8.2 R Code for Metropolis Sampler (Figure 8.3)

Important: This program relies on the mvrnorm function in R, which is part
of the MASS package. The MASS package is part of the typical distribution
of R, and it is probably already installed on your computer. Therefore, the
library(MASS) command in this program will load the MASS package without complaint. If, however, the program balks at that line, you must install the
MASS package. Just uncomment the install.packages command.

(BernTwoMetropolis.R)

```
1     # Use this program as a template for experimenting with the Metropolis
2     # algorithm applied to two parameters called theta1,theta2 defined on the
3     # domain [0,1]x[0,1].
4
5     # Load the MASS package, which defines the mvrnorm function.
6     # If this "library" command balks, you must intall the MASS package:
7     #install.packages("MASS")
8     library(MASS)
9
10    # Define the likelihood function.
11    # The input argument is a vector: theta = c( theta1 , theta2 )
12    likelihood = function( theta ) {
13           # Data are constants, specified here:
14           z1 = 5 ; N1 = 7 ; z2 = 2 ; N2 = 7
15           likelihood = ( theta[1]^z1 * (1-theta[1])^(N1-z1)
16                   * theta[2]^z2 * (1-theta[2])^(N2-z2) )
17           return( likelihood )
18    }
19
20    # Define the prior density function.
21    # The input argument is a vector: theta = c( theta1 , theta2 )
22    prior = function( theta ) {
23           # Here's a beta-beta prior:
24           a1 = 3 ; b1 = 3 ; a2 = 3 ; b2 = 3
25           prior = dbeta( theta[1] , a1 , b1) * dbeta( theta[2] , a2 , b2)
26           return( prior )
27    }
28
29    # Define the relative probability of the target distribution, as a function
30    # of theta.  The input argument is a vector: theta = c( theta1 , theta2 ).
```

```
31  # For our purposes, the value returned is the UNnormalized posterior prob.
32  targetRelProb = function( theta ) {
33          if ( all( theta >= 0.0 ) & all( theta <= 1.0 ) ) {
34                  targetRelProbVal = likelihood( theta ) * prior( theta )
35          } else {
36                  # This part is important so that the Metropolis algorithm
37                  # never accepts a jump to an invalid parameter value.
38                  targetRelProbVal = 0.0
39          }
40          return( targetRelProbVal )
41  }
42
43  # Specify the length of the trajectory, i.e., the number of jumps to try.
44  trajLength = ceiling( 1000 / .9 ) # arbitrary large number
45  # Initialize the vector that will store the results.
46  trajectory = matrix( 0 , nrow=trajLength , ncol=2 )
47  # Specify where to start the trajectory
48  trajectory[1,] = c( 0.50 , 0.50 ) # arbitrary start values of the two param's
49  # Specify the burn-in period.
50  burnIn = ceiling( .1 * trajLength ) # arbitrary number
51  # Initialize accepted, rejected counters, just to monitor performance.
52  nAccepted = 0
53  nRejected = 0
54  # Specify the seed, so the trajectory can be reproduced.
55  set.seed(47405)
56  # Specify the covariance matrix for multivariate normal proposal distribution.
57  nDim = 2 ; sd1 = 0.2 ; sd2 = 0.2
58  covarMat = matrix( c( sd1^2 , 0.00 , 0.00 , sd2^2 ) , nrow=nDim , ncol=nDim )
59
60  # Now generate the random walk. stepIdx is the step in the walk.
61  for ( stepIdx in 1:(trajLength-1) ) {
62          currentPosition = trajectory[stepIdx,]
63          # Use the proposal distribution to generate a proposed jump.
64          # The shape and variance of the proposal distribution can be changed
65          # to whatever you think is appropriate for the target distribution.
66          proposedJump = mvrnorm( n=1 , mu=rep(0,nDim), Sigma=covarMat )
67          # Compute the probability of accepting the proposed jump.
68          probAccept = min( 1,
69                  targetRelProb( currentPosition + proposedJump )
70                  / targetRelProb( currentPosition ) )
71          # Generate a random uniform value from the interval [0,1] to
72          # decide whether or not to accept the proposed jump.
73          if ( runif(1) < probAccept ) {
74                  # accept the proposed jump
75                  trajectory[ stepIdx+1 , ] = currentPosition + proposedJump
76                  # increment the accepted counter, just to monitor performance
77                  if ( stepIdx > burnIn ) { nAccepted = nAccepted + 1 }
78          } else {
79                  # reject the proposed jump, stay at current position
80                  trajectory[ stepIdx+1 , ] = currentPosition
81                  # increment the rejected counter, just to monitor performance
82                  if ( stepIdx > burnIn ) { nRejected = nRejected + 1 }
83          }
```

```
84      }
85
86      # End of Metropolis algorithm.
87
88      #------------------------------------------------------------------------
89      # Begin making inferences by using the sample generated by the
90      # Metropolis algorithm.
91
92      # Extract just the post-burnIn portion of the trajectory.
93      acceptedTraj = trajectory[ (burnIn+1) : dim(trajectory)[1] , ]
94
95      # Compute the mean of the accepted points.
96      meanTraj = apply( acceptedTraj , 2 , mean )
97      # Compute the standard deviations of the accepted points.
98      sdTraj = apply( acceptedTraj , 2 , sd )
99
100     # Display the sampled points
101     par( pty="s" ) # makes plots in square axes.
102     plot( acceptedTraj , type = "o" , xlim = c(0,1) , xlab = bquote(theta[1]) ,
103           ylim = c(0,1) , ylab = bquote(theta[2]) )
104     # Display means and rejected/accepted ratio in plot.
105     if ( meanTraj[1] > .5 ) { xpos = 0.0 ; xadj = 0.0
106     } else { xpos = 1.0 ; xadj = 1.0 }
107     if ( meanTraj[2] > .5 ) { ypos = 0.0 ; yadj = 0.0
108     } else { ypos = 1.0 ; yadj = 1.0 }
109     text( xpos , ypos ,              bquote(
110           "M=" * .(signif(meanTraj[1],3)) * "," * .(signif(meanTraj[2],3))
111           * "; " * N[pro] * "=" * .(dim(acceptedTraj)[1])
112           * ", " * frac(N[acc],N[pro]) * "="
113           * .(signif(nAccepted/dim(acceptedTraj)[1],3))
114           ) , adj=c(xadj,yadj) , cex=1.5   )
115
116
117     # Evidence for model, p(D).
118     # Compute a,b parameters for beta distribution that has the same mean
119     # and stdev as the sample from the posterior. This is a useful choice
120     # when the likelihood function is binomial.
121     a =    meanTraj * ( (meanTraj*(1-meanTraj)/sdTraj^2) - rep(1,nDim) )
122     b = (1-meanTraj) * ( (meanTraj*(1-meanTraj)/sdTraj^2) - rep(1,nDim) )
123     # For every theta value in the posterior sample, compute
124     # dbeta(theta,a,b) / likelihood(theta)*prior(theta)
125     # This computation assumes that likelihood and prior are properly normalized,
126     # i.e., not just relative probabilities.
127     wtd_evid = rep( 0 , dim(acceptedTraj)[1] )
128     for ( idx in 1 : dim(acceptedTraj)[1] ) {
129           wtd_evid[idx] = ( dbeta( acceptedTraj[idx,1],a[1],b[1] )
130                   * dbeta( acceptedTraj[idx,2],a[2],b[2] ) /
131                   ( likelihood(acceptedTraj[idx,]) * prior(acceptedTraj[idx,]) ) )
132     }
133     pdata = 1 / mean( wtd_evid )
134     # Display p(D) in the graph
135     text( xpos , ypos+(.12*(-1)^(ypos)) , bquote( "p(D) = " * .(signif(pdata,3)) ) ,
136           adj=c(xadj,yadj) , cex=1.5 )
```

```
137
138  ### Change next line if you want to save the graph.
139  want_saved_graph = F # TRUE or FALSE
140  if ( want_saved_graph ) { dev.copy2eps(file="BernTwoMetropolis.eps") }
141
142  # Estimate highest density region by evaluating posterior at each point.
143  npts = dim( acceptedTraj )[1] ; postProb = rep( 0 , npts )
144  for ( ptIdx in 1:npts ) {
145      postProb[ptIdx] = targetRelProb( acceptedTraj[ptIdx,] )
146  }
147  # Determine the level at which credmass points are above:
148  credmass = 0.95
149  waterline = quantile( postProb , probs=c(1-credmass) )
150  # Display highest density region in new graph
151  windows()
152  par( pty="s" ) # makes plots in square axes.
153  plot( acceptedTraj[ postProb < waterline , ] , type="p" , pch="x" , col="grey" ,
154        xlim = c(0,1) , xlab = bquote(theta[1]) ,
155        ylim = c(0,1) , ylab = bquote(theta[2]) ,
156        main=paste(100*credmass,"% HD region",sep="") )
157  points( acceptedTraj[ postProb >= waterline , ] ,  pch="o" , col="black" )
158  ## Change next line if you want to save the graph.
159  want_saved_graph = F # TRUE or FALSE
160  if ( want_saved_graph ) { dev.copy2eps(file="BernTwoMetropolisHD.eps") }
```

8.8.3 R Code for BUGS Sampler (Figure 8.6)

Here is the complete program for generating an MCMC sample using BUGS.

(BernTwoBugs.R)
```
1   library(BRugs)           # Kruschke, J. K. (2010). Doing Bayesian data analysis:
2                            # A Tutorial with R and BUGS. Academic Press / Elsevier.
3   #------------------------------------------------------------------------------
4   # THE MODEL.
5
6   modelstring = "
7   # BUGS model specification begins here...
8   model {
9       # Likelihood. Each flip is Bernoulli.
10      for ( i in 1 : N1 ) { y1[i] ~ dbern( theta1 ) }
11      for ( i in 1 : N2 ) { y2[i] ~ dbern( theta2 ) }
12      # Prior. Independent beta distributions.
13      theta1 ~ dbeta( 3 , 3 )
14      theta2 ~ dbeta( 3 , 3 )
15  }
16  # ... end BUGS model specification
17  " # close quote for modelstring
18  # Write model to a file:
19  .temp = file("model.txt","w") ; writeLines(modelstring,con=.temp); close(.temp)
20  # Load model file into BRugs and check its syntax:
21  modelCheck( "model.txt" )
22
23  #------------------------------------------------------------------------------
```

```
24   # THE DATA.
25
26   # Specify the data in a form that is compatible with BRugs model, as a list:
27   datalist = list(
28       N1 = 7 ,
29       y1 = c( 1,1,1,1,1,0,0 ) ,
30       N2 = 7 ,
31       y2 = c( 1,1,0,0,0,0,0 )
32   )
33   # Get the data into BRugs:
34   modelData( bugsData( datalist ) )
35
36   #------------------------------------------------------------------------------
37   # INTIALIZE THE CHAIN.
38
39   modelCompile()
40   modelGenInits()
41
42   #------------------------------------------------------------------------------
43   # RUN THE CHAINS.
44
45   samplesSet( c( "theta1" , "theta2" ) ) # Keep a record of sampled "theta" values
46   chainlength = 10000                     # Arbitrary length of chain to generate.
47   modelUpdate( chainlength )              # Actually generate the chain.
48
49   #------------------------------------------------------------------------------
50   # EXAMINE THE RESULTS.
51
52   theta1Sample = samplesSample( "theta1" ) # Put sampled values in a vector.
53   theta2Sample = samplesSample( "theta2" ) # Put sampled values in a vector.
54
55   # Plot the trajectory of the last 500 sampled values.
56   windows()
57   par( pty="s" )
58   plot( theta1Sample[(chainlength-500):chainlength] ,
59         theta2Sample[(chainlength-500):chainlength] , type = "o" ,
60         xlim = c(0,1) , xlab = bquote(theta[1]) , ylim = c(0,1) ,
61         ylab = bquote(theta[2]) , main="BUGS Result" )
62   # Display means in plot.
63   theta1mean = mean(theta1Sample)
64   theta2mean = mean(theta2Sample)
65   if (theta1mean > .5) { xpos = 0.0 ; xadj = 0.0
66   } else { xpos = 1.0 ; xadj = 1.0 }
67   if (theta2mean > .5) { ypos = 0.0 ; yadj = 0.0
68   } else { ypos = 1.0 ; yadj = 1.0 }
69   text( xpos , ypos ,
70           bquote(
71           "M=" * .(signif(theta1mean,3)) * "," * .(signif(theta2mean,3))
72           ) ,adj=c(xadj,yadj) ,cex=1.5   )
73   dev.copy2eps(file="BernTwoBugs.eps")
74
75   # Plot a histogram of the posterior differences of theta values.
76   thetaDiff = theta1Sample - theta2Sample
77   source("plotPost.R")
```

```
78    windows(7,4)
79    plotPost( thetaDiff , xlab=expression(theta[1]-theta[2]) , compVal=0.0 ,
80              breaks=30 )
81    dev.copy2eps(file="BernTwoBugsDiff.eps")
```

8.8.4 R Code for Plotting a Posterior Histogram

The plotPost.R program, listed here, merely plots a histogram of a sample, with various useful annotations. It was used to generate Figure 8.8, for example. It is used often throughout the remainder of the book. You need to "source" it in R before calling it, of course. Here are the annotations that the program can add to the histogram:

- *The mean (by default) or an estimate of the mode (if specified).* To show the estimate of the mode, instead of the mean, use the argument showMode=T.
- *The estimated HDI of the sample, defaulting to 95% mass.* To specify a different HDI mass (e.g., 90%), use the argument credMass=0.90. The HDI limits are computed by the function described in Section 23.3.2 (HDIofMCMC.R).

  The HDI is marked by a horizontal line and the ends of the HDI are labeled numerically, with the placement of the numerals with respect to the end points of the line governed by a parameter called HDItextPlace. The value of HDItextPlace is the proportion of the numeral that is plotted outside the limits of the HDI. Thus, when HDItextPlace=0.0, the labels fall entirely within the HDI, on both ends, and when HDItextPlace=1.0, the labels fall entirely outside the HDI, on both ends.
- *A comparison value, with the percentage of the distribution below and above that value.* For example, if the values being plotted are differences, a meaningful comparison value would be zero. No comparison value is plotted unless specified by the user (e.g., compVal=0.0).
- *The ROPE, with the percentage of the distribution that falls inside it.* The ROPE is not plotted unless specified by the user (e.g., ROPE=c(-.01,.01)). Notice that the values in the ROPE specification are the actual end points of the ROPE, *not* relative to the compVal. In principle, therefore, you could specify something like ROPE=c(-5,0) and compVal=1.

If the user does not specify a label for the x-axis, using the usual xlab argument, the program defaults to the label "Parameter." Other arguments specified by the user are passed into the histogram function as usual. Because plotPost.R uses the hist() function, it returns what hist() returns, namely, information about the histogram. If plotPost() is called without assigning the result to a variable, then the histogram information is displayed in the command window.

(plotPost.R)

```
1     plotPost = function( paramSampleVec , credMass=0.95 , compVal=NULL ,
2               HDItextPlace=0.7 , ROPE=NULL , yaxt=NULL , ylab=NULL ,
```

```
3          xlab=NULL , cex.lab=NULL , cex=NULL , xlim=NULL , main=NULL ,
4          showMode=F , ... ) {
5    # Override defaults of hist function, if not specified by user:
6    # (additional arguments "..." are passed to the hist function)
7    if ( is.null(xlab) ) xlab="Parameter"
8    if ( is.null(cex.lab) ) cex.lab=1.5
9    if ( is.null(cex) ) cex=1.4
10   if ( is.null(xlim) ) xlim=range( c( compVal , paramSampleVec ) )
11   if ( is.null(main) ) main=""
12   if ( is.null(yaxt) ) yaxt="n"
13   if ( is.null(ylab) ) ylab=""
14   # Plot histogram.
15   par(xpd=NA)
16   histinfo = hist( paramSampleVec , xlab=xlab , yaxt=yaxt , ylab=ylab ,
17                    freq=F , col="lightgrey" , border="white" ,
18                    xlim=xlim , main=main , cex=cex , cex.lab=cex.lab ,
19                    ... )
20   # Display mean or mode:
21   if ( showMode==F ) {
22       meanParam = mean( paramSampleVec )
23       text( meanParam , .9*max(histinfo$density) ,
24             bquote(mean==.(signif(meanParam,3))) , adj=c(.5,0) , cex=cex )
25   } else {
26       dres = density( paramSampleVec )
27       modeParam = dres$x[which.max(dres$y)]
28       text( modeParam , .9*max(histinfo$density) ,
29             bquote(mode==.(signif(modeParam,3))) , adj=c(.5,0) , cex=cex )
30   }
31   # Display the comparison value.
32   if ( !is.null( compVal ) ) {
33       pcgtCompVal = round( 100 * sum( paramSampleVec > compVal )
34                           / length( paramSampleVec )  , 1 )
35       pcltCompVal = 100 - pcgtCompVal
36       lines( c(compVal,compVal) , c(.5*max(histinfo$density),0) ,
37              lty="dashed" , lwd=2 )
38       text( compVal , .5*max(histinfo$density) ,
39             bquote( .(pcltCompVal)*"% <= " *
40                     .(signif(compVal,3)) * " < "*.(pcgtCompVal)*"%" ) ,
41             adj=c(pcltCompVal/100,-0.2) , cex=cex )
42   }
43   # Display the ROPE.
44   if ( !is.null( ROPE ) ) {
45       pcInROPE = ( sum( paramSampleVec > ROPE[1] & paramSampleVec < ROPE[2] )
46                   / length( paramSampleVec ) )
47       ROPEtextHt = .35*max(histinfo$density)
48       lines( c(ROPE[1],ROPE[1]) , c(ROPEtextHt,0) , lty="dotted" , lwd=2 )
49       lines( c(ROPE[2],ROPE[2]) , c(ROPEtextHt,0) , lty="dotted" , lwd=2 )
50       text( mean(ROPE) , ROPEtextHt ,
51             bquote( .(round(100*pcInROPE))*"% in ROPE" ) ,
52             adj=c(.5,-0.2) , cex=1 )
53   }
54   # Display the HDI.
55   source("HDIofMCMC.R")
56   HDI = HDIofMCMC( paramSampleVec , credMass )
```

```
57      lines( HDI , c(0,0) , lwd=4 )
58      text( mean(HDI) , 0 , bquote(.(100*credMass) * "% HDI" ) ,
59          adj=c(.5,-1.9) , cex=cex )
60      text( HDI[1] , 0 , bquote(.(signif(HDI[1],3))) ,
61          adj=c(HDItextPlace,-0.5) , cex=cex )
62      text( HDI[2] , 0 , bquote(.(signif(HDI[2],3))) ,
63          adj=c(1.0-HDItextPlace,-0.5) , cex=cex )
64      par(xpd=F)
65      return( histinfo )
66   }
```

8.9 EXERCISES

Exercise 8.1. [Purpose: To explore a real-world application about the difference of proportions.] Is there a "hot hand" in basketball? This question has been addressed in a frequently cited article by Gilovich, Vallone, & Tversky (1985).The idea of a hot hand is that the success of a shot depends on the success of a previous shot, as opposed to each shot being an independent flip of a coin (or toss of a ball). One way to address this idea is to consider pairs of free throws taken after fouls. If the player has a hot hand, then he should be more likely to make the second shot after a successful first shot than after a failed first shot. If the two shots are independent, however, then the probability of making the second shot after a successful first shot should equal the probability of making the second shot after failing the first shot. Thus, there is a hot hand if the probability of success after a success is better than the probability of success after failure.

During 1980–1982, Larry Bird of the Boston Celtics had 338 pairs of free throws. He was successful on 285 first shots and failed on the remaining 53 first shots. After the 285 successful first shots, he was successful on 251 second shots (and failed on the other 34 second shots). After the 53 failed first shots, he was successful on 48 second shots (and failed on the other 5 second shots). Thus, we want to know if 251/285 (success after success) is different than 48/53 (success after failure).

Let θ_1 represent the proportion of success after a successful shot, and let θ_2 represent the proportion of success after a failed first shot. Suppose we have priors of beta($\theta \mid 30, 10$) on both, representing the belief that we think that professional players make about 75% of their free throws, regardless of when they are made.

(A) Modify the BRugs program of Section 8.8.3 (BernTwoBugs.R) to generate a histogram of credible differences between success-after-success and success-after-failure. Explain what modifications you made to the R code. (Hint: Your result should look like something like Figure 12.2, p. 299).

(B) Based on your results from the previous part, does Larry Bird seem to have a hot hand? In other words, are almost all of the credible differences between success-after-success and success-after-failure well above zero, or is a difference of zero among the credible differences?

Exercise 8.2. [Purpose: To examine the prior in BUGS by omitting references to data.] Reproduce and run the BUGS code that generated the chains in Figure 8.7. Show the sections of code that you modified from the program in Section 8.8.3 (BernTwoBugs.R), and show the resulting graphical output.

Exercise 8.3. [Purpose: To understand the limitations of prior specification in BUGS.] In BUGS, all priors must be specified in terms of probability distributions that BUGS knows about. These distributions include the beta, gamma, normal, uniform, and so on, as specified in the appendix of the BUGS User Manual. There are ways to program novel distributions into BUGS, but explaining how would take us too far afield at this point. Instead, we'll consider how unusual priors can be constructed from the built-in distributions. The left panel of Figure 8.9 shows the prior that results from this model specification (BernTwoFurrowsBugs.R):

```
10    model {
11          # Likelihood. Each flip is Bernoulli.
12          for ( i in 1 : N1 ) { y1[i] ~ dbern( theta1 ) }
13          for ( i in 1 : N2 ) { y2[i] ~ dbern( theta2 ) }
14          # Prior. Curved scallops!
15          x ~ dunif(0,1)
16          y ~ dunif(0,1)
17          N <- 4
18          xt <- sin( 2*3.141593*N * x ) / (2*3.141593*N) + x
19          yt <- 3 * y + (1/3)
20          xtt <- pow( xt , yt )
21          theta1 <- xtt
22          theta2 <- y
23    }
```

Adapt the program in Section 8.8.3 (BernTwoBugs.R) to use this prior, but with N set to 5 instead of 4. (Don't confuse N with $N1$ or $N2$.) Produce a graph of the prior and of the posterior, like those shown in Figure 8.9. This particular furrowed prior would never be used in actual research that I'm aware of; the point is that you can specify unusual priors if you need to.

Exercise 8.4. [Purpose: To understand Metropolis sampling, and to see the importance of tuning the proposal distribution.] For this exercise, assume the prior and data used in Figure 8.1.

(A) In the R code of Section 8.8.2 (BernTwoMetropolis.R), set the standard deviations (both sd1 and sd2) of the proposal distribution to 0.005, and run the program. Does the resulting distribution of sampled values

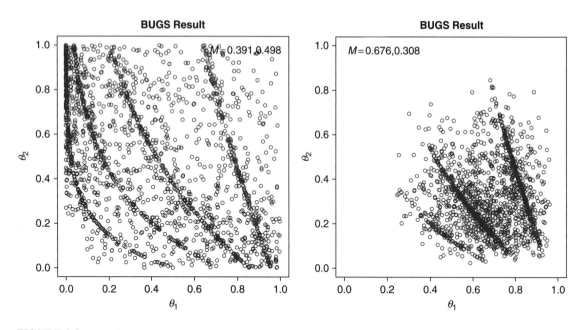

FIGURE 8.9
For Exercise 8.3, a bizarre prior, on the left, specified in BUGS, and the resulting posterior distribution on the right, when using the data from examples in this chapter.

resemble Figures 8.1 or 8.3? What is wrong with the choice of standard deviations?

(B) In the R code of Section 8.8.2 (BernTwoMetropolis.R), set the standard deviations (both sd1 and sd2) of the proposal distribution to 5.0, and run the program. Does the resulting distribution of sampled values resemble Figures 8.1 or 8.3? What is wrong with the choice of standard deviations?

Exercise 8.5. [Purpose: To remember how to do posterior prediction with BUGS.] For this exercise, assume the prior and data used in Figure 8.1. From the posterior, what is the probability that the next flip of the two coins will have $y_1^* = 1$ *and* $y_2^* = 0$? To answer this question, expand the code of Section 8.5 so that it includes posterior predictions. See the example in Section 7.4.2, (p. 143), and simply repeat the same structure for each theta and y value. Include your code with your answer.

CHAPTER 9

Bernoulli Likelihood with Hierarchical Prior

Oh darlin', for love, it's on you I depend.
Well, I s'pose Jack Daniels is also my friend.
But you keep him locked up in the ladies' loo,
S'pose that means his spirits depend on you too.

In the previous chapter, we explored a case in which there were two parameters, and our goal was to estimate each parameter. We had two coins and we wanted

Doing Bayesian Data Analysis: A Tutorial with R and BUGS. DOI: 10.1016/B978-0-12-381485-2.00009-2
© 2011, Elsevier Inc. All rights reserved.

to estimate the bias in each coin. The prior beliefs about each parameter were assumed to be independent of each other, which meant that our prior belief about the bias in one coin had no influence about our prior belief regarding the bias in the other coin. This independence of the parameters meant that the formal specification of the prior belief about either coin's bias had no mention of the other bias.

In this chapter, we explore situations in which there are two or more parameters that do have meaningful dependencies. For example, we may believe that the bias of a coin depends on the characteristics of the factory in which it was minted. We have prior beliefs about the parameter values of the mint, and we have prior beliefs about the dependence of the coin's bias on the minting parameters. Then we flip the coin a few times and observe how many times it comes up heads. The data affect our beliefs about the coin's bias. But, importantly, the data also affect our beliefs about the dependence of the coin's bias on the minting parameters, and the data affect our belief about the minting parameters themselves.

The parameters that directly affect the data are called just that: parameters. But parameters that affect the data indirectly, by affecting beliefs about other parameters, are often called *hyper*parameters. In one respect, there is nothing different about parameters and hyperparameters: They coexist in a multi-dimensional joint parameter space, and we apply Bayes' rule to the joint parameter space to update from prior to posterior beliefs over the joint parameter space. But the dependencies among parameters become useful in at least two respects. First, the dependencies are meaningful for the given application, so it can help to keep that structure intact as we interpret the posterior distribution. Second, the dependencies can motivate relatively efficient Monte Carlo sampling from the posterior. Instead of using a Metropolis algorithm to sample from the joint parameter space, and face the heartbreak of rejected proposals, we might use the dependencies to sample parameters in turn, analogous to Gibbs sampling.

As usual, we consider the scenario of flipping coins. And, as usual, keep in mind that the coin flip is just a surrogate for real-world data involving two outcomes, such as recovery or nonrecovery from a disease after treatment, recalling or not recalling a studied item, choosing candidate A or candidate B in an election, and so on.

9.1 A SINGLE COIN FROM A SINGLE MINT

We begin with a review of the likelihood and prior distribution for our single-coin scenario in order to build new ideas. The bias θ of a coin determines the probability of getting a head, according to the Bernoulli distribution:

$$p(y \mid \theta) = \text{bern}(y \mid \theta)$$
$$= \theta^y (1 - \theta)^{1-y} \tag{9.1}$$

where $y = 1$ for the result "head" and $y = 0$ for the result "tail." We assume independence across flips, so the joint probability of the particular $z = \sum_{i=1}^{N} y_i$ heads out of N flips is $\prod_{i=1}^{N} p(y_i \mid \theta) = \theta^z (1 - \theta)^{N-z}$.

The prior distribution over the biases is denoted $p(\theta)$. For the present example, we suppose that the prior is a beta distribution. As was explained in Chapter 5, the beta distribution has two parameters, a and b, and is defined as $\text{beta}(\theta \mid a, b) = \theta^{a-1}(1 - \theta)^{b-1}/B(a, b)$. To make the parameters of the beta distribution more intuitive, we express them in terms of the corresponding mean μ and sample size K, as described in Section 5.2.1; specifically in Equation 5.5, p. 83. If the mean of our prior belief is μ, and our confidence is reflected by a prior sample size of K, then the corresponding beta parameters are $a = \mu K$ and $b = (1 - \mu)K$. For the purposes of the example, we will treat K as a constant. Because the prior distribution depends on our choice of μ, the prior distribution is a function of μ and can be written

$$p(\theta \mid \mu) = \text{beta}(\theta \mid \mu K, (1 - \mu)K) \tag{9.2}$$

Notice that the magnitude of K is an expression of our prior certainty regarding the dependence of the bias on μ. When K is large, the distribution of θ is narrowly loaded over μ. When K is small, the distribution of θ is widely dispersed around μ. Thus, as K gets large, we are more certain about the form of the dependency of θ on μ.

Now we make the essential expansion of our scenario into the realm of hierarchical models. Instead of specifying a single particular value for μ, we think of μ as taking on many possible values (from 0 to 1), and we specify a probability distribution over those values. This distribution can be thought of as describing the uncertainty in our beliefs about the construction of the mint that manufactured the coin. When μ is large, the mint tends to produce coins with large biases, and when μ is small, the mint tends to produce coins with small biases. Our prior distribution over μ expresses what we believe about how mints are constructed. For the sake of making the example concrete, we suppose that the distribution on μ is again a beta distribution,

$$p(\mu) = \text{beta}(\mu \mid A_\mu, B_\mu) \tag{9.3}$$

where A_μ and B_μ are constants. In this case, we believe that μ is typically near $A_\mu/(A_\mu + B_\mu)$, because that is the mean of the beta distribution, but we believe that the value of μ could be above or below that mean.

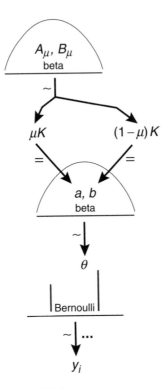

FIGURE 9.1

A model of hierarchical dependencies for data from a single coin. At the bottom of the diagram, the datum y_i for the ith flip depends on the value of the bias parameter θ in a Bernoulli distribution. The arrow has a "~" symbol to indicate that $y_i \sim \text{bern}(y_i \mid \theta)$. The ellipsis next to the arrow denotes the repeated dependency for every flip. Moving up the diagram, we see that the value of θ depends on a beta distribution, which has shape parameters a and b. The a, b values are reparameterized in terms of μ and K. The arrows impinging on a and b are labeled with "=" signs to indicate that these variables have a deterministic dependency, not a probabilistic one. K is a constant set by prior beliefs, expressing how strongly θ depends on μ. The hyperparameter μ expresses the bias of the mint that created the coin, and μ depends on a beta distribution with parameters A_μ and B_μ, which are set by prior beliefs. We simultaneously estimate the two parameters θ and μ.

The scenario is summarized in Figure 9.1. The figure shows the dependency structure among the variables. Downward-pointing arrows denote how higher-level variables generate lower-level variables. For example, because the coin bias θ governs the result of a coin flip, we draw an arrow from the θ-parameterized Bernoulli distribution to the coin flip datum y_i. Thus, when y_i depends on θ, we draw an arrow to y_i from the distribution involving θ. The caption describes the full chain of dependencies. This form of model is referred to as a *hierarchical* model because of the layers of dependencies. When we do Bayesian inference, we are updating our beliefs about both of the parameters θ and μ. Please read the caption for Figure 9.1 now.

Notice what happens to the dependency structure when our prior beliefs about the hyperparameter μ become narrowed to a single value. Suppose that we believe that μ has just one specific value and no other value of μ is possible. This can be visualized as the distribution of $p(\mu)$ being an infinitesimally narrow spike over the specific value of μ that we posit, as if A_μ and B_μ were set to humongous values. Thus, μ is effectively a *constant*, and $p(\theta)$ is a function of two constants, namely, μ and K. In this case, the inclusion of a hyper-level involving μ adds nothing to our inferential process, because we have nothing to infer about μ because we already know the value of μ. Therefore we do not have a hierarchical model any more; instead we have a one-level model of the type we developed in previous chapters. It is only when we have uncertainty about a value that we make inferences about it. When other uncertain values depend on an uncertain value, then we have a chain of dependent uncertainties, formalized by a hierarchial model.

Notice also what happens to the dependency structure when our beliefs about the dependence of θ on μ narrow down to a *deterministic* relationship without any uncertainty. For example, suppose that K becomes infinitely large, so that when μ has a particular value, then θ also has exactly that same value. In this case, the only uncertainty in θ is the uncertainty in μ; there is no additional uncertainty introduced in going from μ to θ. In this case, the inclusion of an intermediate level for θ adds nothing to our inferential process, because we have nothing to infer about θ; we already know that the value of θ is exactly the value of μ. Therefore, we do not have a hierarchical model any more, instead we have a one-level model of the type we developed in previous chapters. *Refrain:* It is only when we have uncertainty about a value that we make inferences about it. When other uncertain values depend on an uncertain value, then we have a chain of dependent uncertainties, formalized by a hierarchial model.

Let's now consider how Bayes' rule applies to this situation. If we treat this situation as simply a case of two parameters, then Bayes' rule is merely $p(\theta, \mu \mid y) = p(y \mid \theta, \mu)p(\theta, \mu)/p(y)$. There are two aspects that are "special" about our present situation. First, the likelihood function does not involve μ, so $p(y \mid \theta, \mu)$ can be rewritten as $p(y \mid \theta)$. Second, because by definition $p(\theta \mid \mu) = p(\theta, \mu)/p(\mu)$, the prior on the joint parameter space can be factored thus: $p(\theta, \mu) = p(\theta \mid \mu)\, p(\mu)$. Therefore, Bayes' rule for our current hierarchical model has the form

$$p(\theta, \mu \mid y) = p(y \mid \theta, \mu)p(\theta, \mu)/p(y)$$
$$= p(y \mid \theta)p(\theta \mid \mu)p(\mu)/p(y) \qquad (9.4)$$

Notice that the three terms of the numerator are given specific expression by our particular example; these specific formulas are summarized in Figure 9.1.

It turns out that direct formal analysis of Equation 9.4 does not yield a simple formula for the *normalized* posterior. We can, nevertheless, use grid approximation to get a thorough picture of what is going on in the present example.

9.1.1 Posterior via Grid Approximation

When the parameter(s) and hyperparameter(s) extend over a finite domain, and there are not too many of them, then we can approximate the posterior via grid approximation. In our present situation, we have the parameter θ and hyperparameter μ that both have finite domains, namely, the interval $[0, 1]$. Therefore, a grid approximation is tractable and the distributions can be readily graphed.

Figure 9.2 shows an example in which the hyperprior distribution has the form of a beta distribution as in Equation 9.3, with $A_\mu = 2$ and $B_\mu = 2$; that is, $p(\mu) = \text{beta}(\mu \mid 2, 2)$. This hyperprior expresses the belief that the mint's μ value is near 0.5, but there is large uncertainty. This beta distribution is shown in the third panel of the upper row of Figure 9.2. The graph is tipped on its side so that the vertical axis is μ; this orientation keeps all the μ axes oriented the same direction in the figure, to facilitate comparison across panels.

The prior distribution on θ—or more precisely, the prior distribution regarding the dependency of θ on μ—is expressed by another beta distribution, as in Equation 9.2 with $K = 100$, whereby $p(\theta \mid \mu) = \text{beta}(\theta, \mu 100, (1 - \mu)100)$. The prior expresses a high degree of certainty that a mint with hyperparameter μ generates coins that have a bias θ close to μ. Two cases of this conditional distribution are shown in the right panel of the second row of Figure 9.2. The upper graph within that panel shows $p(\theta \mid \mu=0.75)$, and the lower graph in that panel shows $p(\theta \mid \mu=0.25)$. You can see that the conditional distributions are fairly tightly centered on 0.75 and 0.25, respectively.

The left and middle panels of the top row of Figure 9.2 show the *joint* prior distribution: $p(\theta, \mu) = p(\theta \mid \mu)p(\mu)$. The contour plot in the middle panel shows a top-down view of the perspective plot in the left panel. As this is a grid approximation, the joint prior $p(\theta, \mu)$ was computed by first multiplying $p(\theta \mid \mu)$ and $p(\mu)$ at every grid point and then, to convert to a discrete probability mass at each grid point, dividing by their sum across the entire grid. The normalized probability masses were then converted to estimates of probability density at each grid point by dividing each probability mass by the area of a grid cell.

The right panel of the top row, showing $p(\mu)$ tipped on its side, is the *marginal* distribution of the prior: If you imagine collapsing (i.e., summing) the joint

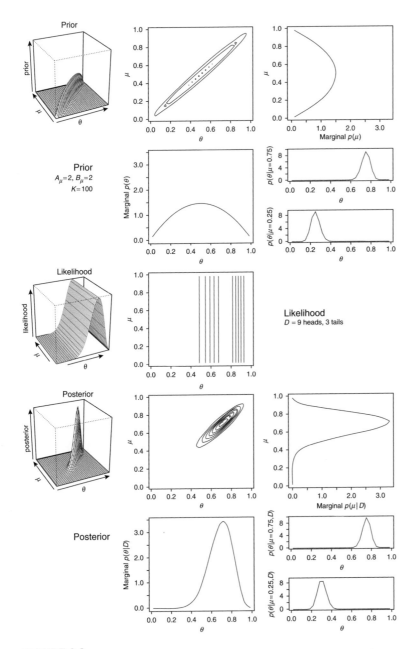

FIGURE 9.2

The prior has low certainty regarding μ but high certainty regarding the dependence of θ on μ. The posterior shows that the distribution of μ has been altered noticeably by the data (see sideways plots of marginal $p(\mu)$), but the dependence of θ on μ has not been altered much (see small plots of $p(\mu|\theta)$). Compare with Figure 9.3, which uses the same data but a different prior.

prior across θ, the pressed flower you end up with is the graph of $p(\mu)$. The scale on the $p(\mu)$ axis is density, not mass, because the the mass at each point of the comb on μ was divided by the width of a comb interval to approximate density.

The right panel of the second row, showing $p(\theta \mid \mu)$, can be thought of as two "slices" through the joint prior. One slice is at the grid value $\mu = 0.75$, and the other slice is at the grid value $\mu = 0.25$. The slices are renormalized, however, so that they are individually proper probability densities that sum to 1.0 over θ.

The middle row of Figure 9.2 shows the likelihood distribution over the parameter space. The data D comprise nine heads and three tails. The likelihood distribution is a product of Bernoulli distributions, $p(D \mid \theta) = \theta^9 (1 - \theta)^3$. Notice in the graph that all the contour lines are parallel to the μ axis and orthogonal to the θ axis. These parallel contours are the graphical signature of the fact that the likelihood function depends only on θ and not on μ.

The posterior distribution in the fourth row of Figure 9.2 is determined by multiplying, at each point of the grid on θ, μ space, the joint prior and the likelihood. The point-wise products are normalized by dividing by the sum of those values across the parameter space.

When we take a slice through the joint posterior at a particular value of μ, and renormalize by dividing by sum of the discrete probability masses in that slice, we get the conditional distribution $p(\theta \mid \mu, D)$. The bottom right panel of Figure 9.2 shows the conditional for two values of μ. Notice in Figure 9.2 that there is not much difference in the graphs of the prior $p(\theta \mid \mu)$ and the posterior $p(\theta \mid \mu, D)$. This is because the prior beliefs regarding the dependency of θ on μ had little uncertainty.

The distribution of $p(\mu \mid D)$, shown in the right panel of the fourth row of Figure 9.2, is determined by summing the joint posterior across all values of θ. This marginal distribution can be imagined as collapsing the joint posterior along the θ axis, with the resulting "pressed flower" being silhouetted in the third panel of the bottom row. Notice that the graphs of the prior $p(\mu)$ and the posterior $p(\mu \mid D)$ are different. The data have had a noticeable impact on beliefs about how μ is distributed because the prior was uncertain and was therefore easily influenced by the data.

As a contrasting case, consider instead what happens when there is high certainty on the prior regarding μ, but low certainty on the prior regarding the dependence of θ on μ. Figure 9.3 illustrates such a case, where $p(\mu) = \text{beta}(\mu \mid 20, 20)$ and $p(\theta \mid \mu) = \text{beta}(\theta \mid \mu 6, (1 - \mu)6)$. The top row, right panel, shows that $p(\mu)$ is sharply peaked over $\mu = 0.5$, but the conditional distributions $p(\theta \mid \mu)$ are broad (second row, right panel). The same data as for Figure 9.2 are used here, so the likelihood graphs look the same in the two

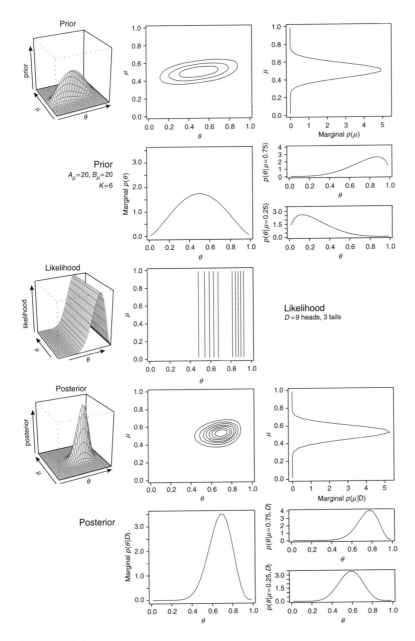

FIGURE 9.3

The prior has high certainty regarding μ but low certainty regarding the dependence of θ on μ. The posterior shows that the distribution of μ has not been altered much by the data (see sideways plots of marginal $p(\mu)$), but the dependence of θ on μ has been altered noticeably (see small plots of $p(\mu \mid \theta)$). Compare with Figure 9.2, which uses the same data but a different prior.

figures. Notice again that the contour lines of the likelihood function are parallel to the μ axis, indicating that μ has no influence on the likelihood. Compare now the prior and the posterior within Figure 9.3. The distribution over μ hardly changes, because it began with high certainty. The distributions $p(\theta \mid \mu, D)$ are different from their priors, however. This is because they began with low certainty, so the data can have a big impact on these distributions. In this case, the data suggest that θ depends on μ in a different way than we initially suspected.

In summary, the data influence (1) our beliefs about the hyperparameter and (2) our beliefs about the dependence of the parameter on the hyperparameter. (The next couple sentences are a challenge to parse, but we hope it is worth the effort.) When the prior on the hyperparameter is uncertain but the prior on the dependence-of-the-parameter-on-the-hyperparameter is highly certain, then the data influence the beliefs about the hyperparameter more than beliefs about the dependence-of-the-parameter-on-the-hyperparameter. This case was illustrated in Figure 9.2. On the other hand, when the prior on the hyperparameter is highly certain but the prior on the dependence-of-the-parameter-on-the-hyperparameter is uncertain, then the data influence the beliefs about the dependence-of-the-parameter-on-the-hyperparameter more than beliefs about the hyperparameter. This case was illustrated in Figure 9.3. In other words, for any aspect of the prior, the more uncertain it is, the more it is affected by the data.

9.2 MULTIPLE COINS FROM A SINGLE MINT

The previous sections considered a scenario in which we flip a *single* coin and make inferences about the bias θ and the hyperparameter μ of the coin. Now we consider an interesting extension: What if we collect data from more than one coin? If each coin has its own distinct bias θ_j, then we are estimating a distinct parameter value for each coin. For now, we assume that all the coins have come from the same mint. This means that we have the same prior belief about μ for all the coins. We also assume that each coin is minted independently of the others. This means that each coin's bias is independent of the others (conditional on μ), in our prior belief distribution.

Okay, you're thinking, I'm willing to entertain these assumptions, but why should I care? What real-world situation (other than minting coins) does this represent? Consider a treatment condition in an experiment, such as administering a certain drug. The drug plays the role of the mint. The subject receiving the treatment plays the role of the coin. The underlying reaction to the drug, by the subject, plays the role of the individual bias. Different subjects will have different individual reactions induced by the drug, but the reactions will be dependent on the overall effect of the drug. We set up the experiment so that subjects don't interact with each other, so we assume that individual biases are independent of each other.[1] We "flip the coin" by measuring Bernoulli responses from the subject. For example, suppose that the drug is supposed to affect memory. We can test memory by giving the subject a list of random words to study and then checking how many words can be recalled several minutes later. Each word is a flip of the coin, and recall or nonrecall corresponds to head or tail.

The scenario is summarized in Figure 9.4. It is much like Figure 9.1, but with one subtle change. Instead of there being a single θ value, there is now a different θ value for each coin, with the bias of the jth coin denoted θ_j. Because the individual flips of the coins come from different coins, the flip results are double-subscripted, such that the ith flip of the jth coin is denoted y_{ji}. Notice that the model involves $J + 1$ parameters, $\theta_1, \ldots, \theta_J$, and μ, all of which are being estimated simultaneously.

Notice what happens to the dependency structure in Figure 9.4 when our beliefs about the hyperparameter μ narrow down to a single value. This can be visualized as the distribution of $p(\mu)$ being an infinitesimally narrow spike over the specific value of μ that we posit. Thus, μ is effectively a *constant*, and $p(\theta_j)$ is a function of two constants, namely, μ and K. In this case, the inclusion of a hyperlevel for μ adds nothing to our inferential process, because we have nothing to infer about μ; we already know the value of μ. Therefore we do not have a hierarchical model any more, instead we have a set of J one-level models of the type we developed in previous chapters. *Cue the choir:* It is only when we have uncertainty about a value that we make inferences about it. When other uncertain values depend on an uncertain value, then we have a chain of dependent uncertainties, formalized by a hierarchial model.

Notice also what happens to the dependency structure in Figure 9.4 when our beliefs about the dependence of θ_j on μ narrow down to an *deterministic* relationship without any uncertainty. For example, suppose that $K \to \infty$, so that when μ has a particular value, then all θ_j's also have exactly that value. In this case, the only uncertainty in θ_j is the uncertainty in μ; there

---

[1] The assumption that biases of coins are independent of each other is actually a bit stronger than the Bayesian mathematics really requires. A weaker assumption goes by the name of *exchangeability* of θ_j values. Exchangeability is a less stringent condition than independence. Exchangeability means that the probability of a sequence of values is the same as the probability of any permutation of that sequence. Independence implies exchangeability, but exchangeability does not imply independence. Be aware that the Bayesian mathematics only requires exchangeability, but the math still works if we make the stronger assumption of independence. Just as we assume each single flip y_j of the jth coin is independently representative of the coin's bias θ_j, we also assume that each individual bias θ_j is independently representative of the mint parameter μ. For further informatiopn about exchangeability, see, for example, Jackman (2009).

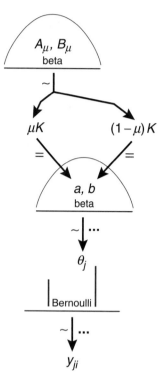

FIGURE 9.4

A model of hierarchical dependencies for data from J coins created independently from the same mint. A datum y_{ji}, from the ith flip of the jth coin, depends on the value of the bias parameter θ_j for the coin. The values of θ_j depend on the value of the hyperparameter μ for the mint that created the coins. The ellipsis on the dependency arrows denotes the repetition of the dependency across flips (for y_{ji}) or coins (for θ_j). The μ parameter has a prior belief distributed as a beta distribution with shape parameters A_μ and B_μ. We simultaneously estimate the $J + 1$ parameters: $\theta_1, \ldots, \theta_J$, and μ. This case is discussed in Section 9.2.

is no additional uncertainty introduced in going from μ to θ_j. In fact, in this case, because $\theta_j = \mu$ and the same μ is being used to model all J coins, we are effectively treating all the coins as if they have the same underlying bias, and the data collapse into simply $\sum_j \sum_{i=1}^{N_j} y_{ji}$ heads out of $\sum_j N_j$ flips. Therefore we do not have a hierarchical model any more, instead we have a one-level model of the type we developed in previous chapters. *All together now:* It is only when we have uncertainty about a value that we make inferences about it. When other uncertain values depend on an uncertain value, then we have a chain of dependent uncertainties, formalized by a hierarchial model.

9.2.1 Posterior via Grid Approximation

As a concrete example, suppose we have two coins from the same mint. We want to estimate the biases θ_1 and θ_2 of the two coins and simultaneously estimate μ of the mint that created them. Figures 9.5 and 9.6 show grid approximations for two different priors.

In Figure 9.5, the prior on μ is gently peaked over $\mu = 0.5$, in the form of a beta($\mu \mid 2, 2$) distribution; that is, $A_\mu = B_\mu = 2$ in the top-level formula of Figure 9.4. The biases of the coins are only weakly dependent on μ according to the prior $p(\theta_j \mid \mu) = \text{beta}(\theta_j \mid \mu \cdot 5, (1 - \mu) \cdot 5)$; that is, $K = 5$ in the middle-level formula of Figure 9.4. The full prior distribution is a joint distribution over three parameters: μ, θ_1, and θ_2. In a grid approximation, the prior is specified as a three-dimensional (3D) array that holds the prior probability at various grid points in the 3D space. The prior probability at point $(\mu, \theta_1, \theta_2)$ is $p(\mu) \, p(\theta_1 \mid \mu) \, p(\theta_2 \mid \mu)$, with exact normalization enforced by summing across the entire grid and dividing by the total.

Because the prior is 3D, it cannot be easily displayed in its entirety. Instead, Figure 9.5 shows various marginal distributions. The top row shows two contour plots, one of the marginal distribution $p(\theta_1, \mu)$, which collapses across θ_2, and the other of the marginal distribution $p(\theta_2, \mu)$, which collapses across θ_1. Also shown is the marginal $p(\mu)$, which collapses across θ_1 and θ_2.

The middle row of Figure 9.5 shows the likelihood function for the data, which comprise 3 heads out of 15 flips of the first coin, and 4 heads out of 5 flips of the second coin. Notice that the contours of the likelihood plot are parallel to the μ axis, indicating that the likelihood does not depend on μ. Notice that the contours are more tightly grouped for the first coin than for the second, which reflects the fact that we have more data from the first coin (i.e., 15 flips versus 5 flips).

The lower two rows of Figure 9.5 show the posterior distribution. Notice that the posterior on θ_1 is centered not far from the proportion $3/15 = 0.2$ in its coin's data, and the posterior on θ_2 is centered not far from the proportion $4/5 = 0.8$ in its coin's data. The posterior on θ_1 has less uncertainty than the posterior on θ_2, as indicated by the density of the contours. Notice also that the posterior on μ has not tightened up much, relative to its prior.

That result should be contrasted with the result in Figure 9.6, which uses the same data with a different prior. In Figure 9.6, the prior on μ is the same gentle peak, but the prior dependency of θ_j on μ is much stronger. The dependency can be seen graphically in the top two panels of Figure 9.6, which show contour plots of the marginals $p(\theta_j, \mu)$. The contours reveal that when θ_j is not close to μ, away from the diagonal, then the probability $p(\theta_j, \mu)$ is small. You can imagine this 3D distribution as spindle shaped or football shaped, tipped diagonally with its ends pointing into opposite corners of a box.

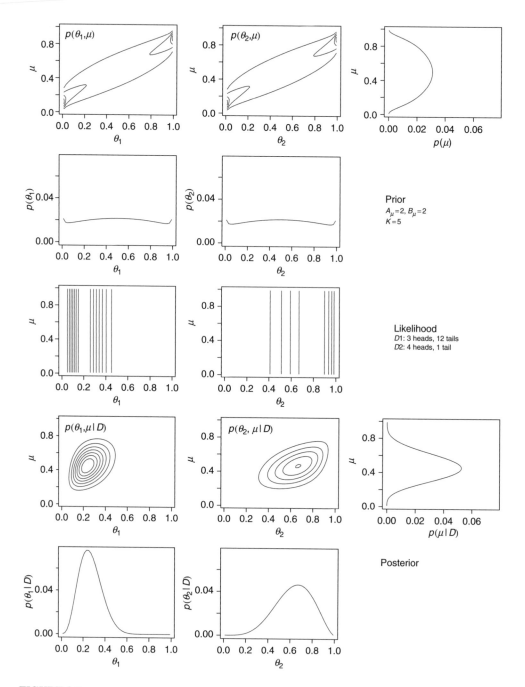

FIGURE 9.5

The prior has only a weak dependency of θ on μ, so the posteriors on θ_1 and θ_2 (bottom row) are weakly influenced by each other's data. Compare with Figure 9.6, which uses the same data but a prior with a strong dependency.

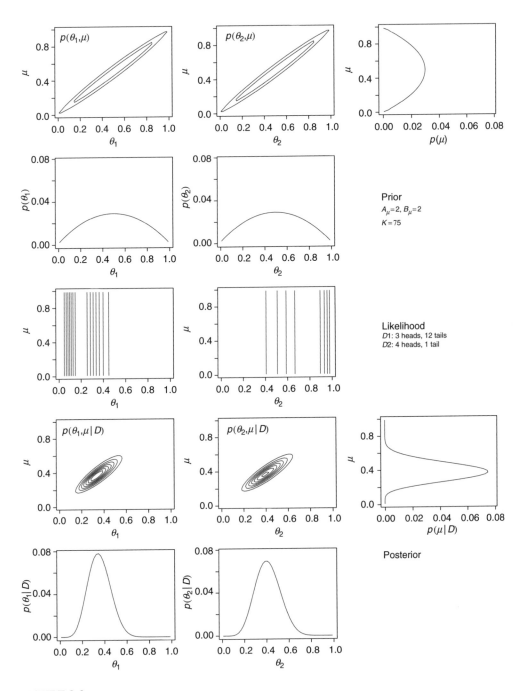

FIGURE 9.6

The prior has a strong dependency of θ on μ, so the posteriors on θ_1 and θ_2 (bottom row) are strongly influenced by each other's data, with θ_2 being pulled toward θ_1 because $N_1 > N_2$. Compare with Figure 9.5, which uses the same data but a prior with a weak dependency.

The plots of the posterior distribution, in the lower rows of Figure 9.6, reveal interesting results. Because the biases and the hyperparameter are being simultaneously estimated, and the biases are strongly dependent on the hyperparameter, the posterior estimates are fairly tightly constrained, especially in comparison with Figure 9.5. Essentially, because the prior emphasizes a relatively narrow spindle within the 3D box, the posterior is restricted to a zone within that spindle. Not only does this cause the posterior to be relatively peaked on all the parameters, it also pulls all the estimates in toward the focal zone. Notice, in particular, that the posterior on θ_2 is peaked around 0.4, far from the proportion $4/5 = 0.8$ in its coin's data! This shift away from the data proportion is caused by the fact that the other coin had a larger sample size, so it has more influence in deciding which part of the prior's spindle is focussed upon.

One of the desirable aspects of using grid approximation to determine the posterior is that we do not rely on any formal analysis of the posterior. Instead, our computer simply keeps track of the values of the prior and likelihood at a large number of grid points and sums over them to determine the denominator of Bayes' rule. Grid approximation can use mathematical formulas for the prior as a convenience for determining the prior values at all those thousands of grid points. What's nice is that we can use, for the prior, any (non-negative) mathematical function we want, without knowing how to formally normalize it, because it will be normalized by the grid approximation. My choice of the priors for this example, summarized in Figure 9.4, was motivated merely by the pedagogical goal of using functions that you are familiar with, not by any formal restriction.

The grid approximation displayed in Figures 9.5 and 9.6 used combs of only 50 points on each parameter (μ, θ_1, and θ_2). This means that the 3D grid had $50^3 = 125,000$ points, which is a size that can be handled easily on an ordinary desktop computer of the early 21st century. It is interesting to remind ourselves that the grid approximation displayed in Figures 9.5 and 9.6 would have been on the edge of computability 50 years ago and would have been impossible 100 years ago.

The number of points in a grid approximation can get rather hefty in a hurry. If we were to expand the example by including a third coin, with its parameter θ_3, then the grid would have $50^4 = 6,250,000$ points, which already strains small computers. Include a fourth coin, and the grid contains more than 312 million points. Grid approximation is not a viable approach to even modestly large problems, which we encounter next.

9.2.2 Posterior via Monte Carlo Sampling

The previous sections have used a simplified model (believe it or not) for the purpose of being able to graphically display the parameter space and gain clear intuitions about how Bayesian inference works. In this section, the first thing we'll do is include one more parameter in the model, to make it more realistic. The previous examples arbitrarily fixed the degree of dependency of θ on μ. The degree of dependency was specified as the value of K, such that when K was large, the individual θ_j values stayed close to μ, but when K was small, the individual θ_j values could spread quite far from μ.

In real situations, we don't know the value of K in advance; instead we let the data inform us regarding its credible values. Intuitively, when the proportions of heads in the different coins are all similar, we have evidence that K is high. But when the proportions of heads in the different coins are diverse, then we have evidence that K is small. Because K will no longer be a constant but will instead be a parameter that we are estimating, we'll call it κ.

Consider the hierarchical model shown in Figure 9.7. This is just like the hierarchy shown in Figure 9.4, except that what was a constant K is now a

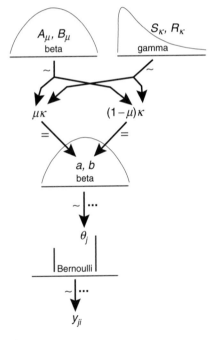

FIGURE 9.7
A model of hierarchical dependencies for data from J coins created independently from the same mint, with the uncertainty of the mint parameterized by μ and κ. The datum y_{ji} from the ith flip of the jth coin depends on the value of the coin's bias parameter θ_j. The values of θ_j depend on the value of the hyperparameters μ and κ for the mint that created the coins. The μ parameter has a prior belief distributed as a beta distribution with shape parameters A_μ and B_μ, whereas the κ parameter has a prior belief distributed as a gamma distribution with shape and rate parameters of S_κ and R_κ. We simultaneously estimate the $J + 2$ parameters, $\theta_1, \ldots, \theta_J$, μ, and κ. This case is discussed in Section 9.2.2.

parameter κ with its own prior distribution. Instead of specifying a single value K for the dependency of θ_j on μ, we allow a distribution of values κ. The prior distribution on κ is expressed mathematically by a *gamma distribution*, which is explained in Figure 9.8. We will be using gamma distributions a lot, just as we've been using beta distributions a lot, so take a look at Figure 9.8 now.

9.2.2.1 Doing It with BUGS

Look again at the hierarchical diagram in Figure 9.7. The arrows in that diagram indicate the dependencies between the variables. Some dependencies are probabilistic; those arrows are labeled with a "~" symbol. Other dependencies are deterministic; those arrows are labeled with a "=" symbol. *The key is that every arrow in the hierarchical diagram has a corresponding statement in the BUGS model specification.* The BUGS model specification is merely a verbal code for the graphical diagram.

Here is a BUGS model specification that corresponds to Figure 9.7 (BernBetaMuKappaBugs.R):

```
11    model {
12        # Likelihood:
13        for ( t in 1:nTrialTotal ) {
14            y[t] ~ dbern( theta[ coin[ t ] ] )
15        }
16        # Prior:
17        for ( j in 1:nCoins ) {
18            theta[j] ~ dbeta( a , b )I(0.0001,0.9999)
19        }
20        a <- mu * kappa
21        b <- ( 1.0 - mu ) * kappa
22        mu ~ dbeta( Amu , Bmu )
23        kappa ~ dgamma( Skappa , Rkappa )
24        Amu <- 2.0
25        Bmu <- 2.0
26        Skappa <- pow(10,2)/pow(10,2)
27        Rkappa <- 10/pow(10,2)
28    }
```

Line 14 corresponds to the arrow to y_{ji} from the Bernoulli distribution. Line 18 corresponds to the arrow to θ_j from the beta distribution.[2] Lines 20 and 21

---

[2] The beta distribution in this BUGS model is specified with *censoring*: theta[j] ~ dbeta(a,b)I (0.0001,0.9999). The suffix "I(lower,upper)" means that the sampled value from the beta distribution must lie between the lower and upper bounds specified. Censoring is only needed because some of the data sets have extreme values that push the parameter values to degenerate extremes. In particular, the data set z = c(1, 1, 1, 1, 5) with N = c(5, 5, 5, 5, 5), used for the example in Figure 9.13, causes BUGS to crash when there is no censoring. Notice that θ_5 in Figure 9.13 loads heavily against 1.0. Unfortunately, there are no certain signatures of when this will be a problem for BUGS, but data in which some $z = N$ or $z = 0$ seem to be especially pernicious.

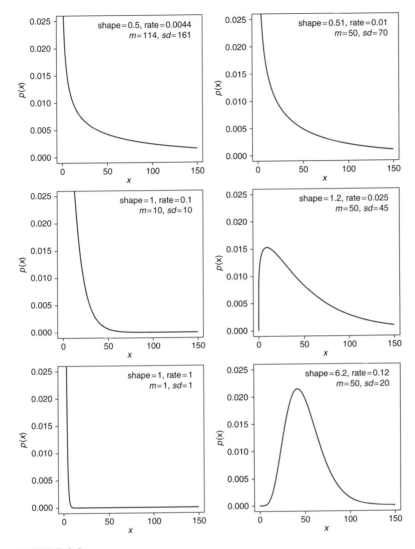

FIGURE 9.8

Examples of the gamma density distribution. The $\text{gamma}(x \mid s, r)$ *distribution* is a probability density for $x \geq 0$, given by $\text{gamma}(x \mid s, r) = \dfrac{r^s}{\Gamma(s)} x^{s-1} e^{-rx}$, where $\Gamma(s)$ is the gamma *function*: $\Gamma(s) = \int_0^\infty dt\, t^{s-1} e^{-t}$. The gamma function is a generalization of the factorial, because for positive integers, $\Gamma(s) = (s-1)!$. In the specification of the distribution, s is called the "shape" parameter and r is called the "rate" (or "inverse scale") parameter. The mean of the gamma distribution is $m = s/r$, and the standard deviation of the gamma distribution is $sd = \sqrt{s}/r$. Hence $s = m^2/sd^2$ and $r = m/sd^2$. In R, the gamma density is provided by dgamma(x,shape=s,rate=r), and the gamma function is provided by gamma(s). Conveniently, BUGS parameterizes the gamma distribution the same way as R, with shape and rate parameters in that order.

correspond to the arrows to *a* and *b* from μ and κ. Finally, lines 22 and 23 correspond to the arrows to μ and κ from their respective beta and gamma distributions.

The remaining lines of the model specification simply specify the values for the constants in the top-level prior distributions, but these constants could be specified directly in the statements of the density distribution. For example, instead of the three statements `mu ~ dbeta( Amu , Bmu )`, `Amu <- 1.0`, and `Bmu <- 1.0`, we could use the single statement `mu ~ dbeta(1.0,1.0)`. The more verbose style is preferred because it explicitly indicates where the hierarchical structure "tops out" at specific constants and because the extra lines more easily allow changing a constant into a stochastic variable if it is meaningful to do so. For example, instead of specifying a constant value, as in `Amu <- 1.0`, we could instead make it a stochastic variable, as in `Amu ~ dgamma(SA,RA)`.

What's amazing is that BUGS figures out how to generate a sample from this model, without requiring us to derive any conditional probabilities or proposal distributions. This leap from model specification to automatic sampling from the posterior deserves highlighting, because it is a huge leap indeed. Going through the effort of deriving a home-grown sampling scheme for each model is arduous and time consuming, even if you do have the mathematical skill to derive it. Then programming its particulars consumes even more effort. Previous drafts of this book included an extensive section that included a mathematical derivation of a hybrid Metropolis-Gibbs sampler for this model. Ultimately its only point was to demonstrate how arduous it was and how easy BUGS is by comparison. Therefore, the derivation has been excluded from this edition. Yeah!

The BUGS model specification used `for` loops that repeat the dependencies for each flip of each coin. The `for` loops implement the ellipsis symbols next to the arrows in Figure 9.7. The loop for the θ_j values, in lines 17 to 19, is an obvious implementation of the diagram: The `for` loop has an index `j` that goes from 1 to the number of coins, `nCoins`. The loop for the y_{ji} values, in lines 13 to 15, uses a technique called *nested indexing*. The index for the loop is denoted `t`, for trial or time. On any given trial, a particular coin is flipped and it has a certain outcome; therefore the data must specify two things for each trial: first, which coin was flipped, and second, what was the outcome. The identity of the coin is specified in a vector called `coin`, and the outcome is specified in a vector called `y`. For example, `coin = c( 1,1,2,2,2 )` and `y = c( 1,0,0,1,1 )` means that the first flip was of coin 1 and it was a head, the second flip was of coin 1 and it was a tail, the third flip was of coin 2 and it was a tail, the fourth flip was of coin 2 and it was a head, and so on. The dependency specified in line 14 of the code indicates that `y[t]` is distributed as a Bernoulli distribution with a θ value for the corresponding coin—that is, `theta[coin[t]]`.

This is called nested indexing because the value of `coin[t]` is used as an index into the vector `theta[]`.

Before showing more complex examples of this BUGS program in action, we will first reproduce the previous examples that were illustrated with grid approximation. This will reassure us that the BUGS sampler is operating correctly, and it will also help us transition from the contour plots of the grid analysis to the scatter plots of MCMC chains.

Please refer back to Figure 9.5, p. 204. In that example, we assumed that the dependency of θ on μ had a fixed value, namely $K = 5$. We will capture that assumption in the present BUGS program by making the prior on κ be a very narrow spike over $\kappa = 5$. This is achieved by setting the mean of its gamma distribution to be 5.0 and the standard deviation of its gamma distribution to be 0.01. The corresponding shape and rate parameter values can be computed as described in the caption for Figure 9.8. The resulting posterior sample is shown in Figure 9.9. Notice that the distribution of κ is extremely narrow: The sampled values are all extremely close to 5.0, as demanded by the spike prior. Now consider the results for the μ and θ_j values, and compare the results with the grid approximation in Figure 9.5, p. 204. Clearly the BUGS-generated sample corresponds closely to the grid approximation.

Figure 9.10, on the other hand, shows what happens when the prior on κ restricts it to values extremely close to 75.0. Compare the result with the grid approximation in Figure 9.6, p. 205. Again we see that the BUGS-generated sample corresponds to the grid approximation.

Now that we are convinced that BUGS is performing properly, we consider some new cases. These continue to be "toy" examples, intended to train our intuition about how Bayesian inference works for this hierarchical prior. But even these toy examples involve too many parameters for grid approximation.

Consider a situation with three coins. We put a prior on κ that is fairly spread out, with a mean of 10.0 and a standard deviation of 10.0, implying shape and rate parameters of 1.0 and 0.1, respectively. Suppose the data from the three coins happen to indicate 5 heads out of 10 flips for every coin. Figure 9.11 shows the resulting posterior distribution. Notice in particular that the posterior mean value of κ is *larger* than its prior mean of 10.0. The estimate of κ has increased from the prior because the results from the three coins are identical. In other words, the lack of variation between coins suggests that the coin biases, θ_j, are strongly dependent on the mint bias μ.

Suppose instead that the results were 1 head out of 10 flips in the first coin, 5 heads out of 10 flips in the second coin, and and 9 heads out of 10 flips in the third coin. Then the results are as shown in Figure 9.12. Notice that the posterior mean value of κ is now *smaller* than its prior mean of 10.0. This estimate

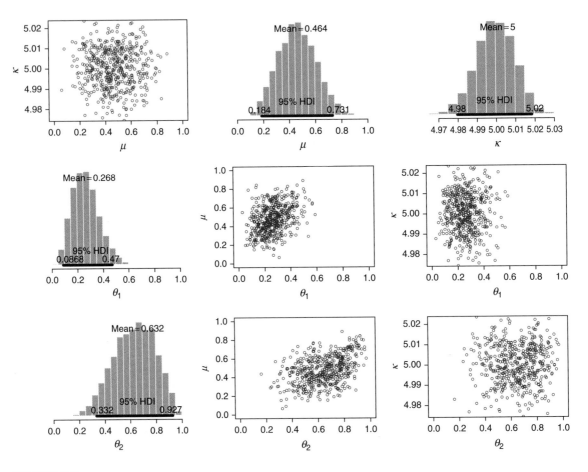

FIGURE 9.9

Posterior sample from BUGS, for model of Figure 9.7, when prior has $\mu \sim \text{beta}(\mu \mid 2, 2)$ and $\kappa \approx 5.0$, and data consist of $N1 = 15$, $z1 = 3$, $N2 = 5$, and $z2 = 4$. Compare with Figure 9.5, p. 204.

of κ has decreased from the prior because the three coins are so different from each other: The huge variation between coins suggests that the coin biases are not strongly dependent on the mint bias.

9.2.3 Outliers and Shrinkage of Individual Estimates

When estimating a bias in an individual coin, the estimate can be affected by the results of the other coins, if there is a belief that the coins depend on a shared mint parameter μ. If many coins yield similar data proportions, then the posterior estimate of the dependence, κ, will tend to be high, and that dependence in turn will tend to yield estimates of the individual biases that more closely resemble the mint parameter μ. This "shrinkage" of individual estimates toward the hyperparameter value is especially evident for outliers.

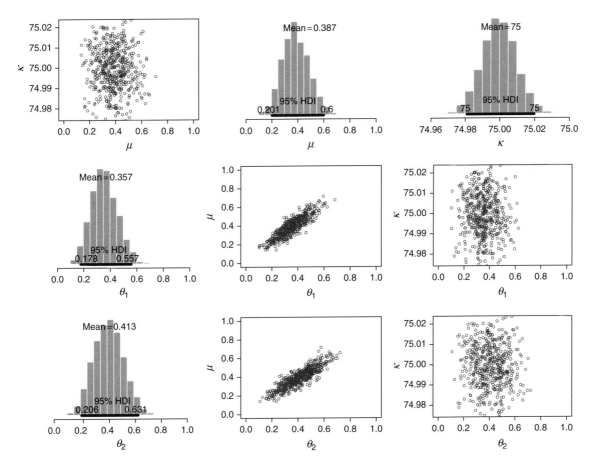

FIGURE 9.10

Posterior sample from BUGS, for model of Figure 9.7, when prior has $\mu \sim \text{beta}(\mu \mid 2, 2)$ and $\kappa \approx 75.0$, and data consist of $N1 = 15$, $z1 = 3$, $N2 = 5$, and $z2 = 4$. Compare with Figure 9.6, p. 205.

Figure 9.13 shows an example with five coins, four of which yield 1/5 heads and one of which yields 5/5 heads. The last coin is an outlier relative to the other four coins. Notice in the top-right panel that the mean of its posterior bias estimate is pulled down quite far from its data proportion of 5/5. The other coins are pulled up from their data proportions of 1/5 relatively less.

The bottom-right panel of Figure 9.13 is a scatter plot of believable combinations of values for θ_5 and κ. The scatter reveals the influence of κ on the estimate of the outlying bias, θ_5. When κ is close to zero, then the believable values of θ_5 tend to be high, close to the actual data proportion. But when κ is larger, then the estimate of θ_5 shrinks toward the typical value of μ.

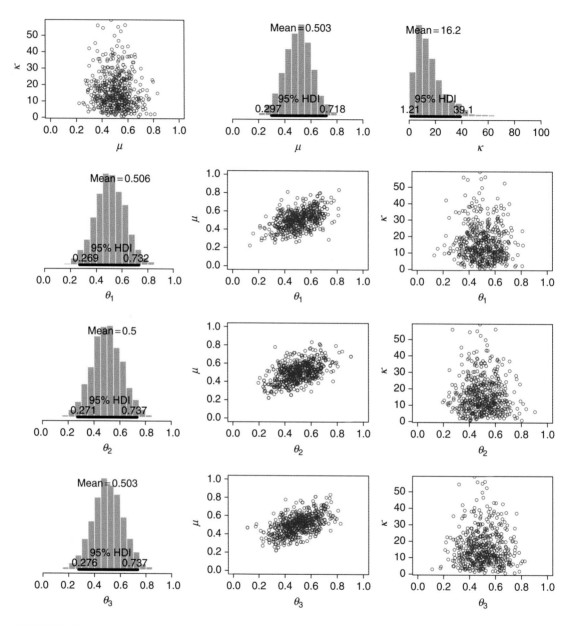

FIGURE 9.11

Posterior sample from BUGS, for model of Figure 9.7, when prior has $\mu \sim \text{beta}(\mu \mid 2, 2)$ and $\kappa \sim \text{gamma}(\kappa \mid 1.0, 0.1)$, and data consist of $N1 = 10$, $z1 = 5$, $N2 = 10$, $z2 = 5$, $N3 = 10$, and $z3 = 5$. The equality of outcome proportions across coins causes the estimate of κ to be high and causes the estimate of μ to be relatively certain. Compare with Figure 9.12.

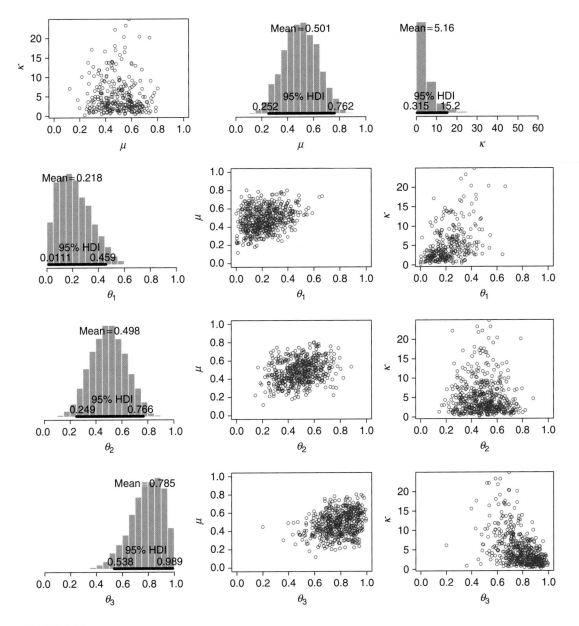

FIGURE 9.12

Posterior sample from BUGS, for model of Figure 9.7, when prior has $\mu \sim \text{beta}(\mu \mid 2, 2)$ and $\kappa \sim \text{gamma}(\kappa \mid 1.0, 0.1)$, and data consist of $N1 = 10$, $z1 = 1$, $N2 = 10$, $z2 = 5$, $N3 = 10$, and $z3 = 9$. The variation of outcome proportions across coins causes the estimate of κ to be low and causes the estimate of μ to be relatively uncertain. Compare with Figure 9.11.

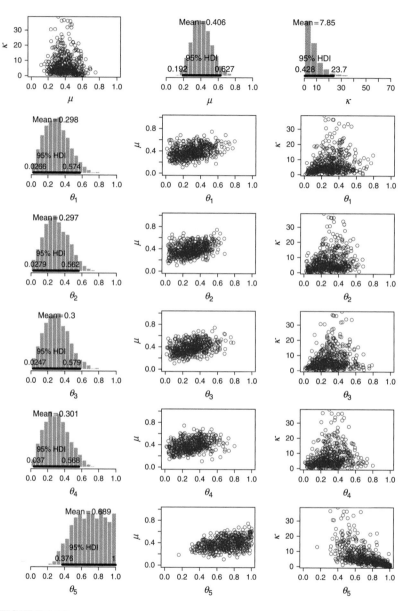

FIGURE 9.13

Posterior sample from BUGS, for model of Figure 9.7, when prior has $\mu \sim \text{beta}(\mu \mid 2, 2)$ and $\kappa \sim \text{gamma}(\kappa \mid 1.0, 0.1)$, and data consist of four coins showing one head in five flips, and one coin showing five heads in five flips. The last coin is an outlier, and the estimate of its bias, θ_5, shows notable shrinkage toward the group average, which is dominated by the mutual consistency of the other coins. The bottom-right panel shows that the shrinkage is stronger when κ is larger. The upper-left panel shows that the estimate of μ is more certain (less variable) when κ is larger.

This shrinkage of the estimates of the individual biases is not a problem with the analysis; in fact, the shrinkage accurately reflects our beliefs: If we believe that all the coins have biases generated by the same mint and that the biases depend on the mint, then the flips from the different coins *should* mutually inform each other's estimated biases. The hierarchical Bayesian analysis is especially nice because it tells us not only the estimates of the biases (θ_j) and the mint parameter (μ), but also the degree of dependence (κ) of the biases on the mint parameter.

9.2.4 Case Study: Therapeutic Touch

Rosa, Sarner, & Barrett (1998) investigated therapeutic touch (TT) among medical practitioners. TT is a technique in which nurses sweep their hands 5 to 10 cm over a patient's body and (claim to) sense depleted or congested areas of the patient's "energy field." The TT practitioners then sweep their hands to "repattern" and smooth the energy field, resulting in healing of the patient. TT has had a notable range of proponents, including some professional organizations (see Rosa, Sarner, & Barrett, 1998, for review).

The crucial prerequisite for TT is sensing of the patient's energy field. Rosa, Sarner, & Barrett (1998) wanted to test practitioners' abilities to achieve this sensing. The testing method was as follows: A practitioner held out her two hands, palms up. The tester put her right hand 5 to 10 cm over one or the other of the practitioner's hands, as determined by the flip of a coin. There was a screen occluding the practitioner's view of her own hands. The practitioner had to guess which of her outstretched hands was being hovered over by the experimenter's hand. Each practitioner was tested for 10 trials.

There were 28 TT practitioners who volunteered to participate (7 were repeated with a several-month separation, so these are counted as distinct subjects). Rosa, Sarner, & Barrett (1998) suggested that the recruitment rate was aided by the fact that the experimenter was a 9-year-old girl (the second author of the article). Results showed that the mean number correct, across practitioners, was 4.39, with the lowest being 1 and the highest being 8. Chance performance is 5 out of 10 correct.

These data are precisely of the form that can be modeled by Figure 9.7. Each practitioner corresponds to a "coin" being flipped 10 times, and the underlying ability of the jth practitioner is denoted θ_j. The practitioners are assumed to be randomly representative of the group of all practitioners, and the group has a mean ability denoted by μ. The dependency of the individual abilities on the group mean is measured by κ.

The prior on the group mean μ was uniform, $\mu \sim \text{beta}(\mu \mid 1, 1)$, thereby putting no prior emphasis on chance. The prior on κ had a mean of 10 and a standard deviation of 10, thereby allowing great variation among practitioners'

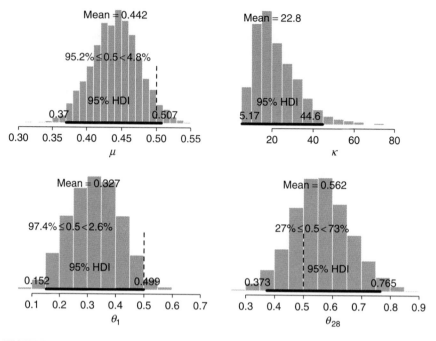

FIGURE 9.14

Posterior distribution for data from Rosa et al. (1998). Prior assumes $\mu \sim \text{beta}(\mu \mid 1, 1)$ and $\kappa \sim \text{gamma}(\kappa \mid 1.0, 0.1)$ (i.e., a prior κ mean of 10.0 and prior κ standard deviation of 10.0).

abilities (by allowing small κ). Note that the prior (and posterior) distribution is a joint distribution on a 30-dimensional space, for parameters $\theta_1, \ldots, \theta_{28}, \mu$, and κ.

Figure 9.14 shows various marginal distributions of the 30-dimensional joint posterior. The posterior on μ (upper-left panel) indicates that the chance value of $\mu = 0.5$ is among the 95% most believable values. The posterior indicates that the most believable group accuracies actually tend to be less than chance correct, meaning that the therapists, if sensing something from the experimenter's hand, were systematically selecting the side *opposite* where the experimenter's hand was.

Figure 9.14 also shows the marginal posterior for κ. The posterior mean κ is larger than the prior mean, indicating that the individual accuracies were more similar to each other than the prior assumed. The two lower panels of Figure 9.14 show the posterior estimates of the individual accuracies for the lowest and highest performing practitioners. In both cases, the 95% HDI spans 0.5, but this is not surprising given that there were only 10 trials per subject. The individual estimates are also affected by shrinkage, of course, as can be discerned from the fact that the 95% HDIs do not even include the proportion-correct actually yielded by the practitioners, which were 0.1 and 0.8, respectively. Although the shrinkage pulls the individual estimates toward the group average, the shrinkage also increases the certainty of the estimate of the group average. In other words, when κ is larger, the distribution of μ gets narrower. This effect on the spread of μ can be seen in the upper-left panel of Figure 9.13, which shows a scatter plot of μ and σ for a similar toy design. Thus, the consistency across subjects, which shrinks individual estimates, also enhances the specificity of the group estimate.

9.2.5 Number of Coins and Flips per Coin

When we collect more data, our estimate of the model parameters becomes more certain. For example, in the previous section's investigation of thera-peutic touch (TT), we know that the sensitivity of the experiment could have been larger if more data were collected. Indeed, if enough data were collected, maybe we could conclude that TT practitioners can, on average, sense the experimenter's hand, albeit only weakly.

The new twist to this old theme is that here we have two ways to get more data: We could include more flips per coin or include more coins. If we have a choice of including more coins or including more flips per coin, which should we choose? If our goal is to estimate the hyperparameters, then the answer is: more coins. For example, suppose you can run an experiment with 250 flips total. You can flip 5 coins 50 times each, or you can flip 50 coins 5 times each. Which should you do? If you want to estimate the minting parameters, go with the 50 coins. The reason is that the larger number of coins puts more constraint on the posterior estimate of μ and κ than the fewer number of coins. The individual coins will have estimates of θ_j that are less specific and more influenced by the other coins, but this is appropriate for the premise of our model: We are presuming that each coin is an independent representative of the *same* mint. Our goal in this case is not so much to estimate the individual coin biases as to estimate the overarching mint parameters. Exercise 9.1 has you explore an example of this issue.

9.3 MULTIPLE COINS FROM MULTIPLE MINTS

9.3.1 Independent Mints

American coins can usually be identified with which mint they came from. For example, recent 1-cent coins minted in Denver have a "D" under the year on their obverse ("head") side, and 1-cent coins minted in Philadelphia have no letter under the year. Suppose we want to estimate the mint parameters μ_m of the different mints. We will assume that the two mints' parameters are inde-pendent of each other, and we will estimate them separately from each other. This amounts to modeling each mint with a structure shown in Figure 9.7 and estimating μ and κ separately for each mint.

This situation arises in real research quite regularly. For example, give each participant a 20-item test, and mark the answers correct or wrong, so each subject has a score in the range 0 to 20. Randomly assign the subjects to one of two conditions for taking the test: noisy environment and quiet environment. We would like to know whether the noise/quiet affected test performance, so we compare the estimates of μ_{noise} and μ_{quiet}. We might also be interested in the values of κ and θ, but our primary interest for this situation is the typical accuracy in the two conditions, as measured by the hyperparamter μ. As another example, suppose we have a new drug for treating some disease. We run a study in which patients at some randomly selected hospitals get the new drug, whereas patients at other randomly selected hospitals get a placebo. At each hospital, we count the number of patients who have recovered. In this case the hospital has the role of the coin, and each patient at the hospital is a "flip" of the coin. We would like to compare μ_{drug} and $\mu_{placebo}$; in other words, we would like to estimate the magnitude of their difference $\mu_1 - \mu_2$.

At this point we'll analyze some real data, as a thorough example of this situation. To understand the experiment, first consider a rectangle with a small vertical segment inside it, like this: . Different instances of this figure have rectangles of different heights, and internal line segments at different lateral positions, like these:

...

In a simple learning experiment, people are shown instances of these rectangles, one at a time, and the people must learn which of two category labels belong to each instance.

For example, it could be that all short rectangles are A's and all tall rectangles are B's. Or it could be that all instances with a left-side line are A's and all instances with right-side lines are B's. These sorts of categorization rules are called *filtration* structures because you can filter out one dimension (line position or rectangle height) but still get the categorization correct. On the other hand, the categorization rule might be: tallest rectangles or rightmost lines are A's, whereas shortest rectangles or leftmost lines are B's. Or tallest rectangles or leftmost lines are A's whereas shortest rectangles or rightmost lines are B's. These sorts of categorization rules are called *condensation* structures because you have to condense information from both dimensions to get the categorization to be correct.

Different theories of learning predict different orderings for the relative difficulty of these category structures. To test the theories, we need to assess how difficult it is to learn the structures and, in particular, estimate differences in difficulties of different category structures.

In an experiment involving these sorts of stimuli and structures, different groups of subjects learned each of the four rules mentioned earlier (Kruschke, 1993). Group 1 learned a filtration structure for which line-segment position was the relevant cue. Group 2 learned a filtration structure for which rectangle height was the relevant cue. Groups 3 and 4 learned condensation structures of the two types mentioned in the previous paragraphs. All four groups saw the same stimuli; the only difference was the category labeling of the stimuli. Each learning trial consisted of the presentation of the figure, the subject guessing the correct label, and then being shown the correct label for that figure. There were 64 learning trials. On each trial, the subject's response could be correct or wrong.

The abstract structure of this experiment can be mapped onto the sort of hierarchical dependencies we've been considering in this chapter. Figure 9.15

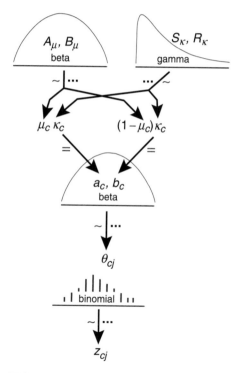

FIGURE 9.15

Hierarchical diagram for model of data from the filtration-condensation experiment, with program in Section 9.5.2 (`FilconBrugs.R`). This diagram is much like Figure 9.7, except for two changes. First, the distribution at the bottom of the hierarchy is binomial instead of Bernoulli, because the data are sums across trials instead of individual trials. Second, the estimated parameters all have a subscript c that denotes the condition from which the data were drawn. In other words, this structure is copied separately for each condition.

shows the dependencies as a hierarchical diagram. We conceive of each trial's response as the flip of a coin, with correct $= 1$ and wrong $= 0$. Across the $N = 64$ trials, the subject gets z of them correct. Each individual subject is a coin with a certain learned bias θ to get the answer correct. (We are *not* modeling the *change* of bias across trials; we are simplifying and modeling each subject as a stationary propensity to be correct, a propensity that has been influenced by the structure of the category labels.) An individual's propensity is a random sample from the group's distribution of propensities, modeled as a beta distribution. The group's typical accuracy is denoted by μ, and the tightness of individual propensities around that group average is denoted by κ. The idea is that the difficulty of the category structure determines the group-average accuracy μ, and each individual is some random variation from that group average. The amount by which individuals tend to deviate from the group average is governed by κ. When κ is large, then the individuals tend to be close to the group average, and when κ is small, then individuals can be spread far from the group average.

Each category structure has its own average difficulty and spread of individual scores around that average. The different groups are denoted by the subscript c. The model assumes that there is no higher level in the structure to express relations across conditions. In other words, the data from one condition have no influence on the data from another condition. This is merely an assumption of the model, which could be changed if it were theoretically meaningful to do so. Our main interest is estimating differences in μ_c across conditions.

The data from this experiment are structured by subject: Each subject was assigned to a particular condition, had a particular number of training trials, and got a particular number of those trials correct. This way of organizing the data is very general, because it allows there to be a different number of subjects in different conditions and allows there to be a different number of training trials for each subject. The filtration/condensation experiment happened to use a balanced design wherein all subjects had the same number of trials and all conditions had the same number of subjects, but it is not necessary for experiments to be balanced in that way. The by-subject structuring of data allows for easy application to unbalanced experiment designs. The data are specified as three vectors: cond[i] specifies the condition, 1 to 4, of the ith subject, N[i] specifies the number of training trials of the ith subject, and z[i] specifies the number of correct responses of the ith subject.

The BUGS model specification has a line for each arrow in the hierarchical diagram of Figure 9.15, as follows (FilconBrugs.R):

```
11   model {
12     for ( subjIdx in 1:nSubj ) {
13       # Likelihood:
14       z[subjIdx] ~ dbin( theta[subjIdx] , N[subjIdx] )
15       # Prior on theta: Notice nested indexing.
```

```
16          theta[subjIdx] ~ dbeta( a[cond[subjIdx]] , b[cond[subjIdx]] )I(0.001,0.999)
17      }
18      for ( condIdx in 1:nCond ) {
19          a[condIdx] <- mu[condIdx] * kappa[condIdx]
20          b[condIdx] <- (1-mu[condIdx]) * kappa[condIdx]
21          # Hyperprior on mu and kappa:
22          mu[condIdx] ~ dbeta( Amu , Bmu )
23          kappa[condIdx] ~ dgamma( Skappa , Rkappa )
24      }
25      # Constants for hyperprior:
26      Amu <- 1
27      Bmu <- 1
28      Skappa <- pow(meanGamma,2)/pow(sdGamma,2)
29      Rkappa <- meanGamma/pow(sdGamma,2)
30      meanGamma <- 10
31      sdGamma <- 10
32  }
```

The model specification uses nested indexing (as we've seen in a previous section) in line 16, which says that the underlying propensity θ for an individual subject is distributed as a beta distribution that has shape parameters a[cond[subjIdx]] and b[cond[subjIdx]] specific to the condition of the that subject. (The beta density is censored with the I(,) suffix to prevent degeneracy problems when the data are extreme; see footnote 2, p. 208.) The complete program is listed in Section 9.5.2 (FilconBrugs.R).

The BUGS model uses a binomial likelihood distribution for total correct, instead of using the Bernoulli distribution for individual trials. (The binomial distribution was mentioned in footnote 1, p. 79, and is explained on p. 267. Although BUGS specifies the binomial density as dbin, R specifies it as dbinom.) This use of the binomial is just a convenience for shortening the program. If the data were specified as trial-by-trial outcomes instead of as total correct, then the model could include a trial-by-trial loop and use a Bernoulli likelihood function.

When the program is run, the result is a large sample of jointly believable parameter values. Notice how many parameter values have been estimated. In every condition, there are 40 θ values for the individual learners, plus a μ value expressing the group average, plus a κ value expressing how tightly the individuals cluster around the group average. Therefore, across the four conditions, the program has estimated 168 parameters! These parameters exist in a joint space of 168 dimensions. Because of this particular model structure, we know that each condition is independent of the others (so the space can be thought of as four distinct 42-dimensional spaces), but in general all the parameters exist in a conjoint space. In particular, believable values of μ_c and κ_c are not necessarily independent of each other, and believable values of θ_{cj} are not independent of the values of κ_c, etc.

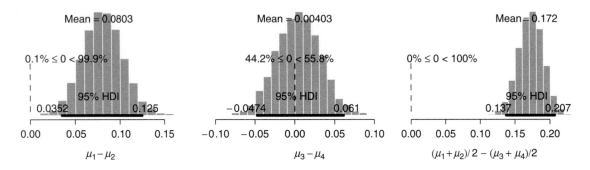

FIGURE 9.16
Histograms of differences in values of the μ parameters, generated by the script FilconBrugs.R (p. 232).

Because the parameter space is so big, we only examine perspectives of the posterior distribution that are theoretically meaningful to consider. Our main interest in this experiment falls at the group level: How big is the difference between groups? We address this question by examining differences in believable μ_c values. Figure 9.16 shows histograms of meaningful differences of μ_c values. The left histogram shows the difference between μ_1, for filtration with line-segment relevant, and μ_2, for filtration with height relevant. The mean difference of posterior mu values is 0.0803, and the 95% HDI goes from 0.0352 to 0.125. Because a difference of zero is not among the 95% most credible values, but is in the tail of credibility, we have a moderately strong belief that the two filtration conditions have different difficulty. The middle histogram shows that there is no credible difference between the two condensation conditions. The right histogram shows that there is overwhelming strength in the belief that filtration, on average, is easier than condensation, on average: *all* of the posterior is well above zero.

9.3.2 Dependent Mints

In some experiment designs, we might assume that the different treatment conditions are totally independent of each other. In the context of minting coins, this is assuming that the parameter values of one mint are completely disconnected from the parameter values of another mint. This assumption of total disconnection was used to analyze the data from the filtration-condensation experiment in the previous section (e.g., Figure 9.16).

On the other hand, we could treat the mint parameters as being mutually informative, perhaps because the same governmental agency oversees the creation of all the mints. Therefore, we can establish a prior distribution of beliefs regarding possible values of the mint parameters, with the mth mint's μ_m and κ_m parameters being representative of the overarching distribution of governmentally constrained mint parameters.

To make these ideas more concrete, consider again the filtration-condensation experiment. In the previous section's analysis, we estimated each condition's parameters completely separately from the other conditions. But this assumption of complete informational insulation between conditions might be losing some useful information. In particular, consider the κ_c parameters, which indicate how tightly the individual subjects' accuracies cluster around the group average μ_c. If individual variation from the group mean is caused only by random influences that are not affected by the group's treatment (i.e., category structure), then the variation seen in each group should be informative for the estimate of κ_c in other groups.

We could express this mutuality of κ_c in different ways. One way is to assume that all conditions have the *same* κ value. The left side of Figure 9.17 shows a hierarchical diagram for this assumption. The diagram is only subtly different from the one in Figure 9.15: Occurrences of κ have no subscript c, and the dependency arrow pointing at κ has no ellipsis. Again, what this means conceptually is that the random spread of individual accuracies around the group mean accuracy is the same for all groups. In other words, the category structure influences the mean accuracy for the group, but the category structure does not influence the random variation of individuals around that mean accuracy.

So far, we have considered two extreme possibilities: Either the κ_c's are completely insulated from each other, or they are identical. There is a middle ground. We could have prior beliefs that the κ_c's might be influenced by the category structures, but the degree of that influence is something we estimate from the data. We conceive of the the κ_c values as coming from an overarching distribution, and the dispersion of that distribution is estimated by considering how much the group variances actually differ from each other. The right side of Figure 9.17 shows a hierarchical diagram for this assumption. Notice that the gamma distribution for the κ_c values no longer has its shape and rate parameters as preset constants. Instead, the shape of the gamma distribution is estimated from the data. The diagram shows that the shape and rate parameters of the gamma distribution are first reexpressed in terms of the mean μ_γ and the standard deviation σ_γ. The mean and standard deviation, in turn, are given uniform prior distributions over a suitable (positive) range. Although this diagram may seem confusing at first, the idea is actually simple: If the data indicate that individual variation within each group is nearly identical across groups, then the estimated κ_c values should be near each other and the distribution of κ_c values should be narrow. The narrowness of the distribution of κ_c values is captured by the estimated standard deviation σ_γ. In particular, the overarching distribution of κ_c values will provide shrinkage of their estimates to the extent that the data suggest it. For example, if three of the four conditions have similar within-group variation but the fourth condition has a somewhat outlying within-group variation, then the estimate of the fourth

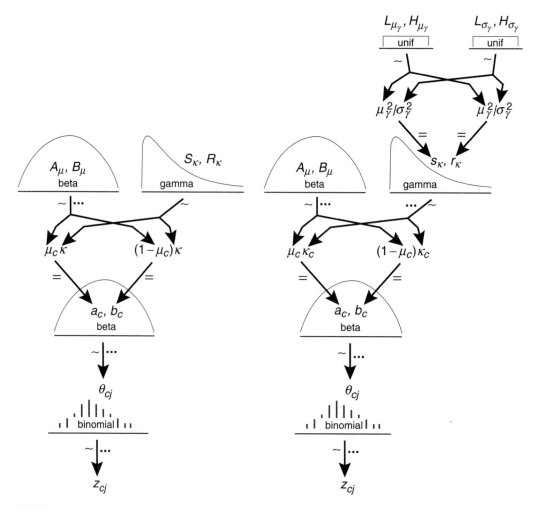

FIGURE 9.17

Left: A hierarchical diagram for a model that constrains the κ of every condition to be the same value. Notice that there is no subscript c on κ and no ellipsis on the dependency arrow leading to κ. *Right*: A hierarchical diagram for a model that constrains the κ_c values to come from the same overarching distribution, namely, gamma($\kappa_c \mid s_\kappa, r_\kappa$), but the parameters of the overarching distribution are estimated by considering data from all the conditions. BUGS code for this model is presented on p. 236. Compare with Figure 9.15, for which the κ_c values come from the same overarching distribution, but the parameters of which are fixed and uninfluenced by the data from the other conditions.

group's κ_c will be pulled toward the overall average within-group variation. On the other hand, if the groups have very different within-group variations, then only large values of σ_γ will be credible in the posterior.

Because of the simplicity of the BRugs software, it is easy to program this sort of hierarchy and generate samples from the posterior distribution. In Exercise 9.2

you will modify the BUGS program for the filtration-condensation analysis and implement these variations of assumptions about κ_c. Importantly, you will examine the indirect effects on estimates of μ_c, which are of primary applied interest for the research. (Analogous higher-level estimation will be simultaneously applied to the μ_c parameters in Exercise 13.4, p. 349). Another benefit of implementing the more elaborate hierarchial model is that we see a clear case in which extensive burn-in and thinning is required. In Exercise 9.3, you will see this graphically.

9.3.3 Individual Differences and Meta-Analysis

In experiments with human participants, one of the striking results is vast variation between people. Seat 10 people at a simple response-time task, and you will get 10 different mean response times. These various results across people are referred to as *individual differences*. We could estimate parameters for each individual completely insulated from estimates of other individuals. Alternatively, we could believe that the individual results are all taken from a common overarching distribution, because all the individuals were of the same species. We could specify the dependence of individual performance parameters on the overarching distribution, and then our estimates of the individual parameters would be influenced by results from other individuals, via the codependence on the hyperparameter. We've seen several examples of this mutual influence via higher-level dependencies, such as so-called shrinkage.

So far in this book we've been focused on binary-scaled data (e.g., heads versus tails). For an example of modeling individual differences in this vein but when data are measured on a continuous scale, see Rouder & Lu (2005). For an interesting approach that uses an infinite dimensional parameter space to entertain many possible groupings of individuals, see Navarro et al. (2006). For a different approach that uses Bayesian model comparison to select groupings for individuals, see Lee & Webb (2005). The next chapter of this textbook explains how model comparison can be thought of as hierarchical modeling.

Hierarchical models can also be used for *meta-analysis*. The idea is that different replications of an experiment are independent representatives of an overarching distribution of effect magnitudes. Each experiment's data informs a posterior distribution regarding its own first-level parameters, but the other experiments' data also influence the estimates because of their dependency on the overarching parameter. The estimate of the overarching parameter indicates the result of meta-analysis. See Section 5.6 of Gelman et al. (2004) for an example of Bayesian meta-analysis, but using normal priors on continuous data instead of beta priors on binomial data as we've been using here.

9.4 SUMMARY

This chapter introduced the notion of hierarchical dependency using a simple two-parameter model, shown in Figures 9.2 and 9.3. This simple example showed quite graphically the notion of parameterized dependency: One of the parameters explicitly described the degree of dependency of the data on the other parameter. From that conceptual foundation, the chapter introduced MCMC approximations, applying them to the simple cases first so that you can see that the MCMC methods do in fact yield the same results as the grid-based methods.

The power and scope of MCMC methods applied to hierarchical models was demonstrated in the analysis of data from the filtration-condensation experiment. In those analyses, a parameter was estimated for every subject, plus two parameters for every condition, plus in some cases other overarching parameters. Despite the number of parameters, MCMC methods yield rapid and stable results.

9.5 R CODE

9.5.1 Code for Analysis of Therapeutic-Touch Experiment

This program was used to generate Figures 9.13, 9.14, and the like.

(BernBetaMuKappaBugs.R)

```
1    graphics.off()
2    rm(list=ls(all=TRUE))
3    library(BRugs)              # Kruschke, J. K. (2010). Doing Bayesian data analysis:
4                                # A Tutorial with R and BUGS. Academic Press / Elsevier.
5    #-------------------------------------------------------------------------------
6    # THE MODEL.
7
8    # Specify the model in BUGS language, but save it as a string in R:
9    modelString = "
10   # BUGS model specification begins ...
11   model {
12       # Likelihood:
13       for ( t in 1:nTrialTotal ) {
14           y[t] ~ dbern( theta[ coin[ t ] ] )
15       }
16       # Prior:
17       for ( j in 1:nCoins ) {
18           theta[j] ~ dbeta( a , b )I(0.0001,0.9999)
19       }
20       a <- mu * kappa
21       b <- ( 1.0 - mu ) * kappa
22       mu ~ dbeta( Amu , Bmu )
23       kappa ~ dgamma( Skappa , Rkappa )
```

```
24       Amu <- 2.0
25       Bmu <- 2.0
26       Skappa <- pow(10,2)/pow(10,2)
27       Rkappa <- 10/pow(10,2)
28   }
29   # ... BUGS model specification ends.
30   " # close quote to end modelString
31
32   # Write the modelString to a file, using R commands:
33   .temp = file("model.txt","w") ; writeLines(modelString,con=.temp) ; close(.temp)
34   # Use BRugs to send the model.txt file to BUGS, which checks the model syntax:
35   modelCheck( "model.txt" )
36
37   #------------------------------------------------------------------------------
38   # THE DATA.
39
40   # Demo data for various figures in the book:
41    N = c( 5, 5, 5, 5, 5 ) # c( 10, 10, 10 )  # c( 15, 5 ) # c( 5, 5, 5, 5, 5 )
42    z = c( 1, 1, 1, 1, 5 ) # c(  1,  5,  9 )  # c(  3, 4 ) # c( 1, 1, 1, 1, 5 )
43   # Therapeutic touch data:
44   #   z = c(1,2,3,3,3,3,3,3,3,3,4,4,4,4,4,5,5,5,5,5,5,5,6,6,7,7,7,8)
45   #   N = rep(10,length(z))
46   # Convert z,N to vectors of individual flips.
47   # coin vector is index of coin for each flip.
48   # y vector is head or tail for each flip.
49   # For example,
50   #   coin = c( 1, 1, 2, 2, 2 )
51   #   y    = c( 1, 0, 0, 0, 1 )
52   # means that the first flip was of coin 1 and it was a head, the second flip
53   # was of coin 1 and it was a tail, the third flip was of coin 2 and it was a
54   # tail, etc.
55   coin = NULL ; y = NULL
56   for ( coinIdx in 1:length(N) ) {
57       coin = c( coin , rep(coinIdx,N[coinIdx]) )
58       y = c( y , rep(1,z[coinIdx]) , rep(0,N[coinIdx]-z[coinIdx]) )
59   }
60   nTrialTotal = length( y )
61   nCoins = length( unique( coin ) )
62   dataList = list(
63       y = y ,
64       coin = coin ,
65       nTrialTotal = nTrialTotal ,
66       nCoins = nCoins
67   )
68
69   # Use BRugs commands to put the data into a file and ship the file to BUGS:
70   modelData( bugsData( dataList ) )
71
72   #------------------------------------------------------------------------------
73   # INTIALIZE THE CHAIN.
74
75   nChains = 3
76   modelCompile( numChains = nChains )  # BRugs tells BUGS to compile the model.
```

```
77   modelGenInits() # BRugs tells BUGS to randomly initialize the chains.
78
79   #-------------------------------------------------------------------------------
80   # RUN THE CHAINS.
81
82   # Run some initial steps without recording them, to burn-in the chains:
83   burninSteps = 1000
84   modelUpdate( burninSteps )
85   # BRugs tells BUGS to keep a record of the sampled values:
86   samplesSet( c( "mu" , "kappa" , "theta" ) )
87   nPerChain = 1000
88   modelUpdate( nPerChain , thin=10 )
89
90   #-------------------------------------------------------------------------------
91   # EXAMINE THE RESULTS.
92
93   # Check for mixing and autocorrelation:
94   source("plotChains.R")
95   plotChains( "mu" , saveplots=F )
96   plotChains( "kappa" , saveplots=F )
97   plotChains( "theta[1]" , saveplots=F )
98
99   # Extract the posterior sample from BUGS:
100  muSample = samplesSample( "mu" ) # BRugs gets sample from BUGS
101  kappaSample = samplesSample( "kappa" ) # BRugs gets sample from BUGS
102  thetaSample = matrix( 0 , nrow=nCoins , ncol=nChains*nPerChain )
103  for ( coinIdx in 1:nCoins ) {
104      nodeName = paste( "theta[" , coinIdx , "]" , sep="" )
105      thetaSample[coinIdx,] = samplesSample( nodeName )
106  }
107
108  # Make a graph using R commands:
109  source("plotPost.R")
110  if ( nCoins <= 5 ) { # Only make this figure if there are not too many coins
111  windows(3.2*3,2.5*(1+nCoins))
112  layout( matrix( 1:(3*(nCoins+1)) , nrow=(nCoins+1) , byrow=T ) )
113  par(mar=c(2.95,2.95,1.0,0),mgp=c(1.35,0.35,0),oma=c( 0.1, 0.1, 0.1, 0.1))
114  nPtsToPlot = 500
115  plotIdx = floor(seq(1,length(muSample),length=nPtsToPlot))
116  kPltLim = signif( quantile( kappaSample , p=c(.01,.99) ) , 4 )
117  plot( muSample[plotIdx] , kappaSample[plotIdx] , type="p" , ylim=kPltLim ,
118          xlim=c(0,1) , xlab=expression(mu) , ylab=expression(kappa) , cex.lab=1.5 )
119  plotPost( muSample , xlab="mu" , xlim=c(0,1) , main="" , breaks=20 )
120  plotPost( kappaSample , xlab="kappa" , main="" , breaks=20 , HDItextPlace=.6 )
121  for ( coinIdx in 1:nCoins ) {
122      plotPost( thetaSample[coinIdx,] , xlab=paste("theta",coinIdx,sep="") ,
123              xlim=c(0,1) , main="" , breaks=20 , HDItextPlace=.3 )
124      plot( thetaSample[coinIdx,plotIdx] , muSample[plotIdx] , type="p" ,
125              xlim=c(0,1) , ylim=c(0,1) , cex.lab=1.5 ,
126              xlab=bquote(theta[.(coinIdx)]) , ylab=expression(mu) )
127      plot( thetaSample[coinIdx,plotIdx] , kappaSample[plotIdx] , type="p" ,
128              xlim=c(0,1) , ylim=kPltLim , cex.lab=1.5 ,
129              xlab=bquote(theta[.(coinIdx)]) , ylab=expression(kappa) )
```

```
130    }
131    dev.copy2eps(file=paste("BernBetaMuKappaBugs",paste(z,collapse=""),".eps",sep=""))
132    } # end if ( nCoins <= ...
133
134    ## Uncomment the following if using therapeutic touch data:
135    #windows(7,5)
136    #layout( matrix( 1:4 , nrow=2 , byrow=T ) )
137    #par(mar=c(2.95,2.95,1.0,0),mgp=c(1.35,0.35,0),oma=c( 0.1, 0.1, 0.1, 0.1) )
138    #plotPost( muSample , xlab="mu" , main="" , breaks=20 , compVal=0.5 )
139    #plotPost( kappaSample , xlab="kappa" , main="" , breaks=20 , HDItextPlace=.1 )
140    #plotPost( thetaSample[1,] , xlab="theta1" , main="" , breaks=20 , compVal=0.5 )
141    #plotPost( thetaSample[28,] , xlab="theta28" , main="" , breaks=20 , compVal=0.5 )
142    #dev.copy2eps(file="BernBetaMuKappaBugsTT.eps")
```

9.5.2 Code for Analysis of Filtration-Condensation Experiment

The beta density on line 16 is suffixed with "I(0.001,0.999)," which means that the sampled value is censored at 0.001 below and 0.999 above. This is necessary because the sampled theta value is used as an argument in dbin on line 14, and when the theta value gets too extreme, the dbin function crashes on some data points. This is especially a problem for condition 1 of the experiment, in which the accuracies are very high and the sampled theta values also tend to be close to 1.0. This issue was first mentioned in footnote 2, p. 208.

An important method for chain initialization is used here. Instead of letting BUGS randomly generate initial values from the prior, the chains are started "intelligently" so that they are likely to be in the midst of the posterior. This sort of initialization reduces burn-in time, but, more important, it assures that BUGS will run successfully instead of balking at parameter values that are too far afield from the true posterior. The chain initialization uses the data to determine reasonable parameter values that may be in the vicinity of posterior credible values. For example, the mu values are initialized at the means of each group. The group means are computed by using the aggregate(dataVec , list(groupVec) , mean) command in R, which cleverly extracts the data in dataVec according to the group codes in groupVec and applies the mean to each set of extracted data. The initial kappa values are computed via Equation 5.6 from the standard deviation of each group. Once these reasonable values are computed, they are sent to BUGS using the genInitList function that is defined on lines 77 to 89. *The genInitList function returns a list of the named variables and their initial values. The function is called with the BRugs commands* modelInits(bugsInits(genInitList)). This method for initialization will be used frequently for complex models.

(FilconBrugs.R)

```
1   graphics.off()
2   rm(list=ls(all=TRUE))
3   fileNameRoot="FilconBrugs" # for constructing output filenames
4   library(BRugs)              # Kruschke, J. K. (2010). Doing Bayesian data analysis:
5                               # A Tutorial with R and BUGS. Academic Press / Elsevier.
6   #------------------------------------------------------------------------
7   # THE MODEL.
8
9   modelstring = "
10  # BUGS model specification begins here...
11  model {
12     for ( subjIdx in 1:nSubj ) {
13        # Likelihood:
14        z[subjIdx] ~ dbin( theta[subjIdx] , N[subjIdx] )
15        # Prior on theta: Notice nested indexing.
16        theta[subjIdx] ~ dbeta( a[cond[subjIdx]] , b[cond[subjIdx]] )I(0.001,0.999)
17     }
18     for ( condIdx in 1:nCond ) {
19        a[condIdx] <- mu[condIdx] * kappa[condIdx]
20        b[condIdx] <- (1-mu[condIdx]) * kappa[condIdx]
21        # Hyperprior on mu and kappa:
22        mu[condIdx] ~ dbeta( Amu , Bmu )
23        kappa[condIdx] ~ dgamma( Skappa , Rkappa )
24     }
25     # Constants for hyperprior:
26     Amu <- 1
27     Bmu <- 1
28     Skappa <- pow(meanGamma,2)/pow(sdGamma,2)
29     Rkappa <- meanGamma/pow(sdGamma,2)
30     meanGamma <- 10
31     sdGamma <- 10
32  }
33  # ... end BUGS model specification
34  " # close quote for modelstring
35  # Write model to a file:
36  writeLines(modelstring,con="model.txt")
37  # Load model file into BRugs and check its syntax:
38  modelCheck( "model.txt" )
39
40  #------------------------------------------------------------------------
41  # THE DATA.
42
43  # For each subject, specify the condition s/he was in,
44  # the number of trials s/he experienced, and the number correct.
45  # (These lines intentionally exceed the margins so that they don't take up
46  # excessive space on the printed page.)
47  cond = c(1,1,1,1,1,1,1,1,1,1,1,1,1,1,1,1,1,1,1,1,1,1,1,1,1,1,1,1,1,1,1,1,1,1,1,1,1,1,
        1,1,1,1,2,2,2,2,2,2,2,2,2,2,2,2,2,2,2,2,2,2,2,2,2,2,2,2,2,2,2,2,2,2,2,2,2,2,2,2,
        2,2,2,2,2,2,2,3,3,3,3,3,3,3,3,3,3,3,3,3,3,3,3,3,3,3,3,3,3,3,3,3,3,3,3,3,3,3,3,3,
        3,3,3,3,3,3,3,4,4,4,4,4,4,4,4,4,4,4,4,4,4,4,4,4,4,4,4,4,4,4,4,4,4,4,4,4,4,4,4,4,
        4,4,4,4,4,4,4,4,4,4)
```

```
48  N = c(64,64,64,64,64,64,64,64,64,64,64,64,64,64,64,64,64,64,64,64,64,64,64,64,
        64,64,64,64,64,64,64,64,64,64,64,64,64,64,64,64,64,64,64,64,64,64,64,64,64,
        64,64,64,64,64,64,64,64,64,64,64,64,64,64,64,64,64,64,64,64,64,64,64,64,64,
        64,64,64,64,64,64,64,64,64,64,64,64,64,64,64,64,64,64,64,64,64,64,64,64,64,
        64,64,64,64,64,64,64,64,64,64,64,64,64,64,64,64,64,64,64,64,64,64,64,64,64,
        64,64,64,64,64,64,64,64,64,64,64,64,64,64,64,64,64,64,64,64,64,64,64,64,64,
        64,64,64,64,64)
49  z = c(45,63,58,64,58,63,51,60,59,47,63,61,60,51,59,45,61,59,60,58,63,56,63,64,64,
        60,64,62,49,64,64,58,64,52,64,64,64,62,64,61,59,59,55,62,51,58,55,54,59,57,58,
        60,54,42,59,57,59,53,53,42,59,57,29,36,51,64,60,54,54,38,61,60,61,60,62,55,38,
        43,58,60,44,44,32,56,43,36,38,48,32,40,40,34,45,42,41,32,48,36,29,37,53,55,50,
        47,46,44,50,56,58,42,58,54,57,54,51,49,52,51,49,51,46,46,42,49,46,56,42,53,55,
        51,55,49,53,55,40,46,56,47,54,54,42,34,35,41,48,46,39,55,30,49,27,51,41,36,45,
        41,53,32,43,33)
50  nSubj = length(cond)
51  nCond = length(unique(cond))
52
53  # Specify the data in a form that is compatible with BRugs model, as a list:
54  datalist = list(
55   nCond = nCond ,
56   nSubj = nSubj ,
57   cond = cond ,
58   N = N ,
59   z = z
60  )
61
62  # Get the data into BRugs:
63  # Function bugsData stores the data file (default filename is data.txt).
64  # Function modelData loads data file into BRugs (default filename is data.txt).
65  modelData( bugsData( datalist ) )
66
67  #-------------------------------------------------------------------------------
68  # INTIALIZE THE CHAINS.
69
70  nChain = 3
71  modelCompile( numChains=nChain )
72
73  if ( F ) {
74    modelGenInits() # often won't work for diffuse prior
75  } else {
76   # initialization based on data
77   genInitList <- function() {
78     sqzData = .01+.98*datalist$z/datalist$N
79     mu = aggregate( sqzData , list(datalist$cond) , "mean" )[,"x"]
80     sd = aggregate( sqzData , list(datalist$cond) , "sd" )[,"x"]
81     kappa = mu*(1-mu)/sd^2 - 1
82     return(
83       list(
84         theta = sqzData ,
85         mu = mu ,
86         kappa = kappa
87       )
88     )
```

```
89        }
90      for ( chainIdx in 1 : nChain ) {
91        modelInits( bugsInits( genInitList ) )
92      }
93    }
94
95    #------------------------------------------------------------------------------
96    # RUN THE CHAINS.
97
98    burninSteps = 2000
99    modelUpdate( burninSteps )
100   samplesSet( c("mu","kappa","theta","a","b") )
101   nPerChain = ceiling(5000/nChain)
102   modelUpdate( nPerChain , thin=10 )
103
104   #------------------------------------------------------------------------------
105   # EXAMINE THE RESULTS.
106
107   # Check for convergence, mixing and autocorrelation:
108   source("plotChains.R")
109   sumInfo = plotChains( "mu" , saveplots=T , fileNameRoot )
110   sumInfo = plotChains( "kappa" , saveplots=F )
111   sumInfo = plotChains( "theta[1]" , saveplots=F )
112
113   # Extract parameter values and save them.
114   mu = NULL
115   kappa = NULL
116   for ( condIdx in 1:nCond ) {
117     mu = rbind( mu , samplesSample( paste("mu[",condIdx,"]",sep="") ) )
118     kappa = rbind( kappa , samplesSample( paste("kappa[",condIdx,"]",sep="") ) )
119   }
120   save( mu , kappa , file=paste(fileNameRoot,"MuKappa.Rdata",sep="") )
121   chainLength = NCOL(mu)
122
123   # Histograms of mu differences:
124   windows(10,2.75)
125   layout( matrix(1:3,nrow=1) )
126   source("plotPost.R")
127   plotPost( mu[1,]-mu[2,] , xlab=expression(mu[1]-mu[2]) , main="" ,
128           breaks=20 , compVal=0 )
129   plotPost( mu[3,]-mu[4,] , xlab=expression(mu[3]-mu[4]) , main="" ,
130           breaks=20 , compVal=0 )
131   plotPost( (mu[1,]+mu[2,])/2 - (mu[3,]+mu[4,])/2 ,
132           xlab=expression( (mu[1]+mu[2])/2 - (mu[3]+mu[4])/2 ) ,
133           main="" , breaks=20 , compVal=0 )
134   dev.copy2eps(file=paste(fileNameRoot,"MuDiffs.eps",sep=""))
135
136   # Scatterplot of mu,kappa in each condition:
137   windows()
138   muLim = c(.60,1) ; kappaLim = c( 4.0 , 40 ) ; mainLab="Posterior"
139   thindex = round( seq( 1 , chainLength , len=300 ) )
140   plot( mu[1,thindex] , kappa[1,thindex] , main=mainLab ,
141       xlab=expression(mu[c]) , ylab=expression(kappa[c]) , cex.lab=1.75 ,
```

```
142          xlim=muLim , ylim=kappaLim , log="y" , col="red" , pch="1" )
143    points( mu[2,thindex] , kappa[2,thindex] , col="blue" , pch="2" )
144    points( mu[3,thindex] , kappa[3,thindex] , col="green" , pch="3" )
145    points( mu[4,thindex] , kappa[4,thindex] , col="black" , pch="4" )
146    dev.copy2eps(file=paste(fileNameRoot,"Scatter.eps",sep=""))
```

9.6 EXERCISES

Exercise 9.1. [Purpose: To investigate research design—more coins versus more flips per coin.] In Section 9.2.5, p. 219, it was argued that if the goal of the research is to get a good estimate of the group average μ, then it is better to collect data from more coins than to collect more flips per coin. This exercise has you generate simulated data to bolster this conclusion.

(A) Use the R code of Section 9.5.1 (BernBetaMuKappaBugs.R) for this exercise. In the data section of that program, comment out the lines that specify N and z. Insert the following lines:

```
ncoins = 50 ; nflipspercoin = 5
muAct = .7 ; kappaAct = 20
thetaAct = rbeta( ncoins ,muAct*kappaAct ,(1-muAct)*kappaAct )
z = rbinom( n=ncoins ,size=nflipspercoin ,prob=thetaAct )
N = rep( nflipspercoin , ncoins )
```

Explain in words what that code does. This is important; if you don't understand this code, the rest of the exercise will not make much sense. (Hint: It's generating random data, for specific "actual" parameter values; explain in detail.)

(B) At the bottom of the program, UNcomment the lines that plot the posteriors of muSample, kappaSample, and thetaSample[1,]. (Don't plot thetaSample[28,], because it doesn't exist.) You should also UNcomment the windows() and layout(...) command, so that the plots don't overwrite each other. Run the program a few times and include the graphs of one *typical* run in your write-up.

(C) Now change the data-generation code so that the number of coins is 5 (instead of 50) and the number of flips per coin is 50 (instead of 5). Run the program a few times and include the graphs of one *typical* run in your write-up.

(D) Is the posterior estimate of μ more certain for 5 coins or for 50 coins? Is the posterior estimate of θ_1 more certain for 5 coins (50 flips per coin) or for 50 coins (5 flips per coin)? Is it better to use more coins or more flips per coin if the goal is to estimate μ?

Exercise 9.2. [Purpose: To examine the effect of different assumptions about across-group constraints. Specifically, different assumptions about κ_c in the analysis of data from the filtration-condensation experiment.] For this exercise, we will perform alternative analyses of the data from the filtration-condensation experiment described in Section 9.3.1. You will adapt the code listed in Section 9.5.2 (FilconBrugs.R), which implements the hierarchical model diagrammed in Figure 9.7, to implement the alternative hierarchical models diagrammed in Figure 9.17.

(A)   The left side of Figure 9.17 shows a model in which the same κ value is used for all groups simultaneously. The idea here is that accuracies of individuals in each group depend on the group mean accuracy, and we are going to estimate the magnitude of that dependency of individuals on the group mean, *but* we assume that whatever the degree of dependency is, it is the same in every group. This assumption can be thought of as saying that the category structure (e.g., filtration or condensation) affects the mean accuracy of the group, but individual variation from the mean accuracy is caused only by other factors that are the same across groups, not by the category structure. To implement this assumption in the program, do the following: In the model specification, because κ does not depend on the group (i.e., the condition), move the line that specifies the distribution of κ *outside* the for loop that cycles through the conditions. Moreover, because there is only one κ, remove the index from κ, (i.e., change kappa[condIdx] to kappa). Then, in the initialization of the chains, make sure that kappa is initialized to a single value instead of a vector of four values. To do this, set kappa to the mean of the four condition kappas. Now run the modified program. (Hint: For an example of results, see Figures 9.18 and 9.19, and compare the results with those in Figure 9.16.) In your answer: (1) Report the modified model-specification section of your code; (2) Show the graph of the estimated μ differences; (3) Answer this question: *Why* is the 95% HDI of the $\mu_1 - \mu_2$ differences farther away from zero than in the original analysis?

(B)   The right side of Figure 9.17 shows a model in which the κ_c values for the different groups come an overarching distribution, the parameters of which are estimated by considering all the groups. The idea here is that accuracies of individuals in each group depend on the group mean accuracy, and we are going to estimate the magnitude of that dependency of individuals on group mean, *but* we assume that whatever the degree of dependency is, it will tend to be similar across groups, and we let the data inform our estimate of that similarity. To implement this assumption in the program, do the following: First, starting with the original program, in the model specification, change the lines that specify the mean and standard deviation of the gamma distribution, from constants to uniform distributions. Use ranges from 0.01 to 30. For example, change meanGamma <- 10 to meanGamma~dunif(0.01,30), and do the

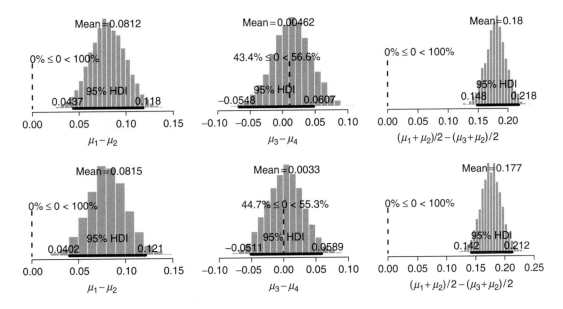

FIGURE 9.18

For Exercise 9.2. Upper row shows results when κ is constrained to be equal for all groups, as in the hierarchical diagram of the *left* side of Figure 9.17. Lower row shows results when κ has its distribution across groups estimated, as in the hierarchical diagram of the *right* side of Figure 9.17. Compare these results with those in Figure 9.16.

same for sdGamma. The resulting model specification could look like this (FilconCoKappaBrugs.R):

```
11   model {
12       for ( subjIdx in 1:nSubj ) {
13           # Likelihood:
14           z[subjIdx] ~ dbin( theta[subjIdx] , N[subjIdx] )
15           # Prior on theta: Notice nested indexing.
16           theta[subjIdx] ~ dbeta( a[cond[subjIdx]] , b[cond[subjIdx]] )I(0.001,0.999)
17       }
18       for ( condIdx in 1:nCond ) {
19           a[condIdx] <- mu[condIdx] * kappa[condIdx]
20           b[condIdx] <- (1-mu[condIdx]) * kappa[condIdx]
21           # Hyperprior on mu and kappa:
22           mu[condIdx] ~ dbeta( Amu , Bmu )
23           kappa[condIdx] ~ dgamma( Skappa , Rkappa )
24       }
25       # Constants for hyperprior:
26       Amu <- 1
27       Bmu <- 1
28       Skappa <- pow(meanGamma,2)/pow(sdGamma,2)
29       Rkappa <- meanGamma/pow(sdGamma,2)
30       meanGamma ~ dunif( 0.01 , 30 )
31       sdGamma ~ dunif( 0.01 , 30 )
32   }
```

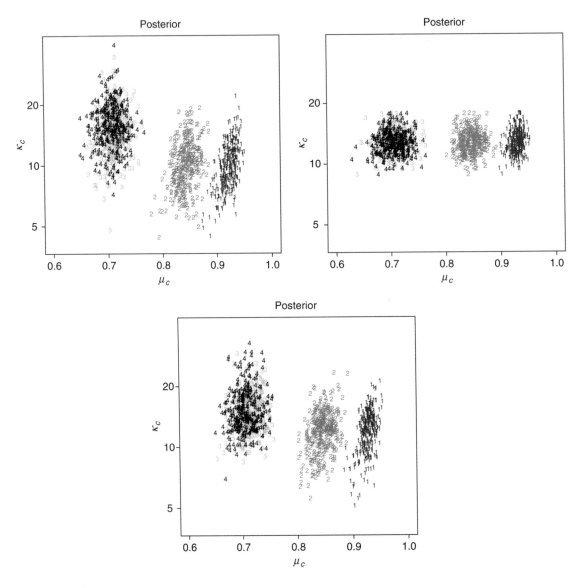

FIGURE 9.19

For Exercise 9.2. Scatterplots of μ_c and κ_c values for the four conditions of the filtration-condensation experiment. A plotted numeral indicates the condition. *Upper left panel:* All κ_c values separately estimated, as in model of Figure 9.15. *Lower panel:* κ_c values mutually informed by estimated overarching distribution, as in model of right side of Figure 9.17. *Upper right panel:* κ_c values constrained to be the same, as in model of left side of Figure 9.17.

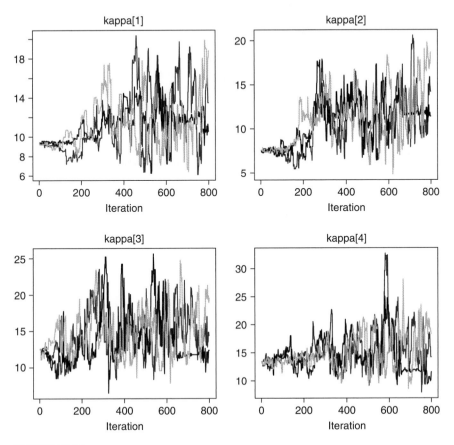

FIGURE 9.20

For Exercise 9.3. Markov chains for the kappa parameters of the four conditions, *during the initial burn-in trials, and with no thinning.* Notice that it takes a few hundred trials for the chains to migrate to their stable regions, which is why burn-in is needed. Notice also that chains change only gradually and are autocorrelated (and can even get "stuck" as is seen here in trials 600 to 800), which is why thinning is needed.

Second, be sure that the initialization of the chains give initial values to meanGamma and sdGamma; such as mean(kappa[]) and sd(kappa[]), respectively, where kappa[] are data-derived kappas for the four groups. Now run the modified program. (Hint: For an example of results, see Figures 9.18 and 9.19, and compare the results with those in Figure 9.16). In your answer: (1) Report the modified model-specification section of your code; (2) Show the graph of the estimated μ differences; (3) Answer this question: *Why is the 95% HDI of the $\mu_1 - \mu_2$ differences farther away from zero than in the original analysis, but not as far away from zero as when assuming the same κ for all conditions?*

Exercise 9.3. [Purpose: To see, graphically, the importance of burning-in the chains, and the meaning of autocorrelation.] This exercise uses the BUGS program you created in the last part of the previous exercise (i.e., the model for the right side of Figure 9.17). You might find it useful to refer to Section 23.2, p. 623, when doing this exercise.

Set the burn-in to zero and the thinning to zero. In other words, don't burn-in at all, and remove the the thin argument. Have the model go for 800 steps. What do the graphs of the chains look like, in particular for kappa? See, for example, Figure 9.20. Are the chains thoroughly overlapping? Include a relevant graph and discuss both burn-in and autocorrelation.

CHAPTER 10

Hierarchical Modeling and Model Comparison

CONTENTS

The magazine model comparison game
Leaves all of us wishing that we looked like them.
But they have mere fantasy's bogus appeal,
'Cause none obeys fact or respects what is real.[1]

10.1 MODEL COMPARISON AS HIERARCHICAL MODELING

Consider again the simple hierarchical dependency in which a single coin with bias θ depends on a hyperparameter μ. This dependency was diagrammed in Figure 9.1 (p. 194). An example of Bayesian updating for this structure, using

---

[1] Did you notice "Bayes' factor" sounded out in the verse?

Doing Bayesian Data Analysis: A Tutorial with R and BUGS. DOI: 10.1016/B978-0-12-381485-2.00010-9
© 2011, Elsevier Inc. All rights reserved.

specific prior constants, was plotted in Figure 9.2 (p. 197). Please take a look at those figures now.

In that previous scenario, the hyperparameter μ was assumed to have any possible value in the interval $[0, 1]$. Figure 10.1 shows a case in which the hyperparameter μ is instead allowed only *two* values: $\mu = 0.25$ and $\mu = 0.75$. We can think of this as being a prior with two ridges, one ridge over each of the possible values. The hyperprior has put equal prior probability on the two values of μ, as can be seen by the equal heights of the spikes in the marginal hyperprior distribution, $p(\mu)$, shown in the top-right panel.

The dependence of θ on μ is determined as before by Equation 9.2, in this case with $K = 12$ (arbitrarily). Therefore, when $\mu = 0.25$, the prior distribution over θ is beta($\theta, 3, 9$), and when $\mu = 0.75$, the prior distribution over θ is beta($\theta, 9, 3$). These distributions can be seen as the profiles of the vertical ridges in the perspective plot of the prior in Figure 10.1, and in the plots of $p(\theta \mid \mu)$ in the right panel of the second row.

In this example, we flip a coin $N = 9$ times and observe $z = 6$ heads. The likelihood is shown in the middle row of Figure 10.1, and the posterior is obtained, as always, by multiplying the likelihood and the prior, point-by-point across the parameter space.

The posterior shows two ridges, like the prior, because anywhere the prior is zero, the posterior must also be zero. The shape of each ridge is just the updated beta distributions: When $\mu = 0.25$, the posterior distribution over θ is beta($\theta, 3 + 6, 9 + 3$), and when $\mu = 0.75$, the posterior distribution over θ is beta($\theta, 9 + 6, 3 + 3$). These can be seen in the conditional distributions in the lower-right panel.

The posterior also reveals how much we believe in each candidate value of μ, as can be seen in the graph of $p(\mu \mid D)$ in the right panel of the fourth row of Figure 10.1. (The scale on the $p(\mu)$ axis is not meaningful in this case, because it is a grid-based estimate of density, but in this case $p(\mu)$ is really a discrete probability mass at each of the two values of μ.) Notice that the spike over $\mu = 0.75$ has grown much taller, and the ratio of $p(\mu = 0.75 \mid D)$ to $p(\mu = 0.25 \mid D)$ is roughly 26 to 4, which is 6.5 to 1.

This comparison of beliefs in two values of μ is tantamount to comparison of beliefs in two models at the level of θ, namely, the model beta($\theta \mid 3, 9$) versus the model beta($\theta \mid 9, 3$). When summing the posterior distribution over θ, we are effectively computing, at each value of $\mu = \mu^*$, the evidence for the model $\mu = \mu^*$: $p(D \mid \mu) = \int d\theta \, p(D \mid \theta) p(\theta \mid \mu)$. Exercise 10.1 (p. 259) has you explore this in a little more detail, and it would be good for you to take a look at it now.

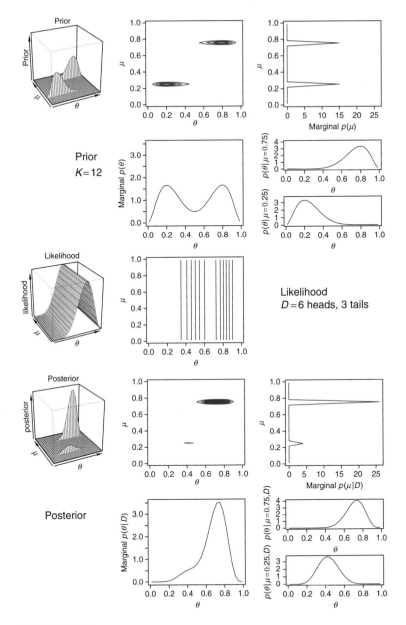

FIGURE 10.1

A prior with only two values of the hyperparameter μ is tantamount to model comparison at the level of θ. The marginal $p(\mu \mid D)$ indicates the posterior beliefs in the two values of μ.

10.2 MODEL COMPARISON IN BUGS

The example of Figure 10.1 showed a case in which two models were formed by two discrete values of a continuous hyperparameter. More generally, different models can be thought of as dependent on a *categorical* hyperparameter, which merely indexes the models. When the indexical hyperparameter has value j, then the jth model, with its parameters and priors, is used to estimate the data. With the models represented by values of the categorical hyperparameter, the posterior probability of each model is just the posterior probability of each value of the categorical hyperparameter. This framework allows us to set up Markov chain Monte Carlo (MCMC) chains that jump from model to model, as well as jump from parameter value to parameter value within models. MCMC chains that jump between models are sometimes called *transdimensional* because different models can have different numbers of parameters (i.e., different dimensionality).

10.2.1 A Simple Example

We'll continue with the example of Figure 10.1 to illustrate the basic idea in BUGS. What we'll do is define two possible priors for theta, with the priors indexed by a hyperparameter called modelIndex. The indexical hyperparameter is itself a random variable, estimated along with the parameter theta.

A straightforward way to implement this scheme is shown by this BUGS model specification (BernBetaModelCompBrugs.R):

```
8    model {
9        # Likelihood:
10       for ( i in 1:nflips ) {
11           y[i] ~ dbern( theta )   # y[i] distributed as Bernoulli
12       }
13       # Prior distribution:
14       theta ~ dbeta( aTheta , bTheta ) # theta distributed as beta density
15       aTheta <- muTheta * kappaTheta
16       bTheta <- (1-muTheta) * kappaTheta
17       # Hyperprior:
18       muTheta <- muThetaModel[ modelIndex ]
19       muThetaModel[1] <- .75
20       muThetaModel[2] <- .25
21       kappaTheta <- 12
22       # Hyperhyperprior:
23       modelIndex ~ dcat( modelProb[] )
24       modelProb[1] <- .5
25       modelProb[2] <- .5
26   }
```

The model specification starts with the likelihood, which is just the familiar Bernoulli distribution using theta as the parameter. The prior on theta is then specified as a beta density, with the parameters of the beta density defined in terms of a hyperparameter called muTheta, which is simply the mean of the beta distribution of theta.

It's at this point that the new stuff begins, at line 17, commented "Hyperprior." The hyperparameter muTheta has a different value depending on the model. If it's model 1, then muTheta has value 0.75, but if it's model 2, then muTheta has value 0.25. This assignment is accomplished by specifying a vector muTheta-Model[] with two components having values 0.75 and 0.25. The parameter muTheta is assigned one of the values from the vector muThetaModel[] depending on the model index.

The next lines, 23 to 25, specify the prior on the model index itself. The model index is named modelIndex. It is a random variable, distributed according to a *categorical density* function dcat. The dcat distribution takes a J-component vector as its argument and generates a value $1, 2, \ldots, J$ with the probability specified in the $1^{st}, 2^{nd}, \ldots, J^{th}$ component of the vector. The code specifies the prior probabilities on the two model indices to be the same: 0.5. This means that we believe each model is equally probable before we collect new data.

Figure 10.2 shows the output of the program. Compare it with the posterior in the lower-right panel of Figure 10.1. In particular, notice that the posterior probability that the model index has value 1 is 0.868, which is roughly 6.5 times larger than the probability it has value 2. This result matches the result

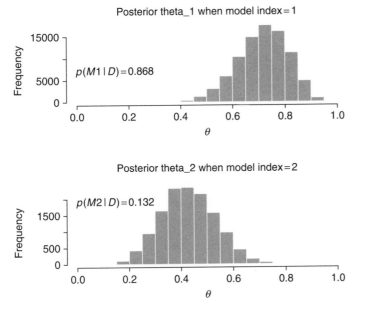

FIGURE 10.2
Output of BUGS model described in Section 10.2.1. Compare with the posterior in the lower-right panel of Figure 10.1, p. 243.

shown in Figure 10.1 (and computed exactly in Exercise 10.1). It is important to notice that the `theta` chain produced by the BRugs code contains interspersed values for both models. At steps in the chain when `model Index = 1`, the `theta` value represents the posterior for model 1, but at steps in the chain when `model Index = 2`, the `theta` value represents the posterior for model 2.

The simple approach taken in this example is fine in principle but works in practice only for the most elementary cases like this one. One aspect that made this example simple was that the same parameter (`theta`) was used for both models, and only the hyperprior constants differed between models. More generally, however, different models can have different parameters altogether. In such situations, the approach taken earlier can run into problems because BUGS can get stuck sampling from one model for a long time before it jumps to the other model. The next section illustrates this situation and one way to address the problem.

10.2.2 A Realistic Example with "Pseudopriors"

Consider again the experiment that involved learning of filtration or condensation structures, introduced in Section 9.3.1 (p. 219). There were four different category structures, a.k.a. conditions, for learning. In the initial BRugs code for that example, each condition was allowed its own mean (μ) and certainty (κ) hyperparameter values for the beta distribution from which individual participant biases were generated. The model therefore had a different θ parameter for each individual, four μ parameters (one for each condition), and four κ parameters (one for each condition). This model was diagrammed in Figure 9.15, p. 221.

Alternatively, we also considered a model that assumed the same certainty value for all four conditions. This model was diagrammed in the left side of Figure 9.17, p. 226. This assumption says that all participants are equally dependent on their condition's μ parameter, even though the μ parameter may differ between groups. In terms of the learning task, the assumption is that the category structure affects the mean difficulty of learning, but not the variability of that difficulty across participants. This model has just one κ_0 parameter, instead of four different κ parameters. Importantly, this single κ_0 parameter is distinct from any of the four condition κ's of the previous model, and therefore it is given a distinctive subscript here and denoted `kappa0` in the BUGS model specification:

(FilconModelCompBrugs.R)

```
10   model {
11     for ( i in 1:nSubj ) {
12       # Likelihood:
13       nCorrOfSubj[i] ~ dbin( theta[i] , nTrlOfSubj[i] )
14       # Prior on theta: Notice nested indexing.
15       theta[i] ~ dbeta( aBeta[ CondOfSubj[i] ] ,
```

```
16                        bBeta[ CondOfSubj[i] ] )I(0.0001,0.9999)
17          }
18       # Hyperprior on mu and kappa:
19       kappa0 ~ dgamma( shapeGamma , rateGamma )
20       for ( j in 1:nCond ) {
21          mu[j] ~ dbeta( aHyperbeta , bHyperbeta )
22          kappa[j] ~ dgamma( shapeGamma , rateGamma )
23       }
24       for ( j in 1:nCond ) {
25          aBeta[j] <- mu[j]     * ((kappa[j]*(2-mdlIdx))+(kappa0*(mdlIdx-1)))
26          bBeta[j] <- (1-mu[j]) * ((kappa[j]*(2-mdlIdx))+(kappa0*(mdlIdx-1)))
27          # BUGS equals(,) won't work here, for no apparent reason.
28          # Took me hours to isolate this problem (argh!). So, DO NOT use:
29          # aBeta[j] <- mu[j]     * (kappa[j]*equals(mdlIdx,1)+kappa0*equals(mdlIdx,2))
30          # bBeta[j] <- (1-mu[j]) * (kappa[j]*equals(mdlIdx,1)+kappa0*equals(mdlIdx,2))
31       }
32       # Constants for hyperprior:
33       aHyperbeta <- 1
34       bHyperbeta <- 1
35       shapeGamma <- 1.0
36       rateGamma <- 0.1
37       # Hyperprior on model index:
38       mdlIdx ~ dcat( modelProb[] )
39       modelProb[1] <- .5
40       modelProb[2] <- .5
41    }
```

The model specification begins with the binomial likelihood for each individual's data. Then each individual theta[i] value is distributed as a beta distribution with parameters aBeta and bBeta. On lines 25 to 26 something new happens. The aBeta and bBeta values depend on the value of the model index, mdlIdx. For example, when mdlIdx is 1, then aBeta[j] <- mu[j] * kappa[j], but when mdlIdx is 2, then aBeta[j] <- mu[j] * kappa0. The program accomplishes this conditional assignment by a contorted algebraic manipulation because the BUGS interpreter has limited abilities; see the comment on line 27.

Figure 10.3 illustrates the output. The top panel shows the MCMC chain for the model index. The most obvious feature is that there is a much higher probability for model 2 than model 1. In other words, despite the fact that model 1 has distinct certainties for each condition and can therefore fit the data better in terms of maximum likelihood, it is *not* the more believable model. The simpler model 2, with only one certainty for all conditions, has noticeably higher posterior probability. The reason that model 1 suffers is that the prior believability of its extra parameters gets diluted over a high-dimensional space. Even though model 1 has particular parameter values with a higher likelihood than model 2, the prior believability of those parameter values is low. If the data were different, such that some conditions had huge variance and other conditions had tiny variance, then the more complex model 1 might have higher posterior probability because the simpler model wouldn't fit the data very well.

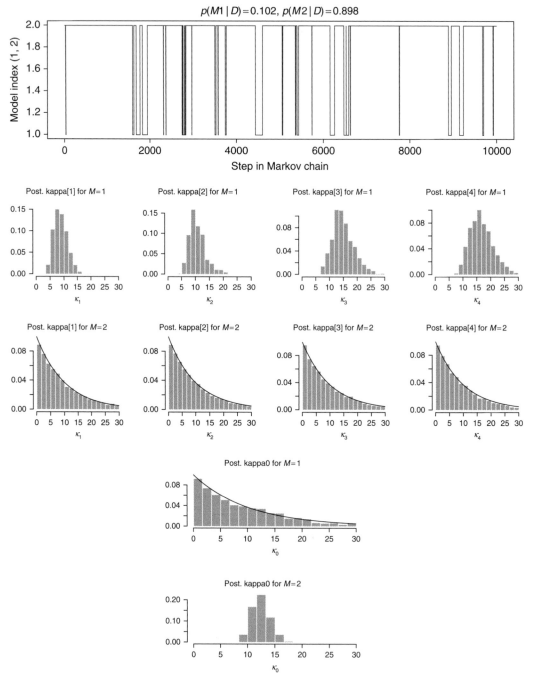

FIGURE 10.3
Output from model comparison with*out* using pseudopriors. Compare with Figure 10.4.

For the actual data, the variances within the four conditions are similar enough that the simpler model wins.

The middle rows of Figure 10.3 show distributions of the κ_j parameter values, sampled in the MCMC chain. There are two rows of histograms. The upper row shows the four κ_j parameters when the model index is 1. The lower row shows the four κ_j parameters when the model index is 2. Even though the κ_j are not used to model the data when the model index is 2, BUGS goes ahead and generates values for them anyway; BUGS generates a value for every parameter at every step in the chain. Notice that the distributions in the two rows are quite different! When the model index is 1, the data are being modeled by the four κ_j values, so the sampled values are constrained by the data. But when the model index is 2, the κ_j values are not used in the model; they are floating unconstrained by data. Therefore, the sampled values can only reflect the prior distribution. The curves superimposed on the histograms show the prior distribution that was specified in the model; notice that the sample histograms match the specified prior very closely.

The bottom section of Figure 10.3 shows the distribution of kappa0 for the two model indices. The kappa0 parameter is used to model the data when the model index is 2 but kappa0 is unconstrained by the data when the model index is 1. The histograms reflect this situation: When the model index is 1, the histogram matches the prior distribution.

Look again at the MCMC chain for the model index in the top panel of Figure 10.3. Although it might not be obvious, the model index doesn't change back and forth as often as it should if it were really sampling independently from the posterior probabilities of the models. The model index tends to linger at one value or the other. This is, essentially, a problem of *autocorrelation* in the chain. This malingering can be problematic because it implies that the chain must be *very* long before we have a good estimate of the relative probabilities of the models. In principle, all we have to do is wait a long time, but in practice, we might not be able to wait long enough and we don't know in advance how long is long enough. So we should try to fix this autocorrelation problem.

Why does the model index get stuck? Recall that the way the chain jumps away from its current model is by considering a proposed set of parameter values in the *other* model. What are the proposed parameters in the other model? For the model that is not currently constrained by the data, BUGS has generated random values from the model's prior. Such random values are usually not very good fits to the data; therefore the chain won't often jump to the other model.

How can the problem be solved? One approach is to define "pseudopriors" for the models that are used when the models are not constrained by the data (Carlin & Chib, 1995). These pseudopriors merely stand in for the real prior when the model is not being used to actually fit the data. The pseudoprior merely keeps the MCMC chain "in the ballpark" of the posterior, so that proposed parameter values are reasonable. The challenge then becomes how to specify the pseudoprior. A simple solution is to make the pseudoprior mimic the posterior. I did this in the following code by using gamma densities with the same mean and standard deviation as the posterior kappa distributions simulated in Figure 10.3. The following code shows details, discussed after the listing:

(FilconModelCompPseudoPriorBrugs.R)

```
10   model {
11       for ( i in 1:nSubj ) {
12           # Likelihood:
13           nCorrOfSubj[i] ~ dbin( theta[i] , nTrlOfSubj[i] )
14           # Prior on theta: Notice nested indexing.
15           theta[i] ~ dbeta( aBeta[ CondOfSubj[i] ] ,
16                             bBeta[ CondOfSubj[i] ] )I(0.0001,0.9999)
17       }
18       # Hyperprior on mu and kappa:
19       kappa0 ~ dgamma( shk0[mdlIdx] , rak0[mdlIdx] )
20       for ( j in 1:nCond ) {
21           mu[j] ~ dbeta( aHyperbeta , bHyperbeta )
22           kappa[j] ~ dgamma( shk[j,mdlIdx] , rak[j,mdlIdx] )
23       }
24       for ( j in 1:nCond ) {
25           aBeta[j] <- mu[j]     * ((kappa[j]*(2-mdlIdx))+(kappa0*(mdlIdx-1)))
26           bBeta[j] <- (1-mu[j]) * ((kappa[j]*(2-mdlIdx))+(kappa0*(mdlIdx-1)))
27           # BUGS equals(,) won't work here, for no apparent reason.
28           # Took me hours to isolate this problem (argh!). So, DO NOT use:
29           # aBeta[j] <- mu[j]     * (kappa[j]*equals(mdlIdx,1)+kappa0*equals(mdlIdx,2))
30           # bBeta[j] <- (1-mu[j]) * (kappa[j]*equals(mdlIdx,1)+kappa0*equals(mdlIdx,2))
31       }
32       # Constants for hyperprior:
33       aHyperbeta <- 1
34       bHyperbeta <- 1
35
36       # Actual priors:
37       shP <- 1.0 # shape for prior
38       raP <- 0.1 # rate for prior
39       # shape, rate kappa0[ model ]
40       shk0[2] <- shP
41       rak0[2] <- raP
42       # shape kappa[ condition , model ]
43       shk[1,1] <- shP
44       shk[2,1] <- shP
45       shk[3,1] <- shP
46       shk[4,1] <- shP
47       # rate kappa[ condition , model ]
48       rak[1,1] <- raP
49       rak[2,1] <- raP
50       rak[3,1] <- raP
51       rak[4,1] <- raP
52
53       # Pseudo priors:
54       # shape, rate kappa0[ model ]
```

```
55        shk0[1] <- 54.0
56        rak0[1] <- 4.35
57        # shape kappa[ condition , model ]
58        shk[1,2] <- 11.8
59        shk[2,2] <- 11.9
60        shk[3,2] <- 13.6
61        shk[4,2] <- 12.6
62        # rate kappa[ condition , model ]
63        rak[1,2] <- 1.34
64        rak[2,2] <- 1.11
65        rak[3,2] <- 0.903
66        rak[4,2] <- 0.748
67
68        # Hyperprior on model index:
69        mdlIdx ~ dcat( modelProb[] )
70        modelProb[1] <- .5
71        modelProb[2] <- .5
72    }
```

As before, the kappa parameters are distributed as gamma distributions. But the shape and rate constants for those gamma distributions now depend on the model index. Line 19 of the code says that the gamma distribution has shape parameter shk0[mdlIdx] and rate parameter rak0[mdlIdx]. These values are assigned in lines 40 to 41 and 55 to 56. Those lines state that when the model index is 2, use the real prior, but when the model index is 1, use the pseudoprior.

The shape and rate constants for the gamma pseudopriors were set so that the gamma distributions had the same mean and standard deviation as the posterior sample of kappa values. For example, consider the parameter kappa0. In the initial run without pseudopriors, a reasonable posterior sample was obtained for model index 2 (i.e., when kappa0 is actually constrained by the data, as shown in the bottom panel of Figure 10.3). That posterior sample had a mean m and a standard deviation s. The conversion from mean and standard deviation to shape and rate parameters was explained in the caption of Figure 9.8. The conversion yields shape and rate values specified in lines 55 to 56 of the code.

Figure 10.4 shows the results from the use of pseudopriors. The top panel shows that the chain for the model index jumps back and forth much more often than before (compare with Figure 10.3). Therefore the sampled relative probabilities of model indices is much more trustworthy. This reduction of autocorrelation in the chain of model indices is the primary advantage of using pseudopriors. If you compare the posterior that does *not* use pseudopriors (Figure 10.3) with the posterior that *does* use pseudopriors (Figure 10.4), you will see that the probability of model-index 2 is slightly different. Which probability is more trustworthy? The one using pseudopriors, because its chain of model-index values is less autocorrelated (i.e., less stagnant and clumpy).

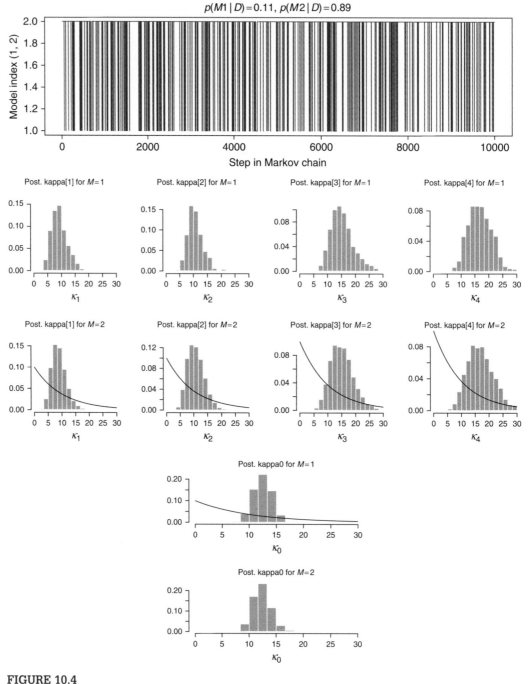

FIGURE 10.4

Output from model comparison *with* using pseudopriors. Compare with Figure 10.3.

The middle and lower panels of Figure 10.4 show the sampled κ values for each model index. Notice in particular that the sampled values of kappa[j], when the model index is 2, are distributed according to the pseudoprior, not according to the real prior indicated by the superimposed curve. The pseudoprior histograms do, in fact, resemble the true posterior historgrams, as intended. Analogous remarks apply to the histograms of kappa0 in the lower portion of Figure 10.4. To recapitulate, the pseudopriors are not being used to model the data; they are used only to keep the MCMC chain of the unused parameters in the vicinity of believable values until the chain jumps over and uses them again, at which point the true prior is invoked.

10.2.3 Some Practical Advice When Using Transdimensional MCMC with Pseudopriors

When doing the first run without pseudopriors, do not bother programming a separate model that uses only the real priors. All you need is the program that can accommodate the pseudopriors, which you run iteratively with different specifications for the pseudopriors. The first time you run it, the pseudopriors are specified as the real priors. This first run yields estimates of the posteriors for the parameters when their model index applies. With those estimates, pseudopriors can be calculated and then the program can be run again with the modified pseudopriors. You can do this repeatedly if needed, each time tuning the pseudoprior to better match the posterior that has been more accurately sampled because of the most recent tuning of the pseudoprior.

When one model is much less believable than another, the unbelievable model rarely gets visited in the chain, so there are few points in the sample to represent its parameter values. To compensate for this imbalance, you can arbitrarily set the the model-index *prior* probabilities to compensating values. For example, suppose you do an initial run with the model-index priors set with modelProb[1] <- .5 and modelProb[2] <- .5, with which you find that model 1 has a posterior probability of only 0.02. You can then rerun the sampler with the model priors set to modelProb[1] <- .98 and modelProb[2] <- .02, so that model 1 gets visited a lot more frequently. This will give you a better sample of the posteriors for creating reasonable pseudopriors. When you fiddle with the model priors, beware that the relative posterior probabilities of the models is the Bayes factor *times the ratio of the prior probabilities* (see Equation 4.6, p. 58). The posterior probabilities of the model indices are affected by the prior probabilities, although the Bayes factor is not. If you are interested in the Bayes factor, convert the obtained posterior probabilities of the models by multiplying by the reciprocal of the priors. Exercise 10.3 provides an example in which the different-kappa model is strongly preferred.

10.3 MODEL COMPARISON AND NESTED MODELS

The model comparison of the previous section indicated that a model with a single κ_0 parameter for all four groups was about eight times more believable than a model with distinct κ_1, κ_2, κ_3, and κ_4 parameters for the four groups. Given this analysis, do we then believe that the four groups really have exactly the same κ value? Asked a different way, if we wanted to estimate the kappa values of the four groups, what would that estimate look like? Some people might answer that the single κ_0, shared by all four groups, is the best estimate. But that is not my position. Instead, the best estimates of the distinct kappa parameters are the separate estimates, as shown in Figure 10.4. The posterior distribution on the four distinct parameters indicates a large degree of over-lap, but that does *not* mean that the parameter values are literally the same; it means merely that the differences are small compared to our uncertainty about the values. Most likely, the parameter values for the groups really are different, because the treatments received by the groups really are different, but the differences are small and our data are not sufficient to confidently detect the differences among the values.

Instead of doing a model comparison between a model with four distinct κ values and a model with one shared κ value, we could just look at the posterior distribution of the four distinct κ values. At every step in the chain, we compute the differences of the various group's κ values and examine whether the differences tend to be near zero. Figure 10.5 shows the differences between κ values of the four groups. For every comparison, the HDI of the difference includes zero, which means that no two κ values are credibly different. Thus,

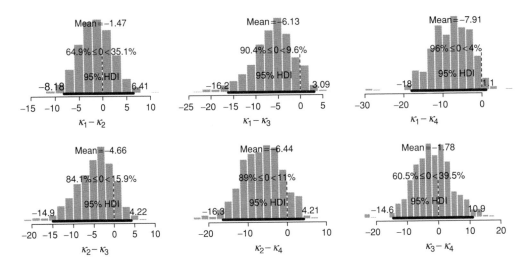

FIGURE 10.5
Differences of posterior κ values for the four groups.

by simply considering the posterior on the distinct κ values, we have arrived at a conclusion similar to the model comparison: The κ values are not credibly different. However, unlike the model comparison that favors the single shared κ parameter, we have retained all the information about posterior estimates of the distinct κ parameters.

The model involving a single shared κ value is a restricted version of the full model involving four distinct κ values. We get to the restricted model from the full model by demanding that all the distinct κ values are equal: $\kappa_1 = \kappa_2 = \kappa_3 = \kappa_4 = \kappa_0$. We say that the restricted model is *nested* within the full model and that the model comparison of the previous section was a *nested model comparison*.

Consider nested model comparison in general. Suppose a particular nested model comparison favors a restricted model in which parameters that are distinct in the full model are set equal in the restricted model. Does this mean that we should believe that the parameters in the full model are truly not different? In general, no. The full model is still (usually) the richer and more accurate description, but the posterior beliefs over the full model will indicate that the distinct parameters are not credibly different, while also providing estimates of the distinct parameters. When we are testing "generic" models and believe that there really are distinct parameter values, and the restricted model is just a simplification of convenience, then it is not advisable to use model comparison for the nested models. The reason is that, even if the restricted model ends up having the higher posterior probability, we do not actually believe the restriction. I used the nested model comparison in the previous section for pedagogical purposes. For a clear example of how model comparison on nested models can lead us astray, see Section 12.2.2, p. 307. There it is shown that nested model comparison can strongly prefer the restricted model, even though estimates of the different parameters show clear differences!

Is there ever a situation in which nested model comparison is appropriate? Yes, when the restricted model is genuinely believable for theoretical reasons. In this case, when we have a viable theory that asserts that a specific restriction is true, then it makes sense to test that restriction as a unique model. But if the restriction is merely a simplification of convenience, without genuine theoretical motivation, then it is more meaningful to examine the posterior distribution of the full model, rather than the restricted model that we do not actually believe.

Another way of expressing this distinction is formally. Denote the full model as M_F and denote the restricted model as M_R. We know from Bayes' rule that

$$\frac{p(M_R \mid D)}{p(M_F \mid D)} = \underbrace{\frac{p(D \mid M_R)}{p(D \mid M_F)}}_{\text{BF}} \frac{p(M_R)}{p(M_F)} \tag{10.1}$$

Essentially I am arguing that even if the Bayes factor (BF) strongly favors the restricted model, our posterior belief in the restricted model is, nevertheless, very small if our prior belief in the restricted model is very small. If the restricted model is merely a simplification of convenience, then our prior belief in the restricted model is essentially nil, so our posterior belief in the restricted model remains small regardless of the BF. It is only when the prior on the restricted model is nontrivial that the posterior on the restricted model can be large. The prior on the restricted model is nontrivial only in situations when a viable theory asserts it could be true.

In conclusion, model comparison for nested models should be undertaken only when it is truly meaningful to do so; it should not be undertaken routinely and automatically as "the" way to assess parameter values. And when nested model comparison is conducted, the parameter estimates in the unrestricted models should be examined for coherence with the conclusion from the nested model comparison, because the unrestricted model might show credible differences among parameters even if the restricted model "wins" a model comparison.

10.4 REVIEW OF HIERARCHICAL FRAMEWORK FOR MODEL COMPARISON

The upper panel of Figure 10.6 shows the general hierarchical framework for model comparison. The data are denoted by y, at the bottom of the hierarchy. The data depend on parameters θ in model 1 and on parameters ϕ in model 2. And the degree to which each model space is used depends on the model index ι (Greek letter "iota"). When estimating parameters, we are estimating θ, ϕ, and ι in a high-dimensional parameter space, namely, the conjoint space spanned by all combinations of those parameters, called the "product space" of the parameters. In terms of an MCMC trajectory, this means that at every step in the chain, there are values specified for all the parameters. In particular, at any point in the chain when $\iota = 1$, values for both θ and ϕ are generated, and at any point in the chain when $\iota = 2$, values for both ϕ and θ are generated. After the chain is generated, we only use the θ values for steps when $\iota = 1$, and we only use the ϕ values for steps when $\iota = 2$. In principle, it does not matter what the θ and ϕ values are when they are not being used. But in practice, to make the MCMC steps efficient, we use pseudopriors that keep the unused values in the appropriate region of parameter space.

The lower panel of Figure 10.6 shows a special case of the upper panel, for which the models differ only in their priors on the model parameter. In other words, the two models involve the same parameter θ and the same likelihood function $p(y \mid \theta)$ and differ only in the priors, $p(\theta \mid \iota=1)$ and $p(\theta \mid \iota=2)$. In this situation, we are estimating θ and ι. There is only one θ to estimate, even

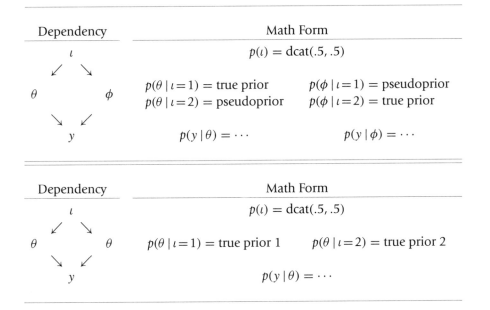

FIGURE 10.6

Hierarchical framework for model comparison. *Upper panel*: When the models involve different parameters, hence different likelihood functions. *Lower panel*: When the models involve the same likelihood function and differ only in the prior on the parameter. The lower panel shows two copies of parameter θ merely for the purpose of comparison with the upper panel. There is really only one θ parameter in this case. For both panels, the 50-50 constants for the prior on the model index are arbitrary.

though the lower panel of Figure 10.6 shows two copies of θ merely for purposes of comparison with the upper panel. There is no extra dimensionality introduced by the second model, because the second model does not have different parameters. When generating steps in a MCMC trajectory through the parameter space, every step generates a usable value for θ; each step's θ value is used for whichever model is specified by ι on that step. Thus, there is no need for pseudopriors on unused parameter samples, because there are no unused parameter samples. We encountered this situation back in the simple example accompanying Figure 10.2, p. 245.

10.4.1 Comparing Methods for MCMC Model Comparison

Section 10.2.2 provided an example of model comparison using transdimensional MCMC with pseudopriors. Notice that the method yielded the relative posterior probabilities of the models without computing the individual values of $p(D|M1)$ and $p(D|M2)$. If you want the values of those evidences, you can use the method explained in Section 7.3.3. By computing $p(D|M1)$ and $p(D|M2)$ explicitly, it is then easy to compute the Bayes factor and the posterior probabilities of the models.

Both methods for model comparison (i.e., transdimensional MCMC with pseudopriors and approximating $p(D|M)$ directly) require us to find an approximation to the posteriors. Is there an advantage of one method over the other? The direct method yields values for $p(D|M)$, which cannot be obtained from transdimensional MCMC. But we rarely need the absolute magnitude of $p(D|M)$, because it is meaningful primarily only relative to the evidence of another model. Transdimensional MCMC is perhaps more difficult to program, because we must carefully index the models and get both models to peacefully coexist in the same code. On the other hand, transdimensional MCMC is conceptually useful for thinking about the models as alternatives in a broader space of possible models. For our purposes, there is not a clear advantage of one method over the other. If you do both, and the results come out the same, you have increased confidence that you didn't make a mistake in your programming! For more in-depth comparisons, see, for example, Han & Carlin (2001).

10.4.2 Summary and Caveats

This chapter's main message regarding model comparison is that different models can be thought of as varying on a higher-level categorical parameter that indexes the models. Estimates of this indexical parameter provide an estimate of the posterior believability of the models. Thus, model comparison is "really" just estimation at a higher level. In BUGS, the model index can be set up as a random variable, and its probability can be approximated like any other variable. To do it efficiently, however, we had to apply a trick of using pseudopriors.

Bayesian model comparison is especially useful when comparing models that have genuine prior believability. If we instead start with a model or models that have little if any prior believability, then the Bayesian model comparison is an exercise in meaninglessness, as the result will only reveal which of the unbelievable models is least unbelievable. In particular, sometimes a mathematically convenient but otherwise arbitrary restriction on a full model, such as setting distinct group parameters to a single shared value, has essentially no prior believability. In this situation, the different groups are believed in advance to have at least slightly different parameter values, because they receive different treatments, by definition. Even if we conduct a model comparison and the restricted model is favored by the Bayes factor, we still do not believe the restricted model (as was discussed in the context of Equation 10.1, p. 255). Instead, the posterior parameter estimates on the full model reveal our beliefs, including the fact that some of the parameters are not very different relative to the uncertainty in those parameter values. As mentioned earlier, for a clear example of how model comparison on nested models can lead us astray, see Section 12.2.2, p. 307. Only sometimes does a restricted model have genuine prior believability. In these cases, Bayesian model comparison on nested models makes sense.

Bayesian model comparison is especially useful for comparing non-nested models, which involve distinct accounts of the data, using different parameters and different likelihood functions. *The resulting Bayes factor can be very sensitive to the priors used within each model, however* (e.g., Liu & Aitkin, 2008). To see why the posterior probabilities of the models can be so sensitive to the priors on the parameters, recall that the Bayes factor is just the ratio of the evidences for the two models. And, recall that the evidence for a model can be strongly affected by the prior chosen for the model. Indeed, we've considered many cases of model comparison in which all we've done is use different priors for the same likelihood function, and found radically different evidences emerge. Therefore, when comparing two models that have different parameters, the priors for the two models must be established using comparable criteria. One way to establish appropriate priors for the two models is by using informed priors instead of uninformed, "automatic," convenience priors. One approach is to use a set of previous data, or plausible and audience-agreeable fictitious data, that is, small but large enough to overwhelm any vague, primordial prior. This way, the priors for both models are at least "in the neighborhood" of appropriate parameter values, instead of being diluted over vast regions of parameter space that are only considered because of mathematical form. In any case, an important step in model comparison is checking the robustness of the conclusion when the priors in the models are changed in reasonable ways. For an example, see Exercise 10.2, which is highly recommended.

10.5 EXERCISES

Exercise 10.1. [Purpose: To see how first-level model comparison is like hierarchical modeling with just two values of a hyperparameter.]

(A) Use the R code of Section 5.5.1 (`BernBeta.R`) (p. 91) to compare the model $M1 : \text{beta}(\theta \mid 3, 9)$ and the model $M2 : \text{beta}(\theta \mid 9, 3)$, when the observed data are $z = 6$ heads in $N = 9$ flips. What is the Bayes factor for the two models?

(B) Verify the values of $p(D)$ by using Equation 5.10, p. 88.

(C) From the right panel of the fourth row of Figure 10.1 (p. 243), visually estimate $p(\mu = 0.75 \mid D)$ and $p(\mu = 0.25 \mid D)$, and compute their ratio. Does that ratio nearly match the Bayes factor that you computed in the first two parts? (It should.)

Exercise 10.2. [Purpose: To see that the prior can strongly affect the outcome of a model comparison.] In this exercise we consider a "toy" model comparison to illustrate how the priors on the parameters in two models can strongly affect the outcome of the comparison.

The models: Consider two models for the bias of a coin. For both models, the probability of getting a "head" is a Bernoulli distribution of the bias θ, but each model has a different expression for determining the value of θ in terms of a different parameter:

M1 : $\theta = 1/(1 + \exp(-\nu))$, where ν is any real value ("ν" is Greek "nu"), and
M2 : $\theta = \exp(-\eta)$, where η is non-negative ("η" is Greek "eta").

(You can easily graph those two function in R if you want to see what they look like. Figure 14.6, p. 375, shows variations of model M1.) For model 1, we put a normal prior on ν, and for model 2, we put a gamma prior on η.

The data: Suppose we flip the coin and find that we get 8 heads out of 30 flips.

(A) Suppose the prior on ν is normal with mean 0 and *precision* 0.1, whereas the prior on η is gamma with shape 0.1 and rate 0.1. What are the resulting posterior probabilities of the models (if the prior on the models is 50/50)? (Hint: See Figure 10.7 for the posterior on the parameters and the model probabilities.) The following chunks of code may be helpful. Notice that BUGS parameterizes the normal density in terms of mean and *precision*, where precision is defined as the reciprocal of variance, (i.e., 1 over the squared standard deviation). The motivation for using precision is explained in Section 15.1.1. Beware that although BUGS parameterizes dnorm by mean and precision, R parameterizes dnorm by mean and standard deviation. See lines 18 and 19 of the code below for the specification of the priors on the parameters:

(ToyModelComp.R)

```
10  model {
11     for ( i in 1:nFlip ) {
12        # Likelihood:
13        y[i] ~ dbern( theta )
14     }
15     # Prior
16     theta <- ( (2-mdlIdx) * 1/(1+exp( -nu )) # theta from model index 1
17              + (mdlIdx-1) * exp( -eta ) )    # theta from model index 2
18     nu ~ dnorm(0,.1)      # 0,.1   vs  1,1
19     eta ~ dgamma(.1,.1)   # .1,.1  vs  1,1
20     # Hyperprior on model index:
21     mdlIdx ~ dcat( modelProb[] )
22     modelProb[1] <- .5
23     modelProb[2] <- .5
24  }
```

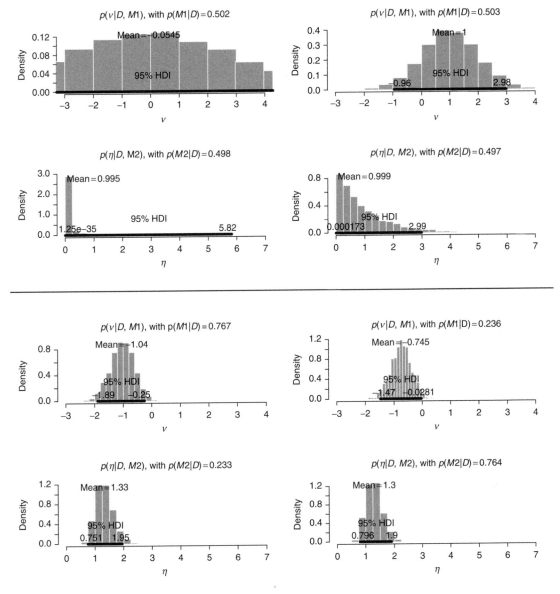

FIGURE 10.7
For Exercise 10.2, which does model comparison for different priors within each model. *Upper two rows,*
above the black line: Prior distributions as sampled from BUGS. Although the titles for these upper two rows
suggest that these are distributions given D, they are in fact priors because D is empty. *Lower two rows,*
below the black line: Posterior distributions. *Left column:* Results for priors of part A. *Right column:* Results
for priors of part B. Notice that the probability of the models is displayed in the title of each distribution.

(ToyModelComp.R)

```
36    N = 30
37    z = 8
38    datalist = list(
39        y = c( rep(1,z) , rep(0,N-z) ) ,
40        nFlip = N
41    )
```

(B) Suppose the prior on v is normal with mean 1 and precision 1, while the prior on η is gamma with shape 1 and rate 1. What are the resulting posterior probabilities of the models (if the prior on the models is 50/50)? Which model is preferred? (Hint: See Figure 10.7 for the posterior on the parameters and the model probabilities.

(C) Why does the choice of prior within each model (when the prior on $p(M1)$ and $p(M2)$ remains unchanged) have such a big influence on the posterior of the model believabilities? (Hint: Notice in Figure 10.7 that the posterior on the parameter values is in roughly the same place regardless of the prior—that is, the location of the posterior on the parameter values is dominated by the data. But notice that the prior probabilities at those posterior-favored values are rather different, and remember that $p(D|M) = \int d\phi\, p(D|\phi, M)\, p(\phi|M)$.)

(D) Discuss how we might decide which choice of priors on v and η would put the priors on an equal playing field. (Technically, in this case, the priors can be reparameterized to yield exact equivalence, but that's not the answer being sought for this exercise. In general, different models cannot be reparameterized into equivalence. Your discussion should address what to do in the general case, when models cannot be reparameterized to equivalence, but be motivated by considering the present toy example.) Here's one scheme to consider: Suppose we have some modest pilot data, or even some audience-agreeable plausible fictitious data, that in 7 flips there were 2 heads. To generate a prior for the subsequent new data, we start each model with an extremely vague proto-prior and update it with the modest pilot data. For example, suppose that the proto-prior on v is normal with mean zero and precision of 0.0001 (i.e., standard deviation of 100), and the proto-prior on η has rate = 0.01 and shape = 0.01. Notice that both of the resulting priors (i.e., the extremely vague proto-priors updated with the modest pilot data) are now reasonably "in the ballpark" of realistic data, and therefore the models are starting on a more equal playing field.

Exercise 10.3. [Purpose: To gain hands-on experience with the pseudopriors approach.] For this exercise, we'll consider a situation very similar to the filtration-condensation experiment that has been used repeatedly in recent pages, and was introduced in Section 9.3.1. The new experiment also involved people learning to categorize shapes. There were again four different groups, but this time the groups differed on the type of structural *shift* from an initial category structure to a new category structure (Kruschke, 1996). Theories of learning are concerned with this type of situation because different theories predict different types of transfer from one phase to the next. Our goal in this exercise is to determine whether the κ values for the four groups are equivalent, just as we asked this question of the filtration-condensation data in Section 10.2.2, p. 246.

Modify the program FilconModelCompPseudoPriorBrugs.R for this exercise. Do the following:

- In the data section of the program, delete the filtration-condensation data. At that point in the program, load the relevance-shift data instead. Type in `load("Kruschke1996CSdatsum.Rdata")`. You must have that data file in the same folder as your program, of course.
- In the model-specification, change the prior probability of model 1 to only 0.003, and the prior probability of model 2 to 0.997. This is done because it turns out that model 1 is far more believable than model 2, so to get any samples at all from model 2, its prior probability must be set very high.

(A) Run the program. Notice that the pseudopriors are not well matched to the posterior distribution. Change the pseudopriors so that they have the same mean and standard deviation as the posterior. Show this section of your modified code (i.e., the lines that set the shape and rate constants of the pseudopriors). (Hint: Use `mean(kappa1sampleM1)` and `sd(kappa1sampleM1)` to get the mean and standard deviation of the posterior. Then use the identities described in Figure 9.8 to determine the shape and rate parameters of a gamma distribution with that mean and standard deviation.)

(B) Rerun the program with the better-tuned pseudopriors that you determined in the previous part. Show histograms of the kappa distributions like those in Figure 10.4. (Hint: Your histograms in the top and bottom rows should be similar, because the pseudoprior is supposed to mimic the actual posterior. You will find that the histogram for κ_1 falls largely above 30; that is, it exceeds the right limit of the x axis as it is presently set.)

(C) Are the kappa values of the four groups the same or different? Answer this question two ways, as follows. First, what is the value of the Bayes factor in favor of the different-kappa model? Be careful to take into account that the BUGS simulation used priors that were not 50-50. (Hint: It's about 600 to 1. Explain how to get that value.) Does the Bayes factor alone tell us anything about *which* groups are different? (The correct answer to that last question is no. Briefly explain why.) Second, considering only the different-kappa model, what are the distributions of the differences

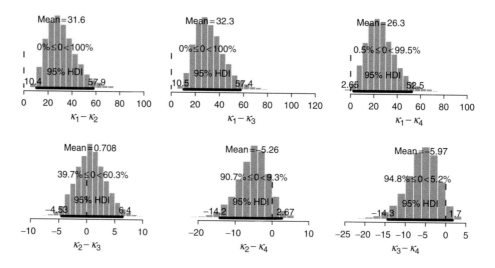

FIGURE 10.8

For Exercise 10.3. The first group has larger kappa than the other groups, which do not have credibly different kappa values.

between kappas for model 1? Specifically, which kappas are different from which other kappas? Does the Bayes factor tell us anything that the kappa estimates do not? (The correct answer to that last question is no. Briefly explain why.) (Hint: Your histograms of kappa differences should look something like Figure 10.8).

CHAPTER 11

Null Hypothesis Significance Testing

CONTENTS

Doing Bayesian Data Analysis: A Tutorial with R and BUGS. DOI: 10.1016/B978-0-12-381485-2.00011-0
© 2011, Elsevier Inc. All rights reserved.

My baby don't value what I really do.
She only imagines who else might come through.
She'll only consider my worth to be high
If she can't conceive of some much nicer guys.

In the previous chapters we have seen a thorough introduction to Bayesian inference involving a Bernoulli likelihood function. It is appropriate now to compare Bayesian inference with 20th-century *null hypothesis significance testing* (NHST) of binomial (i.e., dichotomous valued) data. In NHST, the goal of inference is to decide whether a particular value of a model parameter can be rejected. For example, we might want to know whether a coin is fair, which in NHST becomes the question of whether we can reject the hypothesis that the bias of the coin has the specific value 0.5.

To make the logic of NHST concrete, suppose we have a coin that we want to test for fairness. We decide that we will conduct an experiment wherein we flip the coin $N = 26$ times, and we observe how many times it comes up heads. If the coin is fair, it should usually come up heads about 13 times out of 26 flips. Only rarely will it come up with far fewer or far greater than 13 heads. Suppose we now conduct our experiment: We flip the coin $N = 26$ times and we happen to observe $z = 8$ heads. All we need to do is figure out the probability of getting that few heads if the coin were truly fair. If the probability of getting so few heads is sufficiently tiny, then we doubt that the coin is truly fair.

Notice that this reasoning depends on the notion of repeating the intended experiment, because we are computing the probability of getting 8 heads if we were to repeat an experiment with $N = 26$. In other words, we are figuring out the probability of getting 8 heads relative to the space of all possible outcomes when $N = 26$. Why do we restrict consideration to $N = 26$? Because that was the intention of the experimenter.

The problem with NHST is that the interpretation of the observed outcome depends on the space of possible outcomes when the experiment is repeated. Why is that a problem? Because the definition of the space of possible outcomes depends on the intentions of the experimenter. If the experimenter intended to flip the coin exactly $N = 26$ times, then the space of possibilities is all samples with $N = 26$. But if the experimenter intended to flip the coin for one minute (and merely happened to make 26 flips during that time), then the space of possibilities is all samples that could occur when flipping the coin for one minute. Some of those possibilities would have $N = 26$, but some would have $N = 23$, and some would have $N = 32$, and so on. On the other hand, the experimenter might have intended to flip the coin until observing 8 heads, and it just happened to take 26 flips to get there. In this case, the space of possibilities is all samples that have the 8th head as the last flip.

Notice that for any of those intended experiments (fixed N, fixed time, or fixed z), the actually observed data are the same: $z = 8$ and $N = 26$. But the probability of the observed data is different relative to each experiment space. The space of possibilities is determined by what the experimenter had in mind while flipping the coin.

Do the observed data depend on what the experimenter had in mind? We certainly hope not! A good experiment is founded on the principle that the data are insulated from the experimenter's intentions. The coin "knows" only that it was flipped 26 times, regardless of what the experimenter had in mind while doing the flipping. Therefore our conclusion about the coin should not depend on what the experimenter had in mind while flipping it.

This chapter explains some of the gory details of NHST, to bring mathematical rigor to the preceding comments and to bring rigor mortis to NHST. You'll see how NHST is committed to the notion that the covert intentions of the experimenter are crucial to interpreting the data, even though the data are not supposed to be influenced by the covert intentions of the experimenter.

11.1 NHST FOR THE BIAS OF A COIN

11.1.1 When the Experimenter Intends to Fix N

Now for some of the mathematical details of NHST. Suppose we intend to flip a coin $N = 26$ times and we happen to observe $z = 8$ heads. This result seems to suggest that the coin is biased, because the result is less than the 13 heads that we would expect to get from a fair coin. But someone who is skeptical about the claim that the coin is biased, a defender of the null hypothesis that the coin is fair, would argue that the seemingly biased result could have happened merely by chance from a genuinely fair coin. Because a "false alarm" (rejection of a null hypothesis when it is really true) is considered to be costly in scientific practice, we decide that we will only reject the null hypothesis if the probability that it could generate the result is very small, conventionally less than 5%. In other words, to reject the null hypothesis, we need to show that the probability of getting something as extreme as $z = 8$, when $N = 26$, is less than 5%.

What is the probability of getting a particular number of heads when N is fixed? The answer is provided by the *binomial probability distribution*, which states that the probability of getting z heads out of N flips is

$$p(z \mid N, \theta) = \binom{N}{z} \theta^z (1 - \theta)^{N-z} \tag{11.1}$$

where the notation $\binom{N}{z}$ will be defined later. The binomial distribution is derived by the following logic. Consider any specific sequence of N flips with z heads. The probability of that specific sequence is simply the product of the individual flips, which is the product of Bernoulli probabilities $\prod_i \theta^{y_i} (1 - \theta)^{1-y_i} = \theta^z (1 - \theta)^{N-z}$, which we first saw in Section 5.1, p. 78. But there are many different specific sequences with z heads. Let's count how many ways there are. Consider allocating z heads to N flips in the sequence. The first head could go in any one of the N slots, the second head could go in any one of the remaining $N - 1$ slots, the third head could go in any one of the remaining $N - 2$ slots, and so on until the zth head could go in any one of the remaining $N - (z - 1)$ slots. Multiplying those possibilities together means that there are $N \cdot (N - 1) \cdot \dots \cdot (N - (z - 1))$ ways of allocating z heads to N flips. As an algebraic convenience, notice that $N \cdot (N - 1) \cdot \dots \cdot (N - (z - 1)) = N! / (N - z)!$. In this counting of the allocations, we've counted different orderings of the same allocation separately. For example, putting the first head in the first slot and the second head in the second slot was counted as a different allocation than putting the first head in the second slot and the second head in the first slot. In the space of possible outcomes, there is no meaningful difference in these allocations, because they both have a head in the first and second slots. Therefore we get rid of this duplicate counting by dividing out by the number of ways of permuting the z heads among their z slots. The number of permutations of z items is $z!$. Putting this all together, the number of ways of allocating z heads among N flips, without duplicate counting of equivalent allocations, is $N! / [(N - z)! \, z! \,]$. This factor is also called the number of ways of choosing z items from N possibilities, or "N choose z" for short, and is denoted $\binom{N}{z}$. Thus, the overall probability of getting z heads in N flips is the probability of any particular sequence of z heads in N flips times the number of ways of choosing z slots from among the N possible flips. The product appears in Equation 11.1. An illustration of a binomial probability distribution is provided in the right panel of Figure 11.1, for $N = 26$ and $\theta = 0.5$. Notice that the abscissa ranges from $z = 0$ to $z = 26$, because in $N = 26$ flips it is possible to get anywhere from no heads to all heads.

The binomial probability distribution in Figure 11.1 is also called a *sampling distribution*. This terminology stems from the idea that any set of N flips is a representative sample of the behavior of the coin. If we were to repeatedly run experiments with a fair coin, such that in every experiment we flip the coin exactly N times, then, in the long run, the probability of getting each possible z would be the distribution shown in Figure 11.1. To describe it carefully, we would call it "the probability distribution of the possible sample outcomes," but that's usually abbreviated as "the sampling distribution."

The left side of Figure 11.1 shows the null hypothesis. It shows the probability distribution for the two states of the coin. According to the null hypothesis, the coin is fair, whereby $p(y = \text{heads}) = \theta = 0.5$. The two panels in the figure are connected by an implication arrow to denote the fact that when the sample size N is fixed, the sampling distribution on the right is implied.

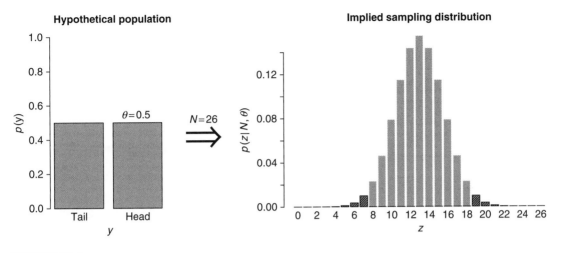

FIGURE 11.1
Binomial sampling distribution (Eqn. 11.1) when $p(y=\text{head}) = \theta = 0.5$ and N is fixed by the experimenter's intention. The total probability of the dark bars does not exceed 5%.

Our goal, as you'll recall, is to determine whether the probability of getting the observed result, $z = 8$, is tiny enough that we can reject the null hypothesis. By using the binomial probability formula in Equation 11.1, we determine that the probability of getting *exactly* $z = 8$ heads in $N = 26$ flips is 2.3%. Figure 11.1 shows this probability as the height of the bar over $z = 8$. However, we do not want to determine the probability of only the actually observed result. After all, for large N, *any* specific result z can be improbable. For example, if we flip a fair coin $N = 1000$ times, the probability of getting exactly $z = 500$ heads is only 2.5%, even though $z = 500$ is precisely what we would expect if the coin were fair.

Therefore, instead of determining the probability of getting exactly the result z from the null hypothesis, we determine the probability of getting z *or a result even more extreme than what we would expect*. The reason for considering more extreme outcomes is this: If we reject the null hypothesis because the result z is too far from what we would expect, then any other potential result that has an even more extreme value would also cause us to reject the null hypothesis. Therefore we want to know the probability of getting the actual outcome *or an outcome more extreme* relative to what we expect. This total probability is referred to as "the p value."[1] If this p value is less than a critical amount, then we reject the null hypothesis.

---

[1] The p value defined at this point is the "one-tailed" p value, because it sums the extreme probabilities in only one tail of the sampling distribution. In practice, the one-tailed p value is multiplied by 2 to get the two-tailed p value. We consider both tails of the sampling distribution because the null hypothesis could be rejected if the outcome were too extreme in either direction. This is explained later in the main text.

The critical probability is conventionally set to 5%. In other words, we will reject the null hypothesis whenever the total probability of the observed z or an outcome more extreme is less than 5%. Notice that this decision rule will cause us to reject the null hypothesis 5% of the time *when the null hypothesis is true*, because the null hypothesis itself generates those extreme values 5% of the time, just by chance. The critical probability, 5%, is the proportion of false alarms that we are willing to tolerate in our decision process. We set the critical z values such that the false alarm rate is no greater than 5%.

We also have to be careful to consider both directions of deviation from what we would expect. If we flip a coin and it comes up heads almost all the time, we suspect that it is biased. But if the coin comes up heads almost never, we also suspect that it is biased. Therefore, we have to establish the range of all possible extreme values, high *or* low, that would cause us to reject the null hypothesis. We let half the false alarms be due to high values and half be due to low values. Because we want the total false alarm rate to be no greater than 5%, we will reject the null hypothesis only when z is so *high* that it would reach or exceed that value less than 2.5% of the time by chance alone, *or* when z is so *low* that it would be that small or smaller by chance only 2.5% of the time. Figure 11.1 shows these extreme values of z as the darkly shaded bars in the tails of the distribution. The total probability mass of these bars does not exceed 2.5% in either tail. If the actually observed z falls among any of these darkly shaded extreme values, we reject the null hypothesis. For our specific situation, where the experimenter intended $N = 26$, we need a value of z that is 19 or greater, or 7 or less, to reject the hypothesis that $\theta = 0.5$.

Here's the conclusion for our particular case. The actual observation had $z = 8$, and so we would *not* reject the null hypothesis that $\theta = 0.5$. In NHST parlance, we would say that the result "has failed to reach significance." This does not mean we *accept* the null hypothesis; we merely suspend judgment regarding rejection of this particular hypothesis. Notice that we have not determined any degree of belief in the hypothesis that $\theta = 0.5$. The hypothesis might be true or might be false; we suspend judgment.

It is worth reiterating how this conclusion was reached: We considered the space of all possible outcomes if the intended experiment were repeated, and we determined the probabilities of extreme outcomes in this space of possibilities. We then examined whether the one actually observed outcome fell into the extreme zones of the space of possible outcomes.

11.1.2 When the Experimenter Intends to Fix z

Suppose that the experimenter did not intend to stop flipping when N flips were reached. Instead, the intention was to stop when z heads were reached. This scenario can happen in many real-life situations; for example, widgets on an assembly line can be checked for defects until z defective widgets are identified. In this situation, z is fixed in advance and N is the random variable. We don't talk about the probability of getting z heads out of N flips; we instead talk about the probability of taking N flips to get z heads.

What is the probability of taking N flips to get z heads? To answer this question, consider this: We know that the Nth flip is the zth head, because that is what signalled us to stop flipping. Therefore, the previous $N - 1$ flips had $z - 1$ heads in some random sequence. The probability of getting $z - 1$ heads in $N - 1$ flips is $\binom{N-1}{z-1}\theta^{z-1}(1 - \theta)^{N-z}$. The probability that the last flip comes up heads is θ. Therefore, the probability that it takes N flips to get z heads is

$$p(N \mid z, \theta) = \binom{N - 1}{z - 1}\theta^{z-1}(1 - \theta)^{N-z} \times \theta$$

$$= \binom{N - 1}{z - 1}\theta^{z}(1 - \theta)^{N-z}$$

$$= \frac{z}{N}\binom{N}{z}\theta^{z}(1 - \theta)^{N-z} \qquad (11.2)$$

Figure 11.2 shows an example of this probability distribution. This distribution is sometimes called the "negative binomial." Notice that values of N start at z and rise to infinity, because it takes at least z flips to get z heads, and it might take a huge number of flips to finally get the zth flip.

If the coin is biased to come up heads *rarely*, then it will take a *large* number of flips until we get z heads. If the coin is biased to come up heads *frequently*,

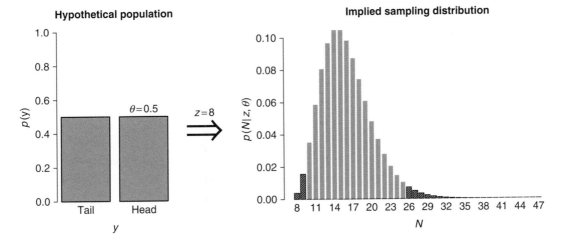

FIGURE 11.2
Sampling distribution of N when $p(y=\text{head}) = \theta = 0.5$ and z is fixed (Eqn. 11.2). The total probability of the dark bars does not exceed 5%.

then it will take a *small* number of flips until we get z heads. Figure 11.2 shows the values of observed N for which the probability of getting that result, or something more extreme, is less than 2.5% in each tail. These extreme values are marked as dark bars in the sampling distribution.[2] If the observed N falls in either of these extreme tails, we reject the null hypothesis.

Here is the conclusion for our specific example: The actual observation is N = 26, which falls in the extreme tail of the sampling distribution, and therefore we *reject* the null hypothesis. In other words, in the space of possible outcomes, the null hypothesis predicts that it is very rare for a fair coin to need 26 flips to get 8 heads; so rare, in fact, that we reject the null hypothesis. Notice that while we have rejected the null hypothesis, we still have no particular degree of disbelief in it, nor do we have any particular degree of belief in any other hypothesis. All we know is that the actual observation lies in an extreme end of the space of possibilities if the intended experiment were repeated.

11.1.3 Soul Searching

Let's summarize the situation. We watch the experimenter flip the coin. We see the same results as the experimenter, and we observe z = 8 heads out of N = 26 flips. According to NHST, if the intention of the experimenter was to stop when N = 26, then we do *not* reject the null hypothesis. If the intention of the experimenter was to stop when z = 8, then we *do* reject the null hypothesis. In other words, for us to draw a conclusion from the data, we need to know the experimenter's intentions. Exercise 11.1 shows you other examples of this dependence of the decision on the experimenter's intentions.

Notice that the actual observed events are the same regardless of how the experimenter decided to stop flipping coins; in either case we observe z heads in N flips. An outside observer of the flipping experiment, who is not privy to the covert intentions of the flipper, simply sees N flips, of which z were heads. It could be that the flipper intended to flip N times and then stop. Or it could be that the flipper intended to keep flipping until getting z heads. Or it could be that the flipper intend to flip for one minute. Exercise 11.3 explores this case of a fixed duration for an experiment. Or it could be that the flipper intended to flip until some other number of flips, but was interrupted by some outside factor, such as funding running out.

In real research, the actual reason for stopping is often neither because a pre-planned N was reached nor because a preplanned z was reached, nor because time ran out. Instead, real researchers will sometimes monitor the data as they are collected and do "preliminary" analyses on the data collected so far. If the

---

[2]The total probability in the right tail of Figure 11.2 is an infinite sum. It is easily computed by considering the finite complement to its left. In particular, $\sum_{n=26}^{n\to\infty} p(n\,|\,z,\theta) = 1 - \sum_{n=z}^{n=26-1} p(n\,|\,z,\theta)$.

currently collected data show significance, then data collection stops. If the data are close to significance, then data collection continues. With these intentions, the probability of getting significance is inflated because there is a chance of rejecting the null at every step along the way. In particular, if the experimenter *intended* to allow additional data to be collected after the preliminary inspection, *but did not end up collecting additional data*, the true probability of falsely rejecting the null is still inflated because the *potential* data space is larger, and there are more opportunities for rejecting the null. Exercise 11.4 shows you the details.

The solution to this mess is simple. All we have to do, to determine whether or not to reject the null hypothesis, is search the soul of the experimenter, to discover his or her true intentions about the experiment. Thus, when an experimenter reports his or her results, she or he can sign an affidavit of intent, or testify before Congress under oath. Or perhaps advances in brain scanning will one day give us objective measures of subconscious intent. Then NHST will be on solid ground. Right? Wrong.

In all of these scenarios, *the coin itself has no idea what the flipper's intention is*, and the propensity of the coin to come up heads does not depend on the intentions of the flipper. Indeed, we carefully design experiments to insulate the coins from the intentions of the experimenter. Therefore, our inference about the coin should not depend on the intentions of the experimenter.

A defender of NHST might be tempted to argue that I'm quibbling over trivial differences in the critical values. The critical values for the two cases I described earlier are nearly equal. Unfortunately, different intentions do not always lead to small differences in critical values. For example, when experimenters check their data after every flip of the coin to see if the result so far is "significant" by fixed-N critical values, the false alarm rate sky rockets. If you check at every flip to see if conventional 5% critical values have been exceeded, then the actual false alarm rate with 10 flips is 5.5%, with 20 flips it's 10.7%, with 30 flips it's 14.9%, with 40 flips it's 15.4%, and with 50 flips the true false alarm rate is 17.1%. In other words, if you are willing to flip the coin up to 50 times, and along the way you check at every flip to see if you can reject the hypothesis that the coin is fair, using critical values that are supposed to keep the false alarm rate down to 5% or less, then you actually have a 17.1% chance of falsely rejecting the hypothesis even when the coin is truly fair. You have to change the critical values quite a lot if you intend to check after every flip. Another situation in which critical values change dramatically is when experimenters intend to make multiple comparisons across different conditions in an experiment, as will be discussed later. Thus, we are not quibbling over tiny differences in critical values. Depending on the intentions, the critical values can change dramatically.

More fundamentally, the argument, that if any intention leads to nearly the same critical values then it's okay to use intentions, still fully admits that experimenter intentions influence the interpretation of data. It's like arguing that we shouldn't worry about the butchers putting their fingers on the scale, because no matter which butcher does it, the cheating is about the same. Admitting that experimenter intention influences the interpretation of data contradicts a basic premise of the data collection, that experimenter intentions have no influence on the data.

11.1.4 Bayesian Analysis

The Bayesian interpretation of data does not depend on the covert intentions of the data collector. In general, for data that are independent across trials, the probability of the conjoint set of data is simply the product of the probabilities of the individual outcomes. Thus, for $z = \sum_{i=1}^{N} y_i$ heads in N flips, the likelihood is $\prod_{i=1}^{N} \theta^{y_i}(1 - \theta)^{1-y_i} = \theta^z(1 - \theta)^{N-z}$, regardless of the experimenter's private reasons for collecting those data. The likelihood function captures everything we assume to influence the data. In the case of the coin, we assume that the bias of the coin is the only influence on its outcome, and that the flips are independent. The Bernoulli likelihood function completely captures those assumptions.

In summary, the NHST analysis and conclusion depend on the covert intentions of the experimenter, because those intentions define the space of all possible (unobserved) data. This dependence of the analysis on the experimenter's intentions conflicts with the opposite assumption that the experimenter's intentions have no effect on the observed data. The Bayesian analysis operates only with the actual data obtained and does not depend on the space of possible unobserved data.

11.2 PRIOR KNOWLEDGE ABOUT THE COIN

 Suppose that we are not flipping a coin, but we are flipping

a flat-headed nail. In a social science setting, this is like asking a survey question about left- or right-handedness of the respondent, which we know is far from 50/50, as opposed to asking a survey question about male or female sex of the respondent, which we know is close to 50/50. When we flip the nail, it can land with its point touching the ground (which I'll call tails) or it can land balanced on its head with its point sticking up (which I'll call heads). We believe, just by looking at the nail and our previous experience with nails, that it will not come up heads and tails equally often. Indeed, with its narrow head, the nail will very probably come to rest with its point touching the ground (i.e., "tails"). In other words, we have a strong prior belief that the nail is tail-biased. Suppose we flip the nail 26 times and it comes up heads on 8 flips. Is the nail "fair"? Would we use it to determine who gets to kick off at the Super Bowl?

11.2.1 NHST Analysis

The NHST analysis does not care if we are flipping coins or nails. The analysis proceeds the same way as before. To determine whether the nail is biased, we first declare the experimenter's intentions and then compute the probability of getting 8 heads or more if the nail were fair. As we saw in the previous section, if we declare that the intention was to flip the nail 26 times, then an outcome of 8 heads means we do *not* reject the hypothesis that the nail is fair. Let me say that again: We have a nail for which we have a strong prior belief that it is tail-biased. We flip the nail 26 times, and find it comes up heads 8 times. We conclude, therefore, that we cannot reject the null hypothesis that the nail can come up heads or tails 50/50. Huh? This is a *nail* we're talking about. How can you not reject the null hypothesis?

11.2.2 Bayesian Analysis

The Bayesian statistician starts the analysis with an expression of the prior knowledge. We know from prior experience that the narrow-headed nail is biased to show tails, so we express that knowledge in a prior. In a scientific setting, the prior is established by appealing to publicly accessible and reputable previous research. In our present toy example involving a nail, suppose that we represent our prior beliefs by a fictitious previous sample that had 95% tails in a sample size of 20. That translates into a beta($\theta \mid 2, 20$) prior distribution. If we wanted to go through the trouble, we could instead derive a prior from established theories regarding the mechanics of such objects, after making physical measurements of the nail such as its length, diameter, mass, rigidity, and so on. In any case, to make the analysis convincing to an audience of peers, the prior must be agreeable to that audience. Suppose that the agreed prior for the nail is beta($\theta \mid 2, 20$), then the posterior distribution is beta($\theta \mid 2 + 8, 20 + 18$), as shown in the right side of Figure 11.3. The posterior beliefs clearly do not include the nail being fair.

The differing inferences for a coin and a nail make good intuitive sense. Our posterior beliefs about the bias of the object *should* depend on our prior knowledge of the object: 8 heads in 26 flips of a narrow-headed nail *should* leave us with a different opinion than 8 heads in 26 flips of a coin. For additional details and a practical example, see Lindley & Phillips (1976).

11.2.2.1 *Priors Are Overt and Should Influence*

Some people might assert that prior beliefs are just as mysterious as the experimenter's intentions. But this assertion is wrong. Prior beliefs are not capricious and idiosyncratic. Prior beliefs are overt, explicitly debated, and consensual. A Bayesian analyst might have personal priors that differ from what most

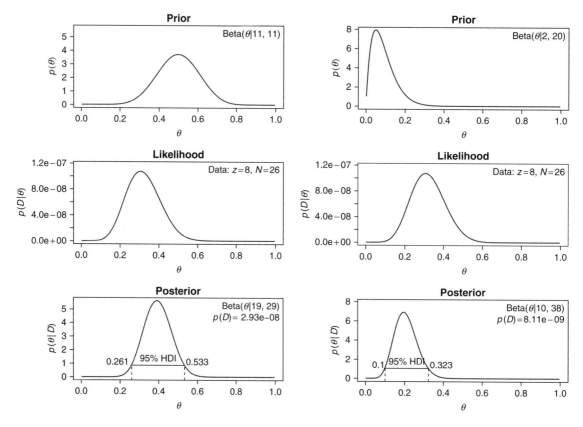

FIGURE 11.3

Posterior HDI for the bias of a Bernoulli process, when the prior assumes a fair coin *(left column)* or tail-strong nail *(right column)*.

people think, but if the analysis is supposed to convince an audience, then the analysis must use priors that the audience finds palatable. It is the job of the Bayesian analyst to make cogent arguments for the particular prior that is used. The research will not get published if the reviewers and editors think that that prior is untenable. Perhaps the researcher and the reviewers will have to agree to disagree about the prior, but even in that case the prior is an explicit part of the argument, and the analysis should be run with both priors in order to assess the robustness of the posterior. Science is a cumulative process, and new research is presented always in the context of previous research. A Bayesian analysis acknowledges this obvious fact, but NHST ignores it.

Some people might wonder, if subjective priors are allowed for Bayesian analyses, then why not allow subjective intentions for NHST? Because the subjective intentions in the data collector's mind do not influence the data and therefore should not influence the analysis. Subjective prior beliefs, on the other hand, are not about how beliefs influence the data but about how the data influence beliefs: Prior beliefs are the starting point from which we move in the light of new data.

Bayesian analysis tells us how much we should change our beliefs relative to our prior beliefs. Bayesian analysis does not tell us what our prior beliefs should be. Nevertheless, the priors are overt, public, and cumulative. Bayesian analysis provides an intellectually coherent method for determining the degree to which beliefs should change, and the conclusion is influenced by exactly what it should be influenced by, namely, the priors and the observed data. The conclusion is not influenced by what it should not be influenced by, namely, the experimenter's covert intention while gathering the data.

11.3 CONFIDENCE INTERVAL AND HIGHEST DENSITY INTERVAL

11.3.1 NHST Confidence Interval

The primary goal of NHST is determining whether a particular "null" value of a parameter can be rejected. One can also ask what *range* of parameter values would not be rejected. This range of non-rejectable parameter values is called the *confidence interval*. (There are different ways of defining an NHST confidence interval; this one is conceptually the most general and coherent with NHST precepts.) The 95% confidence interval consists of all values of θ that would not be rejected by a (two-tailed) significance test that allows 5% false alarms.

For example, in a previous section we found that $\theta = 0.5$ would not be rejected when $z = 8$ and $N = 26$, for a flipper who intended to stop when $N = 26$. The question is, which other values of θ would we not reject? Figure 11.4 shows the sampling distribution for different values of θ. The upper row shows the case of $\theta = 0.144$, for which the sampling distribution has $z = 8$ snug against the upper rejection tail. In fact, if θ is nudged any smaller, the rejection tail includes $z = 8$, which means that smaller values of θ can be rejected. The lower row of Figure 11.4 shows the case of $\theta = 0.517$, for which the sampling distribution has $z = 8$ snug against the lower rejection tail. If θ is nudged any larger, the rejection tail includes $z = 8$, which means that larger values of θ can be rejected. In summary, the range of θ values we would not reject is $\theta \in [0.144, 0.517]$. This is the 95% confidence interval when $z = 8$ and $N = 26$, for a flipper who intended to stop when $N = 26$. Exercise 11.2 has you examine this "hands-on."

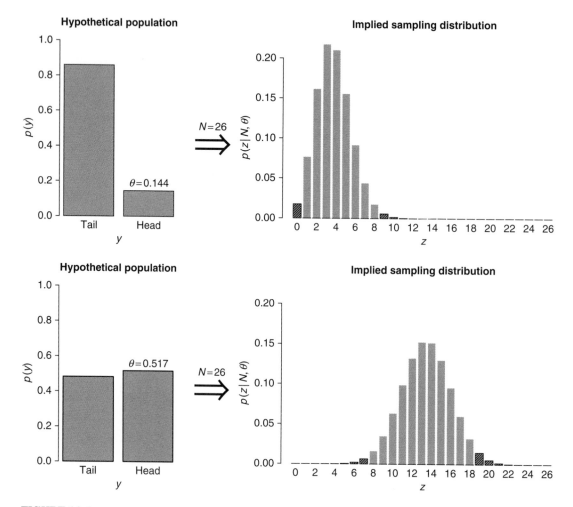

FIGURE 11.4
Confidence interval when $z = 8$ and $N = 26$, with N fixed by the experimenter's intention. Upper row shows $\theta = 0.144$, which is the lowest for which $z = 8$ is not in the rejection tail. Lower row shows $\theta = 0.517$, which is the highest for which $z = 8$ is not in the rejection tail. The NHST confidence interval is, therefore, $\theta \in [0.144, 0.517]$.

We can also determine the confidence interval for the experimenter who intended to stop when $z = 8$. Figure 11.5 shows the sampling distribution for different values of θ. The upper row shows the case of $\theta = 0.144$, for which the sampling distribution has $N = 26$ snug against the lower rejection tail. In fact, if θ is nudged any smaller, the rejection tail includes $N = 26$, which means that smaller values of θ can be rejected. The lower row of Figure 11.5 shows the case

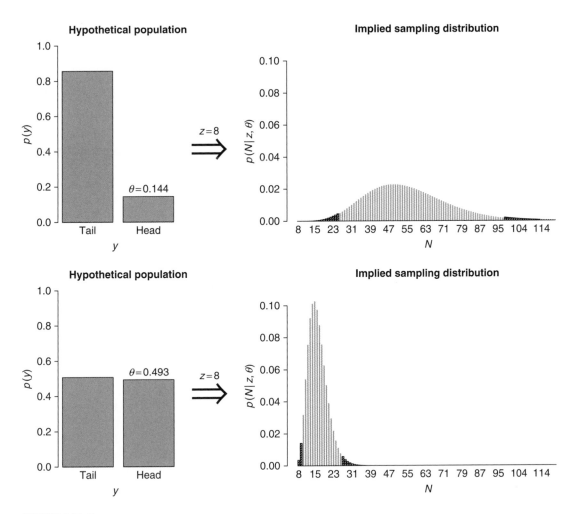

FIGURE 11.5
Confidence interval when $z = 8$ and $N = 26$, with z fixed by the experimenter's intention. Upper row shows $\theta = 0.144$, which is the lowest for which $N = 26$ is not in the rejection tail. Lower row shows $\theta = 0.493$, which is the highest for which $N = 26$ is not in the rejection tail. The NHST confidence interval is, therefore, $\theta \in [0.144, 0.493]$.

of $\theta = 0.493$, for which the sampling distribution has $N = 26$ snug against the upper rejection tail. If θ is nudged any larger, the rejection tail includes $N = 26$, which means that larger values of θ can be rejected. In summary, the range of θ values we would not reject is $\theta \in [0.144, 0.493]$. This is the 95% confidence interval when $z = 8$ and $N = 26$, for a flipper who intended to stop when $z = 8$.

We have just seen that the NHST confidence interval depends on the covert intentions of the experimenter. When the intention was to stop when $N = 26$, then the range of biases that would not be rejected is $\theta \in [0.144, 0.517]$. But when the intention was to stop when $z = 8$, then the range of biases that would not be rejected is $\theta \in [0.144, 0.493]$ (the fact that the lower ends of the confidence intervals are the same is merely an accidental coincidence for this case). The confidence interval depends on the experimenter's intentions, because those intentions dictate the space of possible unobserved data relative to which the actually observed data are judged. If the experimenter had other intentions, such as flipping for a fixed duration, then the confidence interval would be yet something different. Thus, the interpretation of the NHST confidence interval is as convoluted as the interpretation of NHST itself, because the confidence interval is merely the significance test conducted at every candidate value of θ.

The confidence interval tells us something about the probability of extreme unobserved data values that we might have gotten if we repeated the experiment according to the covert intentions of the experimenter. But the confidence interval tells us little about the believability of any particular θ value, which is what we want to know.

11.3.2 Bayesian HDI

A concept in Bayesian inference that is somewhat analogous to the NHST confidence interval is the highest density interval (HDI), which was introduced in Section 3.3.5, p. 40. The 95% HDI consists of those values of θ that have at least some minimal level of posterior believability, such that the total probability of all such θ values is 95%.

Let's consider the HDI when we flip a coin and observe $z = 8$ and $N = 26$. Suppose we have a prior informed by the fact that the coin appears to be authentic, which we express here, for illustrative purposes, as a beta($\theta \mid 11, 11$) distribution. The left side of Figure 11.3 shows that the 95% HDI goes from $\theta = 0.261$ to $\theta = 0.533$. These limits span the 95% most believable values of the bias. Moreover, the posterior density shows exactly how believable each bias is. In particular, we can see that $\theta = 0.5$ is within the 95% HDI, which we might use as a criterion if we are forced to categorically declare whether or not fairness is credible.

There are at least three advantages of the HDI over an NHST confidence interval. First, the HDI has a direct interpretation in terms of the believabilities of values of θ. The HDI is explicitly about $p(\theta \mid D)$, which is exactly what we want to know. The NHST confidence interval, on the other hand, has no direct relationship with what we want to know; there's no clear relationship between the probability of rejecting the value θ and the believability of θ. Second, the HDI has no dependence on the intention of the experimenter during data collection. The NHST confidence interval, in contrast, tells us about probabilities of data relative to what might have been if we replicated the experimenter's covert intentions. Third, the HDI is responsive to the analyst's prior beliefs, as it should be. The Bayesian analysis indicates how much the new data should alter our beliefs. The prior beliefs are overt and publicly decided. The NHST analysis, on the contrary, is ignorant of, and unresponsive to, the accumulated prior knowledge of the scientific community.

11.4 MULTIPLE COMPARISONS

In most experiments there are multiple conditions or treatments. Recall, for example, the experiment that investigated learning of four different category structures, two of which were "filtration" structures and two of which were "condensation" structures (Section 9.3.1, p. 219). Recall that the participants saw a rectangle with an interior vertical line segment. On different trials, the rectangle had different heights, and the interior vertical segment had different lateral positions. People had to learn which figures indicated which of two category labels. For example, some people were trained with labels such that tall rectangles indicated category A, whereas short rectangles indicated category B. This was a case of a filtration condition because the lateral position could be filtered out of consideration; only the height mattered for correct classification. Other people were trained with labels such that tallest rectangles or rightmost line segments indicated category A, whereas other figures indicated category B. This was a case of a condensation condition because both dimensions of variation had to be considered and condensed into a single categorical response. These conditions were studied because different theories of learning predict that some conditions should be easier to learn than others. Therefore, the goal of data analysis is to determine how different the learning performance is across the four conditions.

When comparing multiple conditions, the constraint in NHST is to keep the overall false alarm rate down to the desired level (e.g., 5%). Abiding by this constraint depends on the number of comparisons that are to be made, which in turn depends on the intentions of the experimenter. In a Bayesian analysis, however, there is just one posterior distribution over the parameters that describe the conditions. That posterior distribution is unaffected by the intentions of the experimenter, and the posterior distribution can be examined from multiple perspectives, as suggested by insight and curiosity. The next two sections expand on NHST and Bayesian approaches to multiple comparisons. I will often use the terms "condition" and "group" interchangeably.

11.4.1 NHST Correction for Experimentwise Error

When there are multiple groups, it often makes sense to compare each group to every other group. With four groups, for example, there are six different pairwise comparisons we can make (e.g., groups 1 versus 2, 2 versus 3, 1 versus 3, etc.). In NHST, we have to take into account which comparisons we intend to run for the whole experiment. The problem is that each comparison involves a decision with the potential for false alarm. Suppose we set a criterion for rejecting the null such that each decision has a "per-comparison" (PC) false alarm rate of α_{PC} (e.g., 5%). Our goal is to determine the overall false alarm rate when we conduct several comparisons. To get there, we do a little algebra. First, suppose the null hypothesis is true, which means that the groups are identical, and we get apparent differences in the samples by chance alone. This means that we get a false alarm on a proportion α_{PC} of replications of a comparison test. Therefore, we do *not* get a false alarm on the complementary proportion $1 - \alpha_{PC}$ of replications. If we run c independent comparison tests, then the probability of not getting a false alarm on *any* of the tests is $(1 - \alpha_{PC})^c$. Consequently, the probability of getting at least one false alarm is $1 - (1 - \alpha_{PC})^c$. We call that probability of getting at least one false alarm, across all the comparisons in the experiment, the "experimentwise" false alarm rate, denoted α_{EW}. Here's the rub: α_{EW} is greater than α_{PC}. For example, if $\alpha_{PC} = 0.05$ and $c = 6$, then $\alpha_{EW} = 1 - (1 - \alpha_{PC})^c = 0.26$. Thus, even when the null hypothesis is true, and there are really no differences between groups, if we conduct six independent comparisons, we have a 26% chance of rejecting the null hypothesis for at least one of the comparisons. Usually not all comparisons are structurally independent of each other, so the false alarm rate does not increase so rapidly, but it does increase whenever additional comparison tests are conducted.

One way to keep the experimentwise false alarm rate down to 5% is by reducing the permitted false alarm rate for the individual comparisons, that is, setting a more stringent criterion for rejecting the null hypothesis in individual comparisons. One often-used re-setting is the *Bonferonni correction*, which sets $\alpha_{PC} = \alpha_{EW}^{desired}/c$. For example, if the desired experimentwise false alarm rate is 0.05, and there are 6 comparisons planned, then we set each individual comparison's false alarm rate to 0.05/6. This is a conservative correction, because the actual experimentwise false alarm rate will usually be much less than $\alpha_{EW}^{desired}$.

There are many different corrections available to the discerning NHST aficionado (e.g., Maxwell & Delaney, 2004, Ch. 5). Not only do the correction factors depend on the structural relationships of the comparisons, but the correction factors also depend on whether the analyst intended to conduct the comparison before seeing the data, or was provoked into conducting the comparison only after seeing the data. If the comparison was intended in advance, it is called a *planned* comparison. If the comparison was thought of only after seeing a trend in the data, it is called a *post hoc* comparison. Why should it matter whether a comparison is planned or post hoc? Because even when the null hypothesis is true, and there are no real differences between groups, there will always be a highest and lowest random sample among the groups. If we don't plan in advance which groups to compare, but do compare whichever two groups happen to be farthest apart, we have an inflated chance of declaring groups to be different that aren't truly different.

The point, for our purposes, is not which correction to use. The point is that the NHST analyst must make some correction, and the correction depends on the number and type of comparisons that the analyst *intends* to make. This creates a problem because two analysts can come to the same data but draw different conclusions because of the variety of comparisons that they find interesting enough to conduct and what provoked their interest. The creative and inquisitive analyst, who wants to conduct many comparisons either because of deep thinking about implications of theory or because of provocative unexpected trends in the data, is penalized for being thoughtful. A large set of comparisons can be conducted only at the cost of using a more stringent threshold for each of the comparisons. The uninquisitive analyst is rewarded with an easier criterion for achieving significance. This seems to be a counterproductive incentive structure: You have a higher chance of getting a "significant" result, and getting your work published, if you feign narrow-mindedness under the pretense of protecting the world from false alarms.

To make this concrete, consider again the filtration/condensation experiment from Section 9.3.1, p. 219. The theory relating category structure to learning difficulty predicts that the filtration structures should be easier than the condensation structures, that the two condensation structures should be approximately equally difficult, and that the two filtration structures might be somewhat different in difficulty. Theory implies, therefore, three planned comparisons. But what if the analyst was less thoughtful, or took a more broad-brush approach, and planned only one comparison: The average filtration versus the average condensation. This single comparison would indeed address the primary theoretical issue, without worrying about ancillary nuances. The broad-brusher would be rewarded with a less stringent criterion for the test to achieve significance. On the other hand, suppose that upon seeing the data, the detail-oriented analyst discovers that the slower of the two filtration groups is not much faster than the faster of the two condensation groups. The two groups should therefore be compared. The analyst can treat this as a post hoc comparison, or the analyst can realize that it would have made sense to plan to compare each of the filtration groups individually against each of the condensation groups. After all, it's just as post hoc to notice that the slower of the filtration groups is clearly much faster than the faster of the two condensation groups, and decide therefore *not* to compare them. So we might as well be honest about it, and realize that the comparisons should have been planned in the first place. All this leaves the NHST analyst walking on the quicksand of soul-searching. Was the comparison truly planned or post hoc? Did the analyst commit premeditated exclusion of comparisons that should have been planned, or was the analyst merely superficial, or was the exclusion post hoc? This problem is not solved by picking a story and sticking to it, because it still presumes that the analyst's intentions should influence the data interpretation.

11.4.2 Just One Bayesian Posterior No Matter How You Look at It

The data from an experiment or from an observational study are carefully collected so to be totally insulated from the experimenter's intentions regarding subsequent tests. Indeed, the data should be uninfluenced by the presence or absence of any other condition or subject in the experiment! For example, it doesn't matter to an individual in a filtration group whether or not the experiment includes the other filtration group, or the condensation groups, or other conditions, or how many subjects there are in the groups. Moreover, the data are uninfluenced by the experimenter's intentions regarding the other groups and sample size. So why should our interpretation of the data depend on the experimenter's intentions if the data themselves are not influenced by the experimenter's intentions?

In a Bayesian analysis, the interpretation of the data is indeed *un*influenced by the experimenter's intentions. A Bayesian analysis yields a posterior distribution over the parameters of the model. The posterior distribution is the complete implication of the data. The posterior distribution can be examined in as many different ways as the analyst deems interesting; various comparisons of groups are merely different perspectives on the posterior distribution.

For example, in the case of the filtration-condensation experiment, the Bayesian analysis yields a posterior distribution over a high-dimensional parameter space, which includes the four μ_j parameters that describe the learning accuracies of the four conditions. (The other parameters were the individual learning biases, denoted θ_{ji}, and the four κ_j parameters that described how strongly the individual accuracies depended on the condition's μ_j.) Let's collapse across the other parameters and focus on the four conditions' μ_j parameters. If we want to determine whether group 1 tends to be more accurate than group 2, we determine how much of the posterior distribution has $\mu_1 > \mu_2$. Figure 9.16 showed a histogram of that difference, and that figure is repeated as Figure 11.6 for convenience. The left panel shows the posterior distribution of $\mu_1 - \mu_2$, from which we can ascertain whether a difference of zero is credible. The posterior distribution also tells us the believability of each candidate difference of μ's.

The other two panels of Figure 11.6 show other comparisons of the μ_j parameters. Those histograms merely summarize the posterior distribution from other perspectives. The posterior distribution itself is unchanged by how we look at it. We can examine any other comparison of μ_j parameters without worrying about what motivated us to consider it, because the posterior distribution is unchanged by those motivations.

In summary, the Bayesian posterior distribution is appropriately *insensitive* to the experimenter's covert intentions to compare or not compare various groups. The Bayesian posterior also directly tells us the believabilities of the magnitudes of differences, unlike NHST, which tells us only about whether a difference is extreme in a space of possibilities determined by the experimenter's intentions.

11.4.3 How Bayesian Analysis Mitigates False Alarms

No analysis is immune to false alarms, because randomly sampled data will occasionally contain accidental coincidences of outlying values. Bayesian analysis eschews the use of p values as a criterion for decision making, however, because the probability of false alarm depends dramatically on the experimenter's intentions. Bayesian analysis instead accepts the fact that the posterior is the best inference we can make, given the observed data and the prior beliefs.

How, then, does a Bayesian analysis address the problem of false alarms? By incorporating prior knowledge into the structure of the model. Specifically, if we know that different groups have some overarching commonality, even if their specific treatments are different, we can nevertheless model the different group parameters as having been drawn from an overarching distribution that expresses the commonality. An example of this was described in the right side of Figure 9.17, p. 226, where the group κ_c parameters were modeled by an overarching distribution. If several of the groups yield similar data, this similarity

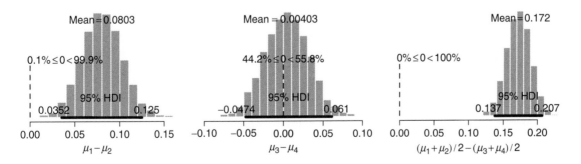

FIGURE 11.6
Histograms of differences in values of the mu parameters, generated by the script FilconBrugs.R (p. 232). (This figure repeats Figure 9.16.)

informs the overarching distribution, which in turn implies that any outlying groups should be estimated to be a little more similar than they would be otherwise. In other words, just as there can be shrinkage of individual estimates toward the group central tendency (recall Section 9.2.3, p. 212), there can be shrinkage of group estimates toward the overall central tendency. The shrinkage protects against accidental outliers and false alarms (e.g., Berry & Hochberg, 1999; Gelman, 2005; Gelman, Hill, & Yajima, 2009; Lindquist & Gelman, 2009; Meng & Dempster, 1987). This shrinkage is not an arbitrary "correction" like those applied in NHST. The shrinkage is a rational consequence of the prior knowledge expressed in the model structure. Section 18.2, p. 502, provides additional discussion and examples in the context of metric variables, which in NHST are analyzed with t-tests and ANOVA.

11.5 WHAT A SAMPLING DISTRIBUTION *IS* GOOD FOR

I hope to have made it clear that sampling distributions aren't as useful as posterior distributions for making inferences about hypotheses from a set of observed data. The reason is that sampling distributions tell us the probabilities of possible data if we run an intended experiment given a particular hypothesis, rather than the believabilities of possible hypotheses given that we have a particular set of data. Nevertheless, sampling distributions are appropriate and useful for other applications. Two of those applications are described in the following sections.

11.5.1 Planning an Experiment

Until this point in the book, I have emphasized analysis of data that have already been obtained. But a crucial part of conducting research is planning the study before actually obtaining the data. When planning research, we have a hypothesis about how the world might be, and we want to gather data that will inform us about the viability of that hypothesis. Typically we have some notion already about the experimental treatments or observational settings, and we want to plan how many observations we'll probably need to make or how long we'll need to run the study to have reasonably reliable evidence one way or the other.

For example, suppose that my theory suggests a coin should be biased with $\theta = 0.60$. Perhaps the coin represents a population of voters, hence flipping the coin means polling a person in the population, and the outcome heads means preference for candidate A. The theory regarding the bias may have come from previous polls regarding political attitudes. We would like to plan a survey of the population that will give us precise posterior beliefs about the true preference for candidate A. Suppose our intended survey will sample people until we obtain $z = 100$ people in favor of candidate A. By simulating the experiment over and over, using the hypothesized $\theta = 0.60$, we can generate expected data, and then derive a Bayesian posterior distribution for every set of simulated data. For every posterior distribution, we determine some measure of accuracy, such as the width of the 95% HDI. From many simulated experiments, we get a sampling distribution of HDI widths. From the sampling distribution of HDI widths, we can decide whether $z = 100$ typically yields high enough accuracy for our purposes. If not, we repeat the simulation with a larger z. Once we know how big z needs to be to get the accuracy we seek, we can decide whether or not it is feasible to conduct such a study.

Notice that we used the intended experiment to generate a space of possible data in order to anticipate what is likely to happen *when the data are analyzed with Bayesian methods*. For any single set of data (simulated or actual), we recognize that the individual data points in the set are insulated from the intentions of the design, and we conduct a Bayesian analysis of the data set. The use of a distribution of possible sample data, from an intended experiment, is perfectly appropriate here because it is exactly the implications of this hypothetical data distribution that we want to find out about.

The issues of research design are explored in greater depth in Chapter 13, which is entirely devoted to this topic.

11.5.2 Exploring Model Predictions (Posterior Predictive Check)

A Bayesian analysis only indicates the *relative* veracities of the various parameter values or models under consideration. The posterior distribution only tells us which parameter values are relatively less bad than the others. The posterior does not tell us whether the least bad parameter values are actually any good.

For example, suppose we believe that a coin is a heavily biased trick coin, and it either comes up heads 99% of the time or comes up tails 99% of the time; we just don't know which direction of bias it has. Now we flip the coin 40 times and it comes up heads 30 of those flips. It turns out that the 99%-head model has a far bigger posterior probability than the 99%-tail model. But it is also the case that the 99%-head model is a terrible model of a coin that comes up heads 30 out of 40 flips!

One way to evaluate whether the least unbelievable parameter values are any good is via a posterior predictive check. A posterior predictive check is an inspection of patterns in simulated data that are generated by typical posterior parameters values. Back in Exercise 5.8, p. 98, we explored an example of a posterior predictive check, and another example appeared in Section 7.4.2, p. 143. The idea of a posterior predictive check is as follows: If the posterior parameter values really are good descriptions of the data, then the predicted data from the model should actually "look like" real data. If the patterns in the predicted data do not mirror the patterns in the actual data, then we are motivated to invent models that can produce the patterns of interest.

This use of the posterior predictive check is suspiciously like null hypothesis significance testing: We start with a hypothesis (i.e., the least unbelievable parameter values), and we generate simulated data as if we were repeating our intended experiment over and over. Then we see if the actual data are typical or atypical in the space of simulated data. If we were to go further and determine critical values for false alarm rates and then reject the model if the actual data fall in its extreme tails, then we would indeed be doing NHST. But we don't go that far. Instead, the goal of the posterior predictive check is to drive intuitions about the qualitative manner in which the model succeeds or fails, and about what sort of novel model formulation might better capture the trends in the data. Once we invent another model, then we can use Bayesian methods to quantitatively compare it with the other models.

11.6 EXERCISES

R Hint: For many of these exercises, you may find it helpful to use R's `dbinom(x,size,prob)` function, where x corresponds to z (a vector from 0 to N) and `size` corresponds to N (a constant) in Equation 11.2. R also has a function for the negative binomial, `dnbinom(x,size,prob)`. Be careful if you use the negative binomial density, because the argument x corresponds to $N-z$ (a vector starting with 0) and the argument `size` corresponds to z (a constant) in Equation 11.2.

Exercise 11.1. [Purpose: To determine critical values for a two-tailed test, conduct the test, and notice the dependence of the conclusion on the intended stopping rule.]

Suppose we flip a coin $N = 17$ times. Our goal is to determine what values for the number of heads would be extreme enough to reject the hypothesis that the coin is fair. We want the total probability of false alarm to be less than 5%. In other words, if the null hypothesis is really true, we will mistakenly reject it less than 5% of the time. Therefore, we desire critical values z_{low} and z_{high} for the number of observed heads, such that $p(z \leq z_{low} | N = 17, \theta = 0.5) < 0.025$ and $p(z \geq z_{high} | N = 17, \theta = 0.5) < 0.025$. This is called a two-tailed test, because extreme values in either tail of the distribution can reject the null hypothesis.

(A) What is the value of z_{low}? Explain how you got your answer. Hint: Try
```
cumsum( dbinom( 0:17 , 17 , .5 ) ) < .025
```
and carefully explain what that does!

(B) What is the value of z_{high}? Explain how you got your answer. Hint: Try
```
cumsum( dbinom( 17:0 , 17 , .5 ) ) < .025
```
and carefully explain what that does!

(C) Suppose we flip the coin 17 times and get 4 heads. How many heads would we expect to get if the coin were fair? Can we reject the null hypothesis "at the 0.05 level" (which means, with two-tailed false alarm rate less than 0.05)?

(D) *New scenario:* We have a six-sided die, and we want to know whether the probability that the six-dotted face comes up is fair. Thus, we are considering two possible outcomes: six dots or not six dots. If the die is fair, the probability of the six-dot face is 1/6. Suppose we roll the die $N = 45$ times, intending to stop at that number of rolls. What are the values of z_{low} and z_{high}, using a two-tailed false alarm rate of 5%? Explain how you got your answer. Suppose we get 3 six-dot rolls. Can we reject the null hypothesis that the die is fair? (See the upper panel of Figure 11.7 for guidance.)

(E) Suppose we roll the die until we get 3 six-dot outcomes. It takes 45 rolls. (Notice this is the same result as the previous part.) How many six-dot outcomes would we expect to get if the coin were fair? Can we reject the null hypothesis that the die is fair, at the 0.05 level? (See the lower panel of Figure 11.7 for guidance.)

Exercise 11.2. [Purpose: To determine NHST confidence intervals and notice that they depend on the experimenter's intention.] Suppose we flip a coin $N = 26$ times and observe $z = 8$ heads. Assume that the intention was to stop when $N = 26$.

(A) Show that when $\theta = 0.144$, the probability of $z \geq 8$ is just over 2.5%, but for smaller values of θ, the probability of $z \geq 8$ is less than 2.5%. Hint: Try
```
for( theta in seq( .140 , .150 , .001 ) ) {
    show( c( theta , sum( dbinom( 8:26 , 26 , theta ) ) ) )
}
```
and explain carefully what that does.

(B) Show that when $\theta = 0.517$, the probability of $z \leq 8$ is just over 2.5%, but when θ exceeds 0.517, then $p(z \leq 8)$ drops below 2.5%.

(C) What is the 95% confidence interval for θ? (Just summarize the results of the previous two parts.)

(D) Suppose that the intention was to stop when $z = 8$ heads. Explain how to determine the NHST confidence interval, and include R code as appropriate. (Compare to Figure 11.5.)

Exercise 11.3. [Purpose: To determine the p value if a coin is flipped for a fixed period of time instead of for a fixed number of flips.] (For another example of NHST for fixed-duration samples, see Kruschke, 2010a.)

An experimenter is investigating whether there are more conservatives or liberals in her subject pool. She recruits 46 subjects and finds that 30 of them are liberal. We are interested in testing the null hypothesis that the population is split 50-50; (i.e., $\theta = 0.5$).

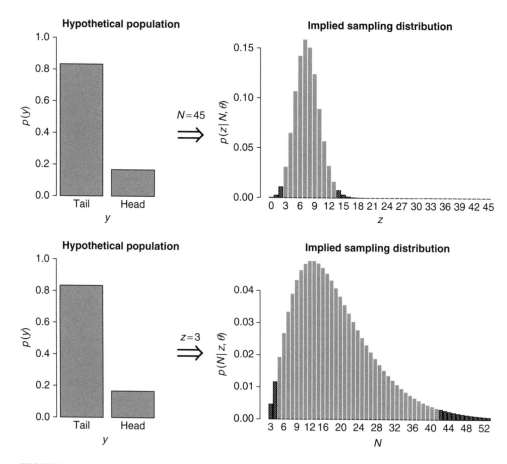

FIGURE 11.7

For use with latter parts of Exercise 11.1. *Upper panel*: Binomial sampling distribution (Eqn. 11.1) when $p(y=\text{head}) = \theta = 1/6$ and N is fixed at 45. The domain of z in the sampling distribution is integers from 0 to N. *Lower panel*: Sampling distribution (Eqn. 11.2) when $p(y=\text{head}) = \theta = 1/6$ and z is fixed. The domain of z in the sampling distribution is integers from z to ∞.

(A) If we assume that the experimenter intended to stop when $N = 46$, what is the probability of getting 30 or more liberals in a sample according to the null hypothesis? Do we reject the null hypothesis? Hint: `sum( dbinom( 30:46 , 46 , .5 ) ) < 0.025`

(B) We ask the experimenter why she chose $N = 46$. She replies that she didn't; in fact, she chose to run the experiment for 1 week, and she just happened to get 46 subjects. (This is typical in real-world research. There is nothing wrong with this procedure because we are assuming that every person polled is independent of every other person polled and uninfluenced by the poller's intentions.) With these intentions regarding duration, we cannot assume that N is fixed at 46, and therefore the space of possible experiment outcomes is much larger.

We know that if we repeated the experiment (i.e., recruited subjects for 1 week), then we would get $N \approx 46$, but not necessarily exactly $N = 46$. We will assume that the sample size N is distributed as a Poisson variable: $p(N \mid \mu) = \frac{1}{N!} e^{-\mu} \mu^N$. (Graphs of the Poisson distribution appear in Figure 22.1, p. 602.) A Poisson process is often used to model the arrival of customers in a queue (e.g., Sadiku & Tofighi, 1999). The Poisson distribution has mean μ. The Poisson distribution indicates the probability of getting N events during a unit time interval when the mean rate of events per unit time interval is μ. Because we observed 46 subjects in 1 week, we will set $\mu = 46$. The Poisson distribution $p(N \mid \mu = 46) = \frac{1}{N!} e^{-46} 46^N$ has highest probability around 46, indicating that usually a week will yield 46 subjects, but sometimes more and sometimes less. In R, the Poisson probability is given by dpois(N , mu).

We seek to determine the probability of getting a *proportion* of heads that is as large or larger than the observed proportion $30/46 = 0.652$, when recruiting subjects for a week, if the null hypothesis were true. We will compute the answer in the following manner. Suppose we obtain N subjects during the week. The probability of getting N is the Poisson probability $p(N \mid \mu = 46) = \frac{1}{N!} e^{-46} 46^N$. We use the binomial distribution to determine the probability of getting $z \geq \frac{30}{46} N$ liberals from the null hypothesis, with $p(z \mid N, \theta = .5) = \binom{N}{z} 0.5^z (1 - 0.5)^{N-z}$. Thus, the probability of the outcome z, N is $p(N \mid \mu = 46) \, p(z \mid N, \theta = 0.5)$. We accumulate this probability across all values of N and all values of $z \geq \frac{30}{46} N$. The accumulation over N is an infinite sum, so we approximate it by summing over a large range of N, beyond which the contribution by larger values of N is negligible. In R, we can execute this computation as follows (BinomNHSTpoissonrate.R)

```
1   z_obs = 30 ; N_obs = 46
2   nulltheta = .5
3   tail_prob = 0   # Zero initial value for accumulation over possible N.
4   for ( N in 1 : (3*N_obs) ) {   # Start at 1 to avoid /0. 3*N_obs is arbitrary.
5     # Create vector of z values such that z/N >= z_obs/N_obs
6     zvec = (0:N)[ (0:N)/N >= z_obs/N_obs ]
7     tail_prob = tail_prob + (
8                 dpois( N , N_obs ) * sum( dbinom( zvec , N , nulltheta ) ) )
9   }
10  show( tail_prob )
```

Is the probability of getting the observed proportion less than 2.5%? Do we reject the null hypothesis? How does this compare with the conclusion from the previous part of the exercise?

(C) Repeat the previous two parts, for $z = 26$ and $N = 39$.

Exercise 11.4. [Purpose: To determine the false alarm rate for a two-tiered data collection process (compare to Berger & Berry, 1988).]

Suppose an experimenter plans to collect data based on a two-tier stopping criterion. The experimenter will collect an initial batch of data with $N = 30$ and then do a NHST. If the result is not significant, then an additional 15 subjects' data will be collected, for a total N of 45. Suppose the researcher intends to use the standard critical values for determining significance at both the $N = 30$ and $N = 45$ stages. Our goal is determine the actual false alarm rate for this two-stage procedure, and to ponder what the mere intention of doing a second phase implies for interpreting the first stage, even if data collection stops with the first stage.

(A) For $N = 30$, what are z_{low} and z_{high}, assuming a two-tailed false alarm rate of .05 or less? (See Exercise 11.1.) (Hint: The answer is 9 and 21; your job is to explain how to get that answer.)

(B) For $N = 45$, what are z_{low} and z_{high}, assuming a two-tailed false alarm rate of .05 or less? (See Exercise 11.1.) (Hint: The answer is 15 and 30; your job is to explain how to get that answer.)

For the next parts of the exercise, consider Table 11.1. Each cell of Table 11.1 corresponds to a certain outcome from the first 30 flips and a certain outcome from the second 15 flips. A cell is marked by a dagger, †, if it has a result for the first 30 flips that would reject the null hypothesis. A cell is marked by a double dagger, ‡, if it has a result for the total of 45 flips that would reject the null hypothesis. For example, the cell with 10 heads from the first 30 flips and 1 head from the second 15 flips is marked with a ‡ because the total number

Table 11.1 Possible Results for Two-Stage NHST

Second 15 Flips	First 30 Flips									
	0	...	9	10	...	15	16	...	20	21	...	30
0	†, ‡	...	†, ‡	‡	...	‡	-	...	-	†	...	†, ‡
1	†, ‡	...	†, ‡	‡	...	-	-	...	-	†	...	†, ‡
⋮												
5	†, ‡	...	†, ‡	‡	...	-	-	...	-	†	...	†, ‡
6	†, ‡	...	†, ‡	-	...	-	-	...	-	†	...	†, ‡
7	†, ‡	...	†	-	...	-	-	...	-	†	...	†, ‡
8	†, ‡	...	†	-	...	-	-	...	-	†	...	†, ‡
9	†, ‡	...	†	-	...	-	-	...	-	†, ‡	...	†, ‡
10	†, ‡	...	†	-	...	-	-	...	‡	†, ‡	...	†, ‡
⋮												
15	†, ‡	...	†	-	...	‡	‡	...	‡	†, ‡	...	†, ‡

of heads for that cell, $10 + 1 = 11$, is less than 15 (which is z_{low} for $N = 45$). That cell has no single dagger, †, because getting 10 heads in the first 30 flips is not extreme enough to reject the null.

(C) Denote the number of heads in the first 30 flips as z_1, and the number of heads in the second 15 flips as z_2. Explain why it it true that the z_1, z_2 cell of the table has conjoint probability equal to `dbinom(z1,30,.5) *` `dbinom(z2,15,.5)`.

(D) What is the sum of the probabilities of all the cells that contain a † (whether or not it contains a ‡)? Explain how you got your answer! (Hint: The answer is not greater than 0.05. See also the Hint at the end of the exercise).

(E) What is the sum of the probabilities of all the cells that contain a ‡ (whether or not it contains a †)? Explain how you got your answer! (Hint: The answer is not greater than 0.05. See also the Hint at the end of the exercise.)

(F) What is the sum of the probabilities of all the cells that contain *either* a † or a ‡? *Note: This is the false alarm rate for the two-stage design, because these are all the ways you would decide to reject the null even when it's true.* Explain how you got your answer! (Hint: The answer *is* greater than 0.05. See also the Hint at the end of the exercise.)

(G) Suppose that the researcher intends to run an experiment using this two-stage stopping criterion. She collects the first 30 flips and finds 8 heads. She therefore rejects the null hypothesis and reports that $p < 0.05$. Is that correct? Explain. (Hint: The answer is no, it's not correct, because the design of the experiment included a larger potential sampling space.)

(H) Whenever we run an experiment and get a result that trends away from the null expectation, but isn't quite significant, it's natural to consider collecting more data. We saw in the previous part that even intending to collect more data, but not actually doing it, inflates the false alarm rate. Doesn't the fact that we always consider collecting more data mean that we always have a much higher false alarm risk than we pretend we do? Doesn't the actual false alarm rate of an experiment depend on the maximal number of data points we'd be willing to collect over the course of our lifetimes?

(HINT: Here is some R code for solving the above. Notice you can try other values of $N1$ and $N2$, which produce even more dramatic results.)

(NHSTtwoTierStoppingExercise.R)
```
1   # For NHST exercise regarding two-tier testing.
2
3   N1 = 30         # Number of flips for first test. Try 17.
4   N2 = 15         # Number of _additional_ flips for second test. Try 27 or 50.
5
6   theta = .5      # Hypothesized bias of coin.
```

```
7   FAmax = .05   # False Alarm maximum for a single test.
8   NT = N1 + N2  # Total number of flips.
9
10  # Determine critical values for N1:
11  # EXPLAIN what each function does and why, including
12  # dbinom, cumsum, which, max, and (0:N)[...]
13  loCritN1 = (0:N1)[ max( which( cumsum( dbinom(0:N1,N1,theta) ) <= FAmax/2 ) ) ]
14  hiCritN1 = (N1:0)[ max( which( cumsum( dbinom(N1:0,N1,theta) ) <= FAmax/2 ) ) ]
15  # Compute actual false alarm rate for those critical values.
16  # EXPLAIN what this does and why.
17  FA1 = sum( ( 0:N1 <= loCritN1 | 0:N1 >= hiCritN1 ) * dbinom(0:N1,N1,theta) )
18  cat( "N1:",N1 , ", lo:",loCritN1 , ", hi:",hiCritN1 , ", FA:",FA1 , "\n" )
19
20  # Determine critical values for NT:
21  # EXPLAIN what each function does and why, including
22  # dbinom, cumsum, which, max, and (0:N)[...]
23  loCritNT = (0:NT)[ max( which( cumsum( dbinom(0:NT,NT,theta) ) <= FAmax/2 ) ) ]
24  hiCritNT = (NT:0)[ max( which( cumsum( dbinom(NT:0,NT,theta) ) <= FAmax/2 ) ) ]
25  # Compute actual false alarm rate for those critical values.
26  # EXPLAIN what this does and why.
27  FAT = sum( ( 0:NT <= loCritNT | 0:NT >= hiCritNT ) * dbinom(0:NT,NT,theta) )
28  cat( "NT:",NT , ", lo:",loCritNT , ", hi:",hiCritNT , ", FA:",FAT , "\n" )
29
30  # Determine actual false alarm rate for the two-tier test:
31  # EXPLAIN each of the matrices below --- what is in each one?
32  Z1mat = matrix( 0:N1 , nrow=N2+1 , ncol=N1+1 , byrow=TRUE )
33  ZTmat = outer( 0:N2 , 0:N1 , "+" )
34  pZTmat = outer( dbinom( 0:N2 , N2 , theta ) , dbinom( 0:N1 , N1 , theta ) )
35  # EXPLAIN the matrices in computation below.
36  FA1or2 = sum( ( ( ZTmat <= loCritNT | ZTmat >= hiCritNT ) # double dagger matrix
37                  | ( Z1mat <= loCritN1 | Z1mat >= hiCritN1 ) # single dagger matrix
38                  ) * pZTmat )
39  cat( "Two tier FA:" , FA1or2 , "\n" )
```

CHAPTER 12

Bayesian Approaches to Testing a Point (''Null'')Hypothesis

Tell me what character does he possess?
Is he a scoundrel who couldn't care less?
Is he immaculate, squeaky, and clean? I'd
Estimate that he's somewhere in between.

Suppose that you have collected some data, and now you want to answer one of the following questions: Is the coin biased or not? Does the drug work or not? Is there a preference for candidate A or candidate B? In the previous chapter, I argued that the 20th-century way of answering these questions, via null hypothesis significance testing (NHST), has deep problems. This chapter describes Bayesian approaches to the question.

Doing Bayesian Data Analysis: A Tutorial with R and BUGS. DOI: 10.1016/B978-0-12-381485-2.00012-2
© 2011, Elsevier Inc. All rights reserved.

In the context of coin flipping, the question we are asking is whether the bias has some particular value. For example, if we are asking whether the coin is fair, we are asking whether a bias of 0.5 is credible. There are two different ways of formalizing this question in a Bayesian framework. One way to pose the question is to ask whether the value of interest falls among the credible values in the posterior. In particular, we can ascertain whether the value of interest is within the 95% highest density interval (HDI) of the posterior. A different way to pose the question sets up a dichotomy between, on the one hand, a prior distribution that allows *only* the value interest, and, on the other hand, a prior distribution that allows a broad range of all possible values. The posterior believability of the two priors is assessed via Bayesian model comparison.

The chapter explores the two approaches in some detail. The conclusions drawn by the two approaches do not yield the same information, so it is important to choose the approach that is most appropriate to the situation and the question you want to ask of your data.

A full treatment of this issue would involve Bayesian *decision theory*, which takes into account the costs of incorrect decisions and the benefits of correct decisions. For example, we would have to specify the cost of declaring a drug to be credibly better than a placebo when in fact it was not, as well as the cost of declaring a drug to be no better than a placebo when it fact it was. We would also have to specify the benefits of the complementary correct decisions. Moreover, we would have to determine the probability of each of those decision outcomes, so that we could figure out the expected value of the decision rule. Unfortunately, in many situations it is difficult to specify the costs and benefits with much accuracy, and therefore we will not study a full decision-theoretic treatment. See Berger (1985) for more information about Bayesian decision theory. Instead of incorporating costs and benefits of decisions, we will consider two methods that consider only posterior distributions.

12.1 THE ESTIMATION (SINGLE PRIOR) APPROACH

We have used this approach several times already in this book, and it is presented in many other textbooks (e.g., Berry, 1996; Bolstad, 2007; Carlin & Louis, 2009; Gelman et al., 2004; Gelman & Hill, 2007; Lynch, 2007). The decision rule goes roughly like this:

> A parameter value is declared to be *not* credible if it lies outside the 95% HDI of the posterior distribution of that parameter. If a parameter value lies within the 95% HDI, it is said to be among the credible values.

This decision rule will be modified and expanded later, when we get to the notion of a region of practical equivalence in Section 12.1.3. This approach is called the *estimation (single-prior)* method because it addresses the question of credibility by considering the posterior estimate of the parameter, derived from a single, often informed, prior distribution. Let's review this approach in several examples.

12.1.1 Is a Null Value of a Parameter among the Credible Values?

Consider a couple of situations in which we wanted to know whether a coin bias of 0.5 was credible. First, recall Exercise 5.3, p. 96, which described a learning experiment in which people learned to press different response keys for different cue words that appeared on a computer screen. After learning, there were some test cases in which conflicting cue words were displayed together. For each type of conflicting case, the researcher wanted to know if people were responding randomly, or instead had a bias away from 50/50 responding. The (fictitious but realistic) data showed that of 50 respondents, 35 gave response A whereas 15 gave response B. If the researcher begins with a uniform prior, then the posterior is a beta$(\theta \mid 36, 16)$ distribution, which has a 95% HDI from 0.567 to 0.813, as shown in Figure 12.1. Because 0.50 lies outside the 95% HDI, the researcher would conclude that people are not responding randomly (i.e., the "null" bias of 0.50 is not credible).

Another example, in which we considered whether a 50/50 bias was credible, was the investigation of therapeutic touch, in which researchers tested whether therapists could detect the presence of another person's hand several centimeters away. Of primary interest in the analysis was the estimate of the overall accuracy of the group of therapists. The overall accuracy was denoted by the parameter μ (mu), and the posterior estimate of the parameter is shown in the top-left panel of Figure 9.14, p. 218. The 95% HDI goes from 0.370 to 0.507, which includes the chance value of 0.50. The researcher would conclude that the therapists might well have been responding randomly, by 50/50 guessing, but other biases within the HDI are also credible. Notice that the posterior distribution provides an explicit description of our uncertainty in the posterior estimate.

12.1.2 Is a Null Value of a Difference among the Credible Values?

We have also already seen situations in which we wanted to know whether a nonzero *difference* of parameter values was credible. Recall Exercise 8.1, p. 188, which asked whether there is a "hot hand" in basketball, whereby there is a difference between (1) the probability $\theta_{AfterSuccess}$ of successfully making a shot immediately after a previous success and (2) the probability $\theta_{AfterFailure}$ of successfully making a shot immediately after after a previous failure. The posterior estimate of the difference in proportions is shown in Figure 12.2, where it can be seen that the 95% HDI goes from -0.0551 to $+0.109$. The researcher would conclude that there may well be *no* hot hand phenomenon, but nonzero differences within the HDI are credible.

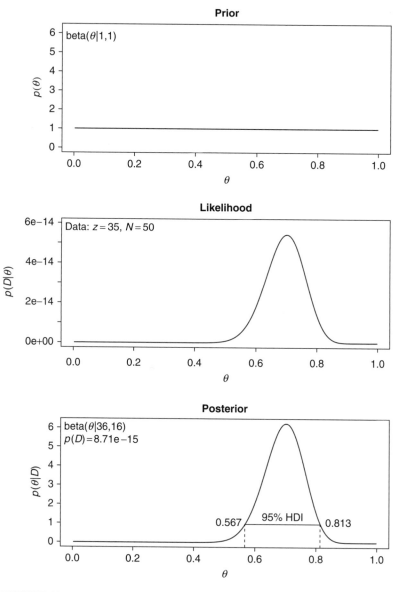

FIGURE 12.1

If there are 35 heads in 50 flips, is a bias of $\theta = 0.5$ among the credible values? According to the 95% HDI, no. (Data from Exercise 5.3, p. 96.)

Another situation in which we wanted to know about a difference of parameter values was the learning experiment involving category structures called filtration or condensation. There were four different category structures, and the parameter describing the accuracy of the jth group was denoted μ_j. Figure 9.16, p. 224, showed the posterior estimate of the differences of the

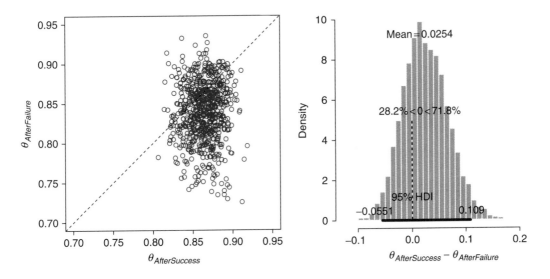

FIGURE 12.2
Posterior of biases from Exercise 8.1, p. 188. The posterior shows that a difference of zero is among the credible values.

parameters. The right panel of that figure shows that the difference between the filtration groups and the condensation groups has a 95% HDI going from 0.137 to 0.207, which suggests that a difference of zero is not credible (i.e., there is highly credible nonzero difference between the types of conditions).

12.1.2.1 Differences of Correlated Parameters
It is important to understand that marginal distributions of single parameters do not reveal whether or not the two parameter values are different. Figure 12.3, in its upper four panels, shows a case in which the posterior distribution for two parameter values has a strong positive correlation. Two of the panels show the marginal distributions of the single parameters. Those two marginal distributions suggest that there is a lot of overlap between the two parameters values. Does this overlap imply that we should not believe that they are very different? No! The histogram of the differences shows that the true difference between parameters is credibly greater than zero, with a difference of zero falling outside the 95% HDI. The upper-left panel shows why: The two parameter values are highly correlated, such that when we believe one parameter value is large, we believe that the other parameter value is also large. Because of this high correlation, the points in the conjoint distribution fall almost all on one side of the line of equality.

Figure 12.3 shows, in its lower four panels, a complementary case. Here, the marginal distributions of the single parameters are exactly the same as before: Compare the histograms of the marginal distributions, for the upper four panels and the lower four panels. Despite the fact that the marginal distributions

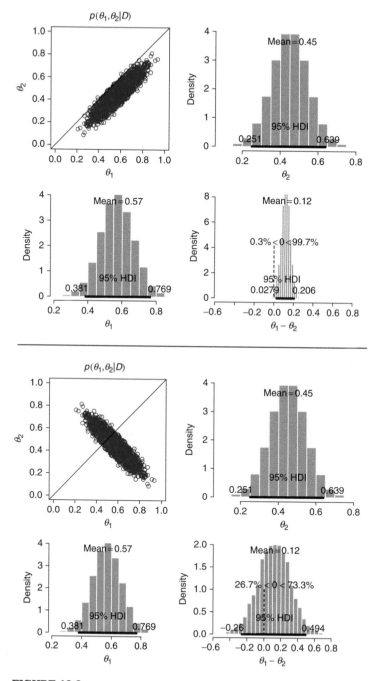

FIGURE 12.3

When there is a positive correlation between parameters, as shown in the upper four panels, the distribution of differences is narrower than when there is a negative correlation, as shown in the lower four panels.

are the same as before, the bottom-right panel reveals that the difference of
parameter values now straddles zero, with a difference of zero firmly in the
midst of the HDI. The plot of the conjoint distribution shows why: The two
parameter values are negatively correlated, such that when we believe one
parameter value is large, we believe that the other parameter value is small.
The negative correlation causes the conjoint distribution to straddle the line of
equality.

In summary, the marginal distributions of single parameters do not indicate
the relationship between the parameter values. The conjoint distribution of the
two parameters might have positive or negative correlation (or even a nonlinear dependency), and therefore the difference of the parameter values should
be explicitly examined.

12.1.3 Region of Practical Equivalence (ROPE)

The estimation approach can be enhanced by including a *region of practical
equivalence* (ROPE), which indicates a small range of values that are considered
to be practically equivalent to the null value for purposes of the particular
application. For example, if we wonder whether a coin is fair, for purposes of
determining which team will kick off at a football game, then we want to know
if the underlying bias in the coin is reasonably close to 0.50, and we don't really
care if the true bias is 0.51 or 0.49, because those values are close enough for
our application. As another example, if we are assessing the efficacy of a drug
versus a placebo, we might only consider using the drug if it improves the
probability of a cure by at least three percentage points, and we would declare
the drug to be counterproductive if it decreases the probability of a cure at all
(i.e., the lower boundary of the ROPE would be zero difference).

Once a ROPE is set, we make decisions according the following rule:

> A parameter value is declared to be *not* credible, or rejected, if its entire
> ROPE lies outside the 95% HDI of the posterior distribution of that
> parameter.

For example, suppose that we want to know whether a coin is fair, and we
establish a ROPE that goes from 0.45 to 0.55. We flip the coin 500 times and
observe 325 heads. If the prior is uniform, the posterior has a 95% HDI from
.608 to .691, which falls completely outside the ROPE. Therefore, we declare
that the null value of 0.5 is rejected for practical purposes.

Because the ROPE and HDI can overlap in different ways, there are different
decisions that can be made. In particular, we can decide to "accept" a null
value:

> A parameter value is declared to be accepted for practical purposes if
> that value's ROPE completely contains the 95% HDI of the posterior of
> that parameter.

For example, suppose that we want to know whether a coin is fair and we establish a ROPE that goes from 0.45 to 0.55. We flip the coin 1000 times and observe 490 heads. If the prior is uniform, the posterior has a 95% HDI from 0.459 to 0.521, which falls completely within the ROPE. Therefore, we declare that the null value of 0.5 is confirmed for practical purposes, because all of the credible values are practically equivalent to the null value.

This use of a ROPE around the null value also implies that if the null value really is true, we will eventually "accept" the null value as the sample size gets large enough. This is because, as the sample size gets larger, the HDI tends to get narrower and closer to the true value. When the sample size gets very large, the HDI is almost certain to be narrow enough, and close enough to the true value, to fall entirely within the ROPE. If we did not use a ROPE around the null value and rejected the null value any time that it falls outside the 95% HDI, then we would incorrectly reject the null in 5% of experiments even if when null value is true. This 5% false alarm rate occurs because random samples of data, when generated by the null value, will have coincidences of outliers, even for large N, that produce a 95% HDI that does not include the null value. Although 5% of those 95% HDIs exclude the null value, the HDIs do get narrower and closer to the null value as N gets large. Therefore, the HDIs do eventually fall within the ROPE around the null, even though 5% of the 95% HDIs exclude the null value itself. As an example of this point: Suppose we establish a ROPE that goes from 0.45 to 0.55. We flip the coin 10,000 times and observe 5200 heads. The 95% HDI goes from 0.510 to 0.530. Despite the fact that the HDI excludes the null value, we accept the null for practical purposes because the HDI falls entirely within the ROPE, which means that all of the credible values are practically equivalent to the null value. We would say that we believe the true bias in the coin is nearly 0.52, but that's close enough to the null value of 0.50 for practical purposes.

How is the size of the ROPE determined? If the application is a domain in which costs and benefits of decisions can be determined, then the full decision-theoretical machinery of expected utility can be brought to bear. In some domains such as medicine, expert clinicians can be interviewed, and their opinions can be translated into a reasonable consensus regarding how big of an effect is useful or important for the application. Otherwise, the ROPE might be established with somewhat arbitrary criteria, bearing in mind the key trade-off: When the ROPE is wider, there is a lower probability of falsely rejecting the null value (i.e., there is a smaller false alarm rate), but there is also a lower probability of rejecting the null value when it is false (i.e., there is a smaller hit rate). For further discussion of the ROPE, under somewhat different appellations of "range of equivalence" and "indifference zone," see, for example, Carlin & Louis (2009); Freedman, Lowe, & Macaskill (1984); Hobbs & Carlin (2008); and Spiegelhalter, Freedman, & Parmar (1994).

It is important to be clear that any discrete declaration about rejecting or accepting a null value does *not* exhaustively capture our beliefs about the parameter values. Our beliefs about the parameter value are described by the full posterior distribution. When making a binary declaration, we have merely compressed all that rich detail into a single bit of information. The broader goal of Bayesian analysis is conveying an informative summary of the posterior, and where the value of interest falls within that posterior. Reporting the limits of an HDI region is more informative than reporting the declaration of a reject/accept decision. By reporting the HDI and other summary information about the posterior, different readers can apply different ROPEs to decide for themselves whether a parameter is practically equivalent to a null value.

12.2 THE MODEL-COMPARISON (TWO-PRIOR) APPROACH

The previous section posed the question whether the null value is among the credible values, in terms of parameter estimation. We started with an informed prior distribution on the parameters, and then examined the posterior distribution on the parameters.

Some researchers prefer instead to pose the question in terms of model comparison. In this framing of the question, the focus is not on estimating the magnitude of the parameter. Instead, the focus is on deciding which of two hypothetical priors is least unbelievable. One hypothetical prior expresses the idea that the parameter value is exactly the null value. The alternative hypothetical prior expresses the idea that the parameter could be anything (or anything but the null value). Notice that neither of these hypothetical priors is informed by prior knowledge. This lack of being informed is often taken as a desirable aspect of the approach, not a defect, because the method is thereby "automatic" insofar as it obviates disputes about prior knowledge. We will see, however, that the model-comparison method can be extremely sensitive to the choice of "uninformed" prior for the alternative hypothesis (recall Figure 10.7, p. 261), and that model comparison is not necessarily meaningful unless both models are viable in the first place.

12.2.1 Are the Biases of Two Coins Equal?

Consider a situation in which we have two coins, and we would like to infer whether their biases are equal. We pose the question as a model comparison, such that one model expresses the "null" hypothesis that the two biases are equal, and the other model expresses the "alternative" hypothesis that the two biases could be any combination. The two models are distinguished by their priors (because the likelihood functions are the same in both models): The null model has zero prior probability on all combinations of biases except

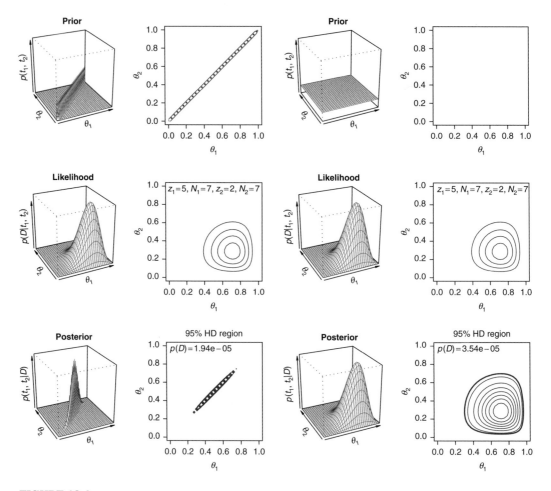

FIGURE 12.4

Two hypothetical priors for model comparison. The left columns show a ridge-shaped prior that allows only points for which $\theta_1 = \theta_2$. The right columns show a uniform prior, in which any combination of theta values is entertained. The top-right contour plot is empty because the surface is exactly horizontal, without any height changes to mark by a contour.

where they are equal, whereas the alternative model has uniform probability on all combinations of biases. The uniform alternative is chosen because it is a convenient expression of noninformed indifference.

Figure 12.4 shows an example of this approach. The "null" model, M_{null}, is shown in the left columns. Notice that this prior is shaped as a ridge along all values for which $\theta_1 = \theta_2$. The prior gives zero probability to any point at which $\theta_1 \neq \theta_2$. Notice also that the height of the ridge is horizontal, meaning that it is not informed by any prior knowledge (e.g., that the coins are probably fair).

The "alternative" model, M_{alt}, is shown in the right columns of Figure 12.4. This prior puts equal probability on all possible combinations of theta values. Notice that this hypothesis also is not informed by any prior knowledge (e.g., that the coins are probably fair).

The lower row of Figure 12.4 shows the posteriors resulting from two hypothetical priors for the same data that have been used in previous examples such as Figure 8.1, p. 161. Notice that the posterior on the null model is a ridge like the prior, except that the extreme ends of the ridge have been attenuated because they are inconsistent with the data.

To do model comparison, we use the implication of Bayes' rule:

$$\frac{p(M_{alt} \mid D)}{p(M_{null} \mid D)} = \underbrace{\frac{p(D \mid M_{alt})}{p(D \mid M_{null})}}_{\text{BF}} \frac{p(M_{alt})}{p(M_{null})} \qquad (12.1)$$

where the ratio marked "BF" is the Bayes factor of the alternative model relative to the null model. The prior beliefs in each model are typically assumed to be equal—that is, $p(M_{null}) = p(M_{alt}) = 0.5$—in the spirit of noninformed priors. The posterior plots in Figure 12.4 display the values of $p(D \mid M)$ for each model. The evidence for the null model is $p(D \mid M_{null}) = 1.94 \times 10^{-5}$, whereas the evidence for the alternative model is $p(D \mid M_{alt}) = 3.54 \times 10^{-5}$. The Bayes factor therefore slightly favors the alternative prior, but not by much. Because the ratio of posterior probabilities is not extreme, we would conclude that either model remains reasonably credible, given the data. If the Bayes factor had turned out to be more extreme, we might decide to declare one or the other prior to be less unbelievable than the other prior.

12.2.1.1 Formal Analytical Solution
The Bayes factor for the null and alternative models can also be computed analytically in this case. The alternative hypothesis has a uniform prior. It is uniform because the alternative hypothesis in this approach is supposed to be an "automatic" conventional prior that expresses a hypothesis complementary to the null hypothesis. A uniform prior on θ_1, θ_2 can be described as a product of beta distributions, namely, beta($\theta_1 \mid 1, 1$) × beta($\theta_2 \mid 1, 1$). Therefore, we can determine an exact value for $p(D \mid M_{alt})$ from Equation 8.5 (p. 160), which becomes

$$p(D \mid M_{alt}) = \frac{B(z_1+1, N_1-z_1+1)\, B(z_2+1, N_2-z_2+1)}{B(1, 1)B(1, 1)}$$

$$= B(z_1+1, N_1-z_1+1)\, B(z_2+1, N_2-z_2+1) \qquad (12.2)$$

because $B(1, 1) = 1$.

We can also derive a formal analytical expression for the evidence for the null hypothesis, $p(D \mid M_{null})$. First, notice this: Because the null hypothesis belief density at any point off the ridge is zero (i.e., $p(\theta_1 \neq \theta_2) = 0$), the double integral for the evidence simplifies to a single integral over $\theta = \theta_1 = \theta_2$:

$$p(D \mid M_{null}) = \iint d\theta_1 d\theta_2 \, p(D \mid \theta_1, \theta_2) p(\theta_1, \theta_1)$$

$$= \int d\theta \, p(D \mid \theta)$$

We now plug in the Bernoulli likelihood function to get

$$p(D \mid M_{null}) = \int d\theta \, p(D \mid \theta)$$

$$= \int d\theta \, \theta^{(z_1)} (1 - \theta)^{(N_1 - z_1)} \theta^{(z_2)} (1 - \theta)^{(N_2 - z_2)}$$

$$= \int d\theta \, \theta^{(z_1 + z_2)} (1 - \theta)^{(N_1 - z_1 + N_2 - z_2)}$$

$$= B(z_1 + z_2 + 1, \, N_1 - z_1 + N_2 - z_2 + 1) \tag{12.3}$$

because, by definition, $B(a, b) \equiv \int d\theta \, \theta^{(a-1)} (1 - \theta)^{(b-1)}$.

The Bayes factor for the alternative hypothesis, relative to the null hypothesis, is then the ratio of Equations 12.2 and 12.3:

$$\frac{p(D \mid M_{alt})}{p(D \mid M_{null})} = \frac{B(z_1 + 1, N_1 - z_1 + 1) \, B(z_2 + 1, N_2 - z_2 + 1)}{B(z_1 + z_2 + 1, \, N_1 - z_1 + N_2 - z_2 + 1)} \tag{12.4}$$

As a concrete example, consider the case presented in Figure 12.4, wherein $z_1 = 5$, $N_1 = 7$, $z_2 = 2$, and $N_2 = 7$. Plugging those values into Equation 12.4 yields $p(D \mid H_{alt})/p(D \mid H_{null}) = 1.82$. The grid approximation in Figure 12.4 displays $p(D \mid H_{alt}) = 3.54 \times 10^{-5}$ and $p(D \mid H_{null}) = 1.94 \times 10^{-5}$, yielding a ratio of 1.82, which is the same as the analytical solution.

12.2.1.2 Example Application

Consider again the question of the "hot hand" in basketball, from Exercise 8.1, p. 188, which asked whether there are more successful shots after a preceding successful shot than after a preceding failed shot. For the case of one famous professional player, there were 251 successes after 285 initial successes, and 48 successes after 53 initial failures.

We can use model comparison to analyze this situation. Using Equation 12.4, we obtain a Bayes factor of 0.128, which, inverted, is a Bayes factor of 7.81 in favor of the null prior. If the priors on the models are 50/50, then the posterior probability of the null model is $p(M_{null} | D) = 88.7\%$, and the posterior probability of the alternative model is $p(M_{alt} | D) = 11.3\%$.

The model comparison suggests that the null model is less unbelievable than the alternative model. But should we conclude, therefore, that there is zero difference between the conditions? Maybe, but not necessarily. Prior belief suggests that it is extremely unlikely that there is literally zero correlation between the first and second shots of a pair of free throws in basketball. Therefore, even though the model comparison suggests that the null prior is less unbelievable than the alternative prior, what we might really want is an estimate of the difference between success after success and success after failure. The model comparison does not provide such an estimate.

On the other hand, the direct estimation of the underlying biases did provide an estimate of the difference, as was shown in Figure 12.2, p. 299. There we saw that a difference of zero was among the credible differences, but we also saw that the mean difference was a bit larger than zero and we saw explicitly that there is moderately large uncertainty in the estimate of the difference. This conclusion from the direct estimate is a more informative inference from the data than the conclusion from model comparison, which states only that the ridge prior is less inconsistent with the data than the uniform prior. It should be remembered that the direct estimate of the difference began with a mildly informed prior, using the knowledge that professional players tend to make about 75% of their free throws, unlike the alternative model's uniform prior in the model comparison. Thus, the posterior from the alternative model is not the same as the posterior from direct estimation.

12.2.2 Are Different Groups Equal?

Suppose we conduct a study that has four conditions. Suppose every participant in every condition gets the same list of 20 words to try to memorize. The ability to recall a word is modeled as a Bernoulli distribution, with probability θ_{ij} for the ith person in the jth condition. The individual recall propensity θ_{ij} depends on hyperparameters, μ_j and κ_j, that describe the overarching recall propensity in each condition, because $\theta_{ij} \sim \text{beta}(\theta_{ij} | \mu_j \kappa_j, (1 - \mu_j)\kappa_j)$. (Abstractly, this is a design just like the four category structures of the filtration-condensation experiment of Section 9.3.1.)

The only difference between the conditions is the type of music being played during learning and recall. For the four groups, the music comes from, respectively, the death-metal band Das Kruschke,[1] Mozart, Bach, and Beethoven. For the four conditions, the mean number of words recalled is 11.85, 9.85, 9.50,

---

[1] To find information regarding Das Kruschke, search www.metal-archives.com. The author has no relation to the band, other than, presumably, some unknown common ancestor many generations in the past. The author was, however, in a garage band as a teenager. That band did not think it was playing death metal, although the music may have sounded that way to the critters fleeing the area.

and 9.60. The fictitious, randomly generated data can be examined in detail by running the program in Section 12.4.1(`OneOddGroupModelComp.R`).

We would like to know whether there is an effect of the type of music on ability to remember words. The most straightforward way to find out is to estimate the parameters and then examine the posterior differences of the parameter estimates. The histograms in Figure 12.5 show the distributions of differences between the μ_j parameters. It can be seen that μ_1 is quite different than μ_3 and μ_4; a difference of zero falls well outside the 95% HDI intervals. From this

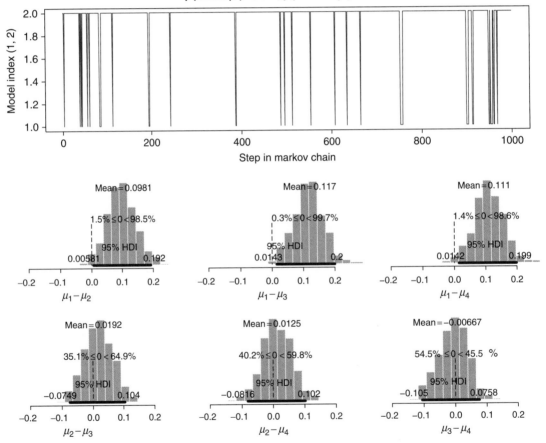

FIGURE 12.5

Upper panel: Results of model comparison shows that the model with a *single* mu parameter ("SameMu") is *preferred* to a model with a separate mu parameter for each group ("DiffMu"). Histograms: Differences of posterior μ_j values for the four groups in the different-mu model. Notice that μ_1 *is credibly different from* μ_3 *and* μ_4.

we would conclude that Das Kruschke enhances memory better than Bach and Beethoven, but perhaps not better than Mozart.

A model-comparison approach addresses the issue a different way. It compares the full model, which has distinct μ_j parameters for the four conditions, against a restricted model, which has a shared μ_0 parameter to describe all the conditions simultaneously. The two models have equal (50/50) prior probabilities. The upper panel of Figure 12.5 shows the results of a model comparison, using transdimensional MCMC with pseudopriors. The shared-μ_0 model is preferred, about 19 to 1 (0.951 to 0.049), over the distinct μ_j model. In other words, from this model comparison we would conclude that there is *no* difference in memory between the groups.

Which analysis should we believe? Is condition 1 different from conditions 3 and 4, or are all the conditions the same? Consider carefully what the model comparison actually says: Given the choice between one shared parameter and four different group parameters, the one-parameter model is less unbelievable. But that does not mean that the one-parameter model is the best possible model. In fact, if a different model comparison is conducted, that compares the one-parameter model against a different model that has one parameter for group 1 and a second parameter that is shared for groups 2 through 4, then the comparison favors the two-parameter model (see for yourself in Exercise 12.1).

In principle, we could consider all possible models formed by partitioning the four groups. For four groups, there are 15 distinct partitions. We could, in principle, put a prior belief on each of the 15 models and then do a comparison of the 15 models (Gopalan & Berry, 1998). From the posterior probabilities of the models, we could ascertain which partition was most believable, and decide whether it is more believable that other nearly-as-believable partitions. (Other approaches have been described by Berry & Hochberg, 1999; Mueller, Parmigiani, & Rice, 2007; Scott & Berger, 2006.) Suppose that we conducted such a large-scale model comparison, and found that the most believable model partitioned groups 2 to 4 together, separate from group 1. Does this mean that we should truly believe that there is zero difference between groups 2, 3, and 4? Not necessarily. If the group treatments are different, such as the four types of music in the present scenario, then there is almost certainly at least some small difference between their outcomes. (In fact, the simulated data do come from groups with all different means.) We may still want to estimate the magnitude of those small differences. An explicit posterior estimate will reveal the magnitude and uncertainty of those estimates. Thus, unless we have a viable reason to believe that different group parameters may be literally identical, an estimation of distinct group parameters will tell us what we want to know, without model comparison.

12.3 ESTIMATION OR MODEL COMPARISON?

12.3.1 What Is the Probability That the Null Value Is True?

There may be situations in which we want to know the probability that the null value is true. Proponents of the model-comparison approach may argue that the approach is especially applicable to these situations because the model comparison yields the posterior probability of the null hypothesis. Unfortunately, the posterior probability of the null is only a relative one, depending completely on the alternative hypothesis that is used in the model comparison. For example, as mentioned earlier for the fictitious memory experiment, a comparison of the null hypothesis (single-parameter) model against the four-parameter model yielded a strong preference *in favor of* the null hypothesis, but a comparison of the null hypothesis against a two-parameter model yielded a preference *against* the null hypothesis.

The posterior probability of the null hypothesis also depends crucially on the prior distribution assumed in the alternative model (e.g., Liu & Aitkin, 2008). Much of the effort in pursuing the model-comparison approach to null-hypothesis testing goes into justifying the form of the alternative-hypothesis prior (e.g., Edwards, Lindman, & Savage, 1963; Gallistel, 2009; Rouder et al., 2009; Wagenmakers, 2007). Often the goal is to establish an "automatic" vague prior for the alternative model that has useful mathematical properties. The benefit of a conventionalized vague prior is that possible disputes regarding the best prior might be ignored by appealing to convention. Unfortunately, none of the automatic vague priors is usually the prior that is most appropriate for model comparison, which is an informed prior. A model comparison is useful if both models are genuinely viable, and viability is enhanced by incorporating prior information. For example, suppose a researcher is interested in investigating whether practitioners of therapeutic touch can detect the presence of another person's hand from a distance of a few centimeters (see Section 9.2.4, p. 217). The null hypothesis is that detection accuracy is at chance: $\theta = 0.5$. Should the alternative hypothesis, which entertains non-chance values, be set at an "automatic" uniform distribution? Proponents of therapeutic touch would probably say not; instead, the prior distribution should incorporate prior knowledge about how well the practitioners might do in this unusual task. The practitioners might argue that accuracy would typically be only a little above chance, perhaps at $\theta = 0.6$, with fair uncertainty, as expressed perhaps in a beta($\theta \mid 9, 6$) distribution.

The estimation approach, when combined with a ROPE, provides a different measure of the probability of the null. This probability is simply the proportion of the posterior distribution that falls within the ROPE. This proportion of the HDI within the ROPE is not the probability that the null value is true;

instead, the proportion is the probability that the parameter is practically equivalent to the null. Obviously, this probability depends enormously on the limits of the ROPE, and it is meaningful only if the limits of the ROPE are meaningful landmarks for the particular application. Finally, the proportion of the HDI within the ROPE is most interpretable when it is large, because the proportion can be small for the trivial reason that the posterior is broad and shallow due to a small data set.

12.3.2 Recommendations

In general, Bayesian model comparison is only useful when both models are genuinely viable. If one or both models has little prior believability, then the Bayes factor and relative posterior believabilities are of little use. For example, suppose that we observe that there are many nicely wrapped gifts strewn across the lawn beside the house. We might hypothesize two models for this observation: One hypothesis is that Santa Claus dropped them from the roof. Another hypothesis is that the Grinch dumped them from the roof. A Bayesian model comparison might prefer the Santa Claus hypothesis 50 to 1 over the Grinch hypothesis, but that does not mean we should believe in Santa Claus. Model comparison requires both models to have prior viability, not just one. To explain the gifts on the lawn, one hypothesis is that Santa Claus lost his grip (i.e., a nonviable hypothesis) and the other hypothesis is that a philosophically impressionable teenager in the household scattered the gifts in a fanatical fit of antimaterialist divestment (i.e., a viable if also unlikely hypothesis). A Bayesian model comparison might prefer the hypothesis of the antimaterialist teenager, 50 to 1, but only by virtue of the fact that the alternative hypothesis was not viable in the first place. (Compare to Section 10.4.2.)

The premise that Bayesian model comparison is most meaningful when both models are genuinely viable implies that "automatic" Bayesian null-hypothesis testing might not be meaningful unless it is carefully applied only to situations in which both the null hypothesis is theoretically viable and the viable alternative hypothesis uses a reasonably informed prior. It is up to the user of the method to decide whether these conditions are met and to carefully interpret the results of the comparison.

In general, Bayesian model comparison is a very useful technique, outside its specific application to null-hypothesis testing, as long as the models are viable, and as long as the models have priors that are equivalently informed. Even if both models are theoretically viable, if one model has a prior distribution on its parameters that is already close to the posterior but the other model has a prior distribution on its parameters that is far from the posterior, then the models are not starting with priors that are equivalently informed. These general points about Bayesian model comparison apply especially strongly to the specific case of Bayesian null-hypothesis testing.

There may be situations that specifically demand a dichotomous decision regarding a null value. In these situations, Bayesian null-hypothesis testing may be called for, but the alternative model and alternative prior should be carefully considered. Otherwise, the estimation approach can be both more informative and easier to implement.

12.4 R CODE

12.4.1 R Code for Figure 12.5

This code is a straightforward modification of the code from Section 10.2.2. Instead of the κ values being either distinct or shared across groups, the μ values are either distinct or shared across groups. In principle, the κ values could (or, indeed, should) simultaneously be either distinct or shared, but here the κ values are always distinct, merely to keep the code more readable. The only other difference from the code from Section 10.2.2 is that fictitious data are generated, whereby the groups have true biases of 0.61, 0.50, 0.49, and 0.51, respectively.

(OneOddGroupModelComp.R)

```
5    #- - - - - - - - - - - - - - - - - - - - - - - - - - - - - - - - - - - - - - - - - - - - - - - - - - - - - - - -
6    # THE MODEL.
7
8    modelstring = "
9    # BUGS model specification begins here...
10   model {
11       for ( i in 1:nSubj ) {
12           # Likelihood:
13           nCorrOfSubj[i] ~ dbin( theta[i] , nTrlOfSubj[i] )
14           # Prior on theta (notice nested indexing):
15           theta[i] ~ dbeta( aBeta[ CondOfSubj[i] ] ,
16               bBeta[ CondOfSubj[i] ] )I(0.0001,0.9999)
16       }
17       # Re-parameterization of aBeta[j],bBeta[j] in terms of mu and kappa:
18       for ( j in 1:nCond ) {
19           # Model 1: Distinct mu[j] each group.  Model 2: Shared mu0 all groups.
20           aBeta[j] <-       ( mu[j]*(2-mdlIdx) + mu0*(mdlIdx-1) )   * kappa[j]
21           bBeta[j] <- ( 1 - ( mu[j]*(2-mdlIdx) + mu0*(mdlIdx-1) ) ) * kappa[j]
22       }
23       # Hyperpriors for mu and kappa:
24       for ( j in 1:nCond ) {
25           mu[j] ~ dbeta( a[j,mdlIdx] , b[j,mdlIdx] )
26       }
27       for ( j in 1:nCond ) {
28           kappa[j] ~ dgamma( shk , rak )
29       }
30       mu0 ~ dbeta( a0[mdlIdx] , b0[mdlIdx] )
31
32       # Constants for hyperprior:
33       # (There is no higher-level distribution of across-group relationships,
```

```
34     # merely to keep the focus here on model comparison.)
35     shk <- 1.0
36     rak <- 0.1
37     aP <- 1
38     bP <- 1
39
40     a0[1] <- .53*400     # pseudo
41     b0[1] <- (1-.53)*400 # pseudo
42
43     a0[2] <- aP # true
44     b0[2] <- bP # true
45
46     a[1,1] <- aP # true
47     a[2,1] <- aP # true
48     a[3,1] <- aP # true
49     a[4,1] <- aP # true
50     b[1,1] <- bP # true
51     b[2,1] <- bP # true
52     b[3,1] <- bP # true
53     b[4,1] <- bP # true
54
55     a[1,2] <- .61*100     # pseudo
56     a[2,2] <- .50*100     # pseudo
57     a[3,2] <- .49*100     # pseudo
58     a[4,2] <- .51*100     # pseudo
59     b[1,2] <- (1-.61)*100 # pseudo
60     b[2,2] <- (1-.50)*100 # pseudo
61     b[3,2] <- (1-.49)*100 # pseudo
62     b[4,2] <- (1-.51)*100 # pseudo
63
64     # Hyperprior on model index:
65     mdlIdx ~ dcat( modelProb[] )
66     modelProb[1] <- .5
67     modelProb[2] <- .5
68     }
69  # ... end BUGS model specification
70  " # close quote for modelstring
71  # Write model to a file:
72  writeLines( text=modelstring , con="model.txt" )
73  # Load model file into BRugs and check its syntax:
74  modelCheck( "model.txt" )
75
76  #------------------------------------------------------------------------------
77  # THE DATA.
78
79  # For each subject, specify the condition s/he was in,
80  # the number of trials s/he experienced, and the number correct.
81  # (Randomly generated fictitious data.)
82  npg = 20  # number of subjects per group
83  ntrl = 20 # number of trials per subject
84  CondOfSubj = c( rep(1,npg) , rep(2,npg) , rep(3,npg) , rep(4,npg) )
85  nTrlOfSubj = rep( ntrl , 4*npg )
86  set.seed(47401)
87  nCorrOfSubj = c( rbinom(npg,ntrl,.61) , rbinom(npg,ntrl,.50) ,
```

```
88                          rbinom(npg,ntrl,.49) , rbinom(npg,ntrl,.51) )
89   nSubj = length(CondOfSubj)
90   nCond = length(unique(CondOfSubj))
91   # Display mean number correct in each group:
92   for ( condIdx in 1:nCond ) {
93       show( mean( nCorrOfSubj[ CondOfSubj==condIdx ] ) )
94   }
95
96   # Specify the data in a form that is compatible with BRugs model, as a list:
97   datalist = list(
98     nCond = nCond ,
99     nSubj = nSubj ,
100    CondOfSubj = CondOfSubj ,
101    nTrlOfSubj = nTrlOfSubj ,
102    nCorrOfSubj = nCorrOfSubj
103  )
104
105  # Get the data into BRugs:
106  modelData( bugsData( datalist ) )
107
108  #- - - - - - - - - - - - - - - - - - - - - - - - - - - - - - - - - - - - - - - - - - - - - - -
109  # INTIALIZE THE CHAINS.
110
111  nchain = 3
112  modelCompile( numChains=nchain )
113  modelGenInits()
114
115  #- - - - - - - - - - - - - - - - - - - - - - - - - - - - - - - - - - - - - - - - - - - - - - -
116  # RUN THE CHAINS.
117
118  burninSteps = 5000
119  modelUpdate( burninSteps )
120  samplesSet( c("mu","kappa","mu0","theta","mdlIdx") )
121  nPerChain = 5000 ; nThin = 10
122  modelUpdate( nPerChain , thin=nThin )
```

12.5 EXERCISES

Exercise 12.1. [Purpose: To investigate model comparison for different partitions of group means.] The program in Section 12.4.1(OneOddGroupModelComp.R) does a model comparison in which M1 has different means for every group and M2 has the same mean for all groups. In this exercise, we consider a different model comparison, with a different partition of the conditions.

(A) Consider a comparison for which M1 has one mean for condition 1 and a second mean for conditions 2 through 4, and M2 has the same mean for all groups. A quick-and-dirty way to program this is by changing the model specification to (OneOddGroupModelCompEx12.1.R)

```
24      for ( j in 1:2 ) {
25          mu[j] ~ dbeta( a[j,mdlIdx] , b[j,mdlIdx] )
26      }
27      mu[3] <- mu[2]
28      mu[4] <- mu[2]
```

Explain in English what the modified code does.

Unfortunately, the modified model structure is not amenable to automatic initialization in BUGS. We can "manually" initialize it this way (OneOddGroupModelCompEx12.1.R):

```
120     genInitList <- function() {
121         sqzData = .01+.98*datalist$nCorrOfSubj/datalist$nTrlOfSubj
122         mu = aggregate( sqzData , list(datalist$CondOfSubj) , "mean" )[,"x"]
123         sd = aggregate( sqzData , list(datalist$CondOfSubj) , "sd" )[,"x"]
124         kappa = mu*(1-mu)/sd^2 - 1
125         return(
126           list(
127             theta = sqzData ,
128             mu = c( mu[1] , mean( mu[2:4] ) , NA , NA ) ,
129             mu0 = mean(mu) ,
130             kappa = kappa ,
131             mdlIdx = 1
132           )
133         )
134     }
135     for ( chainIdx in 1 : nchain ) {
136         modelInits( bugsInits( genInitList ) )
137     }
```

The initialization uses the same technique introduced in Section 9.5.2 (FilconBrugs.R), with only one novel nuance: The last two components of mu are specified as NA because those components are *not* stochastic, but are instead fixed (in the model specification) to equal mu[2].

In general, when initializing chains in BUGS, only stochastic variables have random chains and therefore only the stochastic variables are initialized. Any variable that has a value set by the assignment operator, "<-," has a deterministic, not stochastic value. But sometimes, as in the present application, a vector has some stochastic components and some nonstochastic components. The way to initialize such vectors is to provide a specific value for the stochastic components, and an NA value to the nonstochastic components, as exemplified earlier.

Run the modified program, and show the resulting graphs analogous to those in Figure 12.5. Why are the upper histograms all the same, and why are the lower histograms (e.g., $\mu_2 - \mu_3$) so strange looking?

(B) What should we conclude from the model comparison of the previous part? (Be careful to express your conclusion as a statement about *relative* believabilities.) Should we conclude that the means of conditions 2 through 4 are actually equal?

Exercise 12.2. [Purpose: To estimate a difference, including a ROPE (and the hot hand example with a better prior).]

(A) Consider again the hot hand example from Exercise 8.1, p. 188. In this part of the exercise we establish a better prior: We know from general basketball statistics that professional players tend to make about 75% of their free throws, with some players almost as low as 50% and some players almost as high as 95%. A beta($\theta \mid 16, 6$) distribution nicely captures this prior knowledge. But what we do not know from prior statistics is the difference between success after success and success after failure in pairs of free throws. To be as vague as possible about the difference, we'll put a uniform distribution on the difference. Notice that when the overall success rate is, say, 90%, the difference could be anything from +20% (i.e., 100% versus 80%) to −20% (i.e., 80% versus 100%), but when the overall success rate is, say, 70%, then the difference could be anything from +60% to −60%. The specification of the prior needs to accommodate this range that depends on the overall success. Here is one way to specify this in the BUGS model:

```
theta1 <- mu + deflect
theta2 <- mu - deflect
mu ~ dbeta( 16 , 6 )
delta ~ dbeta( 1 , 1 )
deflect <- (delta-.5)*2 * min(mu,1-mu)
```

The variable `mu` is the overall success rate. The variable `deflect` is the deflection away from `mu` created by a previous success or a previous failure. The value of `deflect` is just a linearly transformed value of the random variable `delta`, which has a uniform prior. Incorporate this code into the model and run without the data, so you can see the prior. Show your complete model specification and a graph of the MCMC sample, which should look much like the upper row of Figure 12.6.

(B) Suppose we establish a ROPE on the difference of success after success and success after failure. We will arbitrarily set it at ±5%. Run the program with the data included, and display the posterior. The `plot-Post.R` function allows specification of a ROPE, with output as shown in Figure 12.6. Include your output graph, and say in English what the ROPE indicates.

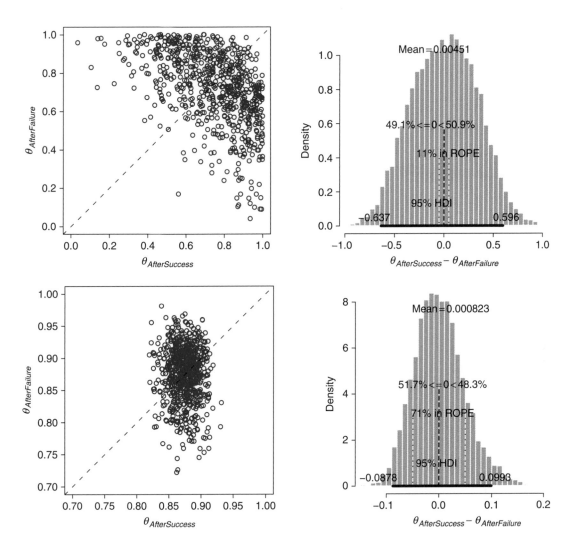

FIGURE 12.6

For Exercise 12.2. *Upper row*: The prior shows that the we believe the overall success rate is between 50% and 95%, but the difference between the two θ's could be anything. *Lower row*: The posterior. Notice the ROPE on the difference, from -0.05 to $+0.05$, along with the display of what percentage of the posterior falls within the ROPE.

Exercise 12.3. [Purpose: To apply the approaches to a real-world example.] The thematic apperception test (TAT) is a method for assessing personality and other aspects of interpersonal attitudes. The subject is shown pictures of people in ambiguous scenes, and the subject is asked to make up a story about what the pictured people are doing. In a study reported by Werner, Stabenau, & Pollin

(1970, Table 4),[2] pictures showing parent-child interactions were shown to women who had children. The stories invented by the mothers were scored for whether or not they expressed "personally involved, child-centered, flexible interactions." Twenty mothers of normal children and 20 mothers of schizophrenic children were each shown 10 pictures. The number of stories, out of 10, that expressed a flexible interaction, were as follows:

> Mothers of normal children: 8, 4, 6, 3, 1, 4, 4, 6, 4, 2, 2, 1, 1, 4, 3, 3, 2, 6, 3, 4.
> Mothers of schizophrenic children: 2, 1, 1, 3, 2, 7, 2, 1, 3, 1, 0, 2, 4, 2, 3, 3, 0, 1, 2, 2.

For purposes of this exercise, we will assume that the prior is informed from previous research, which indicates that the typical number of stories that express flexible interactions for this picture set is around 3 or 4, and can be well described by a beta(3.5,6.5) distribution. This is the distribution of flexible-narrative tendencies across mothers.

(A)   Estimate the difference between TAT scores of the two groups of mothers. Be sure to use the prior knowledge. Is zero among the credible differences? If you assume that the difference has a ROPE from −0.1 to +0.1, is the ROPE excluded from the 95% HDI? (Hints: Set up a hierarchical model in BUGS, using the filtration-condensation example as a guide. The prior applies to the distribution of individual theta[i] values, and the data are distributed as a binomial with bias theta[i].)

(B)   Using model comparison, determine the posterior probability of a model that uses one hyperparameter to describe both groups, relative to a model that uses a different hyperparameter for each group. For this application, the prior on the hyperparameters should *not* be informed, but instead should be extremely vague and broad, because the model comparison is supposed to be "automatic." What is the posterior probability of the different-hyperparameter model? If it is the preferred model, can we use it to estimate the difference between groups? (Answer: Not necessarily, because it does not use the prior knowledge.)

---

[2]The author found these data by perusing the collection by Hand et al. (1994).

CHAPTER 13

Goals, Power, and Sample Size

Just how many times must I show her I care,
Until she believes that I'll always be there?
Well, while she denies that my value's enough,
I'll have to rely on the power of love.[1]

---

[1] The power of "luff"? Sailors know that there's not much power in luffing.

Doing Bayesian Data Analysis: A Tutorial with R and BUGS. DOI: 10.1016/B978-0-12-381485-2.00013-4
© 2011, Elsevier Inc. All rights reserved.

Researchers collect data to achieve a goal. Sometimes the goal is to show that a suspected underlying state of the world is believable; other times the goal is to put a minimal degree of precision on whatever trends are observed. The goal can only be probabilistically achieved, as opposed to definitely achieved, because data are replete with random noise that can obscure the underlying state of the world. Statistical *power* is the probability of achieving the goal of an empirical study, when a suspected underlying state of the world is true. Scientists don't want to waste time and resources pursuing goals that have a small probability of being achieved. In other words, scientists desire power in their experiments.

13.1 THE WILL TO POWER

This section describes a framework for research and data analysis which leads to a more precise definition of power and how to compute it.[2]

13.1.1 Goals and Obstacles

An experiment or observational study can have many possible goals. For example, we might want to show that the rate of recovery for patients who take a drug is higher than the rate of recovery for patients who take a placebo. We might want to show that a coin is biased (i.e., that its tendency to show heads is not equal to 0.5). Goals such as those involve showing that a suspected difference or value really is tenable or, complementarily, showing that a "null" value is not tenable. Other goals do not necessarily have a suspected value in mind. Instead, the goal is merely to put a desired degree of precision on whatever tendencies happen to be observed. For example, when polling a population for preferences of political candidates, the goal is not necessarily to show that any particular candidate is ahead, but to get a precise assessment of the population's preferences.

The various goals of research can be formally expressed in various ways. In this chapter I will focus on two goals, formalized in terms of the highest density interval (HDI).

- *Goal*: Demonstrate that believable parameter values exclude a "null" value.
  - *Formal expression*: Show that the 95% HDI excludes the "null" value, or its region of practical equivalence (ROPE).
- *Goal*: Achieve precision in the estimate of believable values.

---

[2] Regarding the title of this section: Other than the fact that researchers desire statistical power, the notion of statistical power might have profound connections with concepts from Friedrich Nietzsche's work *The Will to Power*. See Exercise 13.1.

—   *Formal expression:* Show that the 95% HDI has some specified maximal width.

In some research, the goal may be to demonstrate that a "null" value is true, rather than false. This goal can also be formalized in terms of precision of the HDI: We want the entire 95% HDI to fall within the ROPE, as described in Section 12.1.3. (p. 301). There are other mathematical formalizations of the various goals, and they will be mentioned later. This chapter focuses on the HDI because of its natural interpretation for purposes of parameter estimation, as explained in the previous chapter.

If we knew the benefits of achieving our goal, and the costs of pursuing it, and if we knew the penalties for making a mistake while interpreting the data, then we could express the results of the research in terms of the long-run expected payoff. When we know the costs and benefits, we can conduct a full decision-theoretic treatment of the situation, and plan the research and data interpretation accordingly (e.g., Chaloner & Verdinelli, 1995; Lindley, 1997). In our applications, we do not have access to those costs and benefits, unfortunately. Therefore we rely on goals such as those outlined above.

The crucial obstacle to the goals of research is that a random sample is only a probabilistic representation of the population from which it came. Even if a coin is actually fair, a random sample of flips will rarely show exactly 50% heads. And even if a coin is biased, it might come up heads 5 times in 10 flips. Drugs that actually work no better than a placebo might happen to cure more patients in a particular random sample. And drugs that truly are effective might happen to show little difference from a placebo in another particular random sample of patients. Thus, a random sample is a fickle indicator of the true value in the underlying world. Whether the goal is showing that a suspected value is or isn't believable or showing that a precise range of values is tenable, random variation is the researcher's bane. Noise is the nemesis.

13.1.2 Power

Because of random noise, the goal of a study can be achieved only probabilistically. The probability of achieving the goal, given the (suspected) true state of the world, is called the *power* of the experiment.

Scientists go to great lengths to try to increase the power of their experiments or observational studies. There are three primary methods by which researchers can increase the chances of detecting an effect. First, we reduce measurement noise as much as possible. For example, if we are trying to determine the cure rate of a drug, we try to reduce other random influences on the patients, such as other drugs they might be stopping or starting, changes in diet or rest, and so on. Reduction of noise and control of other influences is the primary reason for conducting experiments in the lab instead of in the maelstrom of the real world. The second method, by which we can increase the chance of detecting an effect, is to amplify the underlying magnitude of the effect if we possibly can. For example, if we are trying to show that a drug helps cure a disease, we will want to administer as large a dose as possible (assuming there are no negative side effects). In nonexperimental research, in which the researcher does not have the luxury of manipulating the objects being studied, this second method is unfortunately unavailable. Sociologists, economists, and astronomers, for example, are often restricted to observing events that they cannot control or manipulate.

Once we have done everything we can to reduce noise in our measurements and to amplify the effect we are trying to measure, the third way to increase power is to increase the sample size. The intuition behind this method is simple: With more and more measurements, random noises will tend to cancel themselves out, leaving on average a clear signature of the underlying effect. In general, as sample size increases, power increases. Increasing the sample size is an option in most experimental research, and in a lot of observational research (e.g., more survey respondents can be polled), but not in some domains where the population is finite, such as comparative studies of the states or provinces of a nation. In this latter situation, we cannot create a larger sample size, but Bayesian inference is still valid, and perhaps uniquely so (Western & Jackman, 1994).

What we accomplish in this chapter is precise calculations of power. We compute the probability of achieving a specific goal, given (1) a suspected underlying distribution of biases or effects in the population being measured and (2) a specified data-generating process, such as collecting a fixed number of observations. Power calculations are useful for planning an experiment. To plan our research, we conduct "dress rehearsals" before the actual performance. We repeatedly simulate data that we suspect we might get, and conduct Bayesian analyses on the simulated data sets. If the goal is achieved for most of the simulated data sets, then the planned experiment has high power. If the goal is rarely achieved in the analyses of simulated data, then the planned experiment is likely to fail, and we must do something to increase its power.

In general, power can be approximated in the following manner:

1. *Generate a random sample of data points, using the data-generating hypothesis.* The sample should be generated according to how actual data will be gathered in the eventual real experiment. For example, typically it is assumed that the number of data points is fixed at N. It might instead be assumed that data will be collected for a fixed interval of time T, during which data points appear randomly at a known mean rate n/T.
2. *Compute the posterior estimate, using Bayesian analysis with skeptical-audience priors.* The data analysis must be convincing to the audience, and therefore the analysis must use priors that are agreeable to that skeptical audience.

3. *From the posterior estimate, tally whether or not the goal was attained.* The goal could be any of those outlined previously, such as having the 95% HDI exclude a ROPE around the null value or having the 95% HDI be narrower than a desired width.

4. *Repeat above steps many times, to approximate the power.* Power is, by definition, the proportion of times that the goal is attained.

The result of this process is the power: the probability of achieving the goal when using the particular data-sampling procedure. Notice that if the data-sampling procedure uses a fixed sample of size N, then the process determines power as a function of N. If the data-sampling procedure uses a fixed sampling duration T, then the process determines power as a function of T.

Figure 13.1 illustrates the process of computing power. The upper part of the figure shows the flow of information in an actual analysis. The real world generates a single sample of observed data. We use Bayes' rule, starting with a prior for the skeptical audience, to derive an actual posterior distribution. The lower part of the figure shows the flow of information in a power analysis. Instead of the real world, we have a hypothesis about the world. The hypothesis is expressed as a belief distribution over the parameters of the model. In many cases, this hypothesis is a posterior distribution derived from previous research. From the hypothesis, a sample of data is generated. This simulated sample of data is exemplary of data we anticipate if the the hypothesis is true. We then conduct a Bayesian analysis on the anticipated data sample, thereby deriving an anticipated posterior distribution. For the anticipated posterior, we assess whether or not the goal was achieved. We then repeat the simulated data-generation process. The repetition is indicated in Figure 13.1 by the layers of anticipated data samples and anticipated posteriors. Across repetitions, the proportion of times that the goal is achieved is the power. In some simple situations, power can determined exactly, as is done in the program of Section 13.6.1 (`minNforHDIpower.R`). In more realistic situations, power is approximated through simulation (Wang & Gelfand, 2002), as is done in the program of Section 13.6.2 (`FilconBrugsPower.R`).

13.1.3 Sample Size

Power increases as sample size increases (usually). Because gathering data is costly, we would like to know the minimal sample size, or sample duration, that is required to achieve a desired power.

The goal of precision in estimation can always be attained with a large enough sample size. This is because the likelihood of the data, thought of graphically as a function of the parameter, tends to get narrower and narrower as sample size increases. This narrowing of the likelihood is also what causes the data to eventually overwhelm the prior beliefs as sample size increases. As we collect

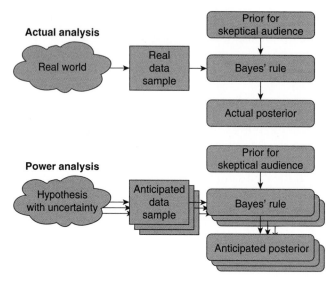

FIGURE 13.1
Upper diagram illustrates the flow of information in an actual Bayesian analysis. Lower diagram illustrates the flow of information in a power analysis. (Adapted with permission from Fig. 5 of Kruschke, 2010b.)

more and more data, the likelihood function gets narrower and narrower (on average), and therefore the posterior gets narrower and narrower. Thus, with a large enough sample, we can make the posterior distribution as narrow (i.e., precise) as we like.

The goal of showing that a parameter value is different from a null value might not be attainable with a high enough probability, however, no matter how big the sample size. Whether or not this goal is attainable with high probability depends on the data-generating distribution. *The best that a large sample can do is exactly reflect the data-generating distribution.* If the data-generating distribution has considerable mass straddling the null value, then the best we can do is get estimates that include and straddle the null value. As a simple example, suppose that we think that a coin may be biased, and the data-generating hypothesis is that $p(\theta = 0.5) = 25\%$, $p(\theta = 0.6) = 25\%$, $p(\theta = 0.7) = 25\%$, and $p(\theta = 0.8) = 25\%$. Because 25% of the simulated data come from a fair coin, the maximal probability of excluding $\theta = 0.5$, even with a huge sample, is 75%.

Therefore, when planning the sample size for an experiment, it is crucial to first decide what a realistic goal is. If there are good reasons to posit a highly certain data-generating hypothesis, perhaps because of extensive previous results, then a viable goal may be to exclude a null value. On the other hand, if the data-generating hypothesis is somewhat vague, then a more reasonable goal is to attain a desired degree of precision in the posterior.

13.1.4 Other Expressions of Goals

There are other ways to express mathematically the goal of precision in estimation. For example, another way of using HDIs was described by Joseph, Wolfson, & du Berger (1995a,b). They considered an "average length criterion," which requires that the *average* HDI width, across repeated simulated data, does not exceed some maximal value L. There is no explicit mention of power (i.e., the probability of achieving the goal) because the sample size is chosen so that the goal is definitely achieved. The goal itself is probabilistic, however, because it regards an average: Although some data sets will have HDI width less than L, many other data sets will not have an HDI width less than L. Another goal considered by Joseph, Wolfson, & du Berger (1995a) was the "average coverage criterion." This goal starts with a specified width for the HDI and requires its mass to exceed 95% (say) on average across simulated data. The sample size is chosen to be large enough to achieve that goal. Again, power is not explicitly mentioned, but the goal is probabilistic: Some data sets will have an L-width HDI mass greater than 95%, and other data sets will not have an L-width HDI mass greater than 95%. Other goals regarding precision are reviewed by Adcock (1997) and by De Santis (2004, 2007). The methods emphasized in this chapter focus on limiting the worst precision, instead of the average precision.

A rather different mathematical expression of precision is *entropy* of a distribution. Entropy describes how spread out a distribution is, hence smaller entropy connotes a more precise distribution. A distribution that consists of an infinitely dense spike, that has an infinitesimally narrow width, has zero entropy. At the opposite extreme, a uniform distribution has maximal entropy. The goal of high precision in the posterior distribution might be re-expressed as a goal of small entropy in the posterior distribution. For an overview of this approach, see Chaloner & Verdinelli (1995). For an introduction to how minimization of expected entropy might be used spontaneously by people as they experiment with the world, see Kruschke (2008). Entropy may be a better measure of posterior precision than HDI width especially in cases of multimodal distributions, for which HDI width is more challenging to determine. I will not further explicate the use of entropy because I think that HDI width is a more intuitive quantity than entropy by which to express precision, at least for most researchers in most contexts.

There are other ways to express mathematically the goal of excluding a null value. In particular, the goal could be expressed as wanting a large Bayes factor (BF) in a model comparison between the spike-null prior and the automatic alternative prior (e.g., Wang & Gelfand, 2002; Weiss, 1997). For example, we might set the desired Bayes factor at a ratio of 19 to 1 (i.e., 0.95 to 0.05). Kass & Raftery (1995) provided some guidelines for choice of criterial Bayes factor. I will not further address this approach, however, because the goal of a criterial

BF for untenable caricatured priors has problems as discussed in the previous chapter. Instead, it will be assumed that the goal of the research is estimation of the parameter values, starting with a viable prior. The resulting posterior is then used to assess whether the goal was achieved.

13.2 SAMPLE SIZE FOR A SINGLE COIN

As our first worked-out example, consider the simplest case: data from a single coin. Perhaps we are polling a population and we want to know if there is a preference for candidate A or candidate B. Or perhaps we want to know if a drug has more than a 50% cure rate. We will go through the steps listed in Section 13.1.2 to compute the exact sample size needed to achieve various degrees of power for different data-generating hypotheses.

13.2.1 When the Goal Is to Exclude a Null Value

Suppose that our goal is to show that the coin is biased. In other words, we want to show that 0.5 is not among the credible values. More specifically, we want to show that the 95% HDI excludes $\theta = 0.50$.

Because of previous research (or just a really strong hunch), we think that the true bias of the coin is very close to $\theta = 0.65$. It is a strong belief, so we'll describe the data-generating hypothesis as a beta distribution with mean 0.65 and certainty based on 2000 fictitious flips of the coin. This puts 95% of the data-generating biases between 0.63 and 0.67.

Next, we generate simulated data from hypothetical coins that have those biases, and tally how often the HDI excludes $\theta = 0.50$. One replication of the process goes like this: First, select a value for the "true" bias in the coin, from the data-generating distribution that is narrowly centered on $\theta = 0.65$. Suppose that the selected value is 0.638. Second, simulate flipping a coin with that bias N times. The simulated data have z heads and $N - z$ tails. The proportion of heads, z/N, will tend to be around .638, but will be higher or lower because of randomness in the flips. Third, using the audience-agreed prior for purposes of data analysis, determine the posterior beliefs regarding θ if z heads in N flips were observed. Tally whether or not the 95% HDI excludes the null value of $\theta = 0.50$. Notice that even though the data were generated by a coin with bias of 0.638, the data might, by chance, show a low proportion of heads, and therefore the 95% HDI might not exclude $\theta = 0.50$. This process is repeated many times to estimate the probability that the null value will be excluded from the HDI. In fact, for this simple situation, the exact power can be computed analytically, as explained in Section 13.6.1 (`minNforHDIpower.R`), p. 335.

Table 13.1 shows the minimal sample size needed for the 95% HDI to exclude $\theta = 0.5$ when flipping a single coin. As an example of how to read the table,

Table 13.1 Minimal Sample Size Required for 95% HDI to Exclude 0.5 When Flipping a Single Coin

Power	Generating Mean θ						
	0.55	0.60	0.65	0.70	0.75	0.80	0.85
0.7	642	148	64	33	18	13	7
0.8	866	191	82	43	26	18	10
0.9	1274	264	111	59	36	24	16

Note. The data-generating distribution is a beta density with mean θ, as indicated by the column header, and with shape parameters of $\theta\kappa$ and $(1-\theta)\kappa$, where $\kappa = 2000$. The audience-agreeable prior is uniform.

suppose you have a data-generating hypothesis that the coin has a bias very near $\theta = 0.65$. This hypothesis is implemented, for purposes of Table 13.1, as a beta distribution with shape parameters of 0.65×2000 and $(1 - 0.65) \times 2000$. The value of 2000 is arbitrary; it's as if the generating mean of 0.65 was based on fictitious previous data containing 2000 flips. The table indicates that if we desire a 90% probability of obtaining a 95% HDI that excludes $\theta = 0.5$, we need a sample size of $N = 111$ (i.e., we need to flip the coin at least 111 times to have a 90% chance that $\theta = 0.5$ falls outside the 95% HDI).

Notice in Table 13.1 that as the generating mean increases, the required sample size decreases. This makes sense intuitively: When the generating mean is large, the sample proportion of heads will tend to be large, and so the HDI will tend to fall toward the high end of the parameter domain. In other words, when the generating mean is large, it doesn't take a lot of data for the HDI to fall well above $\theta = 0.5$. On the other hand, when the generating mean is only slightly above $\theta = 0.5$, then it takes a large sample for the sample proportion of heads to be consistently above 0.5, and for the HDI to be consistently entirely above 0.5.

Notice also, in Table 13.1, that as the desired power increases, the required sample size increases quite dramatically. For example, if the data-generating mean is 0.6, then as the desired power rises from 0.7 to 0.9, the minimal sample size rises from 148 to 264.

13.2.2 When the Goal Is Precision

Not all research has as its goal the exclusion of a particular null value. Sometimes the goal is to establish a precise estimate of the parameter values. Other times, the research may have a null value of interest, but the data-generating hypothesis is too vague for the null value to be excluded most of the time, or, the goal may be to demonstrate that the HDI falls entirely within the ROPE. In these cases, the goal becomes precision of parameter estimation.

Here is an example in which there is a null value of interest, but only a vague data-generating distribution. Suppose you are interested in assessing the preferences of the general population regarding political candidates A and B. In particular, you would like to have high confidence in estimating whether the preference for candidate A exceeds $\theta = 0.5$. A recently conducted poll by a reputable organization found that of 10 randomly selected voters, 6 preferred candidate A and 4 preferred candidate B. If we use a uniform prepoll prior, our postpoll estimate of the population bias is a beta(7,5) distribution. As this is our best information about the population so far, we can use the beta(7,5) distribution as a data-generating distribution for planning the follow-up poll. Unfortunately, a beta(7,5) distribution has a 95% HDI from $\theta = 0.318$ to $\theta = 0.841$, which means that $\theta = 0.5$ is well within the data-generating distribution. How many more people do we need to poll so that 80% of the time we would get a 95% HDI that falls *above* $\theta = 0.5$?

It turns out, in this case, that we can never have a sample size large enough to achieve the goal of 80% of the HDIs falling above $\theta = 0.5$. To see why, consider what happens when we sample a particular value θ from the data-generating distribution (e.g., $\theta = 0.4$). We use that θ value to simulate a random sample of votes. Suppose N for the sample is huge, which implies that the HDI will be very narrow. What value of θ will the HDI focus on? Almost certainly it will focus on the value $\theta = 0.4$ that was used to generate the data. To reiterate, when N is very large, the HDI essentially just reproduces the θ value that generated it. Now recall the data-generating distribution of our example: The beta($\theta|7, 5$) distribution has only about 72% of the θ values above 0.5. Therefore, even with an extremely large sample size, we can get at most 72% of the HDIs to fall above 0.5.

There is a viable alternative goal, however. Instead of trying to reject a particular value of θ, we set as our goal a desired degree of precision in the posterior estimate. For example, our goal might be that the 95% HDI has width less than 0.2, at least 80% of the time. This goal implies that regardless of what values of θ happen to be emphasized by the posterior distribution, the width of the posterior is usually narrow, so that we have attained a suitably high precision in the estimate.

Table 13.2 shows the minimal sample size needed for the 95% HDI to have maximal width of 0.2. As an example of how to read the table, suppose you have a data-generating hypothesis that the coin has a bias roughly around $\theta = 0.6$. This hypothesis is implemented, for purposes of Table 13.2, as a beta distribution with shape parameters of 0.6×10 and $(1 - 0.6) \times 10$. The value of 10 is arbitrary; it's as if the generating mean of 0.6 were based on fictitious previous data containing only 10 flips. The table indicates that if we desire a 90% probability of obtaining an HDI with maximal width of 0.2, we need a sample size of at least 93.

Table 13.2 Minimal Sample Size Required for 95% HDI to Have Maximal Width of 0.2 When Flipping a Single Coin

	Generating Mean θ						
Power	0.55	0.60	0.65	0.70	0.75	0.80	0.85
0.7	91	91	89	86	80	69	58
0.8	92	92	91	90	86	79	67
0.9	93	93	93	92	91	88	79

Note. The data-generating distribution is a beta density with mean θ, as indicated by the column header, and with shape parameters of $\theta\kappa$ and $(1-\theta)\kappa$, where $\kappa = 10$. The audience-agreeable prior is uniform.

Notice in Table 13.2 that as the desired power increases, the required sample size increases only slightly. For example, if the data-generating mean is 0.6, then as the desired power rises from 0.7 to 0.9, the minimal sample size rises from 91 to 93. This is because the distribution of HDI widths, for a given sample size, has a very shunted high tail, and therefore small changes in N can quickly pull the high tail across a threshold such as 0.2. On the other hand, as the desired HDI width decreases (not shown in the table), the required sample size increases rapidly. For example, if the data-generating mean is 0.6 with $\kappa = 10$, and the desired HDI width is 0.1 (instead of 0.2), then the sample size needed for 80% power is 377 (instead of 92).

The R code in Section 13.6.1 (`minNforHDIpower.R`) generated Tables 13.1 and 13.2. You can use the R code to specify your own data-generating distribution, goal, and desired power. You can also specify a ROPE. The R function will tell you the minimal sample size required. Exercise 13.2 has you work through some examples.

13.3 SAMPLE SIZE FOR MULTIPLE MINTS

Consider again the filtration-condensation experiment, described in Section 9.3.1, p. 219, in which people learned four different category structures. Two structures were "filtration" structures in which one stimulus dimension could be ignored but perfect accuracy could be achieved. Two other structures were "condensation" structures in which both stimulus dimensions required processing to achieve perfect accuracy. Some theories of learning predict that filtration should be easier to learn than condensation, and therefore one goal of the experiment was to show that the mean accuracy of people in the filtration structure differed from the mean accuracy of people in the condensation structure. This goal was handily achieved: Using $N = 40$ participants per condition, the 95% HDI on the difference of μ parameters went from 0.137 to 0.207, with 100% of the MCMC μ differences falling well above 0.0; see the rightmost panel of Figure 9.16, p. 224.

Is there a smaller sample size per group that could have been used to achieve high power? And what was the power of detecting the more subtle difference between the two filtration groups, shown in the leftmost panel of Figure 9.16, p. 224? To answer these questions, we pursue the simulation procedure outlined at the beginning of this chapter (Section 13.1.2): Specify a data-generating hypothesis, simulate some data, do a Bayesian data analysis, determine whether the goal is achieved, and repeat many times to estimate the probability of achieving the goal. R code for computing power for this experiment is listed in Section 13.6.2 (FilconBrugsPower.R).

Suppose the goal of the experimenter is to show that the mean of the filtration groups exceeds the mean of the condensation groups. Specifically, we suppose that the goal is achieved if the 95% HDI of $(\mu_1 + \mu_2)/2 - (\mu_3 + \mu_4)/2$ excludes 0.0. It turns out that using merely $N = 6$ in each of the groups achieves over 80% power for this goal! Figure 13.2 shows examples of the posteriors for two runs of simulated data with $N = 6$ per group. The upper row shows a case in which the posterior for $(\mu_1 + \mu_2)/2 - (\mu_3 + \mu_4)/2$ excludes 0.0, as desired. The lower row shows a case in which that goal was not achieved. Across many simulated runs using $N = 6$, 80% achieved the goal. In other words, instead of using $N = 40$, as in the actual experiment, the research could have been conducted with only $N = 6$, and had an 80% chance of inferring a believable difference between filtration and condensation.

This power analysis stems from simulated data that have been generated from moderately certain knowledge, based on results from 40 actual subjects per condition. If we did not have data-generating priors with such high certainty, then our power predictions would not be so high either. For example, if we had data-generating priors on μ_j with the same means as used here, but with wider (more uncertain) distributions, then the sample size would have to be much bigger than $N = 6$ to achieve the same power. This makes intuitive sense: If we are not very certain about what the data will look like, it takes a lot of data to reliably reveal the underlying trends.

Suppose that the goal of the experimenter was, instead, to show that there was a difference between the two filtration groups. Specifically, we suppose that the goal is achieved if the 95% HDI of $\mu_1 - \mu_2$ excludes 0.0. Notice that this goal was not achieved in either simulated run shown in Figure 13.2. It turns out that using $N = 6$ in each of the groups achieves a power of only 4% for this goal. What is the power for this goal if $N = 40$, as used in the actual experiment? It turns out that the power is only 43%! The fact that the *actual* data (Figure 9.16, p. 224) achieved the goal was merely a "lucky" happenstance.

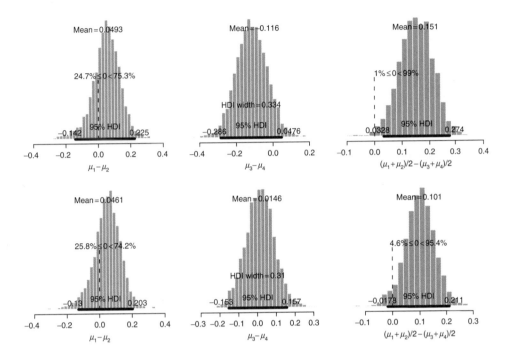

FIGURE 13.2

Posterior distributions from two runs of simulated data, using $N = 6$ per group. *Upper row, right panel:* A case in which the posterior for $(\mu_1 + \mu_2)/2 - (\mu_3 + \mu_4)/2$ excludes 0.0, as desired. *Lower row, right panel:* A case in which that goal is not achieved. Across many simulated runs, over 80% achieved the goal.

Finally, suppose that the goal of the experimenter was to achieve a minimal precision on the difference between the two condensation groups. The motivation for this goal is as follows: Theory suggests that the two condensation groups should not differ much. Therefore the goal cannot be to show that the difference exceeds zero. Instead, the goal should be to achieve some desired degree of precision regarding the difference. Suppose we want the 95% HDI on $\mu_3 - \mu_4$ to have a width less than 0.20. To achieve this goal at least 80% of the time we would need a sample size (per group) of approximately $N = 32$. You can verify this by running the program in Section 13.6.2 (FilconBrugsPower.R).

13.4 POWER: PROSPECTIVE, RETROSPECTIVE, AND REPLICATION

There are different types of power analysis, as shown in Table 13.3. For reference, the top row of the table shows an actual analysis, in which the data generator is the real world, the data sample is observed once, the prior for

Table 13.3 Types of Power Analysis, Including Replication Probability

Analysis Type	Data Generator	Data Sample	Prior for Bayesian Analysis	Posterior
Actual	Real world	Observed once	Skeptical audience	Actual
Prospective Power	Hypothesis	Simulated repeatedly	Skeptical audience	Anticipated
Retrospective Power	Actual posterior	Simulated repeatedly	Skeptical audience	Anticipated
Replication Power	Actual posterior	Simulated repeatedly	Actual posterior	Anticipated

Note. Adapted with permission from Figure 7 of Kruschke, 2010a.

the analysis is designed to be agreeable to a skeptical audience, and the actual posterior distribution is computed.

The second row of Table 13.3 shows a *prospective power* analysis, for which the data generator is a research hypothesis based on theory or results from previous experiments that are not exactly the same as the present experiment. The research hypothesis is expressed as a distribution over model parameters, with a central tendency and uncertainty that captures the researcher's beliefs about what would be the appropriate parameter values if the data of the planned experiment were to appear as it is believed they would. The distribution over the model parameters could be set directly, from theoretical considerations, by the researcher setting the shape constants of the parameter distribution. On the other hand, the data-generating distribution could be set indirectly by first positing data consistent with the theory and then applying Bayesian updating to derive a distribution over parameters that is consistent with the posited data. De Santis (2007) describes how to combine results of previous experiments to construct a data-generating distribution.

Contrast a prospective power analysis with the next row of the table, which shows a *retrospective power* analysis. This refers to a situation in which we have already collected data, and therefore we already have an actual posterior distribution. We want to determine the power of the experiment we ran. To do this, we use the actual posterior as the data generator. This is tantamount to using the data from a posterior predictive check. In other words, at a step in the chain, the parameter values are used to generate simulated data. The simulated data are then analyzed with the same Bayesian model as the actual data, and the posterior is examined for whether or not the goal is achieved. This process is readily programmed in BRugs, as explained in Section 13.6.2 (`FilconBrugsPower.R`). The example of the previous section, regarding the filtration-condensation experiment, was a case of retrospective power analysis.

Finally, suppose that we have already collected some data and we have an actual posterior distribution, and we want to know the probability that we would achieve our goal if we replicated the experiment. In other words, if we were to simply re-collect a new sample of data, what is the probability that we would achieve our goal in the replicated study, also taking into account the first set of data? The bottom row of Table 13.3 indicates how we answer this question of *replication power*. We use the actual posterior derived from the first sample of data as the data generator. But for the analysis prior, we again use the actual posterior from first sample of data, because that is the best-informed prior for the follow-up experiment. An easy way to execute this analysis in BUGS is as follows: Use the actual set of data with a skeptical-audience prior to generate a sample from an actual posterior. Use that actual-posterior sample to generate simulated new data, as if in a posterior predictive check. Then, *concatenate the original data with the novel simulated data* and update the original skeptical-audience prior with the enlarged data set. This technique is tantamount to using the posterior of the original data set as the prior for the novel simulated data. Section 13.6.2 (`FilconBrugsPower.R`) shows BRugs code for this type of analysis.

Computation of replication power is natural in a Bayesian setting, but is difficult or impossible for traditional null-hypothesis significance testing (Miller, 2009). NHST has trouble when addressing replication probability because it has no good way to model a data generator: It has no access to the posterior distribution from the initial analysis.

13.4.1 Power Analysis Requires Verisimilitude of Simulated Data

Power analysis is only useful when the simulated data imitate actual data. We generate simulated data from a descriptive model that has uncertainty in its parameter values, but we assume that the model is a reasonably good description of the actual data. If the model is instead a poor description of the actual data, then the simulated data do not imitate actual data, and inferences from the simulated data are not very meaningful. It is advisable, therefore, to perform a posterior predictive check that the simulated data accurately reflect the actual data.

When simulated data differ from actual data, strange results can arise in power analysis. Consider an analysis of replication probability in which the simulated data are quite different than the actual data. The novel simulated data are combined with the original data to conduct the replication analysis. The combined data set is a broad mixture of two trends (i.e., the actual trend and the different simulated trend), and therefore the estimates of the parameters become more *uncertain* than for the original data alone. It is only when the simulated sample size becomes large, relative to the original sample size, that the simulated trend overwhelms the actual trend, and the replication uncertainty becomes smaller again. If you find in your power analyses that parameter uncertainty initially gets larger as the simulated sample size increases, then you may have a situation in which the model does not faithfully mimic the actual data.

13.5 THE IMPORTANCE OF PLANNING

In the real conduct of research, scientists do not always plan a specific sample size and then stick to it. Nor should they have to. When experiments are set up such that the results from one participant do not affect the next participant, or when observational studies are made such that the results of one observation are independent of any other observation—in other words, when the next coin flip is not influenced by previous flips—then the researcher's intentions do not influence the data production or the data interpretation.

In some situations, data are collected for a particular time period, during which the actual number of participants (coin flips) is a random value that depends on the random rate at which people happen to participate. For example, participants in a typical university psychology experiment might be recruited for a certain number of weeks during a semester. There may be a stable average rate of participation, but the number of participants in any given week is a random number. In this situation, our goal is to determine how many weeks we should plan on recruiting participants so to have a high probability of achieving the goal of the experiment. We can do the power analysis in a manner analogous to the methods of the previous sections. As in those methods, we randomly generate parameter values from our data-generating hypothesis, but the sample size N is a random value that depends on a specified rate of participation and amount of time during which participants are recruited. The sample size could be (but is not necessarily) modeled as a Poisson variable, as was done in Exercise 11.3. We simulate incrementally longer durations until we find the minimal duration that yields a sufficiently high power. Having thereby planned a duration for collecting data (and corresponding approximate number of subjects), we can decide whether or not it is feasible to actually conduct the research. If we do go ahead with the research, we are not stuck with the exact planned duration. We can stop whenever we want, assuming that the data are uninfluenced by our intentions to stop.

The point of the previous paragraphs is to say that sticking to a planned sample size or duration is not crucial to the data analysis, after the data have been collected. Instead, the planning is important for deciding whether or not the research has a reasonable chance of success, before the data are collected.

Conducting a power analysis in advance of collecting data is very important and valuable. Often in real research, a fascinating theory and clever experimental manipulation imply a subtle effect. It can come as a shock to the researcher when power analysis reveals that detecting the subtle effect would take several hundred subjects! But the shock of power analysis is far less than the pain of running dozens of subjects and finding highly uncertain estimates of the sought-after effect.

Power analysis can reduce research pain in other ways. Sometimes in real research, an experiment or observational study is conducted merely to objectively confirm what is anecdotally known to be a strong effect. A researcher may be tempted to conduct a study using the usual large sample size that is typical of related research. But a power analysis may reveal that the strong effect can be easily detected with a much smaller sample size.

Power analysis is also important when proposing research to funding agencies. Proposals in basic research might have fascinating theories and clever experiments, but if the predicted effects are subtle, then reviewers of the proposal may be justifiably dubious, and want to be reassured by a power analysis. Proposals in applied research are even more reliant on power analysis, because the costs and benefits are more immediate and tangible. For example, in clinical research (e.g., medicine, pharmacology, psychiatry, counseling), it can be costly to test patients, and therefore it is important to anticipate the probable sample size or duration.

Although it is important to plan sample size in advance, it can also be important, especially in clinical applications, to monitor data as they are collected and to stop the research as soon as possible. It behooves the researcher to discontinue an experiment as soon as the data clearly indicate a positive or negative outcome: It would be unethical to slavishly continue treating patients with an experimental treatment that is clearly detrimental, and it would be unethical to slavishly continue running patients in a placebo condition when the experimental treatment is clearly having positive effects. If the data are collected in such a way that they are uninfluenced by the experimenter's intentions, then the data collection can be stopped at any time without influencing the interpretation of the data. The decision regarding when to stop collecting data is a topic of much investigation and goes under the name of *Bayesian sequential design*. It will not be discussed further here, but the interested reader is referred to books by Berger (1985) and DeGroot (2004), and various technical articles, for example those cited by Roy, Ghosal, & Rosenberger (2009, p. 427).

13.6 R CODE

13.6.1 Sample Size for a Single Coin

The program described in this section was used to generate Tables 13.1 and 13.2. The program determines the minimal sample size needed to achieve a specified goal with a specified power, when flipping a single coin.

In principle, we could simulate random samples of θ from the generating distribution, and then simulate random samples of coin flips based on the particular values of θ. For each repetition, we would tally whether the goal was achieved. With a large number of repetitions, this procedure would provide a good approximation to the power.

In the present simple situation, however, we can determine mathematically the exact probabilities of each possible outcome. The simulated data are generated by sampling a θ value according to our data-generating prior, and then generating N flips of the coin according to the binomial distribution. Therefore the probability of getting z heads, across repeated sampling from the prior, is

$$p(z|N) = \int_0^1 d\theta\, p(z|N,\theta)\, p(\theta)$$

$$= \int_0^1 d\theta\, \text{binomial}(z|N,\theta)\, \text{Beta}(\theta|a,b)$$

$$= \int_0^1 d\theta \binom{N}{z} \theta^z (1-\theta)^{(N-z)}\, \theta^{(a-1)}(1-\theta)^{(b-1)}/B(a,b)$$

$$= \binom{N}{z} B(z+a, N-z+b)/B(a,b) \tag{13.1}$$

(This probability of possible data is sometimes called the *preposterior marginal distribution of z*; compare to Eqn. 5 of Pham-Gia & Turkkan, 1992.) For each possible outcome, z, we update the audience-agreeable prior to render an audience-agreeable posterior distribution, and then we decide whether the goal has been achieved for that outcome. Because the decision is determined by the outcome z, the probability of the decisions is determined by the probability of the outcomes. Equation 13.1 is implemented in the program on lines 20 to 22, but in logarithmic form to prevent underflow errors.

The function `minNforHDIpower` has several arguments, most of which have default values. Two arguments without defaults are the mean, `genPriorMean`, and effective sample size, `genPriorN`, of the data-generating hypothesis, which is assumed to be a beta density. For example, suppose that the experimenter believes that the actual bias of a coin is very nearly 0.65, and suppose that the belief is based on the equivalent of 2000 fictitious flips of the coin; then `genPriorMean = .65` and `genPriorN = 2000`.

The next two arguments, `HDImaxwid` and `nullVal`, specify the goal of the experiment. One and only one of the arguments must be non-NULL. For example, if the goal is for the HDI to exclude the value of 0.5, then `nullVal = .5` (and `HDImaxwid = NULL`). If the goal is for the HDI to have a maximal width of .2, then `HDImaxwid = .2` (and `nullVal = NULL`). The next argument specifies the limits of the ROPE, if relevant. The function in its present form checks only if the HDI excludes the null value, or ROPE if specified. The function does not check whether the ROPE fully *contains* the HDI.

Finally, the function also has arguments for the audience-agreed prior that is used in the Bayesian analysis. It defaults to a uniform prior, but a different audience-agreed prior can be specified in terms of the mean (`audPriorMean`) and certainty (`audPriorN`) of a beta distribution.

Here is an example of how to use the program:

```
source("minNforHDIpower.R")
minNforHDIpower( genPriorMean=.65 , genPriorN=2000 , nullVal=.5 )
```

The program will compute power for increasing values of sample size, until stopping at N = 82 (as shown in Table 13.1).

(minNforHDIpower.R)

```
1   minNforHDIpower = function( genPriorMean , genPriorN ,
2                       HDImaxwid=NULL , nullVal=NULL , ROPE=c(nullVal,nullVal) ,
3                       desiredPower=0.8 , audPriorMean=0.5 , audPriorN=2 ,
4                       HDImass=0.95 , initSampSize=20 , verbose=T ) {
5       if ( is.null(HDImaxwid) + is.null(nullVal) != 1 ) {
6           stop("One and only one of HDImaxwid and nullVal must be specified.")
7       }
8       # Convert prior mean and N to a,b parameter values of beta distribution.
9       genPriorA = genPriorMean * genPriorN
10      genPriorB = ( 1.0 - genPriorMean ) * genPriorN
11      audPriorA = audPriorMean * audPriorN
12      audPriorB = ( 1.0 - audPriorMean ) * audPriorN
13      # Initialize loop for incrementing sampleSize
14      sampleSize = initSampSize
15      notPowerfulEnough = TRUE
16      # Increment sampleSize until desired power is achieved.
17      while( notPowerfulEnough ) {
18          zvec = 0:sampleSize # All possible z values for N flips.
19          # Compute probability of each z value for data-generating prior.
20          pzvec = exp( lchoose( sampleSize , zvec )
21                          + lbeta( zvec + genPriorA , sampleSize-zvec + genPriorB )
22                          - lbeta( genPriorA , genPriorB ) )
23          # For each z value, compute HDI. hdiMat is min, max of HDI for each z.
24          hdiMat = matrix( 0 , nrow=length(zvec) , ncol=2 )
25          for ( zIdx in 1:length(zvec) ) {
26              z = zvec[zIdx]
27              hdiMat[zIdx,] = HDIofICDF( qbeta , shape1 = z + audPriorA ,
28                                          shape2 = sampleSize - z + audPriorB )
29          }
30          hdiWid = hdiMat[,2] - hdiMat[,1]
31          if ( !is.null( HDImaxwid ) ) {
```

```
32              powerHDI = sum( pzvec[ hdiWid < HDImaxwid ] )
33          }
34          if ( !is.null( nullVal ) ) {
35              powerHDI = sum( pzvec[ hdiMat[,1] > ROPE[2] | hdiMat[,2] < ROPE[1] ]
                    )
36          }
37          if ( verbose ) {
38              cat( " For sample size = ", sampleSize , ", power = " , powerHDI ,
39                  "\n" , sep="" ) ; flush.console()
40          }
41          if ( powerHDI > desiredPower ) {
42              notPowerfulEnough = FALSE
43          } else {
44              sampleSize = sampleSize + 1
45          }
46      } # End while( notPowerfulEnough )
47      # Return the minimal sample size that achieved the desired power.
48      return( sampleSize )
49  } # end of function
```

13.6.2 Power and Sample Size for Multiple Mints

This section explains how to use a BUGS model to estimate power for a given sample size. The first step is to create a working BRugs program for a single set of data, as if the data were from an actual experiment. Make sure that the program has appropriate burn-in and thinning so that the posterior sample is robust and useful. (For a reminder of the issues of burn-in and thinning, see Section 23.2, p. 623.) Once a working BRugs program is established, it is modified and wrapped in a function that gets called repeatedly with different simulated sets of data.

The general scheme for estimating power with a BRugs program is outlined in Figure 13.3. At the top of the code is the BUGS model wrapped in a function called GoalAchievedForSample. The function takes simulated data as input, and returns a true/false value indicating whether or not the goal was achieved. As Figure 13.3 indicates, the BUGS model specification remains unchanged. The data specification is modified, however, to accept the passed-in simulated data rather than the actual data. The MCMC chains are then generated as in the original program, using the burn-in and thinning that is already known to produce well-mixed chains. Finally, when the chains are examined, new commands are included to test whether the goal has been achieved. For example, the code could check whether the 95% HDI excludes the ROPE. The result is a true/false value denoted by the variable goalAchieved, which gets returned by the function.

The lower section of code in Figure 13.3 calls the function defined in the upper section of code. The idea is that we will call the function many times, each time with different simulated data, and check whether the goal is achieved.

```
GoalAchievedForSample = function( simulatedData ) {
    library(BRugs) # needed inside function to re-initialize.
    #-----------------------------------------------------------------
    # Model specification: Unchanged.
    #-----------------------------------------------------------------
    # Data: NEW COMMANDS for using simulatedData.
    #-----------------------------------------------------------------
    # Initialize and run chains: Unchanged.
    #-----------------------------------------------------------------
    # Examine chains: NEW COMMANDS for checking whether goal is achieved.
    # Denote result by true/false variable named goalAchieved.
    return( goalAchieved )
}

nSimulatedExperiments = 500 # An arbitrary large number.
nSuccess = 0 # Initialize counter.
for ( experimentIdx in 1:nSimulatedExperiments ) {
    simulatedData = ... # Create simulated data
    nSuccess = nSuccess + GoalAchievedForSample( simulatedData )
    estPower = nSuccess / experimentIdx
}
```

FIGURE 13.3

Outline of R code for computing power. This outline assumes that there is already a working BRugs script for a single set of data. The working BRugs script is wrapped in a function at the beginning of the code shown here.

The proportion of times that the goal is achieved is the estimated power. The number of times we call the function is denoted nSimulatedExperiments. Then a for loop tallies the number of successes. As the number of simulated experiments increases, the estimate becomes more stable.

We can think of each time through the loop, with a new simulated data set, as the flip of a coin that has a probability of heads equal to the true power of the experiment. We are estimating the bias in that coin (i.e., the power of the experiment) by the observed number of successes. If our prior belief about the power is uniform, then the posterior belief is distributed as beta(nSuccess+1, nSimExperiments-nSuccess+1). The HDI of the estimated power can be computed by calling the function HDIofICDF(qbeta , shape1 = nSuccess+1 , shape2 = nSimExperiments-nSuccess+1).

A complete instantiation of the scheme is shown in the program listing below. It was used to generate Figure 13.2 and the other results mentioned in Section 13.3. The code is modified from the program used to analyze the real data, listed in Section 9.5.2 on p. 232. The program has many embellishments beyond the skeleton outlined in Figure 13.3.

The function, which takes a set of data as input and returns whether or not the goal was achieved, spans lines 5 to 147 of the program. The function has a couple extra arguments that help with plotting graphs of the results, which we might want to see for at least a few sets of data to be sure that the simulations are behaving properly. One extra argument is plotResults, which is simply a true/false flag indicating whether or not to plot the results. Another extra argument is the filename to be used for saving the plots.

The data section, lines 44 to 49, is simpler than in the original program, because the complete datalist is provided from outside the function. The data section merely sends the datalist to BUGS.

In the results sections, beginning at line 89, there are three main changes. First, the chains are not checked for convergence and mixing, because it is assumed that the burn-in and thinning have already been checked for adequacy. (For a reminder of the issues of burn-in and thinning, see Section 23.2, p. 623.) Second, the plot-producing code (lines 104 to 132) is wrapped in an if statement, which causes plots to be produced and saved only if the argument plotResults is true.

The third main change is checking for whether the goal was achieved, as computed in lines 134 to 144. The code checks for *three* goals regarding the μ_c values: (1) Does the HDI on filtration minus condensation exceed zero? (2) Does the HDI on the difference between the two filtration conditions exceed zero? (3) Is the HDI on the difference of condensation conditions of width less than 0.2? The true/false values of these three goals are returned as a vector.

The function is called by code starting at line 150. The code includes the option for computing either *retrospective* or *replication* power, which is specified on line 152. The program then loads the original data, which are used for computing replication power, and the actual posterior sample, which is used for either type of power.

Various constants are specified in lines 170 to 176. In particular, the *simulated* number of subjects per condition is specified in line 170. Also notice that the number of goals computed by the function is specified in line 176. The number of goals is used only to initialize the success counter in line 188.

Because we are simulating results from randomly generated sets of data, and each data set may take a few minutes to run, the complete batch of hundreds of data sets may take many hours to complete. Therefore, we may want to run only a few simulated experiments at a time, or interrupt a run, and save the interim results for future continuation. Also, *be warned that BUGS often decides to crash for no apparent reason on some simulated data sets*, and you will find yourself with interrupted runs whether you intended it or not. For these reasons, the program saves the interim results after every simulated data set, on lines 241 to 242. When the program is invoked, it checks whether there is a previously saved interim result, on lines 178 to 190, and restores those results if they exist.

The loop that repeatedly generates simulated data sets and calls the Bayesian analysis begins on line 193. The simulated data are generated in lines 195 to 209. Each iteration of the loop uses the mu and kappa values from the actual posterior to generate simulated data in each condition. The simulated data are assembled into a datalist in lines 211 to 229. If retrospective power is being computed, then only the newly simulated data are included in the datalist. If replication power is being computed, then the original, actual data are concatenated onto the simulated data. By including the actual data, the model is effectively using the actual posterior as the prior for the new simulated data.

(FilconBrugsPower.R)

```
1    graphics.off()
2    rm(list=ls(all=TRUE))
3
4
5    GoalAchievedForSample = function( datalist , plotResults=F ,
6                                      fileNameRoot="DeleteMe" ) {
7
8    library(BRugs)         # Kruschke, J. K. (2010). Doing Bayesian data analysis:
9                           # A Tutorial with R and BUGS. Academic Press / Elsevier.
10   #------------------------------------------------------------------------------
11   # THE MODEL.
12
13   modelstring = "
14   # BUGS model specification begins here...
15   model {
16       for ( subjIdx in 1:nSubj ) {
17           # Likelihood:
18           z[subjIdx] ~ dbin( theta[subjIdx] , N[subjIdx] )
19           # Prior on theta: Notice nested indexing.
20           theta[subjIdx] ~ dbeta( a[cond[subjIdx]] , b[cond[subjIdx]] )I(0.001,0.999)
21       }
22       for ( condIdx in 1:nCond ) {
23           a[condIdx] <- mu[condIdx] * kappa[condIdx]
24           b[condIdx] <- (1-mu[condIdx]) * kappa[condIdx]
25           # Hyperprior on mu and kappa:
26           mu[condIdx] ~ dbeta( Amu , Bmu )
27           kappa[condIdx] ~ dgamma( Skappa , Rkappa )
28       }
29       # Constants for hyperprior:
30       Amu <- 1
31       Bmu <- 1
32       Skappa <- pow(meanGamma,2)/pow(sdGamma,2)
```

```
33        Rkappa <- meanGamma/pow(sdGamma,2)
34        meanGamma <- 10
35        sdGamma <- 10
36     }
37  # ... end BUGS model specification
38  " # close quote for modelstring
39  # Write model to a file:
40  writeLines(modelstring,con="model.txt")
41  # Load model file into BRugs and check its syntax:
42  modelCheck( "model.txt" )
43
44  #------------------------------------------------------------------------
45  # THE DATA.
46
47  # datalist supplied from outside the function.
48  # Get the data into BRugs:
49  modelData( bugsData( datalist ) )
50
51  #------------------------------------------------------------------------
52  # INTIALIZE THE CHAINS.
53
54  nChain = 3
55  modelCompile( numChains=nChain )
56
57  if ( F ) {
58     modelGenInits() # often won't work for diffuse prior
59  } else {
60     #  initialization based on data
61     genInitList <- function() {
62        sqzData = .01+.98*datalist$z/datalist$N
63        mu = aggregate( sqzData , list(datalist$cond) , mean )[,"x"]
64        sd = aggregate( sqzData , list(datalist$cond) , sd )[,"x"]
65        kappa = mu*(1-mu)/sd^2 - 1
66        return(
67           list(
68              theta = sqzData ,
69              mu = mu ,
70              kappa = kappa
71           )
72        )
73     }
74     for ( chainIdx in 1 : nChain ) {
75        modelInits( bugsInits( genInitList ) )
76     }
77  }
78
79  #------------------------------------------------------------------------
80  # RUN THE CHAINS.
81
82  burninSteps = 1000
83  modelUpdate( burninSteps )
84  cat("Got through burn in...");flush.console()
85  samplesSet( c("mu","kappa","theta","a","b") )
86  nPerChain = ceiling(3000/nChain)
```

```
87    modelUpdate( nPerChain , thin=5 )
88
89    #------------------------------------------------------------------------------
90    # EXAMINE THE RESULTS.
91
92    # Extract chain values from BUGS:
93    mu = NULL
94    kappa = NULL
95    for ( condIdx in 1:nCond ) {
96        nodeName = paste( "mu[" , condIdx , "]" , sep="" )
97        mu = rbind( mu , samplesSample( nodeName ) )
98        nodeName = paste( "kappa[" , condIdx , "]" , sep="" )
99        kappa = rbind( kappa , samplesSample( nodeName ) )
100   }
101   chainLength = NCOL(mu)
102
103   # Display results if desired:
104   if ( plotResults ) {
105     # Histograms of condition (i.e. group) mu differences:
106     windows(12,4)
107     layout( matrix(1:3,nrow=1) )
108     source("plotPost.R")
109     histInfo = plotPost( mu[1,]-mu[2,] , xlab=expression(mu[1]-mu[2]) ,
110                          compVal=0.0 , breaks=30 , main="" )
111     histInfo = plotPost( mu[3,]-mu[4,] , xlab=expression(mu[3]-mu[4]) ,
112                          breaks=30 , main="" )
113     HDIlim = HDIofMCMC( mu[3,]-mu[4,] )
114     text( mean(HDIlim) , .25*max(histInfo$density) , adj=c(.5,0) , cex=1.25 ,
115           bquote( "HDI width = " * .(signif(HDIlim[2]-HDIlim[1],3)) ) )
116     nSubjPerCond = round( datalist$nSubj / datalist$nCond )
117     histInfo = plotPost( (mu[1,]+mu[2,])/2 - (mu[3,]+mu[4,])/2 , compVal=0.0 ,
118                          xlab=expression((mu[1]+mu[2])/2 - (mu[3]+mu[4])/2) ,
119                          breaks=30 , main="" )
120     dev.copy2eps( file = paste( fileNameRoot,"N",nSubjPerCond,"_",expIdx,".eps" ,
121                                 sep="" ) )
122     # Scatterplot of mu, kappa:
123     windows()
124     muLim = c(.60,1) ; kappaLim = c( 2.0 , 50 ) ; mainLab="Posterior"
125     thindex = round( seq( 1 , chainLength , len=300 ) )
126     plot( mu[1,thindex] , kappa[1,thindex] , main=mainLab ,
127         xlab=expression(mu[c]) , ylab=expression(kappa[c]) , cex.lab=1.75 ,
128         xlim=muLim , ylim=kappaLim , log="y" , col="red" , pch="1" )
129     points( mu[2,thindex] , kappa[2,thindex] , col="blue" , pch="2" )
130     points( mu[3,thindex] , kappa[3,thindex] , col="green" , pch="3" )
131     points( mu[4,thindex] , kappa[4,thindex] , col="black" , pch="4" )
132   } # end if plotResults
133
134   # Specify goals and check whether they are achieved:
135   source("HDIofMCMC.R")
136   # Goal is filtration vs condensation 95% HDI exceeds zero:
137   goal1Ach = ( HDIofMCMC( (mu[1,]+mu[2,])/2 - (mu[3,]+mu[4,])/2 )[1] > 0.0 )
138   # Goal is filtration1 vs filtration2 95% HDI exceeds zero:
139   goal2Ach = ( HDIofMCMC( mu[1,]-mu[2,] )[1] > 0.0 )
140   # Goal is condensation1 vs condensation2 95% HDI has width less than 0.2:
```

```
141  HDIlim = HDIofMCMC( mu[3,]-mu[4,] )
142  HDIwidth = HDIlim[2] - HDIlim[1]
143  goal3Ach = ( HDIwidth < 0.2 )
144  goalAchieved = c( goal1Ach , goal2Ach , goal3Ach )
145
146  return( goalAchieved )
147  } # end of function GoalAchievedForSample
148
149  #=========================================================================
150  # Now call the function defined above, using simulated data.
151
152  analysisType = c("Retro","Repli")[1] # specify [1] or [2]
153  fileNameRoot = paste("FilconBrugsPower",analysisType,sep="")
154
155  # Load original data, for use in replication probability analysis:
156  # (These lines intentionally exceed the margins so that they don't take up
157  # excessive space on the printed page.)
158  CondOfSubjOrig = c(1,1,1,1,1,1,1,1,1,1,1,1,1,1,1,1,1,1,1,1,1,1,1,1,1,1,1,1,1,1,1,1,1,
          1,1,1,1,1,1,1,1,2,2,2,2,2,2,2,2,2,2,2,2,2,2,2,2,2,2,2,2,2,2,2,2,2,2,2,2,2,2,2,2,2,2
          ,2,2,2,2,2,2,2,2,3,3,3,3,3,3,3,3,3,3,3,3,3,3,3,3,3,3,3,3,3,3,3,3,3,3,3,3,3,3,3,3,3,3,
          3,3,3,3,3,3,3,3,3,4,4,4,4,4,4,4,4,4,4,4,4,4,4,4,4,4,4,4,4,4,4,4,4,4,4,4,4,4,4,4,4,4,4,4
          ,4,4,4,4,4,4,4,4,4)
159  nTrlOfSubjOrig = c(64,64,64,64,64,64,64,64,64,64,64,64,64,64,64,64,64,64,64,64,
          64,64,64,64,64,64,64,64,64,64,64,64,64,64,64,64,64,64,64,64,64,64,64,64,64,
          64,64,64,64,64,64,64,64,64,64,64,64,64,64,64,64,64,64,64,64,64,64,64,64,64,
          64,64,64,64,64,64,64,64,64,64,64,64,64,64,64,64,64,64,64,64,64,64,64,64,64,
          64,64,64,64,64,64,64,64,64,64,64,64,64,64,64,64,64,64,64,64,64,64,64,64,64,
          64,64,64,64,64,64,64,64,64,64,64,64,64,64,64,64,64,64,64,64,64,64,64,64,64,
          64,64,64,64,64,64,64,64)
160  nCorrOfSubjOrig = c(45,63,58,64,58,63,51,60,59,47,63,61,60,51,59,45,61,59,60,58,63,
          56,63,64,64,60,64,62,49,64,64,58,64,52,64,64,64,62,64,61,59,59,55,62,51,58,55,
          54,59,57,58,60,54,42,59,57,59,53,53,42,59,57,29,36,51,64,60,54,54,38,61,60,61,
          60,62,55,38,43,58,60,44,44,32,56,43,36,38,48,32,40,40,34,45,42,41,32,48,36,29,
          37,53,55,50,47,46,44,50,56,58,42,58,54,57,54,51,49,52,51,49,51,46,46,42,49,46,
          56,42,53,55,51,55,49,53,55,40,46,56,47,54,54,42,34,35,41,48,46,39,55,30,49,27,
          51,41,36,45,41,53,32,43,33)
161  nSubjOrig = length(CondOfSubjOrig)
162  nCondOrig = length(unique(CondOfSubjOrig))
163
164  # Load previously computed posterior mu[cond,step], kappa[cond,step] chains.
165  load( file="FilconBrugsMuKappa.Rdata" )
166  chainLength = NCOL(mu)
167  nCond = NROW(mu) # should be 4, of course
168
169  # SPECIFY NUMBER OF SUBJECTS PER GROUP FOR SIMULATED DATA:
170  nSubjPerCond = 15
171
172  # Specify number of simulated experiments:
173  nSimExperiments = min(200,chainLength)
174  nSubj = nSubjPerCond * nCond # Number of subjects total.
175  nTrlPerSubj = 64 # Number of trials per subject; fixed by design at 64.
176  nGoal=3 # Determined in function above.
177
178  # If previous record of power estimation exists, load it and continue the runs.
```

```
179    filelist = dir( pattern=paste(fileNameRoot,"N",nSubjPerCond,"Result.Rdata",sep="") )
180    if ( length( filelist ) > 0 ) { # if the file already exists...
181        # load  previous expIdx, nSuccess
182        load(paste(fileNameRoot,"N",nSubjPerCond,"Result.Rdata",sep=""))
183        prevExpIdx = expIdx
184        # Use just some of the MCMC steps, distributed among the whole chain:
185        chainIdxVec = round(seq(1,chainLength,len=(prevExpIdx+nSimExperiments)))
186    } else { # ... otherwise, start a new record
187        prevExpIdx = 0
188        nSuccess = rep(0,nGoal) # Initialize success counter.
189        chainIdxVec = round(seq(1,chainLength,len=nSimExperiments))
190    }
191
192    # Simulated experiment loop begins here:
193    for ( expIdx in (1+prevExpIdx):(nSimExperiments+prevExpIdx) ) {
194
195        # Generate random data from posterior parameters
196        chainIdx=chainIdxVec[expIdx]
197        CondOfSubj = sort( rep( 1:nCond , nSubjPerCond ) )
198        nTrlOfSubj = rep( nTrlPerSubj , nSubj )
199        nCorrOfSubj = rep( 0 , nSubj )
200        for ( condIdx in 1:nCond ) {
201            m = mu[condIdx,chainIdx]
202            k = kappa[condIdx,chainIdx]
203            a = m*k
204            b = (1-m)*k
205            # Generate random theta and z values for simulated subjects:
206            thetaVec = rbeta( nSubjPerCond , a , b )
207            zVec = rbinom( n=nSubjPerCond , size=nTrlPerSubj , prob=thetaVec )
208            nCorrOfSubj[ CondOfSubj==condIdx ] = zVec
209        }
210
211        # Put data into list for BUGS program
212        if ( analysisType == "Retro" ) { # retrospective power
213            datalist = list(
214                nCond = nCond ,
215                nSubj = nSubj ,
216                cond = CondOfSubj ,
217                N = nTrlOfSubj ,
218                z = nCorrOfSubj
219                )
220        }
221        if ( analysisType == "Repli" ) { # replication probability
222            datalist = list(
223                nCond = nCond ,
224                nSubj = nSubj + nSubjOrig ,
225                cond = c( CondOfSubj , CondOfSubjOrig ) ,
226                N = c( nTrlOfSubj , nTrlOfSubjOrig ) ,
227                z = c( nCorrOfSubj , nCorrOfSubjOrig )
228                )
229        }
230
231        # Make plots for first 10 simulated experiments:
232        if ( expIdx <= 10 ) { plotRes = T } else { plotRes = F }
```

```
233
234      # Call BUGS program and tally number of successes:
235      nSuccess = nSuccess + GoalAchievedForSample( datalist ,
236                                           plotRes , fileNameRoot )
237      estPower = nSuccess / expIdx
238      cat( "\n*** nSubjPerCond:",nSubjPerCond, ", nSimExp:",expIdx,
239        " , nSuccess:",nSuccess, ", estPower:",round(estPower,2), "\n\n" )
240      flush.console()
241      save( nSuccess , expIdx , estPower ,
242            file=paste(fileNameRoot,"N",nSubjPerCond,"Result.Rdata",sep="") )
243
244    } # end of simulated experiment loop.
```

13.7 EXERCISES

Exercise 13.1. [Purpose: Comic relief.] Read the complete oeuvre of Friedrich Nietzsche, with special attention to his posthumous work, *The Will to Power* (Nietzsche, 1967). Provide a mathematical formalization of the Nietzschian concepts of will and power, using Bayesian probability theory. Show that the notion of statistical power is a special case of formalized Nietzschian power, *and vice versa*. Write your answer in both English and German. Do not submit your answer to the instructor; do post your answer on your personal web blog.

Exercise 13.2. [Purpose: To understand power for flipping a single coin in Tables 13.1 and 13.2.] For this exercise, consider flipping a single coin and inferring its bias.

(A)   Table 13.2 indicates that when the data-generating distribution is vague, with $\kappa = 10$ and $\theta = 0.80$, then 85 flips are needed for an 80% chance of getting the 95% HDI width to be less than 0.2. What is the minimal N needed if the data-generating distribution is certain, with $\kappa = 2000$? Show the command you used, and report the exact power for the smallest N that has power greater than 0.8.

(B)   Regarding the previous part, why might a researcher pursue a goal of precision if the data-generating hypothesis is already precise? (Hint: The audience prior may be different than the data-generating hypothesis. Discuss briefly, perhaps with an example.)

(C)   Table 13.1 indicates that when the data-generating distribution is highly certain, with $\kappa = 2000$ and $\theta = 0.80$, then 18 flips are needed for an 80% chance of getting the 95% HDI to exclude $\theta = 0.5$. What is the minimal N needed if the data-generating distribution is vague, with $\kappa = 2$? Show the command you used, and report the exact power for the smallest N that has power greater than 0.8.

(D)   For the previous part, the goal was for the HDI to exclude the null value (i.e., 0.5). Notice that the goal can be satisfied if the HDI is above the null value *or* if the HDI is below the null value. (1) When the data-generating prior is a beta distribution with $\mu = 0.8$ and $\kappa = 2$, as in the previous part, what proportion of the data-generating biases are greater than the null value? (2) If the goal is for the HDI to fall entirely *above* the null value, what sample size is needed to achieve a power of 0.8? (Hint: Use `minNforHDIpower.R` with the argument `ROPE=c(0,.5)`.) Watch the sample size increase and increase and increase, with the power creeping toward an asymptote. Why does the power never exceed the proportion you computed for (1)?

Exercise 13.3. [Purpose: To determine power for groups of coins, when the goal is precision.] Consider the filtration-condensation experiment summarized in Section 13.3. Suppose we want the 95% HDI on the difference, $\mu_3 - \mu_4$, to have a width less than 0.20. What sample size (N per group) is needed to achieve this goal at least 80% of the time? Determine the answer by running the program in Section 13.6.2 (`FilconBrugsPower.R`) with various values for `nPerGroup`. (Hint: $N \approx 17$; your job is to find the minimal N and discuss how you did it.)

Exercise 13.4. [Purpose: This is a *capstone* exercise that uses real data to review many techniques of the previous chapters, including generating priors in BUGS, checking credibility of null values, estimating retrospective power, and conducting a posterior predictive check.]

This exercise examines a learning experiment that investigated how easy it is for people to learn new category structures after having previously learned an initial structure (Kruschke, 1996). Some new structures had relevant stimulus dimensions that were also previously relevant in the initial structure, while other new structures had relevant stimulus dimensions that were previously *irrelevant* in the initial structure. The initial structure and the subsequently learned structures are outlined in Table 13.4.

The stimuli were simple pictures of freight-train boxcars that had three dimensions: position of "door" (left or right), height of box (short or tall), and color of "wheels" (blank or filled). The initial phase involved a structure in which two dimensions are relevant. For example, in Table 13.4, the initial phase can be described as "If it is short and left doored, or tall and right doored, then it's an X." Notice that wheel color is irrelevant in the initial phase. After learning that classification accurately, people would be trained on one of the four shifts listed in Table 13.4. The Reversal shift reverses all the labels. The Relevant shift uses just one of the previously relevant dimensions. The example in Table 13.4 can be described as "If it's short, then it's an X." The Irrelevant shift uses the one previously irrelevant dimension; the example in Table 13.4 can be described as "If it's blank wheeled, then it's an X." Finally, the Compound shift requires attention to two dimensions, one of which was previously relevant and one of

Table 13.4 Design of Relevance Shift Experiment Reported by Kruschke (1996). Cells indicate category assignment (X or O) of each stimulus. Learners were trained in the Initial Phase and then seamlessly continued into one of the Reversal, Relevant, Irrelevant, or Compound Phases.

Phase	Stimulus							
Initial	X	O	O	X	X	O	O	X
Reversal	O	X	X	O	O	X	X	O
Relevant	X	X	O	O	X	X	O	O
Irrelevant	X	X	X	X	O	O	O	O
Compound	X	O	X	O	O	X	O	X

Note. The actual stimuli had somewhat different proportions than the ones displayed here. The assignment of physical stimuli to abstract structural items was randomly permuted for each subject.

which was previously irrelevant. The example in Table 13.4 can be described as "If it's blank wheeled and left doored, or filled wheeled and right doored, then it's an X."

Various theories of learning predict different degrees of difficulty in learning the new structures after the old structures (see p. 230 of Kruschke, 1996). For example, if learners simply memorize the eight stimuli and their assignments, then the Reversal requires all eight associations to be changed, whereas Relevant, Irrelevant, and Compound shifts all require only four assignments to be changed. Therefore, the reversal should be most difficult to learn, and the other three changes should be equally easy to learn. One of the novel contributions of the experiment design was that it allowed direct comparison of "reversal" shift with "intradimensional" and "extradimensional" shifts, which had not been done previously. Another novel contribution was that the initial phase had dimensions that were relevant to the outcome without being individually correlated with the outcome. Variations of the design have subsequently been used by other researchers (e.g., D. N. George & Pearce, 1999; Oswald et al., 2001).

For each shift, the number of correct responses in the first 32 trials was recorded for each learner. These data are shown in the left column of Figure 13.5, p. 353, and are included in the file Kruschke1996CSdatsum.Rdata. It appears that Reversal shift is easiest (i.e., has highest accuracy), followed by Relevant, Irrelevant, and Compound shifts. The Bayesian analysis will tell us the credibility of those apparent differences between conditions.

(A) [Purpose: To create a BUGS model that has an estimated hyperprior on μ_c and on κ_c.] In the hierarchical diagram on the right side of Figure 9.17, p. 226, the κ_c parameters are drawn from a gamma distribution that has its parameters estimated rather than fixed. For our new model, we want to do the analogous estimation for the μ_c parameters also. Instead of setting A_μ and B_μ as constants, they will be estimated hyperparameters, denoted as $a_\mu = m_\mu{*}k_\mu$ and $b_\mu = (1 - m_\mu){*}k_\mu$, with $m_\mu \sim$ dunif(0, 1) and $k_\mu \sim$ dgamma(1, .1). Here is one way to specify the model (Kruschke1996CSbugs.R):

```
11  model {
12      for ( i in 1:nSubj ) {
13          nCorrOfSubj[i] ~ dbin( theta[i] , nTrlOfSubj[i] )
14          theta[i] ~ dbeta( a[ CondOfSubj[i] ] , b[ CondOfSubj[i] ] )I(0.0001,0.9999)
15      }
16      for ( cond in 1:nCond ) {
17          a[cond] <- mu[cond] * kappa[cond]
18          b[cond] <- (1-mu[cond]) * kappa[cond]
19          mu[cond] ~ dbeta( aMu , bMu )
20          kappa[cond] ~ dgamma( sGamma , rGamma )
21      }
22      aMu <- max( .01 , mMu * kMu )
23      bMu <- max( .01 , (1-mMu) * kMu )
24      mMu ~ dunif(0,1)
25      kMu ~ dgamma(1,.1)
26      sGamma <- max( .005 , pow(muGamma,2)/pow(sigmaGamma,2) )
27      rGamma <- max( .005 , muGamma/pow(sigmaGamma,2) )
28      muGamma ~ dgamma(1,.1)
29      sigmaGamma ~ dgamma(1,.1)
30  }
```

Discuss why we would want to estimate higher-level distributions across the μ_c and κ_c. You may want to mention commonalities across conditions, such as all the learners being the same species, and all learners experiencing the same sort of stimuli on the same computer display. It is also important to discuss *shrinkage* of estimates. Finally, discuss why the particular higher-level distribution might *not* be appropriate.

(B) [Purpose: To check for convergence, mixing, and autocorrelation.] With the data included, run the BUGS model and check for convergence, mixing, and autocorrelation. The model can be initialized automatically, using modelGenInits(), but unfortunately it takes forever for the chains to converge because some are initialized too far away from the mode of the posterior. Instead, manually initialize the chains at reasonable values, as indicated by the data. The same method as was used in Section 9.5.2 (FilconBrugs.R) is used and extended here (Kruschke1996CSbugs.R):

```
77  genInitList <- function() {
78      sqzData = .01+.98*datalist$nCorrOfSubj/datalist$nTrlOfSubj
```

```
79        mu = aggregate( sqzData , list(datalist$CondOfSubj) , "mean" )[,"x"]
80        sd = aggregate( sqzData , list(datalist$CondOfSubj) , "sd" )[,"x"]
81        kappa = mu*(1-mu)/sd^2 - 1
82        mMu = mean( mu )
83        kMu = mMu * (1-mMu) / sd(mu)^2
84        muGamma = mean( kappa )
85        sigmaGamma = sd( kappa )
86        return( list( theta = sqzData ,
87                          mu = mu ,
88                          kappa = kappa ,
89                          mMu = mMu ,
90                          kMu = kMu ,
91                          muGamma = muGamma ,
92                          sigmaGamma = sigmaGamma ) )
93     }
94     for ( chainIdx in 1 : nchain ) {
95        modelInits( bugsInits( genInitList ) )
96     }
```

Run the model and determine reasonable burn-in and thinning. Show the autocorrelation and chain plots for μ_c and κ_c. You may find it useful to refer to Section 23.2, p. 623.

(C)   [Purpose: To examine and interpret the posterior distribution of differences.] Which groups are different from each other, on which parameters? (Hint: See Figure 13.4.)

(D)   [Purpose: To check the robustness of the posterior when the prior is changed in reasonable ways.]   Do the posterior differences change much if the prior changes? (1) Specifically, try this alternative vague prior: Wherever the top-level prior specifies dgamma(1,.1), change it to dgamma(0.1,.1). (2) Also, try this prior that forces all κ_c to be close to 15:

```
muGamma  ~ dgamma(22500,1500)
sigmaGamma  ~ dgamma(25,250)
```

Show your results. Which prior is more reasonable, and why?

(E)   [Purpose: To conduct a posterior predictive check.]   For this part, use the plausible dgamma(1,.1) prior, not the less-plausible other priors explored in the previous part. Conduct a posterior predictive check by generating simulated data from the sampled posterior parameter values. This can be done in R as follows (Kruschke1996CSbugs.R):

```
203    nSimExps = min(500,chainLength) # number of simulated experiments
204    nSubjPerCond = sum( CondOfSubj == 1 ) # number of subjects per condition
205    nTrlPerSubj = nTrlOfSubj[1] # number of trials per subject
206    nCorrOfSubjPredMat = matrix( 0 , nrow=nSimExps , ncol=nSubjPerCond*nCond )
207    CondOfSubjPredMat =  matrix( 0 , nrow=nSimExps , ncol=nSubjPerCond*nCond )
208    nTrlOfSubjPredMat = matrix( 0 , nrow=nSimExps , ncol=nSubjPerCond*nCond )
209    for ( stepIdx in 1:nSimExps ) {
210       for ( condIdx in 1:nCond ) {
211          m = mu[condIdx,stepIdx]
```

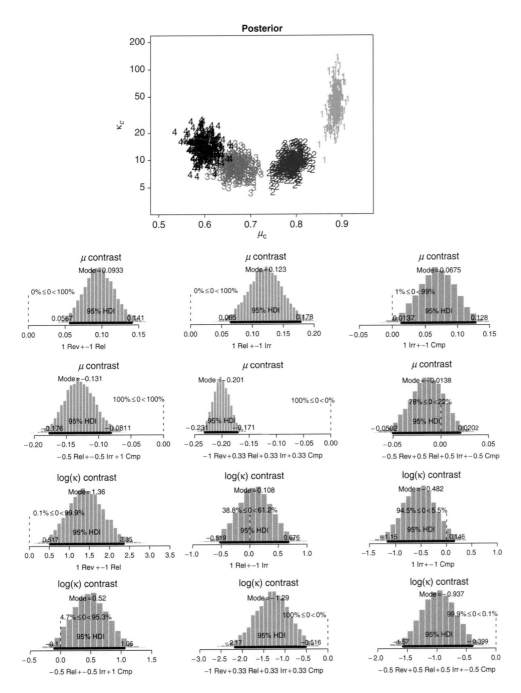

FIGURE 13.4

For Exercise 13.4, Part C. Posterior μ_c and κ_c values and their differences. In the scatter plot, numerals represent a step in the chain for that condition. (Adapted with permission from Figs. 3 and 4 of Kruschke, 2010b.)

```
212          k = kappa[condIdx,stepIdx]
213          a = m*k
214          b = (1-m)*k
215          for ( subjIdx in 1:nSubjPerCond ) {
216              colIdx = (condIdx-1)*nSubjPerCond + subjIdx
217              theta = rbeta( 1 , a , b )
218              nCorrOfSubjPredMat[stepIdx,colIdx] = rbinom( 1 ,
219                              size=nTrlPerSubj , prob=theta )
220              CondOfSubjPredMat[stepIdx,colIdx] = condIdx
221              nTrlOfSubjPredMat[stepIdx,colIdx] = nTrlPerSubj
222          }
223      }
224  }
```

This code puts each randomly generated sample of data into a row of the matrix nCorrOfSubjPredMat. Figure 13.5 shows histograms of simulated data. The preceding code relied on previously extracting the parameter chains from BUGS, as follows (Kruschke1996CSbugs.R):

```
127  mu = NULL
128  kappa = NULL
129  for ( condIdx in 1:nCond ) {
130      nodeName = paste( "mu[" , condIdx , "]" , sep="" )
131      mu = rbind( mu , samplesSample( nodeName ) )
132      nodeName = paste( "kappa[" , condIdx , "]" , sep="" )
133      kappa = rbind( kappa , samplesSample( nodeName ) )
134  }
```

(F) [Purpose: To do a retrospective power analysis.]   Using the method of Figure 13.3, which is exemplified by the code explained in Section 13.6.2 (FilconBrugsPower.R), conduct a retrospective power analysis, and estimate the power of four goals: $\mu_{Rev} - \mu_{Rel} > 0$, $\mu_{Rel} - \mu_{Irr} > 0$, $\mu_{Irr} - \mu_{Com} > 0$, and $\kappa_{Rev} - \kappa_{Rel} > 0$ (condition 1 is Reversal, condition 2 is Relevant, condition 3 is Irrelevant, and condition 4 is Compound). To do this, first analyze the actual data and get a posterior sample of μ_c and κ_c values. Then, wrap the BUGS program into a function, into which you can pass data and from which you check whether each goal is achieved. Then make a loop that goes stepwise through the μ_c, κ_c chain and generates simulated data from those credible parameter values. At each step, pass the simulated data into the BUGS analysis function. Run at least 100 simulated data sets, and report your tally for how many times the goals were achieved. *Cautions and hints*: If each BUGS analysis takes two minutes to run, a batch of a few hundred simulated data sets will take hours. Plan accordingly. Sometimes BUGS will run fine for dozens of simulated data sets, and then inexplicably crash on the next random data set. Therefore, be sure that your results regarding power estimation are saved at the end of each data set. The goal of showing $\mu_{Rev} - \mu_{Rel} > 0$ has a retrospective power of nearly 100%, but at least one of the other goals involving μ_c has retrospective power closer to 60%, which indicates that even with 60

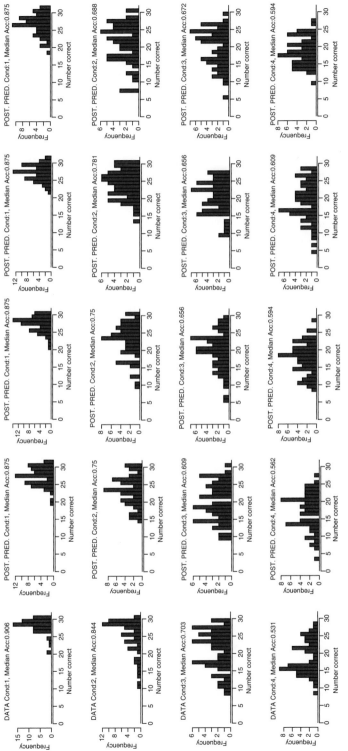

FIGURE 13.5

For Exercise 13.4. Actual data in left column, with examples of posterior-predicted data in other columns. Top row is Reversal shift, second row is Relevant, third row is Irrelevant, and fourth row is Compound. There is a suggestion of bimodality in the data from the Compound shift (lower-left histogram) that is not often generated by simulated posterior-predicted data. Follow-up modeling and empirical work might want to investigate that bimodality if it is deemed theoretically interesting.

subjects per condition, this experiment might have been underpowered. Bayesian analysis does not care whether there are equal numbers of subjects in each condition, so follow-up work could use more subjects in some conditions and fewer subjects in other condtions. The goal of showing $\kappa_{Rev} - \kappa_{Rel} > 0$ has a retrospective power of only about 40%, which is caused largely by shrinkage of κ_c estimates.

(G) [Purpose: To do a replication power analysis.] Suppose you repeat the experiment, using the posterior of the first experiment as the prior for the second experiment. Estimate the probability of achieving these four goals: $\mu_{Rev} - \mu_{Rel} > 0$, $\mu_{Rel} - \mu_{Irr} > 0$, $\mu_{Irr} - \mu_{Com} > 0$, and $\kappa_{Rev} - \kappa_{Rel} > 0$ (condition 1 is Reversal, condition 2 is Relevant, condition 3 is Irrelevant, and condition 4 is Compound). In other words, conduct an analysis of replication probability, regarding the same goals as the previous part. This task is easy to do if you successfully accomplished the previous part. In each step of the chain, instead of using only the simulated data, concatenate the simulated data to the actual data. *Cautions and hints:* The runs will take even longer than in the previous part because of the larger data sets.

Applied to the Generalized Linear Model

CHAPTER 14

Overview of the Generalized Linear Model

Straight and proportionate, deep in your core
All is orthogonal, ceiling to floor.
But on the outside the vines creep and twist
'round all the parapets shrouded in mist.

The previous part of the book explored all the basic concepts of Bayesian analysis applied to a simple likelihood function, namely the Bernoulli distribution. The focus on a simple likelihood function allowed the complex concepts of Bayesian analysis, such as MCMC methods and hierarchical priors, to be developed without interference from additional complications of elaborate likelihood functions with multiple parameters.

Doing Bayesian Data Analysis: A Tutorial with R and BUGS. DOI: 10.1016/B978-0-12-381485-2.00014-6
© 2011, Elsevier Inc. All rights reserved.

In this part of the book, we apply all the concepts to a more complex but versatile model known as the *generalized linear model* (GLM; McCullagh & Nelder, 1989; Nelder & Wedderburn, 1972). This model comprises the traditional "off the shelf" analyses such as *t*-tests, analysis of variance (ANOVA), multiple linear regression, logistic regression, and the like. Because we now know from previous chapters the concepts and mechanisms of Bayesian analysis, we can focus on applications of this versatile model. This chapter is important for understanding subsequent chapters because it lays out the framework for all the models in the remainder of the book.

14.1 THE GENERALIZED LINEAR MODEL (GLM)

To understand the generalized linear model and its many specific cases, we must build up a variety of component concepts regarding relationships between variables and how variables are measured in the first place.

14.1.1 Predictor and Predicted Variables

Suppose we want to predict someone's weight from their height. In this case, weight is the predicted variable and height is the predictor. Or, suppose we want to predict high school grade point average (GPA) from Scholastic Aptitude Test (SAT) score and family income. In this case, GPA is the predicted variable, whereas SAT and income are predictor variables. Or suppose we want to predict the blood pressure of patients who either take drug A or take drug B or take a placebo or merely wait. In this case, the predicted variable is blood pressure, and treatment-group membership is the predictor.

The key mathematical difference between predictor and predicted variables is that the likelihood function expresses the probability of values of the predicted variable as a function of values of the predictor variable. The likelihood function does not describe the probabilities of values of the predictor variable. The value of the predictor variable comes from "outside" the system being modeled, whereas the value of the predicted variable depends on the value of the predictor variable.

Because the predicted variable depends on the predictor variable, at least mathematically in the likelihood function if not causally in the real world, the predicted variable can also be called the *dependent* variable. The predictor variables are sometimes called *independent* variables. The key conceptual difference between an independent and dependent variable is that the value of the dependent variable depends on the value of the independent variable. The term "independent" can be confusing because it can be used strictly or loosely. In experimental settings, the variables that are actually manipulated and set by the experimenter are the independent variables. In this context of experimental manipulation, the values of the independent variables truly are (in principle, at least) independent of the values of other variables, because the experimenter has intervened to arbitrarily set the values of the independent variables. But sometimes a nonmanipulated variable is also referred to as "independent," merely as a way to indicate that it is being used as a predictor variable.

Among nonmanipulated variables, the roles of predicted and predictor are arbitrary, determined only by the interpretation of the analysis. Consider, for example, people's weights and heights. We could be interested in predicting a person's weight from his or her height, or we could be interested in predicting a person's height from his or her weight. Prediction is merely a mathematical dependency, not necessarily a description of underlying causal relationship. Although height and weight tend to covary across people, the two variables are not directly causally related. When a person slouches, thereby getting shorter, she or he does not lose weight. And when a person drinks a glass of water, thereby weighing more, she or he does not get taller.

Just as "prediction" does not imply causation, "prediction" also does not imply any temporal relation between the variables. For example, we may want to predict a person's sex, male or female, from his or her height. Because males tend to be taller than females, this prediction can be made with better than chance accuracy. But a person's sex is not caused by his or her height, nor does a person's sex occur only after that person's height is measured. Thus, we can "predict" a person's sex from his or her height, but this does not mean that the person's sex occurred later in time than his or her height.

In summary, all manipulated independent variables are predictor variables, not predicted. Some dependent variables can take on the role of predictor variables, if desired. All predicted variables are dependent variables. The likelihood function specifies the probability of values of the predicted variables as a function of the values of the predictor variables.

Why we care. We care about these distinctions between predicted and predictor variables because the likelihood function is a mathematical description of the dependency of the predicted variable on the predictor variable. The first thing we have to do in statistical inference is identify the predicted and predictor variables.

14.1.2 Scale Types: Metric, Ordinal, Nominal

Items can be measured on different scales. For example, the participants in a foot race can be measured either by the time they took to run the race, by their placing in the race (first, second, third, etc.), or by the name of the team they represent. These three measurements are examples of metric, ordinal, and nominal scales, respectively (Stevens, 1946).

Examples of *metric* scales include response time (i.e., latency or duration), temperature, height, and weight. Those are actually cases of a specific type of metric scale, called a *ratio* scale, because they have a natural zero point on the scale. The zero point on the scale corresponds to there being a complete absence of the stuff being measured. For example, when the duration is zero, there has been no time elapsed, and when the weight is zero, there is no downward force. Because these scales have a natural zero point, it is meaningful to talk about ratios of amounts being measured, and that is why they are called ratio scales. For example, it is meaningful to say that taking 2 minutes to solve a problem is twice as long as taking 1 minute to solve the problem. On the other hand, the scale of historical time has no known absolute zero. We cannot say, for example, that there is twice as much time in January 2 as there is time in January 1. We can refer to the duration since some arbitrary reference point, but we cannot talk about the absolute amount of time in any given moment. Scales that have no natural zero are called *interval* scales because all we know about them is the amount of stuff in an interval on the scale, not the amount of stuff at a point on the scale. Despite the conceptual difference between ratio and interval scales, we will lump them together into the category of metric scales.

A special case of metric-scaled data is *count* data, also called *frequency* data. For example, the number of cars that pass through an intersection during an hour is a count. The number of poll respondents who say they belong to a particular political party is a count. Count data can only have values that are non-negative integers. Distances between counts have meaning, and therefore the data are metric, but because the data cannot be negative and are not continuous, they are treated with different mathematical forms than continuous, real-valued metric data.

Examples of *ordinal* scales include placing in a race, or rating of degree of agreement. When we are told that, in a race, Jane came in first, Jill came in second, and Jasmine came in third, we only know the order. We do not know whether Jane beat Jill by a nose or by a mile. There is no distance or metric information in an ordinal scale. As another example, many polls have ordinal response scales: Indicate how much you agree with this statement: "Bayesian statistical inference is better than null hypothesis significance testing," with 5 = strongly agree, 4 = mildly agree, 3 = neither agree nor disagree, 2 = mildly disagree, and 1 = strongly disagree. Notice that there is no metric information in the response scale, because we cannot say the difference between ratings of 5 and 4 is the same amount of difference as between ratings of 4 and 3.

Examples of *nominal*, also known as categorical, scales include political party affiliation, the face of a rolled die, and the result of a flipped coin. For nominal scales, there is neither distance between categories nor order between categories. For example, suppose we measure the political party affiliation of a person. The categories of the scale might be Green, Democrat, Republican,

Libertarian, and Other. Although some political theories might infer that the parties fall on some underlying liberal-conservative scale, there is no such scale in the actual categorical values themselves. In the actual categorical labels there is neither distance nor ordering.

In summary, if two items have different nominal values, all we know is that the two items are different (and what categories they are in). On the other hand, if two items have different ordinal values, we know that the two items are different and we know which one is "larger" than the other, but not how much larger. If two items have different metric values, then we know that they are different, which one is larger, and how much larger.

Why we care. We care about the scale type because the likelihood function must specify a probability distribution on the appropriate scale. If the scale has two nominal values, then a Bernoulli likelihood function may be appropriate. If the scale is metric, then a normal distribution may be appropriate as a likelihood function. Whenever we are choosing a model for data, we must answer the question, What kind of scale are we dealing with?

In the following sections, we first consider the case of a metric predicted variable with metric predictors. In that context of all metric variables, we develop the concepts of linear functions and interactions. Once those concepts are established for metric predictors, the notions are extended to nominal predictors.

14.1.3 Linear Function of a Single Metric Predictor

Suppose we have identified one variable to be predicted, which we'll call y, and one variable to be the predictor, which we'll call x. Suppose we have determined that both variables are metric. The next issue we need to address is how to model a relationship between x and y. There are many possible dependencies of y on x, and the particular form of the dependency is determined by the specific meanings and nature of the variables. But in general, across all possible domains, what is the most basic or simplistic dependency of y on x that we might consider? The usual answer to this question is, a linear relationship. A linear function is the generic, "vanilla," off-the-shelf dependency that is used in statistical models. The methods can be generalized to other models when needed.

Linear functions preserve proportionality. If you double the input, then you double the output. If cost of a book is a linear function of the number of pages, then when the number of pages is reduced 10%, the cost should be reduced 10%. If automobile speed is a linear function of gas delivery to the engine, then when you press the pedal 20% further, the car should go 20% faster. Nonlinear functions do not preserve proportionality. For example, in actuality, car speed is not a linear function of gas delivery. At higher and higher speeds, it takes proportionally more and more gas to make the car go faster. Despite the fact that many real-world dependencies are nonlinear, most are at least approximately linear over moderate ranges of the variables. For example, if you have twice the wall area, it takes approximately twice the amount of paint. It is also the case that linear relationships are intuitively prominent (Brehmer, 1974; Hoffman, Earle, & Slovic, 1981; Kalish, Griffiths, & Lewandowsky, 2007). Linear relationships are the easiest to think about: Turn the steering wheel twice as far, and the car should turn twice as sharp. Turn the volume knob 50% higher, the loudness should increase 50%.

The general mathematical form for a linear function of a single variable is

$$y = \beta_0 + \beta_1 x \tag{14.1}$$

When values of x and y that satisfy Equation 14.1 are plotted, they form a line. Examples are shown in Figure 14.1. The value of parameter β_0 is called the y intercept because it is the where the line intersects the y axis when $x = 0$. The left panel of Figure 14.1 shows two lines with different y intercepts. The value of parameter β_1 is called the *slope* because it indicates how much y increases when x increases by 1. The right panel of Figure 14.1 shows two lines with the same intercept but different slopes.

In strict mathematical terminology, the type of transformation in Equation 14.1 is called *affine*. When $\beta_0 \neq 0$, the transformation does not preserve proportionality. For example, consider $y = 10 + 2x$. When x is doubled from $x = 1$ to $x = 2$, y increases from $y = 12$ to $y = 14$, which is not doubling y.

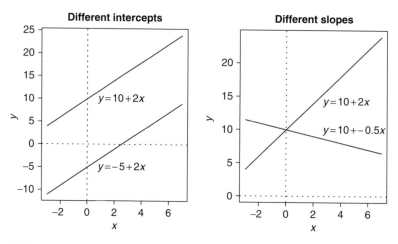

FIGURE 14.1

Examples of linear functions of a single x variable. The left panel shows examples of two lines with the same slope but different intercepts. The right panel shows examples of two lines with the same intercept but different slopes.

Nevertheless, the rate of increase in y is the same for all values of x: Whenever x increases by 1, y increases by 2. Equation 14.1 can be algebraically rearranged so that it does preserve proportionality, as will be shown next.

14.1.3.1 Reparameterization to x Threshold Form

Equation 14.1 can be algebraically rearranged as follows:

$$y = \beta_0 + \beta_1 x$$
$$= \beta_1 \left(x - \underbrace{(-\beta_0/\beta_1)}_{\theta} \right) \tag{14.2}$$

This form of the equation is useful because it explicitly shows the value of the x intercept, also known as the threshold, denoted θ. (Do not confuse this use of the symbol θ with different uses in previous chapters.) The threshold is the value of x when y is zero. This is sometimes also called the x intercept.

The x threshold form preserves proportionality for $x - \theta$. As an example, consider again the case of $y = 10 + 2x$. When changed to x threshold form, it becomes $y = 2(x + 5)$. When x changes from 1 to 2, $x + 5$ changes from 6 to 7, which is an increase of $(7 - 6)/6 = 1/6$. The resulting change in y is from 12 to 14, which is an increase of $(14 - 12)/12 = 1/6$. Thus, a 1/6 increase in $x - \theta$ results in a 1/6 increase in y.

The threshold (i.e., x intercept) is often more meaningful than the y intercept. For example, suppose we are piloting a tugboat upstream on the Mississippi river, and we want to predict how much headway, y, we will gain against the current for a given setting of the throttle, x. Suppose it is the case that $y = -2 + 4x$. This form of the equation indicates that when we apply zero engine power, that is, when $x = 0$, then we lose 2 miles an hour (i.e., $y = -2$). In other words, the y intercept tells us the baseline speed of the river current that we are trying to overcome. What may be more useful to know, however, is the amount of engine power we need to apply in order to overcome the current: How big must x be so that we are just matching the downstream pressure? The answer to this question is the threshold (i.e., the value of x that makes $y = 0$). In our example, wherein $y = -2 + 4x$, the threshold is $\theta = -(-2/4) = 0.5$. In other words, when the throttle is set above the threshold of 0.5, then we make progress upstream because $y > 0$, but when the throttle is set below the threshold of 0.5, then we drift downstream because $y < 0$. Thus, the more intuitive form of the "headway" equation is the x intercept form, $y = 4(x - 0.5)$, because it shows explicitly that our headway is proportional to how much the throttle exceeds 0.5.

Summary of why we care. The likelihood function specifies the form of the dependency of y on x. When y and x are metric variables, the simplest form of dependency, both mathematically and intuitively, is one that preserves proportionality. The mathematical expression of this relation is a so-called *linear function*. The usual mathematical expression of a line is the y intercept form, but often a more intuitive expression is the x threshold form. Linear functions form the core of most statistical models, so it is important to become facile with their algebraic forms and graphical representations.

14.1.4 Additive Combination of Metric Predictors

If we have more than one predictor variable, what function should we use to combine the influences of all the predictor variables? If we want the combination to be linear in each of the predictor variables, then there is just one answer: addition. In other words, if we want an increase in one predictor variable to predict the *same* proportional increase in the predicted variable *for any value of the other predictor variables*, then the predictions of the individual predictor variables must be added.

In general, a linear combination of K predictor variables has the form

$$y = \beta_0 + \beta_1 x_1 + \cdots + \beta_K x_K$$

$$= \beta_0 + \sum_{k=1}^{K} \beta_k x_k \tag{14.3}$$

Figure 14.2 shows examples of linear functions of *two* variables, x_1 and x_2. The graphs show y plotted only over a domain with $0 \le x_1 \le 10$ and $0 \le x_2 \le 10$. It is important to realize that the plane extends from minus to plus infinity, and the graphs only show a small region. Notice in the upper-left panel, where $y = 0 + 1x_1 + 0x_2$, that when $x_1 = 10$, then $y = 10$, regardless of the value of x_2. The plane tilts upward in the x_1 direction, but the plane is horizontal in the x_2 direction. The opposite is true in the upper-right panel: The plane tilts upward in the x_2 direction, but the plane is horizontal in the x_1 direction, because there $y = 0 + 0x_1 + 2x_2$. The lower-left panel shows the two influences added: $y = 0 + 1x_1 + 2x_2$. Notice that the slope in the x_2 direction is steeper than in the x_1 direction. Most important, notice that the slope in the x_2 direction is the same at any specific value of x_1. For example, when $x_1 = 0$, y rises from $y = 0$ to $y = 20$ (i.e., an increase of 20) when x_2 goes from $x_2 = 0$ to $x_2 = 10$. And when $x_1 = 10$, y rises from $y = 10$ to $y = 30$, again an increase of 20, when x_2 goes from $x_2 = 0$ to $x_2 = 10$.

14.1.4.1 Reparameterization to x Threshold Form

For notational convenience, define the length of a vector $\vec{\beta} = \langle \beta_1, \ldots, \beta_K \rangle$ to be $\left\| \vec{\beta} \right\| = \left(\sum_k \beta_k^2 \right)^{1/2}$. This may look complicated, but it's merely the everyday formula for distance from the Pythagorean theorem. Carpenters all memorize a handy special case of this relationship, known as the "3-4-5 rule": When

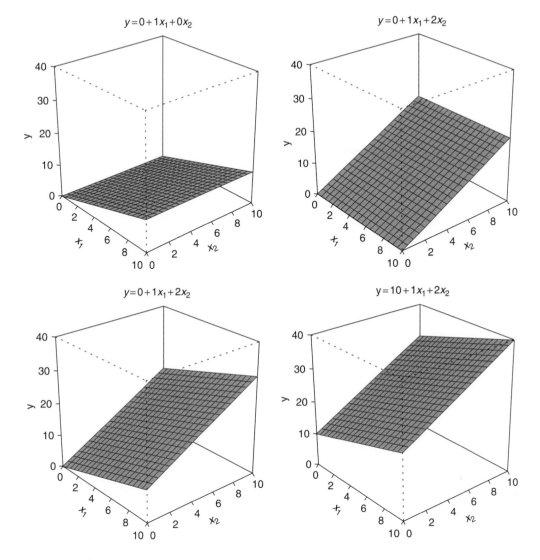

FIGURE 14.2
Examples of linear functions of two variables, x_1 and x_2. *Upper left*: Only x_1 has an influence on y. *Upper right*: Only x_2 has an influence on y. *Lower left*: x_1 and x_2 have an additive influence on y; compare with upper panels. *Lower-right*: Nonzero intercept is added; compare with lower-left panel.

the lengths of the short edges of a right-angled triangle are 3 and 4, then the length of the long edge is exactly 5, because $5 = (3^2 + 4^2)^{1/2}$. (Carpenters actually use the 3-4-5 rule to infer an angle from lengths, rather than infer a third length from two lengths and an angle. Carpenters know that if a triangle has edge lengths of 3, 4, and 5, then the angle between the short edges is exactly 90 degrees.) With this new notation for length, Equation 14.3 can be algebraically reexpressed as

$$
\begin{aligned}
y &= \beta_0 + \sum_{k=1}^{K} \beta_k x_k \\
&= \beta_0 + \left\| \vec{\beta} \right\| \sum_{k=1}^{K} \frac{\beta_k}{\left\| \vec{\beta} \right\|} x_k \\
&= \left\| \vec{\beta} \right\| \left(\sum_{k=1}^{K} \frac{\beta_k}{\left\| \vec{\beta} \right\|} x_k - \underbrace{\left(-\beta_0 \Big/ \left\| \vec{\beta} \right\| \right)}_{\theta} \right)
\end{aligned}
\tag{14.4}
$$

Notice that when there is only a single predictor variable (i.e., when $K = 1$) then $\left\| \vec{\beta} \right\| = |\beta_1|$ and Equation 14.4 reduces to Equation 14.2.

In Equation 14.4, the value of θ is the (Euclidean) length of x when $y = 0$ and when x is in the direction of vector $\langle \beta_1, \ldots, \beta_K \rangle$. In other words, when $\vec{x} = \theta \vec{\beta} \Big/ \left\| \vec{\beta} \right\|$, then $y = 0$. When the length of $\vec{x}$ (in that direction) exceeds the threshold θ, then $y > 0$. The x threshold form in Equation 14.4 can be useful in the context of logistic regression (Chapter 20), but is presented here primarily as a generalization of Equation 14.2.

Summary of section: Predictors are additive if the influence of any single predictor does not depend on the values of the other predictors. The combined influence of two or more predictors can be additive even if the individual influences are nonlinear. But if the individual influences are linear and the combined influence is additive, then the overall combined influence is also linear. The formula of Equation 14.3, or its reparameterization in Equation 14.4, is known as the linear model. It forms the core of many statistical models.

14.1.5 Nonadditive Interaction of Metric Predictors

The combined influence of two predictors does not have to be additive. Consider, for example, a person's self-rating of happiness, predicted from his or her overall health and annual income. It's likely that if a person's health is very poor, then the person is not happy, regardless of his or her income. And if the person has zero income, then the person is probably not happy, regardless of his or her health. But if the person is both healthy and rich, then the person is probably happy (despite celebrated counterexamples in the popular media).

A graph of this sort of nonadditive interaction between predictors appears in the upper-left panel of Figure 14.3. The vertical axis, labeled y, is happiness.

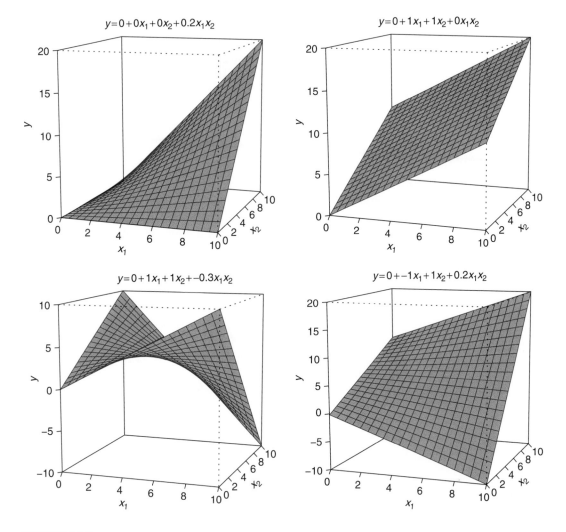

FIGURE 14.3
Multiplicative interaction of two variables, x_1 and x_2. Upper-right panel shows zero interaction, for comparison. Figure 17.8, p. 470, provides additional perspective and insight.

The horizontal axes, x_1 and x_2, are health and income. Notice that if either $x_1 = 0$ or $x_2 = 0$, then $y = 0$. But if both $x_1 > 0$ and $x_2 > 0$, then $y > 0$. The specific form of interaction plotted here is *multiplicative*: $y = 0 + 0x_1 + 0x_2 + .2x_1x_2$. For comparison, the upper-right panel of Figure 14.3 shows a noninterac-tive (i.e., additive) combination of x_1 and x_2. Notice that the graph of the interaction has a twist in it, but the graph of the additive combination is flat.

The lower-left panel of Figure 14.3 shows a multiplicative interaction in which the individual predictors increase the outcome, but the combined variables decrease the outcome. A real-world example of this occurs with some drugs: Individually, each of two drugs might improve symptoms, but when taken together, the two drugs might interact and cause a decline in health. As another example, consider lighter-than-air travel (i.e., ballooning). The levity of a balloon is increased by fire, as in hot air balloons. And the levity of a balloon is increased by hydrogen, as in many early 20th-century blimps and dirigibles. But the levity of a balloon is dramatically decreased by the combination of fire and hydrogen.

The lower-right panel of Figure 14.3 shows a multiplicative interaction in which the direction of influence of one variable depends on the magnitude of the other variable. Notice that when $x_2 = 0$, then an increase in the x_1 variable leads to a decline in y. But when $x_2 = 10$, then an increase in the x_1 variable leads to an increase in y. Again, the graph of the interaction shows a twist and is not flat.

A nonadditive interaction of predictors does not have to be multiplicative. Other types of interaction are possible. The type of interaction is motivated by idiosyncratic theories in different variables in different application domains. Consider, for example, predicting the magnitude of gravitational force between two objects, from three predictor variables: mass of object one, mass of object two, and the distance between the objects. The force is proportional to the product (i.e., multiplication) of their two masses. But the force is proportional to the masses *divided by* the squared distance between them.[1]

14.1.6 Nominal Predictors
14.1.6.1 Linear Model for a Single Nominal Predictor
The previous sections assumed that the predictor was metric. But what if the predictor is nominal, such as political party affiliation or gender? What is the simplest generic model for a metric variable predicted from a nominal variable? The "natural" model has each value of x generate a particular deflection of y away from its baseline level. For example, consider predicting height from sex (male or female). We can consider the overall average height across both sexes as the baseline height. When an individual has the value "male," that adds an upward deflection to the predicted height. When an individual has the value "female," that adds a downward deflection to the predicted height.

Expressing that idea in mathematical notation can get a little tricky. First consider the nominal predictor. We can't represent it appropriately as a single scalar value, such as 1 through 5, because that would mean that level 1

---

[1] Division by squared distance can be thought of as multiplication by the reciprocal of squared distance, but that amounts to reparameterizing the model.

is closer to level 2 than it is to level 5, which is not true of nominal values. Therefore, instead of representing the value of the nominal predictor by a single scalar value x, we will represent the nominal predictor by a vector $\vec{x} = \langle x_1, \ldots, x_J \rangle$, where J is the number of categories that the predictor has. When an individual has level j of the nominal predictor, this is represented by setting $x_j = 1$ and $x_{i \neq j} = 0$. For example, suppose x is sex, with level 1 being male and level 2 being female (so $J = 2$). Then male is represented as $\vec{x} = \langle 1, 0 \rangle$ and female is represented as $\vec{x} = \langle 0, 1 \rangle$. As another example, suppose that the predictor is political party affiliation, with Green as level 1, Democrat as level 2, Republican as level 3, Libertarian as level 4, and Other as level 5. Then Democrat is represented as $\vec{x} = \langle 0, 1, 0, 0, 0 \rangle$, and Libertarian is represented as $\vec{x} = \langle 0, 0, 0, 1, 0 \rangle$. Political party affiliation is being treated here as a categorical label only, with no ordering along a liberal-conservative scale.

Now that we have a formal representation for the nominal predictor variable, we can create a formal representation for the generic model of how the predictor influences the predicted variable. As mentioned earlier, the idea is that there is a baseline level of the predicted variable, and each category of the predictor indicates a deflection above or below that baseline level. We will denote the baseline value of the prediction as β_0. The deflection for the jth level of the predictor is denoted β_j. Then the predicted value is

$$y = \beta_0 + \beta_1 x_1 + \cdots + \beta_J x_J$$
$$= \beta_0 + \vec{\beta} \cdot \vec{x} \qquad (14.5)$$

where the notation $\vec{\beta} \cdot \vec{x}$ is sometimes called the *dot product* of the vectors.

Notice that Equation 14.5 has a form very similar to the basic linear form of Equation 14.1. The conceptual analogy is this: In Equation 14.1 for a metric predictor, the slope β_1 indicates how much y changes when x changes from 0 to 1. In Equation 14.5 for a nominal predictor, the coefficient β_j indicates how much y changes when x changes from neutral to category j.

There is one more consideration when expressing the influence of a nominal predictor as in Equation 14.5: How should the baseline value be set? Consider, for example, predicting height from sex. We could set the baseline height to be zero. Then the deflection from baseline for male might be 5'10" (say), and the deflection from baseline for female might be 5'4" (say). On the other hand, we could set the baseline height to be 5'7". Then the deflection from baseline for male would be +3", and the deflection from baseline for female would be −3". The second way of setting the baseline is the typical way it is done in generic statistical modeling. In other words, the baseline is constrained so that

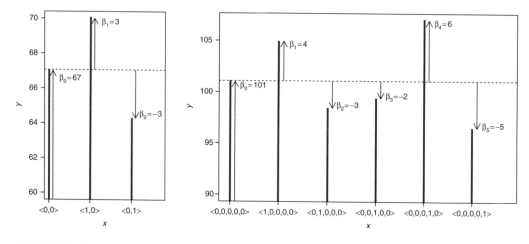

FIGURE 14.4

Examples of a nominal predictor (Equations 14.5 and 14.6). Left panel shows a case with $J = 2$; right panel shows a case with $J = 5$. In each panel, the baseline value of y is on the far left, when all the x components are zero. Notice that the deflections from baseline sum to zero.

the deflections sum to zero across the categories:

$$\sum_{j=1}^{J} \beta_j = 0 \tag{14.6}$$

The expression of the model in Equation 14.5 is not complete without the constraint in 14.6.

Figure 14.4 shows examples of a nominal predictor, expressed in terms of Equations 14.5 and 14.6. The left panel shows a case in which $J = 2$, and the right panel shows a case in which $J = 5$. Notice that the deflections from baseline sum to zero, as demanded by the constraint in Equation 14.6.

14.1.6.2 Additive Combination of Nominal Predictors

Suppose we have two (or more) nominal predictors of a metric value. For example, we might be interested in predicting income as a function of political party affiliation and gender. Figure 14.4 showed examples of each of those predictors individually. What we do now is consider the joint influence of those predictors. If the two influences are merely additive, then the model from Equation 14.5 becomes

$$y = \beta_0 + \vec{\beta}_1 \vec{x}_1 + \vec{\beta}_2 \vec{x}_2 \tag{14.7}$$

$$= \beta_0 + \sum_{j=1}^{J_1} \beta_{1,j} x_{1,j} + \sum_{j=1}^{J_2} \beta_{2,j} x_{2,j}$$

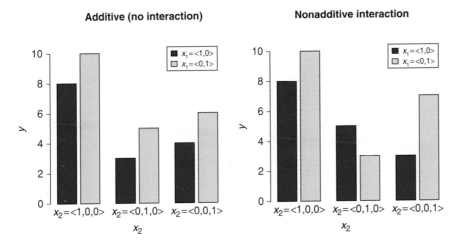

FIGURE 14.5

Combinations of two nominal variables. *Left*: Additive combination. Notice that the difference between adjacent dark and light bars is the same for every level of x_2. *Right*: Nonadditive interaction. Notice that the difference between adjacent dark and light bars is *not* the same for every level of x_2. Figure 19.1, p. 517, provides additional perspective and insight.

with the constraints

$$\sum_{j=1}^{J_1} \beta_{1,j} = 0 \quad \text{and} \quad \sum_{j=1}^{J_2} \beta_{2,j} = 0 \tag{14.8}$$

The left panel of Figure 14.5 shows an example of two nominal predictors that have additive effects on the predicted variable. In this case, the overall baseline is $y = 6$. When $x_1 = \langle 1, 0 \rangle$, there is a deflection in y of -1, and when $x_1 = \langle 0, 1 \rangle$, there is a deflection in y of $+1$. This deflection by x_1 is the same at every level of x_2. The deflections for the three levels of x_2 are $+3$, -2, and -1. These deflections are the same at all levels of x_1. Formally, the left panel of Figure 14.5 is expressed mathematically by the additive combination:

$$y = 6 + \langle -1, 1 \rangle \cdot \vec{x}_1 + \langle 3, -2, -1 \rangle \cdot \vec{x}_2$$

14.1.6.3 *Nonadditive Interaction of Nominal Predictors*

When the predictor variables are nonmetric, it does not even make sense to talk about a multiplicative interaction, because there are no numerical values to multiply. For example, consider predicting annual income from political party affiliation and gender. Both predictors are nominal, so it makes no sense to "multiply" them. But it does make sense to consider nonadditive combination of their influences.

For example, the overall influence of gender is that men, on average, have a higher income than women. The overall influence of political party affiliation is that Republicans, on average, have higher income than Democrats. But it may be that the influences combine nonadditively: Perhaps people who are both Republican and male have a higher average income than would be predicted by merely adding the average income boosts for being Republican and for being male. (This interaction is not claimed to be true; it is being used only as a hypothetical example.)

We need new notation to formalize the nonadditive influence of a combination of nominal values. Just as $\vec{x}_1$ refers to the value of predictor 1, and $\vec{x}_2$ refers to the value of predictor 2, the notation $\vec{x}_{1\times2}$ will refer to a particular *combination* of values of predictors 1 and 2. If there are J_1 levels of predictor 1 and J_2 levels of predictor 2, then there are $J_1 \times J_2$ combinations of the two predictors.

A nonadditive interaction of predictors is formally represented by including a term for the influence of combinations of predictors, beyond the additive influences, as follows: $y = \beta_0 + \vec{\beta}_1 \vec{x}_1 + \vec{\beta}_2 \vec{x}_2 + \vec{\beta}_{1\times2} \vec{x}_{1\times2}$. Whenever the interaction coefficient $\vec{\beta}_{1\times2}$ is nonzero, the predicted value of y is not a mere addition of the separate influences of the predictors.

The right panel of Figure 14.5 shows a graphical example of two nominal predictors that have interactive (i.e., nonadditive) effects on the predicted variable. Notice, in the left pair of bars ($x_2 = \langle 1, 0, 0 \rangle$), that a change from $x_1 = \langle 1, 0 \rangle$ to $x_1 = \langle 0, 1 \rangle$ produces an increase of $+2$ in y, from $y = 8$ to $y = 10$. But for the middle pair of bars ($x_2 = \langle 0, 1, 0 \rangle$), a change from $x_1 = \langle 1, 0 \rangle$ to $x_1 = \langle 0, 1 \rangle$ produces a change of -2 in y, from $y = 5$ to $y = 3$. Thus, the influence of x_1 is *not* the same at all levels of x_2.

An interesting aspect of the pattern in the right panel of Figure 14.5 is that the *average* influences of x_1 and x_2 are the same as in the left panel. Overall, on average, going from $x_1 = \langle 1, 0 \rangle$ to $x_1 = \langle 0, 1 \rangle$ produces a change of $+2$ in y, in both the left and right panels. And overall, on average, for both panels it is the case that $x_2 = \langle 1, 0, 0 \rangle$ is $+3$ above baseline, $x_2 = \langle 0, 1, 0 \rangle$ is -2 below baseline, and $x_2 = \langle 0, 0, 1 \rangle$ is -1 below baseline. The only difference between the two panels is that the combined influence of the two predictors equals the sum of the individual influences in the left panel, but the combined influence of the two predictors does not equal the sum of the individual influences in the right panel.

An interaction between nominal predictors consists of a distinct deflection, for each specific combination of categorical values, away from the additive combination. The magnitude of the interactive deflection is whatever is left

over after the additive effects have been applied to the baseline. The model
that includes an interaction term can be written as

$$y = \beta_0 + \vec{\beta}_1\vec{x}_1 + \vec{\beta}_2\vec{x}_2 + \vec{\beta}_{1\times2}\vec{x}_{1\times2} \qquad (14.9)$$

$$= \beta_0 + \sum_{j=1}^{J_1}\beta_{1,j}x_{1,j} + \sum_{k=1}^{J_2}\beta_{2,k}x_{2,k} + \sum_{j=1}^{J_1}\sum_{k=1}^{J_2}\beta_{1\times2,j,k}x_{1\times2,j,k}$$

with the constraints

$$\sum_{j=1}^{J_1}\beta_{1,j} = 0 \quad \text{and} \quad \sum_{k=1}^{J_2}\beta_{2,k} = 0 \quad \text{and} \quad \sum_{j=1}^{J_1}\beta_{1\times2,j,k} = 0\forall k \quad \text{and} \quad \sum_{k=1}^{J_2}\beta_{1\times2,j,k} = 0\forall j$$

$$(14.10)$$

In these equations, the term $\vec{x}_{1\times2}$ has J_1 times J_2 components, all of which
are zero except for a 1 at the particular combination of levels of x_1 and x_2. This
mysterious and arcane notation will be revealed in all its majestic grandeur in
Chapter 19. For now, the main point is to understand that the term *interaction*
refers to a nonadditive influence of the predictors on the predicted, regardless
of whether the predictors are measured on a nominal scale or a metric scale.

14.1.7 Linking Combined Predictors to the Predicted

Once the predictor variables are combined, they need to be mapped to the predicted variable. This mathematical mapping is called the *(inverse) link* function,
denoted by $f()$ in the following equation:

$$y = f\left(\beta_0 + \beta_1x_1 + \beta_2x_2 + \beta_{1\times2}x_{1\times2}\right) \qquad (14.11)$$

Until now, we have been assuming that the link function is merely the identity
function, $f(x) = x$. For example, in Equation 14.9, y *equals* the linear combination of the predictors; there is no transformation of the linear combination
before mapping the result to y.

Before describing different link functions, it is important to make some clar-
ifications of terminology and corresponding concepts. First, the function $f()$
in Equation 14.11 is usually called the *inverse* link function, because the link
function itself is thought of as transforming the value y into a form that can be
linked to the linear model. I will abuse convention and simply refer to either
$f()$ or $f^{-1}()$ as "the" link function, and rely on context to disambiguate which
direction of linkage is intended. The reason for this terminological sadism is
that the arrows in hierarchical diagrams of Bayesian models will flow from
the linear model toward the data, and therefore it is natural for the functions
to map toward the data, as in Equation 14.11. But repeatedly referring to this
function as the "inverse" link would strain my patience and violate my aes-
thetic sensibilities. Second, the value y that results from the link function

$f(x)$ is not a data value per se. Instead, $f(x)$ is the value of a parameter that expresses some characteristic of the data, usually their mean. Therefore the function $f()$ in Equation 14.11 is sometimes called the *mean* function and is written $\mu = f()$ instead of $y = f()$. I will not use this terminology because most students already think that "mean" means something else, namely the sum divided by N. The fact that y in Equation 14.11 is a parameter value and not a data value will become clear in subsequent sections as we encounter specific cases and examples.

There are situations in which a nonidentity link function is appropriate. Consider, for example, predicting response time (RT) as a function of amount of caffeine consumed. Response time declines as caffeine dosage increases, and therefore a linear prediction of RT from dosage would have a negative slope. This negative slope implies that for a large dosage of caffeine, response time would become negative, which is impossible unless caffeine causes precognition (i.e., foreseeing events before they occur). Therefore, a direct linear function cannot be used for extrapolation to large doses, and we might instead want to use an exponential link function such as $y = \exp(\beta_0 + \beta_1 x)$.

14.1.7.1 The Sigmoid (a.k.a. Logistic) Function

A frequently used link function is the *sigmoid*, also known as the *logistic*:

$$y = \text{sig}(x) = 1/(1 + \exp(-x)) \tag{14.12}$$

Notice the negative sign in front of the x. The sigmoid function ranges between 0 and 1. The sigmoid is nearly 0 when x is large negative, and it is nearly 1 when x is large positive.

For linear combinations of predictors, the sigmoid link function is most conveniently parameterized in x threshold form. For a single predictor variable, the sigmoid link function applied to the linear function of the predictor yields

$$y = \text{sig}(x, \gamma, \theta) = 1/(1 + \exp(-\gamma(x - \theta))) \tag{14.13}$$

where γ, called the *gain*, corresponds to β_1 in Equation 14.2, and where θ, called the *threshold*, corresponds to $-\beta_0/\beta_1$ in Equation 14.2.

Examples of Equation 14.13 (i.e., the sigmoid of a single predictor) are shown in Figure 14.6. Notice that the threshold is the point on the x axis for which $y = 0.5$. The gain indicates how steeply the sigmoid rises through that point.

Figure 14.7 shows examples of a sigmoid of two predictor variables. Above each panel is the equation for the corresponding graph. The equations are parameterized in x threshold form, as in Equation 14.4. In other words, $y = \text{sig}(\gamma(\sum_k w_k x_k - \theta))$, with $(\sum_k w_k^2)^{1/2} = 1$. Notice, in particular, that the coefficients of x_1 and x_2 in the plotted equations do indeed have Euclidean

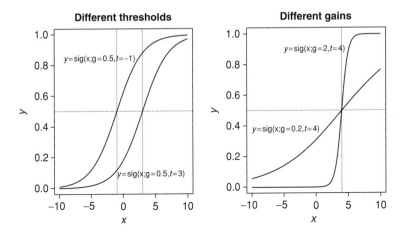

FIGURE 14.6
Examples of sigmoid functions of a single variable. The left panel shows sigmoids with the same gain but different thresholds. The right panel shows sigmoids with the same threshold but different gains.

length of 1.0. For example, in the upper-right panel, $(0.71^2 + 0.71^2)^{1/2} = 1.0$, except for rounding error.

The coefficients of the x variables determine the *orientation* of the sigmoidal *cliff*. For example, compare the two top panels in Figure 14.7, which differ only in the coefficients, not in gain or threshold. In the top-left panel, the coefficients are $w_1 = 0$ and $w_2 = 1$, and the cliff rises in the x_2 direction. In the top-right panel, the coefficients are $w_1 = 0.71$ and $w_2 = 0.71$, and the cliff rises in the positive diagonal direction.

The threshold determines the *position* of the sigmoidal cliff. In other words, the threshold determines the x values for which $y = 0.5$. For example, compare the two left panels of Figure 14.7. The coefficients are the same, but the thresholds (and gains) are different. In the upper-left panel, the threshold is zero, and therefore the midlevel of the cliff is over $x_2 = 0$. In the lower-left panel, the threshold is -3, and therefore the midlevel of the cliff is over $x_2 = -3$.

The gain determines the *steepness* of the sigmoidal cliff. Again compare the two left panels of Figure 14.7. The gain of the upper left is 1, whereas the gain of the lower left is 2.

Terminology: The *logit* function. The inverse of the logistic function is called the *logit* function. For $0 < p < 1$, $\text{logit}(p) = \log\left(p/(1-p)\right)$. It is easy to show (try it!) that $\text{logit}(\text{sig}(x)) = x$, which is to say that the logit is indeed the inverse of the sigmoid. Some authors, and programmers, prefer to express the connection between predictors and predicted in the opposite direction by first

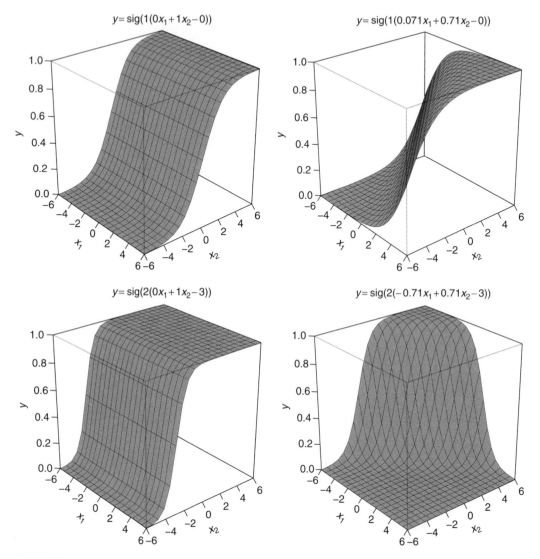

FIGURE 14.7
Examples of sigmoid functions of two variables. Top two panels show sigmoids with the same gain and threshold but different coefficients on the predictors. The left two panels show sigmoids with the same coefficients on the predictors but different gains and thresholds. The lower-right panel shows a case with a negative coefficient on the first predictor.

transforming the predicted variable to match the linear model. In other words, you may see the link expressed either of these ways:

$$y = \text{logistic}\left(\beta_0 + \beta_1 x_1 + \cdots\right)$$
$$\text{logit}(y) = \beta_0 + \beta_1 x_1 + \cdots$$

The two expressions achieve the same result, mathematically. The difference between them is merely a matter of emphasis. In the first expression, the combination of predictors is transformed so it maps onto y expressed in its original scale. In the second expression, y is transformed onto a new scale, and that transformed value is modeled as a combination of predictors.

14.1.7.2 The Cumulative Normal (a.k.a. Phi) Function

Another frequently used link function is the cumulative normal distribution. It is qualitatively similar to the sigmoid or logistic function. Modelers will use the logistic or the cumulative normal depending on mathematical convenience or ease of interpretation. For example, when we consider ordinal predicted variables (in Chapter 21), it will be natural to model the responses in terms of a continuous underlying variable that has normally distributed variability, which leads to using the cumulative normal as a model of response probabilities.

The cumulative normal is denoted $\Phi(x|\mu, \tau)$, where x is a real number and where μ and τ are parameter values, called the *mean* and *precision* of the normal distribution. The parameter μ governs the point at which the cumulative normal, $\Phi(x)$, equals 0.5. In other words, μ plays the same role as the threshold in the logistic sigmoid. The parameter τ governs the steepness of the cumulative normal function at $x = \mu$. The τ parameter plays the same role as the gain parameter in the logistic sigmoid. A graph of a cumulative normal appears in Figure 14.8. For this example, $\mu = 0$, and notice that $\Phi(0) = 0.5$. This means that the area under the normal density to the left of 0 is 0.5.

Terminology: The *probit* function. The inverse of the cumulative normal is called the probit function. ("Probit" stands for "probability unit"; Bliss, 1934). The probit function maps a value p, for $0.0 \le p \le 1.0$, onto the infinite real line, and a graph of the probit function looks much like the logit function. You may see the link expressed either of these ways:

$$y = \Phi\left(\beta_0 + \beta_1 x_1 + \cdots\right)$$
$$\text{probit}(y) = \beta_0 + \beta_1 x_1 + \cdots$$

Traditionally, the transformation of y (in this case, the probit function) is called the *link* function, and the transformation of the linear combination of x (in this case, the Φ function) is called the *inverse* link function. As mentioned before, I abuse the traditional terminology and call either one a link function, relying on context to disambiguate.

14.1.8 Probabilistic Prediction

In the real world, there is always variation in y that we cannot predict from x. This unpredictable "noise" in y might be deterministically caused by sundry factors we have neither measured nor controlled, or the noise might be caused by inherent nondeterminism in y. It does not matter either way because in

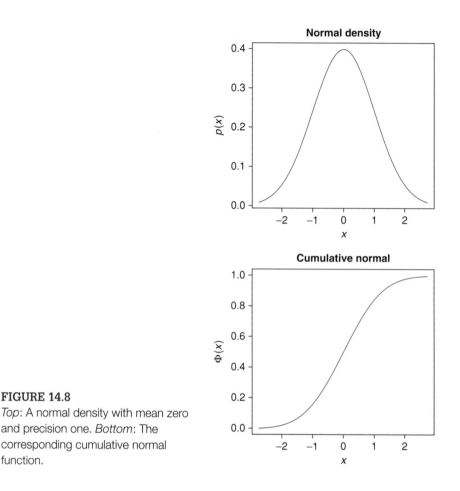

FIGURE 14.8

Top: A normal density with mean zero and precision one. *Bottom*: The corresponding cumulative normal function.

practice the best we can do is predict the *probability* that y will have any particular value, dependent on x. Therefore, we use the deterministic value predicted by Equation 14.11 as the predicted *tendency* of y as a function of the predictors. We do not predict that y is exactly $f\left(\beta_0 + \beta_1 x_1 + \beta_2 x_2 + \beta_{1\times2} x_{1\times2}\right)$ because we would surely be wrong. Instead, we predict that y tends to be near $f\left(\beta_0 + \beta_1 x_1 + \beta_2 x_2 + \beta_{1\times2} x_{1\times2}\right)$.

To make this notion of probabilistic tendency precise, we need to specify a probability distribution for y that depends on $f\left(\beta_0 + \beta_1 x_1 + \beta_2 x_2 + \beta_{1\times2} x_{1\times2}\right)$. To keep the notation tractable, first define $\mu = f\left(\beta_0 + \beta_1 x_1 + \beta_2 x_2 + \beta_{1\times2} x_{1\times2}\right)$. Do not confuse this use of μ with the unrelated μ mentioned in the cumulative normal function. With this notation, we then denote the probability distribution of y as some to-be-specified probability density function, abbreviated as "pdf":

$$y \sim \text{pdf}\left(\mu \ [, \tau, \ldots]\right)$$

The pdf might have various additional parameters, denoted by $\tau, \ldots,$ to specify its shape. Examples are provided in the next section, where all these ideas are brought together.

14.1.9 Formal Expression of the GLM

In general, the likelihood function specifies the probability of each possible predicted value y as a function of the predictor values x_j and various parameter values $\beta, \tau,$ etc. The generalized linear model can be written:

$$\mu = f\left(\beta_0 + \beta_1 x_1 + \beta_2 x_2 + \beta_{1 \times 2} x_{1 \times 2} + \cdots\right) \qquad (14.14)$$

$$y \sim \text{pdf}\left(\mu \ [, \tau, \ldots]\right) \qquad (14.15)$$

The function f in Equation 14.14 is called the *link* function, because it links the combination of predictors to the predicted tendency. The optional parameters $[, \tau, \ldots]$ in Equation 14.15 may be needed for various types of the probability density function (pdf) that describe the probability distribution of y around the tendency μ.

Figure 14.9 shows a random sample of points normally distributed around a line or plane. The upper panel illustrates a case of the generalized linear model of Equations 14.14 and 14.15 in which there is a single predictor x, with $\beta_0 = 10$ and $\beta_1 = 2$. The link function is simply the identity function, $f(\beta_0 + \beta_1 x) = \beta_0 + \beta_1 x$. The probability density function is normal with a standard deviation of 2.0. Profiles of this normal density are superimposed on the graph to make it explicit. Notice that the normal density is always centered on the line that marks the predicted tendency as a function of the predictor.

The lower panel of Figure 14.9 shows a case with two predictor variables. The predictors are combined linearly, with no interaction. The link function is the identity. The probability function is normal with a standard deviation of 4. Each randomly generated point is connected to the underlying linear core by a vertical dotted line, to explicitly indicate the random variation of the point from the plane. The plane marks the predicted tendency as a function of the predictors, and the data values are normally distributed above and below that tendency.

Figure 14.10 shows another case of the GLM. In this case, the points are Bernoulli distributed around a sigmoid function of two predictors, as annotated at the top of the graph. There is a linear combination of predictors, with a sigmoid link function, and a Bernoulli probability function that defines the probability that $y = 1$. The graph shows that values of y can only be 0 or 1, and the sigmoid function defines the probability that y is 1 for particular predictor values. The sigmoidal surface plots the tendency that $y = 1$ as a function of the predictors.

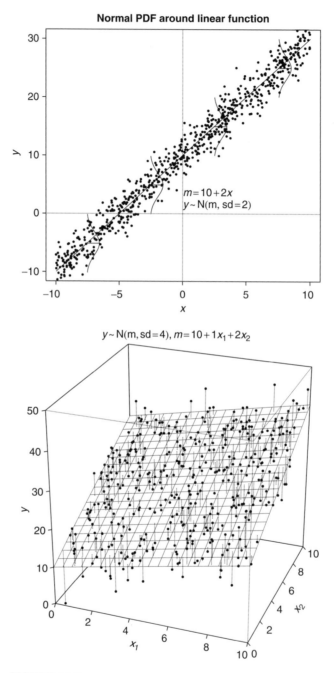

FIGURE 14.9

Examples of points normally distributed around a linear function.

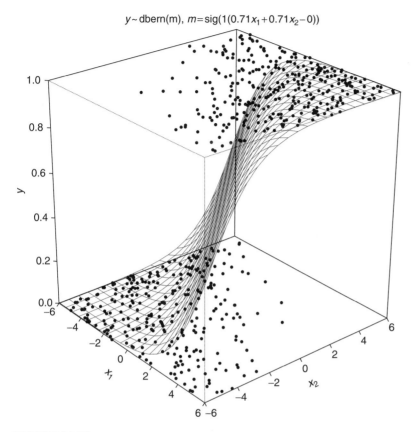

$y \sim \text{dbern(m)}, \ m = \text{sig}(1(0.71 x_1 + 0.71 x_2 - 0))$

FIGURE 14.10
Examples of points Bernoulli distributed around a sigmoid function of two predictors. All the points are either at $y = 1$ or $y = 0$ (intermediate values such as $y = 0.6$ cannot occur).

14.1.10 Two or More Nominal Variables Predicting Frequency

We will also consider the situation in which there are two or more nominal variables used as predictors of a *frequency count*. A frequency count (i.e., how many times something happened) is a special case of a metric scale, but because its values fall at discrete levels, namely, non-negative integers, this situation will have a different sort of likelihood distribution. This type of situation, with nominal predictors and frequency-count predicted values, is often called *contingency table analysis*, and a typical NHST analysis conducts a *chi-square test of independence of attributes*. We explore Bayesian analysis of this situation in Chapter 22.

Here is a brief summary of how contingency tables are analyzed. As a concrete example, suppose we measure political affiliation and religious affiliation of a set of people, and for a sample of people we count how many occurrences there are of each combination. We are interested in analyzing possible relationships between political and religious affiliations. Suppose we conduct a poll for one week. We happen to record 27 people who are Democrats and Unitarians. This observed frequency reflects an underlying rate at which that combination is generated by this sort of poll (i.e., the underlying rate for Unitarian Democrats is roughly 27 people per week). The observed rate (i.e., frequency per unit time) for each combination of nominal values is thought to reflect the true underlying rate at which that combination is generated by the world. We conceive of the observed rate as being a random sample from a true underlying rate denoted by λ. The probability of any particular observed rate, given an underlying rate of λ, is modeled by a Poisson distribution, which is denoted as freq $\sim$ dpois(λ). The Poisson distribution was smuggled into the text back in Exercise 11.3, p. 289, which I'm sure is still as fresh in your memory as a beached fish. Don't worry, the Poisson will be explained again later (Section 22.1.3). The Poisson distribution specifies a probability for each possible observed rate. The Poisson puts highest probabilities on rates near λ.

Our goal is to estimate the underlying rates at which each nominal combination is produced. But more than that, we would like to know if the attributes occur independently of each other or instead covary in some way. For example, if political and religious affiliation are independent, then there should be the same proportion of Unitarians among Democrats as among Republicans. Mathematically, independence means p(Unitarian&Democrat) $=$ p(Unitarian) $\times$ p(Democrat) and p(Unitarian&Republican) $= p$(Unitarian) $\times$ p(Republican) and so on for every combination of attribute values. To shorten the expressions, I'll substitute U for Unitarian and D for Democrat, whereby independence means

$$p(U\&D) = p(U) \times p(D)$$

and so on for every combination of attributes. That expression for probabilities corresponds to the following expression in terms of frequencies:

$$\text{freq}(U\&D)/N = \text{freq}(U)/N \times \text{freq}(D)/N$$

which can be rearranged as

$$\text{freq}(U\&D) = \text{freq}(U) \times \text{freq}(D) \times 1/N$$

Notice that independence is expressed as a multiplicative *product* of attribute influences. But all the models we've considered in this chapter used an additive *sum* of predictor influences. To be able to use our familiar additive models, we'll transform the frequencies by a logarithm, because the logarithm of the product of values equals the sum of the logarithms of the values. In other words,

$$\log(\text{freq}(U\&D)) = \log(\text{freq}(U)) + \log(\text{freq}(D)) + \log(1/N)$$

The notation "log(freq(value))" gets cumbersome, so we'll substitute the notation β_v. Thus,

$$\log(\text{freq}(U\&D)) = \beta_U + \beta_D + \beta_0$$

where β_0 stands in for the constant $\log(1/N)$. Finally, it's unintuitive to talk about the logarithms of frequencies, so we'll exponentiate to get rid of the leading logarithm, yielding:

$$\text{freq}(U\&D) = \exp(\beta_U + \beta_D + \beta_0)$$

To summarize, if independence is true, then the expression above should be true, for every combination of attribute values.

But of course the attributes are usually not independent, and we would like some measure of lack of independence. We already have such a measure in the context of linear models, namely, the interaction term. Thus, we will include an interaction term that estimates deviation from independence. Thus, our model ends up being as follows. For two nominal attributes, we put the observed frequencies in a table, with one attribute's values listed down the rows, and the other attribute's values listed across the columns. The frequency in the rth row and c^{th} column is denoted freq_{rc}, and the underlying rate for that cell is denoted λ_{rc}. The model looks like this:

$$\lambda_{rc} = \exp(\beta_0 + \beta_r + \beta_c + \beta_{r\times c})$$
$$\text{freq}_{rc} \sim \text{dpois}(\lambda_{rc}) \tag{14.16}$$

with the usual constraints (from Equation 14.10)

$$\sum_r \beta_r = 0 \quad \text{and} \quad \sum_c \beta_c = 0 \quad \text{and} \quad \sum_r \beta_{r\times c,r,c} = 0 \ \forall c \quad \text{and} \quad \sum_c \beta_{r\times c,r,c} = 0 \ \forall r$$

The point of this over-fast prelude to contingency table analysis is merely to demonstrate that the core of the model we'll be using is the same as the linear model that was mentioned for metric predicted values. Thus, all the applied analyses we'll see in the remainder of the book are based on the GLM.

14.2 CASES OF THE GLM

Table 14.1, p. 385, displays the various cases of the generalized linear model that are considered in this book. Subsequent chapters of the book progress through the table in reading order: left to right within rows, then top to bottom across rows.

The first row of Table 14.1 lists cases for which the predicted variable is metric. Moving from left to right within this row, the first column indicates a situation in which there is only a single group and the predicted value for the group is simply the mean of the group. In this situation, there is no need to explicitly denote a predictor variable, and instead the mean of the group can be denoted by a single parameter, β_0. This situation corresponds to what classical null hypothesis significance testing (NHST) calls a *single-group t-test*. This case is described in its Bayesian setting in Chapter 15.

Moving to the next column, there is a single metric predictor. This corresponds to the so-called *simple linear regression*, and it is explored in Chapter 16. By inspecting the equation for the GLM in the cell, you can see that the only difference from the previous cell is the inclusion of the predictor x_1 and its coefficient β_1.

Moving rightward to the next column, we come to the scenario involving two or more metric predictors, which corresponds to "multiple regression" and is explored in Chapter 17. By examining the equations for the GLM in the cell, you can see that the basic form is the same but merely with extra terms added for the additional predictors.

The next two columns involve nominal predictors, instead of metric predictors, with the penultimate column devoted to a single predictor and the final column devoted to two or more predictors. The last two columns correspond to what NHST calls *oneway ANOVA* and *multifactor ANOVA*. If that terminology is unfamiliar to you, don't worry, it will be explained in Chapters 18 and 19.

In all the cases in the first row, the link function is the identity, and the probability distribution for the metric predicted values is assumed to be normal. When we move to the second row, however, the predicted variable is dichotomous; therefore the probability distribution for y is a Bernoulli distribution. The link function, which connects the predictors to the probability that $y = 1$, is assumed to be the sigmoid (i.e., logistic function). When the predictors are metric, this situation is generically referred to as "logistic regression" and is discussed in Chapter 20. The case of nominal predictors is also discussed.

The third row of Table 14.1 lists cases for which the predicted variable is ordinal. These cases are considered in Chapter 21. Notice that the link function is the cumulative normal instead of the sigmoid and the ordinal values are generated by a multiple-category generalization of the Bernoulli function, denoted by dcat. The bottom row of Table 14.1 includes cases for which the predicted variable is a frequency count. In this case, the link function is exponential and probability distribution for y is Poisson. Details will be explained at length in

Table 14.1 Cases of the Generalized Linear Model (Single Predicted Variable)

Predicted Variable y	Single Group	Metric		Nominal	
		Single x Factor	Two or More x Factors	Single x Factor	Two or More x Factors
Metric	$\mu = \beta_0$ $y \sim N(\mu, \tau)$	$\mu = \beta_0 + \beta_1 x_1$ $y \sim N(\mu, \tau)$	$\mu = \beta_0 + \beta_1 x_1$ $\quad + \beta_2 x_2$ $\quad + \beta_{1\times2} x_1 x_2$ $y \sim N(\mu, \tau)$	$\mu = \beta_0 + \vec{\beta}_1 \vec{x}_1$ $y \sim N(\mu, \tau)$	$\mu = \beta_0 + \vec{\beta}_1 \vec{x}_1$ $\quad + \vec{\beta}_2 \vec{x}_2$ $\quad + \vec{\beta}_{1\times2} \vec{x}_{1\times2}$ $y \sim N(\mu, \tau)$
Dichotomous	$\mu = \beta_0$ $y \sim \text{dbern}(\mu)$	$\mu = \text{sig}\left(\beta_0 + \beta_1 x_1\right)$ $y \sim \text{dbern}(\mu)$	$\mu = \text{sig}\left(\beta_0 + \beta_1 x_1\right.$ $\quad + \beta_2 x_2$ $\quad \left.+ \beta_{1\times2} x_1 x_2\right)$ $y \sim \text{dbern}(\mu)$	$\mu = \text{sig}\left(\beta_0 + \vec{\beta}_1 \vec{x}_1\right)$ $y \sim \text{dbern}(\mu)$	$\mu = \text{sig}\left(\beta_0 + \vec{\beta}_1 \vec{x}_1\right.$ $\quad + \vec{\beta}_2 \vec{x}_2$ $\quad \left.+ \vec{\beta}_{1\times2} \vec{x}_{1\times2}\right)$ $y \sim \text{dbern}(\mu)$
Ordinal		$\mu = \Phi\left(\beta_0 + \beta_1 x_1\right)$ $\pi_k = \text{thresh}_k(\mu)$ $y \sim \text{dcat}(..., \pi_k, ...)$	$\mu = \Phi\left(\beta_0 + \beta_1 x_1\right.$ $\quad + \beta_2 x_2$ $\quad \left.+ \beta_{1\times2} x_1 x_2\right)$ $\pi_k = \text{thresh}_k(\mu)$ $y \sim \text{dcat}(..., \pi_k, ...)$		
Frequency					$\lambda_{rc} = \exp(\beta_0 + \beta_r$ $\quad + \beta_c$ $\quad + \beta_{r\times c})$ $\text{freq}_{rc} \sim \text{dpois}(\lambda_{rc})$

the forthcoming chapters. The point here is for you to see the overall organization of topics, and to see how all these cases are variations of the same underlying structure.

The table can be expanded with additional columns, but then it gets too big to display easily. Additional columns would include combinations of metric and nominal predictors. It turns out that it is easy in Bayesian models to combine metric and nominal predictors, once you know how to handle metric and nominal predictors individually. In summary, the rows of the table refer to differently scaled predicted values, with their corresponding link functions and pdf's:

y Scale Type	(Inverse) Link Function	pdf
Metric	Identity	Normal
Dichotomous	Sigmoid (a.k.a. logistic)	Bernoulli
Ordinal	Thresholded cumulative normal	Categorical
Frequency	Exponential	Poisson

14.3 EXERCISES

Exercise 14.1. [Purpose: For real-world examples of research, to identify which statistical model is relevant.] For each of the following examples, identify the predicted variable and its scale type, identify the predictor variable(s) and its scale type, and identify the cell of Table 14.1 to which the example belongs.

(A) Guber (1999) examined average performance by public high school students on the Scholastic Aptitude Test (SAT) as a function of how much money was spent per pupil by the state and what percentage of eligible students actually took the exam.

(B) Hahn, Chater, & Richardson (2003) were interested in perceived similarity of simple geometric patterns. Human observers rated pairs of patterns for how similar the patterns appeared, by circling one of the digits 1 to 7 printed on the page, where 1 meant "very dissimilar" and 7 meant "very similar." The authors presented a theory of perceived similarity, in which patterns are perceived to be dissimilar to the extent that it takes more geometric transformations to produce one pattern from the other. The theory specified the exact number of transformations needed to get from one pattern to the other.

(C) R.L. Berger et al. (1988) were interested in the longevity of rats, measured in days, as a function of the rats' diet. One group of rats fed freely, another group of rats had a very low calorie diet.

(D) McIntyre (1994) was interested in predicting the tar content of a cigarette (measured in milligrams) from the weight of the cigarette.

(E) You are interested in predicting the gender of a person, based on the person's height and weight.

(F) You are interested in predicting whether a respondent will agree or disagree with the statement "The United States needs a federal health care plan with a public option," on the basis of the respondent's political party affiliation.

Exercise 14.2. [Purpose: To find *student-relevant* real-world examples of each type of situation in Table 14.1.] For each of the 13 cells of Table 14.1 that is filled with equations, provide an example of research involving that cell's model structure. Do this by finding published articles that describe research with the corresponding structure. The articles do *not* need to have Bayesian data analysis; the articles *do* need to report research that involves the corresponding types of predictor and predicted variables. Because it might be overly time consuming to find published examples of all thirteen types, please find published articles of at least six types. For each example, specify the following:

- The full citation to the published article (see the references of this book for examples of how to cite articles).
- The predictor and predicted variables. Describe their meaning and the type of scale they are. Briefly describe the meaningful context for the varables (i.e., what is the goal of the research).

CHAPTER 15

Metric Predicted Variable on a Single Group

CONTENTS

It's normal to want to fit in with your friends,
Behave by their means and believe all their ends.
But I'll be high tailing it, fast and askew,
Precisely 'cause I can't abide what you do.

In this chapter, we consider a situation in which we have a metric predicted variable that is observed for items from a single group. For example, we could measure the blood pressure (i.e., a metric variable) for people randomly sampled from first-year university students (i.e., a single group). In this case, we might be interested in whether the group's typical blood pressure differs from

Doing Bayesian Data Analysis: A Tutorial with R and BUGS. DOI: 10.1016/B978-0-12-381485-2.00015-8
© 2011, Elsevier Inc. All rights reserved.

the recommended value for people of that age as published by a federal agency. As another example, we could measure the IQ (i.e., a metric variable) of people randomly sampled from everyone self-described as vegetarian (i.e., a single group). In this case, we could be interested in whether this group's IQ differs from the general population's average IQ of 100.

In the context of the generalized linear model (GLM) introduced in the previous chapter, this chapter's situation involves the most trivial case of the linear core of the GLM, namely $\mu = \beta_0$, as indicated in the top-left cell of Table 14.1 (p. 385). The "news" of the present chapter is a detailed look at a particular probability density function, namely the normal distribution, denoted by $y \sim N(\mu, \tau)$ in the top-left cell of Table 14.1. We will explore particular prior distributions for the parameters of the normal distribution, and see how those same density functions can be used in extended hierarchical models for research designs involving repeated measures. We will also provide examples of transforming data so they are approximately normal, so that "off the shelf" normal-likelihood models can be used to describe the data.

15.1 ESTIMATING THE MEAN AND PRECISION OF A NORMAL LIKELIHOOD

A workhorse for the remainder of the book is the normal likelihood function, just as the Bernoulli likelihood was the focus for the previous part of the book. The normal probability density function was introduced in Section 3.3.2.2, p. 35. Please review that section now. The normal distribution specifies the probability density of a value y, given the values of two parameters, the mean μ and standard deviation σ:

$$p(y \mid \mu, \sigma) = \frac{1}{Z} \exp\left(-\frac{1}{2} \frac{(y - \mu)^2}{\sigma^2}\right) \qquad (15.1)$$

where Z is the normalizer (i.e., a constant that makes the probability density integrate to 1). It turns out that $Z = \sigma\sqrt{2\pi}$, but we won't need to use this fact.

To get an intuition for the normal likelihood function, consider three data values $y_1 = 85$, $y_2 = 100$, and $y_3 = 115$, as plotted by large dots in Figure 15.1. The probability (density) of any single datum, given parameter values, is $p(y \mid \mu, \sigma)$ as specified in Equation 15.1. The probability of the whole set of independent data values is $\prod_i p(y_i \mid \mu, \sigma) = p(D \mid \mu, \sigma)$, where $D = \{y_1, y_2, y_3\}$. Figure 15.1 shows $p(D \mid \mu, \sigma)$ for different values of μ and σ. As you can see, there are values of μ and σ that make the data most probable, but other nearby values also accommodate the data reasonably well.

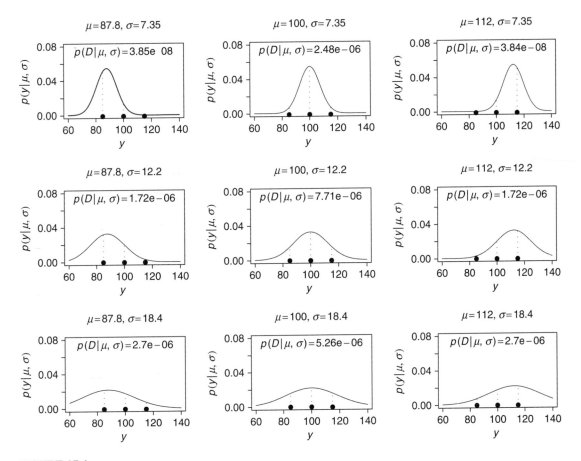

FIGURE 15.1
The likelihood $p(D \mid \mu, \sigma)$ for three data points, $D = \{85, 100, 115\}$, according to a normal likelihood function with different values of μ and σ. The probability density of an individual datum is the height of the dotted line over the point. The probability of the set of data is the product of the individual probabilities. The middle panel shows shows the μ and σ that maximize the probability of the data.

Given a set of data, D, we estimate the parameters with Bayes' rule:

$$p(\mu, \sigma \mid D) = \frac{p(D \mid \mu, \sigma)p(\mu, \sigma)}{\iint d\mu \, d\sigma \, p(D \mid \mu, \sigma)p(\mu, \sigma)} \tag{15.2}$$

Figure 15.1 showed examples of $p(D \mid \mu, \sigma)$ for a particular data set at different values of μ and σ. The prior, $p(\mu, \sigma)$, specifies the believability of each combination of μ, σ values in the two-dimensional conjoint parameter space, without the data. Bayes' rule says that the posterior believability of each combination of μ, σ values is the prior believability times the likelihood, normalized by the evidence. We saw our first examples of Bayes' rule on a two-dimensional parameter space back in Chapter 8, for example, in Figure 8.2, p. 164. Our goal now is to evaluate Equation 15.2, using a normal-density likelihood, for reasonable choices of the prior distribution, $p(\mu, \sigma)$.

15.1.1 Solution by Mathematical Analysis

Purely for the sake of mathematical elucidation, it is convenient to consider the case in which the standard deviation of the likelihood function is fixed at a specific value. In other words, the prior distribution on σ is a spike over that specific value. We'll denote the value as $\sigma = S_y$. With this simplifying assumption, we are only estimating μ because we are assuming perfectly certain prior knowledge about σ.

When σ is fixed, then the prior distribution on μ in Equation 15.2 can be easily chosen to be conjugate to the normal likelihood. (The term *conjugate prior* was defined in Section 5.2, p. 80). It turns out that the product of normal distributions is again a normal distribution; in other words, if the prior on μ is normal, then the posterior on μ is normal. It is easy to derive this fact, as we do next.

Let the prior distribution on μ be normal with mean M_μ and standard deviation S_μ. Then the likelihood times the prior is

$$p(y \mid \mu, \sigma)p(\mu, \sigma) = p(y \mid \mu, S_y)p(\mu)$$

$$\propto \exp\left(-\frac{1}{2}\frac{(y-\mu)^2}{S_y^2}\right)\exp\left(-\frac{1}{2}\frac{(\mu-M_\mu)^2}{S_\mu^2}\right)$$

$$= \exp\left(-\frac{1}{2}\left[\frac{(y-\mu)^2}{S_y^2} + \frac{(\mu-M_\mu)^2}{S_\mu^2}\right]\right)$$

$$= \exp\left(-\frac{1}{2}\left[\frac{S_\mu^2\,(y-\mu)^2 + S_y^2\,(\mu-M_\mu)^2}{S_y^2 S_\mu^2}\right]\right)$$

$$= \exp\left(-\frac{1}{2}\left[\frac{S_y^2 + S_\mu^2}{S_y^2 S_\mu^2}\left(\mu^2 - 2\frac{S_y^2 M_\mu + S_\mu^2 y}{S_y^2 + S_\mu^2}\mu + \frac{S_y^2 M_\mu^2 + S_\mu^2 y^2}{S_y^2 + S_\mu^2}\right)\right]\right)$$

$$= \exp\left(-\frac{1}{2}\left[\frac{S_y^2 + S_\mu^2}{S_y^2 S_\mu^2}\left(\mu^2 - 2\frac{S_y^2 M_\mu + S_\mu^2 y}{S_y^2 + S_\mu^2}\mu\right)\right]\right)$$

$$\times \exp\left(-\frac{1}{2}\left[\frac{S_y^2 + S_\mu^2}{S_y^2 S_\mu^2}\left(+\frac{S_y^2 M_\mu^2 + S_\mu^2 y^2}{S_y^2 + S_\mu^2}\right)\right]\right)$$

$$\propto \exp\left(-\frac{1}{2}\left[\frac{S_y^2 + S_\mu^2}{S_y^2 S_\mu^2}\left(\mu^2 - 2\frac{S_y^2 M_\mu + S_\mu^2 y}{S_y^2 + S_\mu^2}\mu\right)\right]\right) \qquad (15.3)$$

where the transition to the last line was valid because the term that was dropped was merely a constant. This result, believe it or not, is progress. Why? Because we ended up, in the innermost parentheses, with a quadratic expression in μ. Notice that the normal prior is also a quadratic expression in μ. All we have to do is "complete the square" inside the parentheses and do the same trick that got us to the last line of Equation 15.3:

$$p(y \mid \mu, S_y)p(\mu) \propto \exp\left(-\frac{1}{2}\left[\frac{S_y^2 + S_\mu^2}{S_y^2 S_\mu^2}\left(\mu^2 - 2\frac{S_y^2 M_\mu + S_\mu^2 y}{S_y^2 + S_\mu^2}\mu + \left(\frac{S_y^2 M_\mu + S_\mu^2 y}{S_y^2 + S_\mu^2}\right)^2\right)\right]\right)$$

$$= \exp\left(-\frac{1}{2}\left[\frac{S_y^2 + S_\mu^2}{S_y^2 S_\mu^2}\left(\mu - \frac{S_y^2 M_\mu + S_\mu^2 y}{S_y^2 + S_\mu^2}\right)^2\right]\right) \qquad (15.4)$$

Equation 15.4 is the numerator of Bayes' rule. When it is normalized by the evidence in the denominator, it becomes a probability density function. What is the shape of the function? You can see that Equation 15.4 has the same form as a normal distribution on μ, such that the mean is $\frac{S_y^2 M_\mu + S_\mu^2 y}{S_y^2 + S_\mu^2}$ and the standard deviation is $\sqrt{\frac{S_y^2 S_\mu^2}{S_y^2 + S_\mu^2}}$.

That formula is rather unwieldy! It becomes more compact if the normal density is reexpressed in terms of $1/\sigma^2$ instead of σ. The reciprocal of the squared standard deviation is called the *precision* of the normal. When the standard deviation goes down, the precision goes up. A very narrow distribution is highly precise. A wide distribution has low precision. Because the posterior standard deviation is $\sqrt{\frac{S_y^2 S_\mu^2}{S_y^2 + S_\mu^2}}$, the posterior precision is

$$\frac{S_y^2 + S_\mu^2}{S_y^2 S_\mu^2} = \frac{1}{S_\mu^2} + \frac{1}{S_y^2}.$$

In other words—and this is the punch line—the posterior precision is the sum of the prior precision and the likelihood precision.

The posterior mean can also be reexpressed in terms of precisions. The posterior mean is $\frac{S_y^2 M_\mu + S_\mu^2 y}{S_y^2 + S_\mu^2}$, which becomes

$$\frac{1/S_\mu^2}{1/S_y^2 + 1/S_\mu^2}M_\mu + \frac{1/S_y^2}{1/S_y^2 + 1/S_\mu^2}y$$

In other words, the posterior mean is a weighted average of the prior mean and the datum, with the weighting corresponding to the relative precisions of the prior and the likelihood. When the prior is highly precise compared to the likelihood (i.e., when $1/S_\mu^2$ is large compared to $1/S_y^2$), then the prior is weighted heavily and the posterior mean is near the prior mean. But when the prior is imprecise (i.e., very uncertain), then the prior does not get much weight and the posterior mean is close to the datum. We have previously seen this sort of relative weighting of prior and data in the posterior. It showed up in the case of updating a beta prior, back in Equation 5.8, p. 84.

The formulas for the mean and precision of the posterior normal can be naturally extended when there are N values of y in a sample, instead of only a single value of y. The formulas can be derived from the defining formulas, as done previously, but a shortcut can be taken. It is known from mathematical statistics that when a set of values y_i are generated from a normal likelihood function, the mean of those values, $\bar{y}$, is also distributed normally, with the same mean as the the generating mean and with a standard deviation of $\sigma/\sqrt{N}$. Thus, instead of conceiving of this situation as N scores y_i sampled from the likelihood $N(y_i \mid \mu, \sigma)$, we conceive of this as a single score, $\bar{y}$, sampled from the likelihood $N(\bar{y} \mid \mu, \sigma/\sqrt{N})$. Then we just apply the updating formulas we previously derived. Thus, for N scores y_i generated from a normal likelihood $N(y_i \mid \mu, S_y)$ and prior distribution $N(\mu \mid M_\mu, S_\mu)$ on μ, the posterior distribution on μ is also normal with mean

$$\frac{1/S_\mu^2}{N/S_y^2 + 1/S_\mu^2} M_\mu + \frac{N/S_y^2}{N/S_y^2 + 1/S_\mu^2} \bar{y}$$

and precision

$$\frac{1}{S_\mu^2} + \frac{N}{S_y^2}$$

Notice that as the sample size N increases, the posterior mean is dominated by the data mean.

Instead of estimating the μ parameter in the likelihood when σ is fixed, we can estimate the σ parameter when μ is fixed. The situation is also more conveniently expressed in terms of precision: We want to estimate the precision, $1/\sigma^2$, when μ is fixed. It turns out that when μ is fixed, a conjugate prior for the precision is the gamma distribution (e.g., Gelman et al., 2004, p. 50). The gamma distribution was described in Figure 9.8, p. 209. For our purposes, it is not important to review the updating formulas for the gamma distribution in this situation. But it is important to gain an intuition for what a gamma prior on precision means, in terms of the beliefs represented. Consider a gamma distribution that is loaded heavily over very small values, but has a long shallow tail extending over large values. This sort of gamma distribution on precision indicates that we believe most strongly in small precisions, but we admit that large precisions are possible. If this is a belief about the precision of a normal likelihood function, then this sort of gamma distribution expresses a belief that the data will be very spread out, because small precisions imply large standard deviations. If the gamma distribution is loaded over large values of precision, it expresses a belief that the data will be tightly clustered.

Summary. We have assumed that the data are generated by a normal likelihood function, parameterized by a mean μ and standard deviation σ, and denoted $y \sim N(y \mid \mu, \sigma)$. For purposes of mathematical derivation, we made unrealistic assumptions that the prior distribution is either a spike on σ or a spike on μ, in order to make three main points:

1. A natural way to express a prior on μ is with a normal distribution, because this is conjugate with the normal likelihood when its precision is fixed.
2. A natural way to express a prior on the precision $1/\sigma^2$ is with a gamma distribution, because this is conjugate with the normal likelihood when its mean is fixed.
3. The formulas for Bayesian updating of the parameter distribution are more conveniently expressed in terms of precision than standard deviation. Normal distributions will be described sometimes in terms of standard deviation and sometimes in terms of precision, so it is important to glean from context which is being referred to.

A joint prior, on the combination of μ and σ parameter values, can also be specified, in such a way that the posterior has the same form as the prior. We will not pursue these mathematical analyses here, because our purpose is merely to justify and motivate typical expressions for the prior distributions on the parameters, so that they can then be used in MCMC sampling in BUGS. Various other sources describe conjugate priors for the joint parameter space (e.g., Gelman et al., 2004, pp. 78–83).

15.1.2 Approximation by MCMC in BUGS

It is easy to estimate the mean and precision of a set of data in BUGS. Figure 15.2 illustrates the model. The data, y_i, are assumed to be generated by a normal likelihood function with mean μ and precision τ. The prior on μ is assumed to be normal with mean M and precision T. The prior on τ is assumed to be gamma with shape and rate parameters of S and R. We are estimating two parameters, and the prior is over the two-dimensional conjoint parameter space. The prior we have specified assumes that μ and τ are independent.

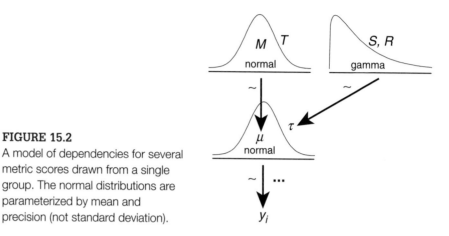

FIGURE 15.2

A model of dependencies for several metric scores drawn from a single group. The normal distributions are parameterized by mean and precision (not standard deviation).

The model of Figure 15.2 is expressed in BUGS as follows

(YmetricXsingleBrugs.R)

```
9    model {
10       # Likelihood:
11       for( i in 1 : N ) {
12           y[i] ~ dnorm( mu , tau ) # tau is precision, not SD
13       }
14       # Prior:
15       tau ~ dgamma( 0.01 , 0.01 )
16       mu ~ dnorm( 0 , 1.0E-10 )
17    }
```

Notice that each arrow in Figure 15.2 has a corresponding specification in the BUGS code. *Notice that BUGS parameterizes* dnorm *by mean and precision, not by mean and standard deviation.* The hyperprior constants in this particular BUGS model are generic and vague; they should be knowledgably informed in real applications.

When learning about or debugging a model, it can be useful to generate fictitious data from known parameter values, and then see how well those parameter values are estimated by the model's posterior distribution. For this purpose, the program in Section 15.4.1 (YmetricXsingleBrugs.R) generates some random data from a normal distribution whose true mean is 100 and standard deviation is 15 (just like IQ scores). The random data themselves have a sample mean and standard deviation somewhere near those generating values. The result of running the BRugs program is shown in Figure 15.3. The posterior precision (i.e., tau) sampled by the BUGS program has been converted to standard deviation (denoted "sigma") via the identity $\sigma = 1/\sqrt{\tau}$. The estimated mean and standard deviation are not far from the generating mean. This reassures us that the program is operating properly.

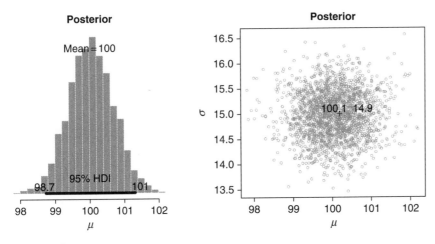

FIGURE 15.3
Posterior distribution from program in Section 15.4.1 (`YmetricXsingleBrugs.R`).

15.1.3 Outliers and Robust Estimation: The *t* Distribution

Figure 15.4 shows examples of the *t* distribution. The *t* distribution was originally invented by Gosset (1908), who used the pseudonym "Student" because his employer (Guinness Brewery) prohibited publication of any research that might be proprietary or imply problems with their product (such as variability in quality). Therefore, the distribution is often referred to as the *Student t* distribution.

In R, the *t* density at *x* is specified by `dt(x,df)`, where `df` is a parameter called the *degrees of freedom*. (The R function `dt` also has an optional argument called `ncp`, the noncentrality parameter. This option is not implemented in the BUGS `dt` function.) The effect of the `df` argument is shown in Figure 15.4. When `df` is small, the *t* distribution has tails that are taller than normal. As `df` approaches infinity, the *t* distribution becomes normal. The `df` parameter

FIGURE 15.4
t distributions, with a normal distribution superimposed for comparison. The *t* distribution has tall tails relative to the normal. The *df* parameter ("degrees of freedom") controls the relative height of the tails.

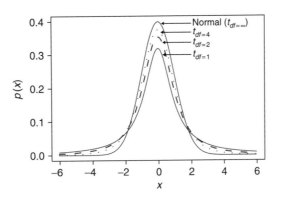

is a continuous value greater than or equal to 1. (Some readers might be familiar with the df parameter as related to sample size when t is a sampling distribution, but the t distribution is not used as a sampling distribution here, and therefore df is not restricted to being an integer.) In BUGS, the t density is specified by dt(mu,tau,df). The mu and tau arguments in BUGS have no equivalent arguments in R. All they do is linearly shift the t distribution, such that dt(mu,tau,df) in BUGS corresponds to sqrt(tau) * dt((x-mu)*sqrt(tau) , df) in R.

Although the t distribution was originally conceived as a sampling distribution for NHST, we will use it instead as a convenient model of data with outliers (as is often done; e.g., Damgaard, 2007; Meyer & Yu, 2000; Tsionas, 2002). Outliers are simply data values that fall unusually far from the model's expected value. Real data often contain outliers, presumably because some extraneous influences have sporadically perturbed the data values. Sometimes these extraneous influences can be identified, and the affected data values can be corrected. But usually we have no way of knowing whether a suspected outlying value was caused by an extraneous influence, or is a genuine reflection of the target being measured. Instead of deleting suspected outliers from the data according to some arbitrary criterion, we retain all the data but use a likelihood function that is less affected by outliers than is the normal distribution.

Figure 15.5 shows examples of how the t distribution is robust against outliers. The curves show the maximum likelihood estimates (MLEs) of the parameters for the t and normal distributions. More formally, for the given data $D = \{x_i\}$, parameter values were found for the normal that maximized $p(D \mid \mu, \tau)$, parameter values were found for the t distribution that maximized $p(D \mid \mu, \tau, df)$, and the curves of those MLEs are plotted with the data. The upper panel of Figure 15.5 shows "toy" data to illustrate that the normal is strongly influenced by an outlier, whereas the t distribution remains centered over the bulk of the data. The lower panel of Figure 15.5 uses realistic data that indicate levels of inorganic phosphorous, measured in milligrams per deciliter, in 177 human subjects aged 65 or older. The authors of the data (Holcomb & Spalsbury, 2005) intentionally altered a few data points to reflect typical transcription errors and to illustrate methods for detecting and correcting such errors. We instead assume that we no longer have access to records of the original individual measurements, and must model the uncorrected data set. The t distribution accommodates the outliers and fits the distribution of data much better than the normal.

The t distribution is useful as a likelihood function for modeling outliers at the level of observed data, but it is also useful for modeling outliers at higher levels in a hierarchical prior. We will encounter such applications when we model individual differences or multiple predictors (e.g., in Section 16.2).

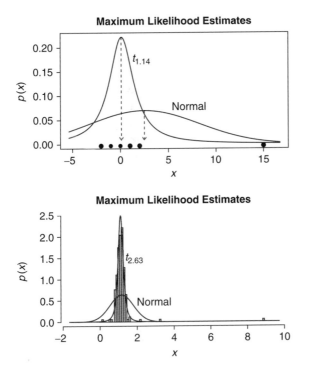

FIGURE 15.5
The maximum likelihood estimates of normal and t distributions fit to the data shown. Upper panel shows "toy" data to illustrate that the normal accommodates an outlier only by enlarging its standard deviation and, in this case, shifting its mean. Lower panel shows actual data to illustrate realistic effect of outliers on estimates of the normal.

15.1.4 When the Data Are Non-normal: Transformations

Suppose there has been discovered a new species of insect that lives in dark caves, and the insects make ultra high frequency sound bursts. One theory posits that the sounds should have an average frequency of 22,000 Hz (cycles per second). An entomologist has measured the frequencies of 200 sound bursts from these insects, and would like to know if a mean of 22,000 Hz is credible.

The upper-left panel of Figure 15.6 shows a histogram of the data, and the upper-right panel shows the posterior distribution of μ from the normal-likelihood model of Figure 15.3. (Assume that the prior was uncontroversial.) According to this posterior, the data *exclude* a μ value of 22,000. Therefore, the entomologist concludes that the 22,000 Hz theory is not credible.

The proponent of the theory says that the theory was misconstrued. Instead of scaling the insect sounds by their raw frequency, they should be scaled by their perceived pitch, which is merely the logarithm of frequency. The natural logarithm of 22,000 is 10. Thus, the theory claims that the sound bursts should have an average log(frequency) of 10. The lower-left panel of Figure 15.6 shows the logarithm of the same data as the upper-left panel. When the normal-likelihood model is applied to these data, the resulting posterior on μ is shown in the lower right of Figure 15.6. (Assume that the prior was uncontroversial.)

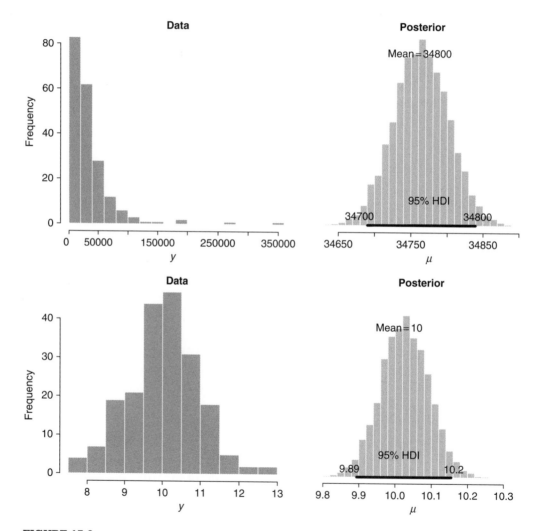

FIGURE 15.6
Upper left shows skewed, non-normal data. Upper right shows the posterior distribution of μ when the data are modeled (inappropriately!) with a normal likelihood. Bottom left shows the same data transformed, via a logarithm. Bottom right shows the posterior distribution of μ when the transformed data are modeled (appropriately) with a normal likelihood.

According to this posterior, the data do *not* exclude a μ value of 10. Indeed, the value of 10 is very near the mode of the posterior distribution.

Which analysis do we trust? Is the theory disconfirmed or upheld? In this case, the theory is clearly upheld because the analysis based on raw frequency is not appropriate. The problem with the analysis based on raw frequency is that the normal-likelihood model assumes that the data are normally distributed, but the raw frequencies are obviously extremely skewed and non-normal. Graphically, the normal-likelihood model in Figure 15.2 shows that the data, y_i, are generated by a normal distribution. The parameters μ and τ are only meaningful to the extent that the normal likelihood actually describes the distribution of the data. Because of the severe skew in the raw-frequency data, the central-tendency parameter μ is being pulled far to the right by the extreme values far above 22,000, for which there are no symmetrically distributed values below 22,000, as is assumed by a symmetric normal distribution. The logarithmically transformed data, on the other hand, are apparently much more normally distributed, and therefore the normal-likelihood model is a viable description of the data.

The moral of the story is this: When using a normal-likelihood model, the data should be at least roughly normally distributed; otherwise the model is a poor description of the data and the model's parameter values may be meaningless. When there is only moderate non-normality in the data, the misrepresentation by the model parameters is often hardly noticeable, but when there is severe non-normality in the data, then the misrepresentation may be substantial.

There are many real-world situations in which data are non-normally distributed. A prominent example is response times. Ask a person to press a button as quickly as possible after a light comes on. Repeat this many times. The person's response times will be strongly skewed toward higher values. This skew makes intuitive sense, because the response time can't be much faster than 100 or 200 milliseconds, but there are many reasons to be slow on some trials. Another example is shown in Figure 15.10, which reveals that the longevity of lab rats on a restricted diet is skewed to the left.

How should we analyze non-normal data? There are two possible solutions. One method perseveres with a normal likelihood function, but transforms the data so that they are nearly normal. There is nothing wrong with transforming data, as long as *all* the data are transformed the same way, and the transformation preserves order. A transformation that preserves order is called *monotonic*. Examples of monotonic transformations include the exponential function, the logarithm (for positive-valued data), and the cubic (i.e., y^3). An example of a nonmonotonic transformation is a sine wave function, when the original data span several cycles of the sine wave. Monotonic transformations merely rescale the data. The measurement scale is arbitrary. Earthquake intensities could be measured in terms of raw energy or in terms of the base-10 logarithm of energy, which is the Richter scale. Temperatures can be measured in terms of the Fahrenheit or Celsius scales. Distances can be measured in terms of the English or metric systems. Acoustic vibrations can be measured by frequency or by perceived pitch, which is, essentially, the logarithm of frequency. The analysis should use whatever transformation makes the data normally distributed, so that the data respect the assumption of the normal-likelihood model and thereby make the parameter estimates meaningful.

The second approach to analyzing non-normal data is to jettison the normal likelihood function, and use a more appropriate likelihood function instead. For example, Rouder Lu, Speckman, Sun, & Jiang (2005) used a Weibull distribution as the likelihood function for modeling skewed response time distributions. This approach requires familiarity with various non-normal probability distributions. Moreover, for use in BUGS, this method also requires that the desired likelihood function can be specified in BUGS. Because of these extra complications, people usually first try the method of transforming data to normal. But a non-normal likelihood function may be preferred when no reasonable data transformation suffices, when transformed scales are too awkward to work with, or when a theory posits a specific non-normal distribution for generating data.

There is one more important caveat when transforming data (or changing likelihood functions): *The prior must be appropriate to the transformed data.* For example, when the insect sounds are measured in terms of raw frequency, then the prior should be appropriate for data in the broad vicinity of 22,000. But when the insect sounds are measured by log(frequency), then the prior should be appropriate for data in the general vicinity of 10.

Exercise 15.2 provides a realistic example in which we are interested in estimating the underlying mean that generated a set of scores. It also asks you to consider the choice of prior, the robustness of the posterior, and whether a normal likelihood seems to be an appropriate model of the data. Exercise 15.3 gives another realistic example in which the data are skewed.

The section previous to this one discussed outliers. Can outliers be "transformed away"? In simple, one-group data sets, the answer, in principle, is yes: We merely fashion a transformation that arbitrarily compresses the extreme tails of the distribution in just the right way. But then the prior should take into account the transformation. Unfortunately, for more typical complex situations, it is difficult to invent a transformation that appropriately compresses all outliers simultaneously. Therefore, transformations are typically used to adjust large-scale skew in the data, and not for outliers. Outliers, on the other hand, are typically addressed by using tall-tail distributions, such as the t distribution.

15.2 REPEATED MEASURES AND INDIVIDUAL DIFFERENCES

In many real-world applications, we have repeated measures from the same individual, although we are mainly interested in the central tendency of the group. For example, we might be interested in the typical blood pressure of employees at a company. We can repeatedly measure the blood pressure of each employee at random different times of day and on random different days. Our overall goal is to estimate the typical blood pressure of the group as a whole, but now we have repeated measurements from the individual subjects.

We assume that the individual subjects have been randomly sampled from the pool of all possible subjects in the population of interest (e.g., the pool of all employees in the company). We also assume that the repeated measurements within each subject are mutually independent. This is sometimes a perilous assumption, because it is probably only approximately true even if we are careful to design the measurement procedure with the assumption in mind. For example, suppose we are measuring someone's blood pressure, with the goal of getting measurements that represent that person's typical blood pressure. We could measure the blood pressure several times in rapid succession, but presumably those measurements would be highly correlated because the person's blood pressure would not change much in only a few seconds' duration. Instead, we should measure blood pressure at random times across hours and days. The goal is to make the repeated measurements far enough apart in time so that we can reasonably assume that the repeated measurements are independent of each other. But we don't want the repeated measurements to be so far apart in time that the person's underlying typical value has changed (e.g., from dramatic change in body fat). Thus, we want the measurements to be far enough apart that they have minimal correlation, but we want the measurements to be close enough together that they represent a stationary underlying propensity. There is no single correct sampling procedure; it depends on theoretical considerations for the specific application domain.

As another example with real data, consider the time between successive failures of the air conditioning system in each aircraft in a fleet of Boeing 720 jet airplanes (Proschan, 1963). The time between successive failures is of interest so that managers can appropriately schedule preventative maintenance and maintain an inventory of spare parts. Thus, the emphasis in this situation is estimation of interfailure duration. We will assume that successive failures of the same air conditioner are independent of each other. One rationale for this assumption is that the repair of a failed system resets the system so that it has no memory for the previous failure. The next failure depends only on other factors, such as the basic design of the air conditioner and the typical weather conditions in which the individual aircraft flies.

The data from 12 aircraft are displayed in Figure 15.7 (these data omit one aircraft that had only two observations, as was done by Hand et al., 1994, set #480). Notice that the data have been transformed from their original scale of days to days$^{1/5}$ (i.e., the fifth root of days). This transformation was necessary because the raw data were extremely skewed and I wanted to use a normal likelihood function, for illustrative purposes (Proschan, 1963, used a more theoretically informed model). A different transformation, such as a logarithm, could have been used instead, but the fifth root yielded nicely symmetric data distributions. Notice that *all* the data were transformed the same way; there were not different transformations applied to different aircraft. As will be explained when the full model is described, the goal of the transformation was to make the data from each individual aircraft be

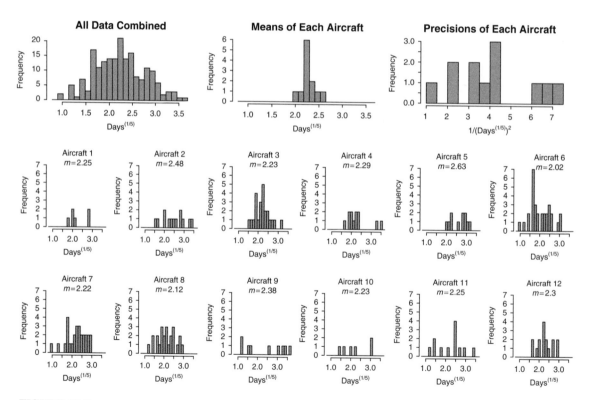

FIGURE 15.7

Number of days between failures of air conditioning systems in aircraft. Notice the scale on the *x*-axis. The lower 12 panels show data from individual aircraft. The upper-left panel shows all those data combined into one histogram. The upper-middle panel shows a histogram of the 12 aircraft means.

approximately normal and to simultaneously make the distribution of aircraft means be approximately normal.

15.2.1 Hierarchical Model

To model this situation, we will assume that data from the jth individual are normally distributed around an underlying individual mean, denoted μ_j. The dispersion of repeated measurements from that individual is described by the precision τ_j. We will further assume that the individual propensities, μ_j, are normally distributed around a group mean μ_G. The dispersion of the individual means around the group mean is described by the precision τ_G. It is these group-level parameters in which we are primarily interested. Because we want to estimate μ_G and τ_G, we need to specify their prior-belief distributions.

Just as the individual propensities μ_j are generated by group-level distributions, the individual precisions τ_j are also generated by group-level distributions. The idea is that the variability of repeated measurements in an individual is dependent on the group they are in. For example, a company might engender employees with relatively stable blood pressures (i.e., all the employees tend to have high precisions in their individual distributions of blood pressure). As another example, an air conditioner manufacturer might build systems that fail erratically (i.e., all the systems tend to have low precisions in their individual distributions of interfailure duration). Therefore, the individual τ_j come from a group-level gamma distribution which has shape and rate parameters we want to estimate from the data. Therefore, the shape and rate parameters of the group-level gamma distribution must themselves be given higher-level prior distributions.

The entire model is illustrated in Figure 15.8. It shows the hierarchical dependency of individual parameters μ_j on the group-level parameters μ_G, τ_G, and the dependency of the individual parameters τ_j on the group-level parameters s_G, r_G. The group-level parameters themselves have higher-level distributions as shown. The higher-level distributions are necessary because it is the group-level parameters we want to estimate; therefore, we must express our prior uncertainty in those group-level parameters by the higher-level distributions. The higher-level distributions for s_G, r_G are reparameterized in terms of the mean (m) and standard deviation (s) of the gamma density for τ_j. This reparameterization is performed merely for convenience in interpreting the values of s_G, r_G.

At the top level, the constants for the hyperpriors must be given specific values. These values must be appropriate for the domain and data at hand. In other words, the hyperprior constants should be at least mildly informed by the situation (or richly informed if audience-agreeable prior information is available). For example, we know from common sense that air conditioners will be manufactured to operate without failure for a duration on the order of

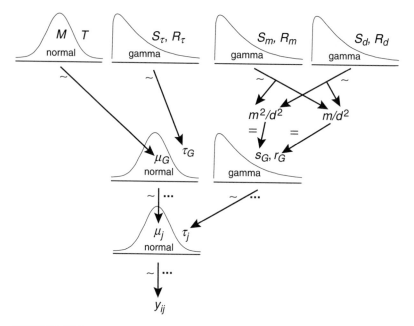

FIGURE 15.8

A model of hierarchical dependencies for repeated measures from each of several subjects drawn from the same group. The ith datum from the jth subject is denoted y_{ij}. It is generated by a normal distribution that has a subject mean of μ_j and subject precision of τ_j. The subject means are generated by a normal distribution centered on the overall mean of μ_G with a precision of τ_G. These group-descriptive parameters, μ_G and τ_G, are themselves being estimated from the data, hence they have prior distributions as shown at the top left of the hierarchy. The subject precisions, τ_j, are generated by a gamma distribution, which has parameters s_G and r_G. These parameters are themselves being estimated, therefore they have prior distributions as shown in the top right of the diagram.

10^2 days, with a range of perhaps 10^0 to 10^4 days. Therefore, the hyperprior on μ_G should reasonably specify a distribution in this range (and not milliseconds or eons!). Moreover, it is important to remember that the air conditioner data were transformed to a different scale, and therefore the hyperprior constants need to be appropriate to the transformed scale.

Before proceeding with parameter estimation, it is important to verify that the basic distributional assumptions of the model are not being blatantly violated. At the lowest level of the model, the normal likelihood function assumes that the data generated by each individual aircraft are normally distributed. A quick look at Figure 15.7 shows that the twelve individual aircraft have data that are reasonably unimodal and symmetric (unlike the raw, untransformed data). At the next higher level in the model, the normal distribution that generates μ_j assumes that the central tendencies of the aircraft are normally distributed. Inspection of the top-middle graph of Figure 15.7 verifies that the 12 aircraft means are unimodal and symmetric. Finally, the gamma distribution that generates the τ_j assumes that the precisions of the 12 aircraft can be described by a gamma distribution. Inspection of the top-right graph of Figure 15.7 verifies that the 12 aircraft precisions may be reasonably described this way.

If there appear to be prominent outliers in the data, then the data might be better modeled by a t distribution than by a normal distribution. As was discussed in connection with Figure 15.5, the t distribution is far less sensitive to outliers than the normal. The novel point here is that outliers might occur at different levels. There might be outliers of interfailure durations within each individual aircraft. These would be visible in the plots of individual aircraft data. Additionally, there might be outliers at the higher level of means for each aircraft. These outliers would be visible in the plot of aircraft means. In the hierarchical model of Figure 15.8, these considerations imply that either or both of the two lower normal distributions, involving μ_j or μ_G, might be replaced with a t distribution.

15.2.2 Implementation in BUGS

It is straightforward to implement the model in BUGS. Every arrow in the dependency diagram of Figure 15.8 has a corresponding specification in the BUGS model (SystemsBrugs.R):

```
9    model {
10       for( i in 1 : Ndata ) {
11           y[i] ~ dnorm( mu[ subj[i] ] , tau[ subj[i] ] )
12       }
13       for ( j in 1 : Nsubj ) {
14           mu[j] ~ dnorm( muG , tauG )
15           tau[j] ~ dgamma( sG , rG )
16       }
17       muG ~ dnorm( 2.3 , 0.1 )
18       tauG ~ dgamma( 1 , .5 )
19       sG <- pow(m,2) / pow(d,2)
20       rG <- m / pow(d,2)
21       m ~ dgamma( 1 , .25 )
22       d ~ dgamma( 1 , .5 )
23    }
```

Notice the use of nested indexing in line 11. In the BUGS model specification, the index i is the overall row of the data matrix, not the ith measure within a subject. Nested indexing is used because the data matrix is structured with each row containing a datum y and the subject number (i.e., aircraft) j from which the datum was obtained. This data structure accommodates the fact that different aircraft contributed different numbers of data points. The complete program is listed in Section 15.4.2 (SystemsBrugs.R).

Aspects of the resulting posterior distribution are shown in Figure 15.9. The left panel shows the marginal distribution on the group-level parameter μ_G, which has a posterior mean of 2.28 days$^{1/5}$, which corresponds to 61.6 days.

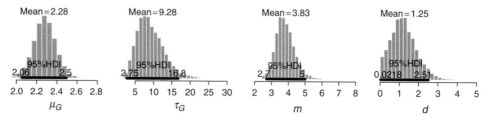

FIGURE 15.9
Posterior distribution for the top-level hyperparameters in Figure 15.8, for the data in
Figure 15.7.

In other words, on average, about two months elapsed between failures of the the air conditioning system on these airplanes. In the second panel, the value of $1/\sqrt{\tau_G}$ indicates the standard deviation across aircraft of the average interfailure duration. The posterior mean τ_G is 9.28, corresponding to a standard deviation of $1/\sqrt{9.28} = 0.33$ days$^{1/5}$. The parameters μ_G and τ_G describe the distribution of aircraft means μ_j, and thus the posterior is saying that among the most credible descriptions of the data distribution in the top-middle panel of Figure 15.7 is a normal distribution with mean 2.28 and standard deviation of 0.33. The remaining posterior marginals in Figure 15.9 show credible values of m and d, for the gamma distribution that describes how the precisions vary across aircraft. The posterior mean m is 3.83 and the posterior mean d is 1.25, which says that among the most credible descriptions of the data distribution in the top-right panel of Figure 15.7 is a gamma distribution with mean of 3.83 and standard deviation of 1.25.

15.3 SUMMARY

This chapter explored the normal likelihood function. The purpose was to explore concepts focused on the normal likelihood when its mean is simply a constant, not involving additional predictors and parameters. With this foundation in place, subsequent chapters can focus on concepts regarding additional predictors. This chapter made several main points:

- When the precision of the normal likelihood function is fixed (not estimated), then the "update rules" for the mean and precision of a normal prior are analytically simple. These formulas also reveal why it is convenient for Bayesians to parameterize a normal distribution by its precision instead of by its standard deviation.
- It is typical to put a normal prior on the mean of a normal likelihood, and a gamma prior on the precision of a normal likelihood, because of considerations of conjugacy for simple situations. Use of these priors is not necessary, however. It is especially easy to program arbitrary priors in BUGS (although not all priors will work equally efficiently for sampling).

- When there are repeated measures within individuals, either the repeated measures within individuals or the means across individuals might be modeled as normal distributions. Again, this hierarchical model can be easily programmed in BUGS.
- Whenever applying a normal likelihood function, it is important to check that the data are reasonably normally distributed (i.e., at least roughly unimodal and symmetric). Otherwise the parameter values may be misleading. If the data are severely non-normal, they might be transformed to normal, or else a non-normal likelihood function could be used instead.
- Data that have prominent outliers might be better modeled with a t distribution than a normal distribution. This idea was introduced in this chapter, but examples of applying it in BUGS wait until subsequent chapters.

15.4 R CODE

15.4.1 Estimating the Mean and Precision of a Normal Likelihood

Notice that the data in this program are randomly generated from a normal distribution in R, on line 30. In R, `dnorm` and `rnorm` are parameterized by mean and standard deviation, unlike BUGS, in which `dnorm` is parameterized by mean and precision.

The program also initializes the chains intelligently so that they start amid the posterior, instead of starting randomly according to the prior. This method of initializing chains was introduced in Section 9.5.2 (`FilconBrugs.R`). Lines 51 to 52 show that the initial value of μ is the mean of the data, and the initial value of τ is the precision of the data. If the prior is fairly vague, and the data are numerous, then the posterior will be near the parameter values that maximize the likelihood of the data. Therefore, we start the chains near those maximum-likelihood values. We still need to allocate a burn-in period, however, because the chains needs to independently diverge from each other and settle into the true posterior.

(YmetricXsingleBrugs.R)

```
1   graphics.off()
2   rm(list=ls(all=TRUE))
3   library(BRugs)          # Kruschke, J. K. (2010). Doing Bayesian data analysis:
4                            # A Tutorial with R and BUGS. Academic Press / Elsevier.
5   #------------------------------------------------------------------------------
6   # THE MODEL.
7   modelstring = "
8   # BUGS model specification begins here...
9   model {
10      # Likelihood:
11      for( i in 1 : N ) {
```

```
12          y[i] ~ dnorm( mu , tau ) # tau is precision, not SD
13        }
14      # Prior:
15      tau ~ dgamma( 0.01 , 0.01 )
16      mu ~ dnorm( 0 , 1.0E-10 )
17    }
18  # ... end BUGS model specification
19  " # close quote for modelstring
20  writeLines(modelstring,con="model.txt")
21  modelCheck( "model.txt" )
22
23  #------------------------------------------------------------------------
24  # THE DATA.
25
26  # Generate random data from known parameter values:
27  set.seed(47405)
28  trueM = 100
29  trueSD = 15
30  y = round( rnorm( n=500 , mean=trueM , sd=trueSD ) ) # R dnorm uses mean and SD
31
32  datalist = list(
33      y = y ,
34      N = length( y )
35  )
36
37  # Get the data into BRugs: (default filename is data.txt).
38  modelData( bugsData( datalist ) )
39
40  #------------------------------------------------------------------------
41  # INTIALIZE THE CHAINS.
42
43  nchain = 3
44  modelCompile( numChains = nchain )
45
46  automaticInit = F # TRUE or FALSE
47  if ( automaticInit ) {
48      modelGenInits()  # automatically initialize chains from prior
49  } else {
50      genInitList <- function() { # manually initialize chains near the data
51          list( mu = mean( datalist$y ) ,
52                tau = 1 / sd( datalist$y )^2 )
53      }
54      for ( chainIdx in 1 : nchain ) {
55          modelInits( bugsInits( genInitList ) )
56      }
57  }
58
59  #------------------------------------------------------------------------
60  # RUN THE CHAINS
61
62  # burn in
63  BurnInSteps = 500
64  modelUpdate( BurnInSteps )
```

```
65   # actual samples
66   samplesSet( c( "mu" , "tau" ) )
67   stepsPerChain = 2000
68   thinStep = 1
69   modelUpdate( stepsPerChain , thin=thinStep )
70
71   #-------------------------------------------------------------------
72   # EXAMINE THE RESULTS
73
74   filenamert = "YmetricXsingleBrugs"
75
76   source("plotChains.R")
77   muSum = plotChains( "mu" , saveplots=F , filenamert )
78   sigmaSum = plotChains( "tau" , saveplots=F , filenamert )
79
80   muSample = samplesSample( "mu" )
81   tauSample = samplesSample( "tau" )
82   sigmaSample <- 1 / sqrt( tauSample ) # Convert precision to SD
83
84   source("plotPost.R")
85   windows()
86   plotPost( muSample , xlab="mu" , breaks=30 , main="Posterior" )
87   dev.copy2eps(file=paste(filenamert,"PostMu.eps",sep=""))
88
89   nPts = length(muSample) ; nPtsForDisplay = min( nPts , 2000 )
90   thinIdx = seq( 1 , nPts , nPts / nPtsForDisplay )
91   windows()
92   plot( muSample[thinIdx] , sigmaSample[thinIdx] , col="gray" ,
93         xlab="mu" , ylab="sigma" , cex.lab=1.5 , main="Posterior" , log="y" )
94   points( mean(muSample) , mean(sigmaSample) , pch="+" , cex=2 )
95   text( mean(muSample) , mean(sigmaSample) ,
96         bquote( .(round(mean(muSample),1)) *"  "* .(round(mean(sigmaSample),1)) ),
97         adj=c(.5,-0.5) )
98   dev.copy2eps(file=paste(filenamert,"PostMuSigma.eps",sep=""))
99
100  #-------------------------------------------------------------------
```

15.4.2 Repeated Measures: Normal Across and Normal Within

Here is the complete program for analyzing the data from Figure 15.7, regarding inter-failure durations of aircraft air conditioners. A glimpse of the posterior was shown in Figure 15.9.

(SystemsBrugs.R)
```
1   graphics.off()
2   rm(list=ls(all=TRUE))
3   library(BRugs)          # Kruschke, J. K. (2010). Doing Bayesian data analysis:
4                           # A Tutorial with R and BUGS. Academic Press / Elsevier.
5   #-------------------------------------------------------------------
6   # THE MODEL.
```

```
7   modelstring = "
8   # BUGS model specification begins here...
9   model {
10      for( i in 1 : Ndata ) {
11          y[i] ~ dnorm( mu[ subj[i] ] , tau[ subj[i] ] )
12      }
13      for ( j in 1 : Nsubj ) {
14          mu[j] ~ dnorm( muG , tauG )
15          tau[j] ~ dgamma( sG , rG )
16      }
17      muG ~ dnorm( 2.3 , 0.1 )
18      tauG ~ dgamma( 1 , .5 )
19      sG <- pow(m,2) / pow(d,2)
20      rG <- m / pow(d,2)
21      m ~ dgamma( 1 , .25 )
22      d ~ dgamma( 1 , .5 )
23  }
24  # ... end BUGS model specification
25  " # close quote for modelstring
26  writeLines(modelstring,con="model.txt")
27  modelCheck( "model.txt" )
28
29  #------------------------------------------------------------------------
30  # THE DATA.
31
32  # Load the aircraft data:
33  load( "Systems.Rdata" ) # loads dataMat
34  nSubj = length( unique( dataMat[,"Aircraft"] ) )
35  # Transform the data:
36  DaysTransf = dataMat[,"Days"]^(1/5)
37  dataMat = cbind( dataMat , DaysTransf )
38  colnames( dataMat ) = c( colnames( dataMat )[1:3] , "DaysTransf" )
39
40  # Specify data, as a list.
41  datalist = list(
42      y = dataMat[,"DaysTransf"] ,
43      subj = dataMat[,"Aircraft"] ,
44      Ndata = NROW(dataMat) ,
45      Nsubj = nSubj
46  )
47  # Get the data into BRugs: (default filename is data.txt).
48  modelData( bugsData( datalist ) )
49
50  #------------------------------------------------------------------------
51  # INTIALIZE THE CHAINS.
52
53  # First, compile the model:
54  nchain = 10
55  modelCompile( numChains = nchain )
56
57  modelGenInits() # works when the priors are not too flat
58
59  #------------------------------------------------------------------------
60  # RUN THE CHAINS
```

```
61
62   # burn in
63   BurnInSteps = 1000
64   modelUpdate( BurnInSteps )
65   # actual samples
66   samplesSet( c( "muG" , "tauG" , "mu" , "tau" , "m" , "d" ) )
67   stepsPerChain = ceiling(10000/nchain)
68   thinStep = 100
69   modelUpdate( stepsPerChain , thin=thinStep )
70
71   #------------------------------------------------------------------------
72   # EXAMINE THE RESULTS
73
74   source("plotChains.R")
75   source("plotPost.R")
76   filenamert = "SystemsBrugs"
77
78   # Examine chains for convergence and autocorrelation:
79   muSum = plotChains( "muG" , saveplots=F , filenameroot=filenamert )
80   tauSum = plotChains( "tauG" , saveplots=F , filenameroot=filenamert )
81   mSum = plotChains( "m" , saveplots=F , filenameroot=filenamert )
82   dSum = plotChains( "d" , saveplots=F , filenameroot=filenamert )
83   mu1Sum = plotChains( "mu[1]" , saveplots=F , filenameroot=filenamert )
84   tau1Sum = plotChains( "tau[1]" , saveplots=F , filenameroot=filenamert )
85
86   # Extract chains from BUGS into R:
87   muGsample = samplesSample( "muG" )
88   tauGsample = samplesSample( "tauG" )
89   mSample = samplesSample( "m" )
90   dSample = samplesSample( "d" )
91   muSample = NULL
92   tauSample = NULL
93   for ( sIdx in 1:nSubj ) {
94       muSample = rbind( muSample , samplesSample( paste("mu[",sIdx,"]",sep="") ) )
95       tauSample = rbind( tauSample , samplesSample( paste("tau[",sIdx,"]",sep="") ) )
96   }
97
98   # Plot the aircraft mu:
99   windows(15,6)
100  layout( matrix( 1:nSubj , nrow=2 , byrow=T ) )
101  for ( sIdx in 1:nSubj ) {
102      plotPost( muSample[sIdx,] , xlab=bquote(mu[.(sIdx)]) )
103  }
104  dev.copy2eps(file=paste(filenamert,"PostMu.eps",sep=""))
105
106  # Plot the aircraft tau:
107  windows(15,6)
108  layout( matrix( 1:nSubj , nrow=2 , byrow=T ) )
109  for ( sIdx in 1:nSubj ) {
110      plotPost( tauSample[sIdx,] , xlab=bquote(tau[.(sIdx)]) , HDItextPlace=.3 )
111  }
112  dev.copy2eps(file=paste(filenamert,"PostTau.eps",sep=""))
113
114  # Plot the hyperdistributions:
```

```
115    windows(15,3)
116    layout( matrix(1:4,ncol=4) )
117    plotPost( muGsample , xlab=expression(mu[G]) , breaks=30 )
118    plotPost( tauGsample , xlab=expression(tau[G]) , breaks=30 )
119    plotPost( mSample , xlab=expression(m) , breaks=30 )
120    plotPost( dSample , xlab=expression(d) , breaks=30 , HDItextPlace=.1 )
121    dev.copy2eps(file=paste(filenamert,"PostHyper.eps",sep=""))
122
123    #-----------------------------------------------------------------------
```

15.5 EXERCISES

Exercise 15.1. [Purpose: To view the prior from BUGS.] For the program in Section 15.4.1 (YmetricXsingleBrugs.R), generate graphs of prior distribution, analogous to Figure 15.3. To do this, comment out the data in the data specification, as explained in Section 8.5.1, p. 174. But also beware of the following: Automatically initialize the chains from the prior, and, when plotting the results, comment out the plotChains commands because the MCMC sigma values are too extreme for some of the BUGS graphics routines.

Exercise 15.2. [Purpose: To explore a realistic example of estimating a single mean, with consideration of priors.] University students who agreed to participate in a problem-solving experiment were also tested for their vocabulary level, using the Wechsler Adult Intelligence Scale Revised (WAIS-R), which is normed to have a general-population mean of 10 and standard deviation of 3. Here are the data from the university students (data from Hand et al., 1994, set #392, p. 322):

14 11 13 13 13 15 11 16 10 13 14 11 13 12 10 14 10 14 16 14 14 11 11 11
13 12 13 11 11 15 14 16 12 17 9 16 11 19 14 12 12 10 11 12 13 13 14 11
11 15 12 16 15 11

(A) Are the data roughly normal, that is, essentially unimodal and symmetrically distributed? (No formal analysis is required here; only an "eyeball" assessment is expected.) Therefore, can a normal likelihood function be applied?

(B) Discuss the rationale for a prior distribution. Justify the constants you choose for the hyperprior. (Hint: The results of the analysis will be presented to a skeptical audience, who may doubt any claims about how prior knowledge of the general population informs an analysis of university students. Therefore, your prior should be very vague and widely dispersed around the general population values.)

(C) Is the mean vocabulary score of the university students credibly different from the general population mean of 10? Report the 95% HDI on μ for the various priors you considered in the previous part.

Exercise 15.3. [Purpose: To observe a realistic example of estimating a mean, with consideration of whether a normal likelihood distribution is appropriate.] Suppose we know that the mean lifetime of rats that eat *ad lib* is 700 days. This value is, in fact, about right for lab rats; see Figure 15.10, which shows data from R.L. Berger et al. (1988), as reported in Hand et al. (1994, data set #242), and which are available in the file RatLives.Rdata. This is an Rdata file, which stores values in compressed format, not text format. To get the data into R, type load("RatLives.Rdata"). It will look like nothing happens when you type that command, but the variables are now loaded into R; you can see the variables that R knows about by typing ls(). You will see listed the two vectors adlibDiet and restrictedDiet. When rats are placed on a restricted diet, their longevity can be affected. Figure 15.10 shows the results.

(A) Using the raw data from the restricted-diet longevities, estimate μ and its 95% HDI. Be sure to report the prior you used. (Hint: The HDI extends from about 917 to 1020 days.)

(B) Is it appropriate to apply a normal likelihood function to these data? Transform the data by squaring the longevities. Estimate μ and its 95% HDI (now in days squared). Be careful to use an appropriate prior! (Hint: The HDI extends from about 967^2 to 1053^2 days.)

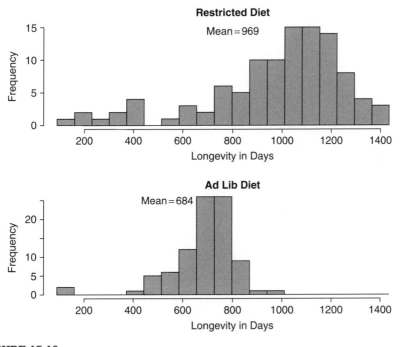

FIGURE 15.10
For Exercise 15.3.

(C) Why is the first estimate lower than the second estimate? Which estimate is more appropriate and why?

Exercise 15.4. [Purpose: To think about specifying an *informed* prior, and non-normal, non-gamma priors.] In the program of Section 15.4.1 (YmetricXsingleBrugs.R), the prior on the group mean uses an extremely small precision. This diffuse prior is silly, because we know that individual IQ scores from the general population should be near 100 with a standard deviation of about 15. Special populations might have IQ scores consistently above or below 100 but almost certainly within, say, 100 points above or below 100. Consider the following expressions of the prior belief on μ:

(1) `mu ~ dnorm( 100 , sd=50 ) # that's sd=50, so`
`                               # tau=pow (50,-2)`
(2) `mu ~ dgamma( 4 , 0.04 )`
(3) `mu ~ dunif( 0 , 200 )`
(4) `mu ~ dgamma( 3 , 0.03 )`

(A) Plot the densities of (1) and (2) on the same graph, superimposed. (Hint:

```
mu = seq( -50 , 300 , length=501 )
plot( mu , dnorm( mu , 100 , 50 ) , type="l" ,
  ylim=c(0,0.01))
lines( mu , dgamma( mu , 4 , 0.04 ) )
```

What do the densities of (1) and (2) have in common? (Hint: Compute the mean and sd of the gamma distribution.) Should either of (1) or (2) be preferred over the other? (Hint: Consider negative IQ scores, which should not be allowed.)

(B) Plot the densities of (3) and (4) on the same graph, superimposed. What do the densities of (3) and (4) have in common? Should one of (3) or (4) be preferred over the other? (Hint: Consider IQ scores greater than 200, which should be allowed.)

In the program of Section 15.4.1 (YmetricXsingleBrugs.R), the prior on the group precision is a general diffuse distribution. This diffuse prior is silly, because we know that individual IQ scores have a standard deviation around 15 (i.e., a precision of $1/15^2 = 0.00444...$) in the general population. The SD might be a bit smaller within special populations.

(C) Suppose we believe that the smallest SD we would find in a specialized population is 5. What is the corresponding precision? (Notice that it is *larger* than $1/15^2$.)

(D) We want to create a gamma distribution for precision that has a mean corresponding to $1/15^2$, such that most of the gamma distribution is less than the precision corresponding to an IQ precision of $1/5^2$, because, as mentioned in the previous part, we don't think that precisions much greater than $1/5^2$ are tenable. Therefore, to be sure that our prior encompasses that maximum but still allows for considerable uncertainty, we will set the standard deviation of the gamma distribution to *half* of the difference between the precisions determined in the previous two parts. What is the value of half of the difference between the precisions determined in the previous two parts? (Hint: The answer is 0.01777 . . . . Explain.)

(E) What are the values of the shape and rate parameters for a gamma distribution that has mean of 0.00444 and standard deviation of 0.01777?

(F) Rerun the program of Section 15.4.1 (`YmetricXsingleBrugs.R`) using a prior for tau from the previous part, and a prior for mu that is the most appropriate from (1)–(4) above. Include the posterior histogram of mu with your answer. Does the conclusion differ noticeably from the diffuse priors?

Exercise 15.5. [Purpose: With repeated measures, to see how the group estimate and individual estimates reflect different sources of variation in the data.]

Consider the data, *y*, with different assignments to subjects, *s*, as follows:

y values:	1	2	3	21	22	23	41	42	43
s, large between subj. var.:	1	1	1	2	2	2	3	3	3
s, small between subj. var.:	1	2	3	1	2	3	1	2	3

The first row shows nine data values. The second row indicates a situation in which the first three data points come from subject 1, the second three data points come from subject 2, and so on. The third row indicates a different situation, in which the first, fourth, and seventh data points come from subject 1, and so forth. The assignment of data to subjects in the second row produces large between-subject variance, whereas the assignment of data to subjects in the third row produces small between-subject variance. *Notice that for both situations, however, the overall mean is the same because the data are the same. We are interested in which situation gives a more precise estimate of the group-level mean.*

Modify the program of Section 15.4.2 (`SystemsBrugs.R`) for use with these data. (No need to be too fancy. For example, in the `datalist`, just type in y = c(1, 2, 3, 21, 22, 23, 41, 42, 43) and subj = c(1,1,1, 2,2,2, 3,3,3).) Be sure to make the prior appropriate for these data. Run the program twice, once for each assignment of data to subjects (using the

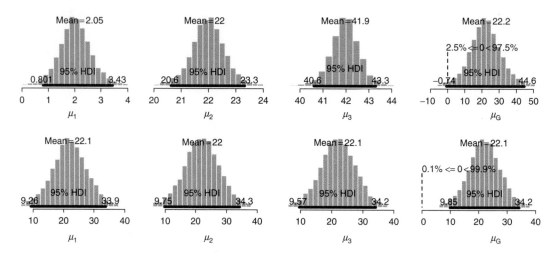

FIGURE 15.11

For Exercise 15.5, involving the same data rearranged across subjects. *Top row*: Large between-subject variance. *Bottom row*: Small between-subject variance. Notice the width of the HDI on the group-level parameter, μ_G.

same prior for both runs). (Hint: Your results should look something like those in Figure 15.11). *Notice that the mean of the posterior of μ_G is essentially the same for both runs, but the HDI widths are quite different. Explain why.*

CHAPTER 16

Metric Predicted Variable with One Metric Predictor

The agri-bank's threatnin' to revoke my lease
If my field's production don't rapid increase.
Oh Lord how I wish I could divine the trend,
Will my furrows deepen, and will my line end?

In this chapter we consider situations such as predicting a person's weight from his or her height, or predicting blood pressure from weight, or predicting income from years of education. In these situations, the predicted variable is metric and the single predictor is also metric. We will describe the relationship between the predicted variable, y, and predictor, x, with a simple linear model and normally distributed residual randomness in y. In terms of the generalized

Doing Bayesian Data Analysis: A Tutorial with R and BUGS. DOI: 10.1016/B978-0-12-381485-2.00016-X
© 2011, Elsevier Inc. All rights reserved.

linear model (GLM), the model assumes that $y \sim N(\mu, \tau)$ with $\mu = \beta_0 + \beta_1 x$. This model appears in Table 14.1 (p. 385) in the first row and second column. This model is often referred to as *simple linear regression*.

16.1 SIMPLE LINEAR REGRESSION

Figure 16.1 shows an example of simulated data generated from the assumed model, with parameter values displayed. At any value of x, the mean predicted value is $\mu = \beta_0 + \beta_1 x$, and the data values y are normally distributed around that mean. For a review of how to interpret the slope, β_1, and intercept, β_0, see Figure 14.1, p. 362.

Note that the model only specifies the dependency of y on x. The model does not say anything about what generates x, and there is no probability distribution assumed for describing x. The x values in the left panel of Figure 16.1 were sampled randomly from a uniform distribution, merely for purposes of illustration, whereas the x values in the right panel of Figure 16.1 were sampled randomly from a bimodal distribution. Both panels show data from the same model of the dependency of y on x.

It is important to emphasize that the model assumes *homogeneity of variance*: At every value of x, the variance of y is the same! This homogeneity of variance

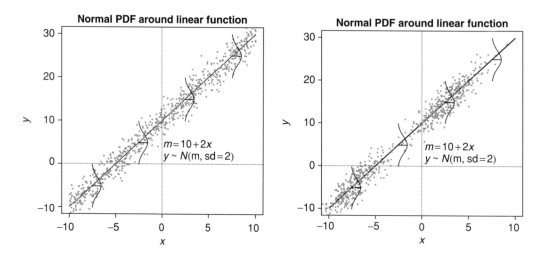

FIGURE 16.1

Examples of points normally distributed around a linear function. (The left panel repeats Figure 14.9, p. 380.) The model assumes that the data y are normally distributed vertically around the line, as shown. Moreover, the variance of y is the same at all values of x. The model puts no constraints on the distribution of x. The right panel shows a case in which x are distributed bimodally, whereas in the left panel the x are distributed uniformly. In both panels, there is homogeneity of variance.

is easy to see in the left panel of Figure 16.1 because the x values are uniformly distributed: The smattering of data points in the vertical, y, direction appears visually to be the same at all values of x. Homogeneity of variance is less easy to identify visually when the x values are not uniformly distributed. For example, the right panel of Figure 16.1 displays data that may appear to violate homogeneity of variance, because the apparent vertical spread of the data seems to be larger for $x = 2.5$ than for $x = 7.5$. Despite this deceiving appearance, the data do respect homogeneity of variance. The reason for the apparent violation is that for regions in which x is sparse, there is far less opportunity for the sampled y values to come from the tails of the normal distribution. In regions where x is dense, there are many opportunities for y to come from the tails.

In applications, the x and y values are provided by some real-world process. In the real-world process, there might or might not be any direct causal connection between x and y. It might be that x causes y, or y causes x, or some third factor causes both x and y, or x and y have no causal connection, or some combination of any or all of those! The simple linear model makes no claims about causal connections between x and y. The simple linear model merely describes a tendency for y values to be linearly related to x values, hence "predictable" from the x values.

As an example, suppose we have measurements of height, in inches, and weight, in pounds, for some randomly selected people. Figure 16.2 shows the height and weight values of the people. The data were generated from the program listed in Section 16.5.1 (HtWtDataGenerator.R), which uses realistic population parameters. The data points in Figure 16.2 do appear to indicate that as height increases, weight also tends to increase. This covariation between height and weight does not imply that one attribute causes the other. When an adult eats a lot of sugary foods or goes on a diet, thereby changing his or her weight, his or her height does not change. Despite the lack of direct causal relationship, the two values do covary, and one can be (imperfectly) predicted from the other.

Our goal is to determine what regression lines are most believable, given the data. In other words, we want to infer what combinations of β_0, β_1, and τ are most believable, given the data. We use Bayes' rule:

$$p(\beta_0, \beta_1, \tau \mid y) = p(y \mid \beta_0, \beta_1, \tau) \, p(\beta_0, \beta_1, \tau) \bigg/ \iiint d\beta_0 \, d\beta_1 \, d\tau \, p(y \mid \beta_0, \beta_1, \tau) \, p(\beta_0, \beta_1, \tau)$$

Analytical forms for the posterior can be obtained for appropriate priors. For example, analogous to the derivation in the previous chapter, if the precision τ is fixed (i.e., if the prior on τ is a spike over a certain value), then normal priors

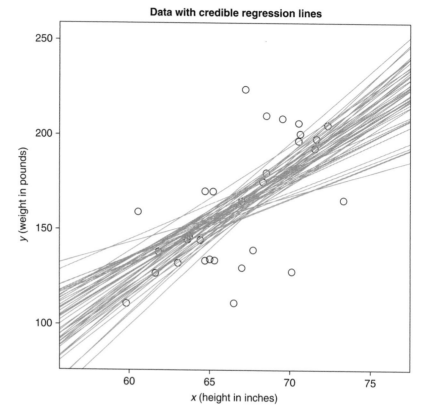

FIGURE 16.2

Data points, with a smattering of believable regression lines superimposed.

on the slope and intercept yield normal posteriors (e.g., see Bolstad, 2007). Fortunately, we do not have to worry much about mathematical derivations because we can let BUGS approximate the posterior. Our job, therefore, is to specify sensible priors and to make sure that BUGS generates a trustworthy posterior sample that is converged and well mixed.

16.1.1 The Hierarchical Model and BUGS Code

There are three parameters in the model; therefore we need to establish a prior distribution on the joint three-dimensional parameter space. To keep things simple, we will assume that the three parameters are independent in our prior beliefs. They would not need to be; for example, we might have a prior belief that the slope and intercept are (anti-)correlated, or we might have a prior belief that the precision and the intercept are correlated. For now, however, we assume independence in the prior, and therefore we can express the prior as three marginal distributions on the separate parameters.

Figure 16.3 shows the hierarchy of dependencies that we will assume. At the bottom of the figure, we see that the data depend on the parameters μ and τ, which describe the mean and precision of a normal distribution. In other words, the data are modeled by a normal likelihood function, as was emphasized in the previous chapter. The value of the precision, τ, depends on our prior belief distribution, shown in the upper right of the figure as a gamma distribution with two shape parameters. The shape parameters are constants that express what we think the precision is likely to be and how uncertain we are in that prior belief.

The mean, μ, of the likelihood function is a linear function of x, as shown in the middle of Figure 16.3. The linear function has two parameters, the intercept β_0 and the slope β_1. The slope parameter is what we are usually most interested in, because it describes the degree of influence between x and y. The values of the slope and intercept depend on the prior beliefs shown in the upper layer of the figure. Both the slope and the intercept here have prior beliefs modeled as normal distributions.

It is useful to compare Figure 16.3 with Figure 15.2, p. 396. The underlying approach is the same for both scenarios, in that there is a normal likelihood for the data, and we set priors on the mean and the precision. The only difference is that Figure 16.3 has one additional prior for the β_1 parameter. Otherwise, the models are identical.

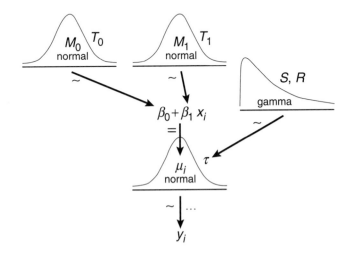

FIGURE 16.3
A model of dependencies for simple linear regression. The parameter β_1 is of primary interest; it describes the slope of the line relating x to y.

Every arrow in Figure 16.3 has a corresponding line in the BUGS model specification (SimpleLinearRegressionBrugs.R):

```
9    model {
10       for( i in 1 : Ndata ) {
11           y[i] ~ dnorm( mu[i] , tau )
12           mu[i] <- beta0 + beta1 * x[i]
13       }
14       beta0 ~ dnorm( 0 , 1.0E-12 )
15       beta1 ~ dnorm( 0 , 1.0E-12 )
16       tau ~ dgamma( 0.001 , 0.001 )
17   }
```

Notice that the constants in the hyperpriors are generic values for a diffuse prior. In real applications, you would probably want to use better informed constants.

16.1.1.1 Standardizing the Data for MCMC Sampling

In principle, we could run the BRugs code on the *raw x, y* data. In practice, however, the attempt often fails. There's nothing wrong with the mathematics or logic; the problem is that believable values of the slope and intercept parameters tend to be tightly correlated, and this narrow diagonal zone of believability is difficult for sampling algorithms to explore.

The right panel of Figure 16.4 shows an example of this type of correlation in believable values. The believable slopes and intercepts are extremely correlated. Sampling from such a tightly correlated distribution is typically difficult to do directly. It is difficult to discover a point in the narrow zone in the first place.

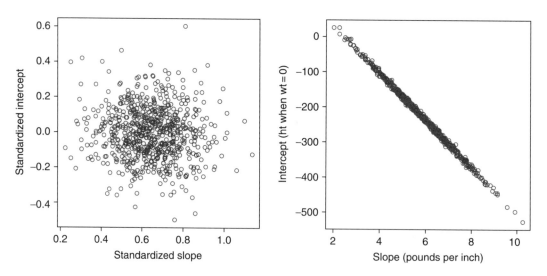

FIGURE 16.4

Slope and intercept values of believable regression lines in standardized and raw scales.

Then, having discovered a viable point, the chain does not move efficiently. Gibbs sampling gets stuck because it keeps "bumping into the walls" (recall the discussion in Section 8.4.2.1, p. 172). Metropolis algorithms often are not clever enough to automatically tune a proposal distribution to match a diagonal posterior.

The reason for the correlation of slope and intercept is evident by examining Figure 16.2. There you can see various believable lines that go through the scatter of points. Notice that if a line has a steep slope, its intercept must be low, but if a line has a smaller slope, its intercept must be higher. Thus, there is a trade-off in slope and intercept for the believable lines.

One of the main tricks used for successful execution of the MCMC sampling is standardizing the data. Standardizing simply means rescaling the data relative to their mean and standard deviation:

$$z(x) = \frac{(x - M_x)}{SD_x} \quad \text{and} \quad z(y) = \frac{(y - M_y)}{SD_y} \tag{16.1}$$

where M_x is the mean of the data x values and SD_x is the standard deviation of the data x values. (Do not confuse M_x and M_y with the constants used in the specification of the priors in the hierarchical diagram.) It is easy to prove, using simple algebra, that the mean of the resulting $z(x)$ values is zero, and the standard deviation of the resulting $z(x)$ values is one, for any initial data set.

Having used BUGS to find slope and intercept values for the standardized data, we then need to convert the parameter values back to the original raw scales. Denote the intercept and slope for standardized data as ζ_0 and ζ_1 (Greek letter "zeta"), and denote the predicted value of y as $\hat{y}$. Then

$$z_{\hat{y}} = \zeta_0 + \zeta_1 z_x \quad \text{by definition of the model}$$

$$\frac{(\hat{y} - M_y)}{SD_y} = \zeta_0 + \zeta_1 \frac{(x - M_x)}{SD_x} \quad \text{from Eqn. 16.1}$$

$$\hat{y} = \underbrace{\zeta_0 SD_y + M_y - \zeta_1 SD_y M_x / SD_x}_{\beta_0} + \underbrace{\zeta_1 SD_y / SD_x}_{\beta_1} x \tag{16.2}$$

Thus, for every believable combination of ζ_0, ζ_1 values, there is a corresponding believable combination of β_0, β_1 values specified by Equation 16.2.

The sampled points in Figure 16.4 were generated by running BUGS on the standardized data, then transforming the results according to Equation 16.2. Notice that the standardized slope and intercept, in the left panel of Figure 16.4, show no noticeable correlation. This is because the believable intercepts tend to be near zero, and that's because the standardized data have means of zero. Thus, even when the slopes of two believable lines differ, their intercepts still hover around zero.

16.1.1.2 Initializing the Chains

Figure 16.4 showed an example of how the believable values in a posterior distribution can occupy a fairly narrow region of parameter space. For the MCMC chain to randomly sample from the posterior, the random walk must first get into the modal region of the posterior in the first place. We might simply start the chain at any point in parameter space, randomly selected from the prior distribution, and wait through the burn-in period until the chain randomly wanders into the bulk of the posterior. Unfortunately, for many real-world situations, this burn-in period can be a *very* long time. Therefore, it helps to initialize the chains near the bulk of the posterior if we can.

If the data set is large and dominates the prior, or if the prior is diffuse, or if the prior is informed from previous results and is reasonably consistent with the new data, then the peak of the likelihood function will be reasonably near the peak of the posterior distribution. Therefore, a reasonable candidate for the initial point of the chain is the maximum-likelihood estimate of the parameters. This heuristic is only useful if we have a simple way of determining the maximum-likelihood estimate. Fortunately, for simple linear regression, we do. When the data are standardized, the maximum-likelihood estimate (MLE) of β_0 is zero, the MLE of β_1 is the correlation (denoted r) of x and y, and the MLE of the precision, τ, is $1/(1 - r^2)$. To get an intuition for the statement about precision, consider what happens when the correlation r approaches 1 (its maximum possible value). As r approaches 1, the x, y data fall very close to a straight line, which implies that the deviation of the data away from the line is very small (recall Figure 16.1). Hence when r is large, σ is small, and hence τ is large, as reflected by the formula $1/(1 - r^2)$.

16.1.2 The Posterior: How Big Is the Slope?

Figure 16.5 shows the posterior distribution of slope values. The standardized and original-scale slopes indicate the same relationship on different scales, and therefore the posterior distributions are identical except for a change of scale. The posterior distribution tells us exactly what we want to know: the believable slopes. We see that weight increases by about 5 or 6 pounds for every 1-inch increase in height. The 95% HDI provides a useful summary of the range of believable slopes.

If we were interested in determining whether the predictor had a nonzero influence on the predicted variable, we might use the decision rules discussed in Section 12.1.3 (p. 301). We may want to establish a ROPE around zero for the predictor and then check whether the entire 95% HDI excludes the ROPE. (Usually, if the true value *is* zero, the HDI will overlap the ROPE, thereby reducing false alarms.)

It would be possible to "test" whether the slope is nonzero by doing a Bayesian comparison of two models: One model would have an arbitrarily diffuse prior

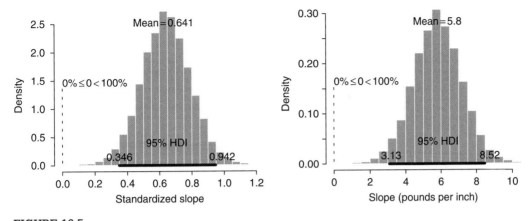

FIGURE 16.5
Posterior distribution of slopes.

on the slope parameter; the other model would have an arbitrarily peaked prior at zero. It would be straightforward, in principle, to set up the model comparison using transdimensional MCMC as described in Section 10.2 (p. 244). But as argued in Section 12.2.2 (p. 307) and elsewhere, all that this model comparison tells us is which of two unbelievable models is less unbelievable. In the present application, the model comparison would favor the model that allows nonzero slopes. But that is not all we want to know, because usually we also want to know what slope values *are* credible. That information is provided by the posterior distribution on the parameter values, often starting with an (mildly) informed prior distribution instead of an arbitrarily diffuse prior of mathematical convenience. Thus, in the present application, there is little to be gained by doing a Bayesian model comparison of null prior against arbitrary diffuse prior.

16.1.3 Posterior Prediction

Linear regression is often used for predicting *y* values from *x* values. For example, we can use the results of the regression to answer the question, If a person were 75 inches tall, what is the probable weight of the person? The Bayesian answer provides the probability of every possible weight *y*, given the height *x* and the previous data: $p(y \mid x, D)$. This distribution of *y* values has uncertainty stemming from the inherent noise τ and also from uncertainty in the estimated values of the regression coefficients.

A simple way to get a good approximation of $p(y \mid x, D)$ is by generating random values of *y* for every step in the MCMC sample of credible parameter values. Thus, at any step in the chain, there are particular values of β_0, β_1, and τ, which we use to generate representative predicted values of *y* according to $y \sim N(\mu = \beta_0 + \beta_1 x, \sigma = 1/\sqrt{\tau})$. We do that at all the steps in the chain and thereby create a large set of predicted y values. From this set, we can compute the mean predicted y value, the HDI of the predicted y values, etc.

A summary of the posterior predictions is displayed in Figure 16.6. Each vertical blue segment, over a particular x value, indicates the extent of the 95% HDI for the distribution of randomly generated y values at that x value. The dash across the middle of the blue segment indicates the mean of the posterior predicted y values.

Consider the limits of the HDIs at each value of x. Notice that the length of the HDIs is larger for high and low x values than for middling x values. To verify this fact, hold a ruler against the HDIs for small, middling, and large values of x. This variation in the length of the HDIs makes sense intuitively: As the x value goes farther from the observed data, the predicted y value should become less certain. Another way of understanding the variation in HDI lengths is by

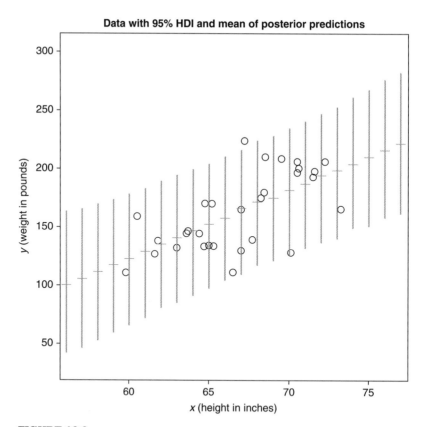

FIGURE 16.6

Data points, with 95% HDIs and means of posterior predictions in y, at selected x values. The HDIs are wider (i.e., taller) for extrapolated x values than for x values in the middle of the observed data.

considering the credible regression lines in Figure 16.2. All of those differently sloped lines go near the bulk of the data in the middle, but the different slopes extend higher or lower at extrapolative values of x. We average across all those lines to generate predicted y values. Therefore, the predicted y values will be more widespread at extrapolative values of x.

If the model of the data is a good model of the data, then believable parameter values for the model ought to generate simulated data that "look like" the real data. What it means for simulated data to "look like" the real data is defined by the analyst, and may be anything that is useful for the application at hand. Loosely speaking, a model is a good model if (1) the data fall mostly within the predicted zone, (2) the data are distributed approximately the way the model says they should be (e.g., normally), and (3) the discrepancies of data from predictions are random and not "systematic."

Posterior prediction is also useful for alerting us to nonlinear trends in data. Figure 16.7 shows a set of data for which the y values have a clear quadratic

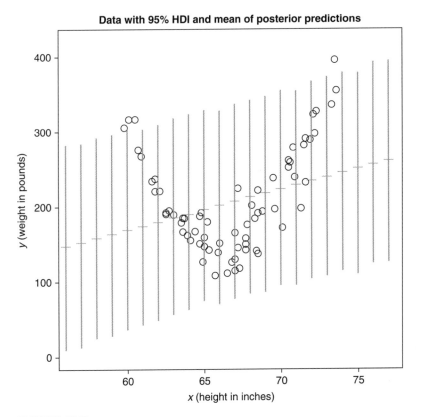

FIGURE 16.7
Data points, with 95% HDIs and means of posterior predictions in y, at selected x values. The data show systematic discrepancy from the linear-model predictions.

(i.e., parabolic) relationship with x, in addition to a linear trend. These data can be entered into the linear regression program, and the resulting posterior prediction HDIs are shown by the vertical blue segments in Figure 16.7. We see that the data do fall mostly within the range of the posterior predictions, but the distribution of the data is *systematically* discrepant from the linear spine of the model: At high and low values of x, the data fall well above the linear spine, but at middling values of x, the data fall well below the linear spine. This sort of systematic discrepancy from the posterior predictions is a clue that the model could be improved.

16.2 OUTLIERS AND ROBUST REGRESSION

Recall from Figure 15.5, p. 399, that the estimated parameter values for a normal likelihood function can be greatly distorted by outliers in the data. This sensitivity also occurs when the normal distribution is used in linear regression. The normal likelihood function demands that the regression line is vertically close to all the data points, because the likelihood value is tiny for points more than about three standard deviations from the line. Consequently, outlying data points can have disproportionate leverage in the estimate of the regression coefficients.

As an example, consider some data regarding 25 brands of cigarette (McIntyre, 1994). For each brand, a cigarette was assessed for its weight, amount of nicotine, amount of tar, and amount of carbon monoxide produced when burned. Many different relationships among these variables might be investigated, but consider in particular predicting the tar content from the weight of the cigarette. These data are plotted in the lower panels of Figure 16.8. The points are rather diversely scattered and do not show an overwhelmingly strong covariation. There is, however, an outlying point that has the heaviest weight and highest tar content (in the upper right of the scatter plot). This outlying point can have a disproportionate influence on the estimated slope (β_1) in the regression model.

The left column of Figure 16.8 shows the results when a normal likelihood function is used (actually, when the likelihood function is a t distribution with large df). The histogram in the upper panel shows that the slope is credibly larger than zero. The plot of posterior predictive HDIs in the lower panel shows that the estimated regression line has been tilted up toward the outlier. The attraction to the outlier is caused by the small tails of the normal likelihood function: For the outlying datum to be "under the umbrella" of the normal likelihood, the line must get fairly close.

The right column of Figure 16.8 shows the results when a t distribution is used as the likelihood function, with a prior on the df parameter that strongly favors

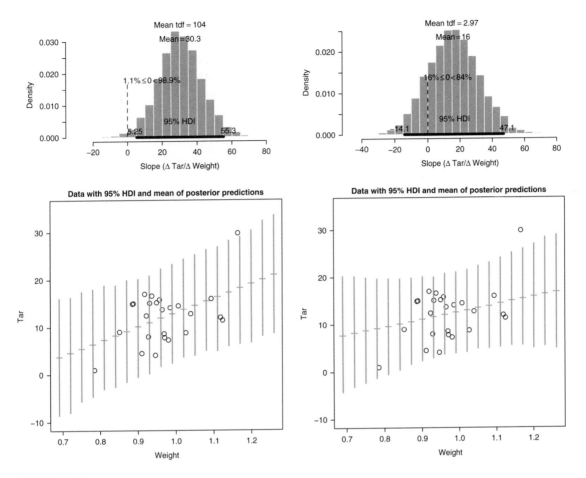

FIGURE 16.8
Data fit with different priors on the *df* parameter of the *t* likelihood function. Left column shows results from prior bias on high *df* values, such that a virtually normal likelihood function is assumed. Right column shows "robust" regression resulting from a strong prior bias on low *df* values.

small values. This prior implies that the likelihood function has the tall tails of a low-*df* *t* distribution, unless the data strongly suggest otherwise. As the figure shows, the 95% HDI of the posterior on the slope contains zero. The lower-right panel shows that the estimated regression line is not slanted as far toward the outlier, because the long tail of the *t* distribution gives the outlier some modest nonzero probability despite being so far from the mean of the *t* distribution. The posterior slope estimate here better reflects the bulk of the points in the scatter plot, and is not so strongly dominated by the outlier.

The BUGS model for robust regression is a simple extension of the model for the normal likelihood. We simply replace the normal distribution with a *t* distribution and include the necessary prior specification. Whereas the normal distribution has two parameters, namely, the mean μ and the precision τ, the *t* distribution has those two plus a third parameter, namely, the degrees of freedom *df*, which can take on real values of 1 or greater (for a reminder of how the *df* parameter works, see Figure 15.4, p. 397). One way to put a prior on the *df* parameter is by sampling a value, denoted udf, from a uniform distribution, and then transforming that value into the range allowed for *df*, which extends from 1 to infinity. One such transformation appears on line 18 of this model specification (SimpleRobustLinearRegressionBrugs.R):

```
9    model {
10       for( i in 1 : Ndata ) {
11          y[i] ~ dt( mu[i] , tau , tdf )
12          mu[i] <- beta0 + beta1 * x[i]
13       }
14       beta0 ~ dnorm( 0 , 1.0E-12 )
15       beta1 ~ dnorm( 0 , 1.0E-12 )
16       tau ~ dgamma( 0.001 , 0.001 )
17       udf ~ dunif(0,1)
18       tdf <- 1 - tdfGain * log(1-udf) # tdf in [1,Inf).
19       # tdfGain specified in data section
20    }
```

Notice that tdf is used as the *df* parameter in the *t* likelihood function specified on line 11. The value of tdf is determined from udf by the transformation tdf <- 1 - tdfGain * log(1 - udf), where tdfGain is a constant that expresses the prior belief in large values of *df*. Figure 16.9 shows a graph of the transformation. The graph shows that when tdfGain is small, the tdf values are close to 1 across almost the entire range of udf. But when tdfGain is large, then the tdf values are large across most of the range of udf.

The prior on the *df* parameter expresses how much we believe there are not outliers in the data. For example, the left column of Figure 16.8 was created by setting tdfGain=100, and the right column of Figure 16.8 was created by setting tdfGain=1. The resulting posteriors are noticeably different.

So, you may ask, which conclusion from Figure 16.8 is correct? Does the tar content of cigarettes increase by about 30 units for every unit increase in weight, as indicated by the nearly normal likelihood, or does the tar content increase by about only half that much, as indicated by the low-*df t* distribution? The answer is, in this case, that it depends on your prior beliefs. If you have a very good reason to believe that tar content should be normally distributed (with the same variance) at every level of weight, then the left column of Figure 16.8 describes your posterior beliefs. But the assumption of normality virtually disavows the possibility of outliers in the data and therefore seems untenable, at least for this application. Therefore, the prior should allow low values of *df*. The dependency of the posterior on the prior suggests that either

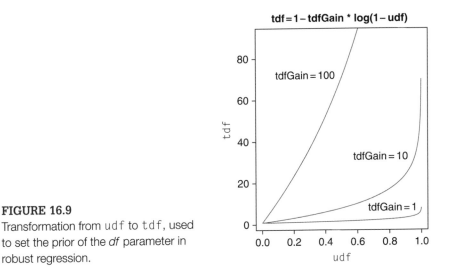

FIGURE 16.9

Transformation from `udf` to `tdf`, used to set the prior of the *df* parameter in robust regression.

more specific prior knowledge should be brought to bear, additional predictors should be included if available, or more data need to be collected.

16.3 SIMPLE LINEAR REGRESSION WITH REPEATED MEASURES

Suppose that for every individual, j, we have multiple observations of x_{ij}, y_{ij} pairs. With these data, we can estimate a linear model for every individual. If we also assume that the individuals are mutually representative of a common group, then we can use the estimates from the individuals to inform estimates of group-level parameters.

One example of this scenario comes from Walker, Gustafson, & Frimer (2007), who measured reading ability scores of children across several years. Thus each child contributed several age and reading-score pairs. A regression line can describe each child's reading ability through time, and higher-level distributions describe the distribution of intercepts and slopes across individuals. By virtue of being linked indirectly through the higher-level distribution, estimates of each individual are mutually informed by other individuals.

As another concrete example of this situation, consider an experiment in which the investigators were interested in how quickly different organs clear themselves of contaminants (Feldman, 1988, reported data from an unpublished experiment by S. B. Weinstock and J. D. Brain). The researchers administered iron oxide particles to rats, because the iron oxide remaining in the body could be assayed noninvasively via magnetometry. Four rats were given intravenous injection of iron oxide, the particles of which are taken up by liver endothelial cells. Four other rats were given the iron oxide by tracheal instillation, so that the particles were taken up by lung macrophages. Although the researchers were specifically interested in comparing the clearance rates of the two groups, we will consider only the "lung" group.

The amount of iron oxide retained in the body, as a percentage of the initially assayed amount, was measured at various times during the following 30 days. The data for the lung group are shown in the top row of Figure 16.10. The retention amount is plotted on a logarithmic scale, so that the retention curves are approximately linear. Notice that all the curves start at a y value of 2.0, because the first measurement establishes the baseline that defines 100%, and $\log_{10}(100) = 2.0$. Some curves rise above the initial value, which presumably does not indicate spontaneous creation of iron, but instead indicates measurement noise, either at the initial or subsequent times. The graphs indicate a reduction through time. The goal of our analysis is to determine what the believable rates of reduction are, given the data. We may also want to know if the apparent reduction really is believably nonzero.

To model this situation, each subject's data set is described by an individual linear regression, and the regression coefficients of the individuals are, in turn, modeled by group-level distributions. The group-level distributions are controlled by parameters that describe the central tendency and variability of the group, and it is these group-level parameters in which we are primarily interested.

Figure 16.11 shows the hierarchy of dependencies. At the lowest level, we just see a simple linear regression for each individual, with the same components as Figure 16.3. The regression coefficients for the jth individual, namely, β_{0j}, β_{1j}, and τ_j, in turn come from distributions that describe the group. For example, the individual slope coefficients, β_{1j}, are assumed to come from a normal distribution, with mean μ_{1G} and precision τ_{1G}. We are interested in estimating those group-level parameters; therefore each is given a prior distribution at the top level of the diagram.

Every arrow in Figure 16.11 has a corresponding line in the BUGS model specification (SimpleLinearRegressionRepeatedBrugs.R):

```
8    model {
9        for( r in 1 : Ndata ) {
10            y[r] ~ dnorm( mu[r] , tau[ subj[r] ] )
11            mu[r] <- b0[ subj[r] ] + b1[ subj[r] ] * x[r]
12        }
13        for ( s in 1 : Nsubj ) {
14            b0[s] ~ dnorm( mu0G , tau0G )
15            b1[s] ~ dnorm( mu1G , tau1G )
16            tau[s] ~ dgamma( sG , rG )
17        }
18        mu0G ~ dnorm(0,.01)
```

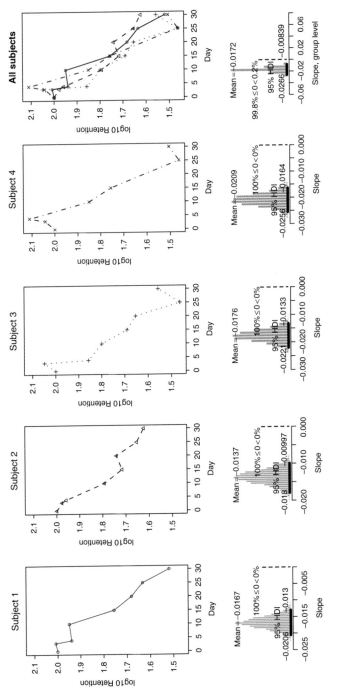

FIGURE 16.10

Upper row shows data from Feldman (1988). Lower row shows believable values of the slopes.

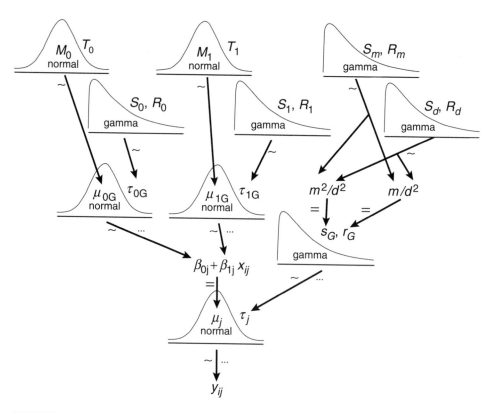

FIGURE 16.11

A model of dependencies for repeated scores from N subjects drawn independently from the same group. The slope for the jth subject is β_{1j}. Across subjects, it is distributed normally, with mean μ_{1G} and precision τ_{1G} (i.e., across subjects the variance of the slopes is $1/\tau_{1G}$).

```
19      tau0G ~ dgamma(.1,.1)
20      mu1G ~ dnorm(0,.01)
21      tau1G ~ dgamma(.1,.1)
22      sG <- pow(m,2)/pow(d,2)
23      rG <- m/pow(d,2)
24      m ~ dgamma(1,.1)
25      d ~ dgamma(1,.1)
26   }
```

The constants in the top-level priors are generically vague because of laziness during programming. More thoughtful priors may be desirable.

The results of running the program are shown in the lower row of Figure 16.10. The results match closely those reported in the non-Bayesian analysis of Feldman (1988), who reported -0.0173 for the group slope. The results also show that every individual slope is believably different from zero, as is the estimate of the group average slope.

The four rats in this experiment had remarkably similar clearance rates. In many realms of research, there is much more variation between subjects. Suppose that different rats had different clearance rates, with some faster and some slower than the rats reported here, such that the group-average clearance rate was about the same. What aspects of the posterior would be most affected by the increase in variability of individual slopes? Would the certainty of individual slope estimates be much affected? Would the certainty of the group-average slope be much affected? Exercise 16.1 has you systematically alter the data to answer these questions. It is also a good intuition-stretching exercise to consider the influence of differing individual intercepts. In particular, without changing the individual slopes, but merely by changing the individual intercepts, the estimate of the group-average slope can be affected. See Exercise 16.2.

The data in this example were well behaved, insofar as there were no blatantly outlying data points. Specifically, within each individual, all the values descended in a reasonably linear trend. And, across subjects, all the individual slopes were reasonably similar to each other, with no subjects who had clearance rates much faster or much slower than the others. With a larger sample, however, outlying subjects might be encountered. And there is always the chance that device malfunctions or transcription errors could be inadvertently introduced into individual data points. Whenever there is reason to believe that outliers may be lurking in the data, t distributions can be substituted for the normal distributions at the appropriate level. Exercise 16.3 shows an example of outlying slopes and intercepts, involving income and family size. A generalization of the model described in this section can be found in Section 19.1.4.1, p. 525, regarding analysis of covariance.

16.4 SUMMARY

This chapter explained Bayesian simple linear regression, where "simple" means it involves a single predictor. There were several main points:

- Prediction of y from x does not imply or assume any particular causal or temporal relation between y and x.
- The usual model for linear regression assumes homogeneity of variance: The variance of y is the same at all values of x.
- It is typical, but not necessary, to put normal priors on the intercept and slope parameters in the linear regression model.
- To make MCMC sampling efficient, it is helpful to standardize the y and x data, and to initialize the chains at the MLE values.

- Posterior predictions are useful for predicting *y* values from *x* values, especially when extrapolating or interpolating to novel *x* values. Posterior predictions are also useful for checking whether there are systematic discrepancies of the data from the predictions of the model.
- It is straightforward to extend the model to situations with repeated measures, in which every individual has data estimated by simple linear regression, and overarching parameters describe the distribution of slope, intercept, and precision parameters of the group.
- When there are outliers in repeated measures from an individual, or in the distribution of parameters across individuals, a *t* distribution may be used instead of a normal distribution.

The next chapter introduces additional predictors into the model and explores the concept of interaction across predictors.

16.5 R CODE

16.5.1 Data Generator for Height and Weight

(HtWtDataGenerator.R)

```
1   HtWtDataGenerator = function( nSubj , rndsd=NULL ) {
2   # Random height, weight generator for males and females. Uses parameters from
3   # Brainard, J. & Burmaster, D. E. (1992). Bivariate distributions for height and
4   # weight of men and women in the United States. Risk Analysis, 12(2), 267-275.
5   # Kruschke, J. K. (2010). Doing Bayesian data analysis:
6   # A Tutorial with R and BUGS. Academic Press / Elsevier.
7
8   require(MASS)
9
10  # Specify parameters of multivariate normal (MVN) distributions.
11  # Men:
12  HtMmu = 69.18
13  HtMsd = 2.87
14  lnWtMmu = 5.14
15  lnWtMsd = 0.17
16  Mrho = 0.42
17  Mmean = c( HtMmu , lnWtMmu )
18  Msigma = matrix( c( HtMsd^2 , Mrho * HtMsd * lnWtMsd ,
19                     Mrho * HtMsd * lnWtMsd , lnWtMsd^2 ) , nrow=2 )
20  # Women cluster 1:
21  HtFmu1 = 63.11
22  HtFsd1 = 2.76
23  lnWtFmu1 = 5.06
24  lnWtFsd1 = 0.24
25  Frho1 = 0.41
26  prop1 = 0.46
27  Fmean1 = c( HtFmu1 , lnWtFmu1 )
28  Fsigma1 = matrix( c( HtFsd1^2 , Frho1 * HtFsd1 * lnWtFsd1 ,
29                      Frho1 * HtFsd1 * lnWtFsd1 , lnWtFsd1^2 ) , nrow=2 )
30  # Women cluster 2:
```

```
31   HtFmu2 = 64.36
32   HtFsd2 = 2.49
33   lnWtFmu2 = 4.86
34   lnWtFsd2 = 0.14
35   Frho2 = 0.44
36   prop2 = 1 - prop1
37   Fmean2 = c( HtFmu2 , lnWtFmu2 )
38   Fsigma2 = matrix( c( HtFsd2^2 , Frho2 * HtFsd2 * lnWtFsd2 ,
39                        Frho2 * HtFsd2 * lnWtFsd2 , lnWtFsd2^2 ) , nrow=2 )
40
41   # Randomly generate data values from those MVN distributions.
42   if ( !is.null( rndsd ) ) { set.seed( rndsd ) }
43   datamatrix = matrix( 0 , nrow=nSubj , ncol=3 )
44   colnames(datamatrix) = c( "male" , "height" , "weight" )
45   maleval = 1 ; femaleval = 0 # arbitrary coding values
46   for ( i in 1:nSubj ) {
47       # Flip coin to decide sex
48       sex = sample( c(maleval,femaleval) , size=1 , replace=TRUE , prob=c(.5,.5) )
49       if ( sex == maleval ) { datum = mvrnorm(n = 1, mu=Mmean, Sigma=Msigma ) }
50       if ( sex == femaleval ) {
51           Fclust = sample( c(1,2) , size=1 , replace=TRUE , prob=c(prop1,prop2) )
52           if ( Fclust == 1 ) { datum = mvrnorm(n = 1, mu=Fmean1, Sigma=Fsigma1 ) }
53           if ( Fclust == 2 ) { datum = mvrnorm(n = 1, mu=Fmean2, Sigma=Fsigma2 ) }
54       }
55       datamatrix[ i , ] = c( sex , round( c( datum[1] , exp( datum[2] ) ) , 1 ) )
56   }
57
58   return( datamatrix )
59   } # end function
```

16.5.2 BRugs: Robust Linear Regression

(SimpleRobustLinearRegressionBrugs.R)

```
1    graphics.off()
2    rm(list=ls(all=TRUE))
3    library(BRugs)              # Kruschke, J. K. (2010). Doing Bayesian data analysis:
4                                # A Tutorial with R and BUGS. Academic Press / Elsevier.
5    #-------------------------------------------------------------------------
6    # THE MODEL.
7    modelstring = "
8    # BUGS model specification begins here...
9    model {
10       for( i in 1 : Ndata ) {
11           y[i] ~ dt( mu[i] , tau , tdf )
12           mu[i] <- beta0 + beta1 * x[i]
13       }
14       beta0 ~ dnorm( 0 , 1.0E-12 )
15       beta1 ~ dnorm( 0 , 1.0E-12 )
16       tau ~ dgamma( 0.001 , 0.001 )
17       udf ~ dunif(0,1)
18       tdf <- 1 - tdfGain * log(1-udf) # tdf in [1,Inf).
19       # tdfGain specified in data section
```

```
20    }
21    # ... end BUGS model specification
22    " # close quote for modelstring
23    writeLines(modelstring,con="model.txt")
24    modelCheck( "model.txt" )
25
26    #------------------------------------------------------------------------
27    # THE DATA.
28
29    cigData = read.csv(file="McIntyre1994data.csv")
30    nSubj = NROW(cigData)
31    x = cigData[,"Wt"]
32    xName="Weight"
33    y = cigData[,"Tar"]
34    yName="Tar"
35
36    # Re-center data at mean, to reduce autocorrelation in MCMC sampling.
37    # Standardize (divide by SD) to make initialization easier.
38    xM = mean( x ) ; xSD = sd( x )
39    yM = mean( y ) ; ySD = sd( y )
40    zx = ( x - xM ) / xSD
41    zy = ( y - yM ) / ySD
42
43    # Specify data, as a list.
44    tdfGain = 1 # 1 for low-baised tdf, 100 for high-biased tdf
45    datalist = list(
46      x = zx ,
47      y = zy ,
48      Ndata = nSubj ,
49      tdfGain = tdfGain
50    )
51    # Get the data into BRugs:
52    modelData( bugsData( datalist ) )
53
54    #------------------------------------------------------------------------
55    # INTIALIZE THE CHAINS.
56
57    nchain = 3
58    modelCompile( numChains = nchain )
59
60    genInitList <- function() {
61        r = cor(x,y)
62        list(
63            beta0 = 0 ,    # because data are standardized
64            beta1 = r ,       # because data are standardized
65            tau = 1 / (1-r^2) , # because data are standardized
66            udf = 0.95 # tdf = 4
67        )
68    }
69    for ( chainIdx in 1 : nchain ) {
70        modelInits( bugsInits( genInitList ) )
71    }
72
73    #------------------------------------------------------------------------
```

```
74    # RUN THE CHAINS
75
76    # burn in
77    BurnInSteps = 100
78    modelUpdate( BurnInSteps )
79    # actual samples
80    samplesSet( c( "beta0" , "beta1" , "tau" , "tdf" ) )
81    stepsPerChain = ceiling(10000/nchain)
82    thinStep = 10
83    modelUpdate( stepsPerChain , thin=thinStep )
84
85    #------------------------------------------------------------------
86    # EXAMINE THE RESULTS
87
88    source("plotChains.R")
89
90    fname = paste("SimpleRobustLinearRegressionBrugsTdfGain",tdfGain,
91                  sep="")
92    #beta0Sum = plotChains( "beta0" , saveplots=F , filenameroot=fname )
93    #beta1Sum = plotChains( "beta1"    , saveplots=F , filenameroot=fname )
94    #tauSum   = plotChains( "tau"      , saveplots=F , filenameroot=fname )
95    #tdfSum   = plotChains( "tdf"      , saveplots=F , filenameroot=fname )
96
97    # Extract chain values:
98    tdfSamp = samplesSample( "tdf" )
99    tdfM = mean( tdfSamp )
100   z0 = samplesSample( "beta0" )
101   z1 = samplesSample( "beta1" )
102   zTau = samplesSample( "tau" )
103   zSigma = 1 / sqrt( zTau ) # Convert precision to SD
104
105   # Convert to original scale:
106   b1 = z1 * ySD / xSD
107   b0 = ( z0 * ySD + yM - z1 * ySD * xM / xSD )
108   sigma = zSigma * ySD
109
110   # Posterior prediction:
111   # Specify x values for which predicted y's are needed:
112   xRang = max(x)-min(x)
113   yRang = max(y)-min(y)
114   limMult = 0.25
115   xLim= c( min(x)-limMult*xRang , max(x)+limMult*xRang )
116   yLim= c( min(y)-limMult*yRang , max(y)+limMult*yRang )
117   yLim = c(-10,35)
118   xPostPred = seq( xLim[1] , xLim[2] , length=20 )
119   # Define matrix for recording posterior predicted y values at each x value.
120   # One row per x value, with each row holding random predicted y values.
121   postSampSize = length(b1)
122   yPostPred = matrix( 0 , nrow=length(xPostPred) , ncol=postSampSize )
123   # Define matrix for recording HDI limits of posterior predicted y values:
124   yHDIlim = matrix( 0 , nrow=length(xPostPred) , ncol=2 )
125   # Generate posterior predicted y values.
126   # This gets only one y value, at each x, for each step in the chain.
127   for ( chainIdx in 1:postSampSize ) {
```

```
128        yPostPred[,chainIdx] = rnorm( length(xPostPred) ,
129                            mean = b0[chainIdx] + b1[chainIdx] * xPostPred ,
130                            sd = rep( sigma[chainIdx] , length(xPostPred) ) ) )
131    }
132    source("HDIofMCMC.R")
133    for ( xIdx in 1:length(xPostPred) ) {
134        yHDIlim[xIdx,] = HDIofMCMC( yPostPred[xIdx,] )
135    }
136
137    # Display believable beta0 and b1 values
138    windows()
139    par( mar=c(4,4,1,1)+0.1 , mgp=c(2.5,0.8,0) )
140    #layout( matrix(1:2,nrow=1) )
141    thinIdx = seq(1,length(b0),length=700)
142    #plot( z1[thinIdx] , z0[thinIdx] , cex.lab=1.75 ,
143    #       ylab="Standardized Intercept" , xlab="Standardized Slope" )
144    plot( b1[thinIdx] , b0[thinIdx] , cex.lab=1.75 ,
145           ylab="Intercept" , xlab="Slope" )
146    dev.copy2eps(file=paste(fname,"SlopeIntercept.eps",sep=""))
147
148    # Display the posterior of the b1:
149    source("plotPost.R")
150    windows(7,4)
151    par( mar=c(4,4,1,1)+0.1 , mgp=c(2.5,0.8,0) )
152    #layout( matrix(1:2,nrow=1) )
153    #histInfo = plotPost( z1 , xlab="Standardized slope" , compVal=0.0 ,
154    #                        breaks=30   )
155    histInfo = plotPost( b1 , main=bquote("Mean tdf"==.(signif(tdfM,3))) , cex.main=2 ,
156                        xlab=bquote("Slope (" * Delta * .(yName) / Delta * .(xName)
157                                        * ")") , compVal=0.0 , breaks=30   )
158    dev.copy2eps(file=paste(fname,"PostSlope.eps",sep=""))
159
160    # Display data with believable regression lines and posterior predictions.
161    windows()
162    par( mar=c(3,3,2,1)+0.5 , mgp=c(2.1,0.8,0) )
163    # Plot data values:
164    plot( x , y , cex=1.5 , lwd=2 , col="black" , xlim=xLim , ylim=yLim ,
165           xlab=xName , ylab=yName , cex.lab=1.5 ,
166           main="Data with credible regression lines" , cex.main=1.33   )
167    # Superimpose a smattering of believable regression lines:
168    for ( i in seq(from=1,to=length(b0),length=50) ) {
169        abline( b0[i] , b1[i] , col="grey" )
170    }
171    dev.copy2eps(file=paste(fname,"DataLines.eps",sep=""))
172
173    # Display data with HDIs of posterior predictions.
174    windows()
175    par( mar=c(3,3,2,1)+0.5 , mgp=c(2.1,0.8,0) )
176    # Plot data values:
177    #yLim= c( min(c(yHDIlim,y)) , max(c(yHDIlim,y)) )
178    plot( x , y , cex=1.5 , lwd=2 , col="black" , xlim=xLim , ylim=yLim ,
179           xlab=xName , ylab=yName , cex.lab=1.5 ,
180           main="Data with 95% HDI & Mean of Posterior Predictions" , cex.main=1.33   )
```

```
181    # Superimpose posterior predicted 95% HDIs:
182    segments( xPostPred, yHDIlim[,1] , xPostPred, yHDIlim[,2] , lwd=3, col="grey" )
183    points( xPostPred , rowMeans( yPostPred ) , pch="+" , cex=2 , col="grey" )
184    dev.copy2eps(file=paste(fname,"DataPred.eps",sep=""))
185
186    #-------------------------------------------------------------------------
```

16.5.3 BRugs: Simple Linear Regression with Repeated Measures

(SimpleLinearRegressionRepeatedBrugs.R)

```
1     graphics.off()
2     rm(list=ls(all=TRUE))
3     library(BRugs)                 # Kruschke, J. K. (2010). Doing Bayesian data analysis:
4                                    # A Tutorial with R and BUGS. Academic Press / Elsevier.
5     #-------------------------------------------------------------------------
6     # THE MODEL.
7     modelstring = "
8     model {
9         for( r in 1 : Ndata ) {
10            y[r] ~ dnorm( mu[r] , tau[ subj[r] ] )
11            mu[r] <- b0[ subj[r] ] + b1[ subj[r] ] * x[r]
12        }
13        for ( s in 1 : Nsubj ) {
14            b0[s] ~ dnorm( mu0G , tau0G )
15            b1[s] ~ dnorm( mu1G , tau1G )
16            tau[s] ~ dgamma( sG , rG )
17        }
18        mu0G ~ dnorm(0,.01)
19        tau0G ~ dgamma(.1,.1)
20        mu1G ~ dnorm(0,.01)
21        tau1G ~ dgamma(.1,.1)
22        sG <- pow(m,2)/pow(d,2)
23        rG <- m/pow(d,2)
24        m ~ dgamma(1,.1)
25        d ~ dgamma(1,.1)
26    }
27    " # close quote for modelstring
28    writeLines(modelstring,con="model.txt")
29    modelCheck( "model.txt" )
30
31    #-------------------------------------------------------------------------
32    # THE DATA.
33
34    # Data from H. A. Feldman, 1988, Table 4, p. 1731.
35    # Columns are "group" , "subjID" , "time" , "retention"
36    source("Feldman1988Table4data.R")
37    # Remove missing data:
38    includeRowVec = is.finite( Feldman1988Table4data[,"retention"] )
39    dataMat = Feldman1988Table4data[ includeRowVec , ]
40    # Retain only the Group 1 (lung) data:
```

```
41    dataMat = dataMat[ dataMat[,"group"]==1 , ]
42    # Convert to log10(retention):
43    dataMat[,"retention"] = log10( dataMat[,"retention"] )
44    # Column names and plot labels
45    yColName = "retention" ; yPlotLab = "log10 Retention"
46    xColName = "time" ; xPlotLab = "Day"
47    subjColName = "subjID" ; subjPlotLab = "Subject"
48    fname = "SimpleLinearRegressionRepeatedBrugs"
49
50    if ( F ) { # change to T to use income data instead of contam.retention data.
51      # Data from http://www.census.gov/hhes/www/income/statemedfaminc.html
52      # Downloaded Dec. 06, 2009.
53      load("IncomeFamszState.Rdata") # loads IncomeFamszState
54      dataMat = IncomeFamszState
55      yColName="Income" ; yPlotLab = "Income"
56      xColName="Famsz" ; xPlotLab="Family Size"
57      subjColName="State" ; subjPlotLab="State"
58      fname = "IncomeFamszState"
59    }
60
61    # Extract data info to pass to BUGS:
62    Ndata = NROW(dataMat)
63    subj = as.integer(factor(dataMat[,subjColName]))
64    Nsubj = length(unique(subj))
65    x = as.numeric(dataMat[,xColName])
66    y = as.numeric(dataMat[,yColName])
67
68    # Re-center data at mean, to reduce autocorrelation in MCMC sampling.
69    # Standardize (divide by SD) to make initialization easier.
70    xM = mean( x ) ; xSD = sd( x )
71    yM = mean( y ) ; ySD = sd( y )
72    zx = ( x - xM ) / xSD
73    zy = ( y - yM ) / ySD
74
75    # Specify data, as a list.
76    datalist = list(
77      Ndata = Ndata ,
78      Nsubj = Nsubj ,
79      subj = subj ,
80      x = zx ,
81      y = zy
82    )
83    # Get the data into BRugs:
84    modelData( bugsData( datalist ) )
85
86    #-----------------------------------------------------------------------------------
87    # INTIALIZE THE CHAINS.
88
89    nchain = 3
90    modelCompile( numChains = nchain )
91
92    genInitList <- function() {
93      b0 = b1 = tau = rep(0,length=Nsubj)
94      for ( sIdx in 1:Nsubj ) {
```

```
95          yVec = datalist$y[datalist$subj==sIdx]
96          xVec = datalist$x[datalist$subj==sIdx]
97          lmInfo = lm( yVec ~ xVec )
98          b0[sIdx] = lmInfo$coef[1]
99          b1[sIdx] = lmInfo$coef[2]
100          tau[sIdx] = length(yVec) / sum(lmInfo$res^2)
101      }
102      mu0G = mean(b0)
103      tau0G = 1/sd(b0)^2
104      mu1G = mean(b1)
105      tau1G = 1/sd(b1)^2
106      m = mean(tau)
107      d = sd(tau)
108      list( b0=b0 , b1=b1 , tau=tau ,
109            mu0G=mu0G , tau0G=tau0G ,
110            mu1G=mu1G , tau1G=tau1G ,
111            m=m , d=d )
112  }
113  for ( chainIdx in 1 : nchain ) {
114      modelInits( bugsInits( genInitList ) )
115  }
116
117  #-------------------------------------------------------------------------
118  # RUN THE CHAINS
119
120  # burn in
121  BurnInSteps = 500
122  modelUpdate( BurnInSteps )
123  # actual samples
124  samplesSet( c( "b0","b1","tau" , "mu0G","tau0G", "mu1G","tau1G", "m","d" ) )
125  stepsPerChain = ceiling(5000/nchain)
126  thinStep = 100 # 40 or more
127  modelUpdate( stepsPerChain , thin=thinStep )
128
129  #-------------------------------------------------------------------------
130  # EXAMINE THE RESULTS
131
132  source("plotChains.R")
133  source("plotPost.R")
134
135  # Check convergence and autocorrelation:
136  checkConvergence = T  # check this first time through, examine m,d,tau0G,tau1G
137  if ( checkConvergence ) {
138      # check a few selected chains
139      b01Sum   = plotChains( "b0[1]"  , saveplots=F , filenameroot=fname )
140      b11Sum   = plotChains( "b1[1]"  , saveplots=F , filenameroot=fname )
141      tau1Sum  = plotChains( "tau[1]" , saveplots=F , filenameroot=fname )
142      mu0GSum  = plotChains( "mu0G"   , saveplots=F , filenameroot=fname )
143      tau0GSum = plotChains( "tau0G"  , saveplots=F , filenameroot=fname )
144      mu1GSum  = plotChains( "mu1G"   , saveplots=F , filenameroot=fname )
145      tau1GSum = plotChains( "tau1G"  , saveplots=F , filenameroot=fname )
146      mSum     = plotChains( "m"      , saveplots=F , filenameroot=fname )
147      dSum     = plotChains( "d"      , saveplots=F , filenameroot=fname )
148  }
```

```
149
150    # Extract chain values for subsequent examination:
151    zmu0Gsamp = samplesSample( "mu0G" )
152    zmu1Gsamp = samplesSample( "mu1G" )
153    zb0samp = NULL
154    zb1samp = NULL
155    for ( subjIdx in 1:Nsubj ) {
156        zb0samp = rbind( zb0samp , samplesSample( paste("b0[",subjIdx,"]",sep="") ))
157        zb1samp = rbind( zb1samp , samplesSample( paste("b1[",subjIdx,"]",sep="") ))
158    }
159
160    # Convert to original scale:
161    mu0Gsamp = zmu0Gsamp * ySD + yM - zmu1Gsamp * ySD * xM / xSD
162    mu1Gsamp = zmu1Gsamp * ySD / xSD
163    b1samp   = zb1samp * ySD / xSD
164
165    # Display believable intercept and slope values
166    windows(10,5.5)
167    par( mar=c(4,4,1.75,1)+0.1 , mgp=c(2.5,0.8,0) )
168    layout( matrix(1:2,nrow=1) )
169    thinIdx = round(seq(1,length(mu0Gsamp),length=700))
170    plot( zmu1Gsamp[thinIdx] , zmu0Gsamp[thinIdx] , cex.lab=1.75 ,
171          ylab="Standardized Intercept" , xlab="Standardized Slope" )
172    plot( mu1Gsamp[thinIdx] , mu0Gsamp[thinIdx] , cex.lab=1.0 ,
173          ylab=paste("Intercept (",yPlotLab," when ",xPlotLab," =0)",sep="") ,
174          xlab=paste("Slope (change in",yPlotLab,"per unit",xPlotLab,")") )
175    dev.copy2eps(file=paste(fname,"SlopeIntercept.eps",sep=""))
176
177    # Make graphs of data and corresponding believable slopes:
178    windows(12,6)
179    par( mar=c(4,4,1.75,1)+0.1 , mgp=c(2.5,0.8,0) )
180    layout(matrix(c(1:5,1:5,6:10),nrow=3,byrow=T))
181    xlims = c( min( dataMat[,xColName] ) ,  max( dataMat[,xColName] ) )
182    ylims = c( min( dataMat[,yColName] ) ,  max( dataMat[,yColName] ) )
183    sIdVec = unique( dataMat[,subjColName] )
184    # Plot data of individual subjects:
185    nSubjPlots = 4 # number of representative subject plots to make
186    subjIdxVec = round(seq(1,length(sIdVec),length=nSubjPlots))
187    for ( sIdx in subjIdxVec ) {
188        rVec = ( dataMat[,subjColName] == sIdVec[sIdx] )
189        plot( dataMat[rVec,xColName] , dataMat[rVec,yColName] , type="o" ,
190              ylim=ylims , ylab=yPlotLab , xlab=xPlotLab , cex.lab=1.5 ,
191              pch=sIdx%%26 , lty=sIdx , main=bquote(.(subjPlotLab) * " " * .(sIdx)) ,
192              cex.main=1.75 )
193    }
194    # Plot data of all subjects superimposed
195    plot( NULL,NULL, xlab=xPlotLab,xlim=xlims , ylab=yPlotLab,ylim=ylims ,
196          cex.lab=1.5 , main=paste("All ",subjPlotLab,"s",sep="") , cex.main=1.75 )
197    for ( sIdx in 1:length(sIdVec) ) {
198        rVec = ( dataMat[,subjColName] == sIdVec[sIdx] )
199        lines( dataMat[rVec,xColName] , dataMat[rVec,yColName] ,
200               lty=sIdx , pch=sIdx%%26 , type="o")
201    }
202    # Plot histograms of corresponding posterior slopes:
```

```
203  for ( sIdx in subjIdxVec ) {
204      histInfo = plotPost( blsamp[sIdx,] , xlab="Slope" , compVal=0.0 , breaks=30 ,
205                  HDItextPlace=0.9 )
206  }
207  histInfo = plotPost( mu1Gsamp , xlab="Slope, Group Level" , compVal=0.0 ,
208                  breaks=30 , HDItextPlace=0.9 )
209  dev.copy2eps(file=paste(fname,"Data.eps",sep=""))
210
211  #- - - - - - - - - - - - - - - - - - - - - - - - - - - - - - - - - - - - - - - - - - - - - - - - - - - - - - - - - - -
```

16.6 EXERCISES

Exercise 16.1. [Purpose: To see the influence of individual slope differences on the estimate of the group-average slope.]

The data shown in Figure 16.10 indicate that all subjects had similar rates of decline in retention, and therefore the estimate of the group average is fairly certain. In this exercise, we change the data so that the individual slopes differ more dramatically and examine the effect on the estimate of the group average.

(A) Alter the data as follows: In the program listed in Section 16.5.3 (SimpleLinearRegressionRepeatedBrugs.R), just before the data are renamed from dataMat to x and y (at about line 65), subtract .030x from subject 1, subtract .015x from subject 2, do nothing to subject 3, and add .015x to subject 4. Here is an example of code for subject 1:

```
subjRowVec = ( dataMat[,subjColName] == 1 )
dataMat[ subjRowVec , yColName ] = ( dataMat   [ subjRowVec , yColName ]
                         - .030 * dataMat[ subjRowVec , xColName ] )
```
Repeat and modify for the remaining subjects. Run the program and include the plot of the data. (See Figure 16.12.) The data curves for the four subjects should have four very different slopes.

(B) Relative to the original data, has the posterior mean of the group slopes gotten farther away from zero or closer to zero? Include the histogram of the posterior in your write-up. Are all the individual slopes believably different from zero (according to the 95% HDI)? Is the group slope believably different from zero (according to the 95% HDI)? *Why is the group-level slope, which is now farther away from zero on average, less believably different from zero than in the original data?*

Exercise 16.2. [Purpose: To see the influence of differences in *individual intercepts* on the estimate of *group-average slope*.]

The data shown in Figure 16.10 all start at 2.0 because that is $\log_{10}(100)$, and the data were measured as *percentage* of original value. Suppose that the data were kept in their raw magnitudes instead of converted to percentage of first

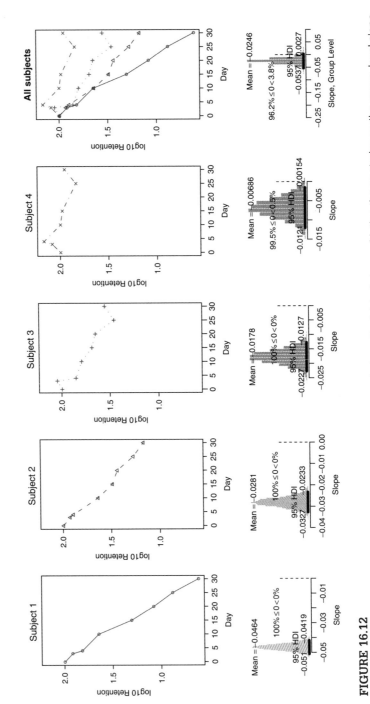

FIGURE 16.12

For Exercise 16.1. Contaminant-retention data with greater variation of individual slopes. Notice the posterior on the group-level slope parameter. Compare with Figure 16.10.

magnitude. This would merely change the intercepts of the individual data curves, without changing their slopes, because $\log(ky) = \log(k) + \log(y)$. In this exercise we find out whether this change would have any effect on the estimate of the group slope.

(A) Alter the data as follows: In the program listed in Section 16.5.3 (SimpleLinearRegressionRepeatedBrugs.R), right *before* the data are converted to log10 on line 42, insert this code:

```
for ( subjIdx in 1:4 ) {
rowIdx = ( dataMat[, "subjID"] == subjIdx )
dataMat[rowIdx, "retention"] = (dataMat[rowIdx, "retention"]
  * 10^(subjIdx-1))
}
```

Run the program and include the plot of the data. (See Figure 16.13.) The data curves for the four subjects should have four very different intercepts.

(B) Relative to the original data, are the estimates of the individual slopes different? Relative to the original data, is the posterior mean of the estimated *group* slope different? Include the histogram of the posterior of the group slope. Is the group slope believably different from zero (according to the 95% HDI)? *Why is the group-level slope less believably different from zero, compared to the original data?* (Hint: The individual intercepts affect the certainty of the group-average intercept. The group-average intercept trades off with the group-average slope; consider scatter plots of mu0G and mu1G.)

Exercise 16.3. [Purpose: To observe real data for repeated measures of individual regression, with an outlying individual and nonlinear trend.] Suppose we are interested in whether families with more members have higher incomes. The U.S. Census Bureau has published data that indicate the median family income as a function of number of persons in the family, for all 50 states and the District of Columbia and Puerto Rico. The data are plotted in Figure 16.14.

(A) Run the program of Section 16.5.3 (SimpleLinearRegressionRepeatedBrugs.R) with the income data. Examine the data section of the program, and you will find that the necessary lines of code are already available. The program should generate a figure much like Figure 16.14, p. 451.

(B) There is suggestion of outliers in these data. One curve (for Puerto Rico) falls barely above 20,000, which is far lower than all the others. This suggests an outlier for the distribution of intercepts. Some single data points fall far from the individual linear trends. For example, the six-person family in the District of Columbia has an income of only about 30,000, whereas the five-person family has an income of more

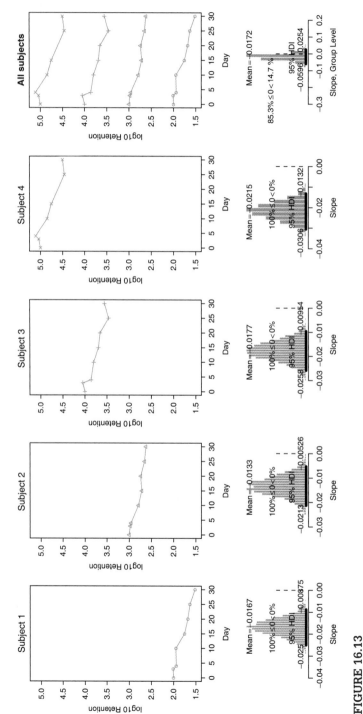

FIGURE 16.13

For Exercise 16.2. Contaminant-retention data with greater variation of individual *intercepts*. Notice the posterior on the group-level slope parameter. Compare with Figure 16.10.

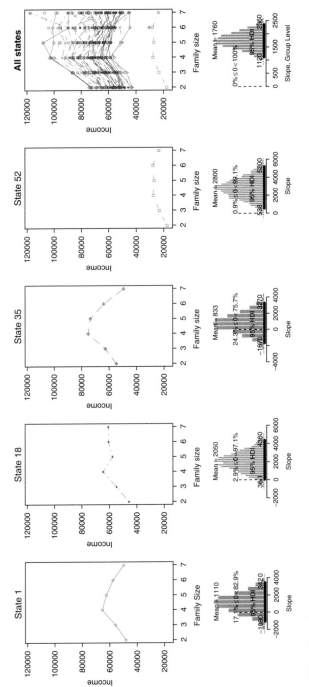

FIGURE 16.14

For Exercise 16.3. The top-right graph shows data from 50 states plus Puerto Rico and the District of Columbia, and the first four panels show data from four randomly selected individual states. The lower row shows marginal posterior distribution on the slope parameters.

than 80,000. This suggests an outlier for points around linear trends. Finally, some individual slopes might seem quite steep compared to others. For example, the income in Hawaii rises about $50,000 from two-person to seven-person families. This increase might or might not be an outlier relative to other states. Which distributions in the hierarchical model of Figure 16.11 should be changed to t distributions to address these outliers? Change the model specification to accommodate these t distributions. Use the example in Section 16.5.2 (`SimpleRobustLinearRegressionBrugs.R`) as a guide. Report your model specification in your write-up. Show the posterior estimate of the intercept of Puerto Rico for small-*df* t bias and for large-*df* t bias. Show the posterior estimate of the slope of Hawaii for small-*df* t bias and for large-*df* t bias.

(C)   The data also suggest a nonlinear trend in the data. Incomes appear to rise for two-, three-, and four-person families, but they then level off and decline as family size gets larger. Include in the original (non-t) model another term that can capture "quadratic curvature" in the income level: $\mu = \beta_0 + \beta_1 x + \beta_2 x^2$. The prior on β_2 is analogous to the prior on β_1. Is the group-average estimate of curvature credibly nonzero?

CHAPTER 17

Metric Predicted Variable with Multiple Metric Predictors

When I was young two plus two equaled four, but
Since I met you things don't add up no more.
My keel was even before I was kissed, but
Now my predictions all come with a twist.

In this chapter we are concerned with situations such as predicting a person's college grade point average (GPA) from his or her high school GPA and scholastic aptitude test (SAT) score. Another such situation is predicting a person's blood pressure from his or her height and weight. In these situations,

Doing Bayesian Data Analysis: A Tutorial with R and BUGS. DOI: 10.1016/B978-0-12-381485-2.00017-1
© 2011, Elsevier Inc. All rights reserved.

the value to be predicted is on a metric scale, and there is more than one predictor, each of which is also on a metric scale.

We will consider models in which the predicted variable is an additive combination of predictors, all of which have proportional influence on the prediction. This kind of model is called *multiple linear regression*, and it is listed in Table 14.1 (p. 385) in the first row and third column. We will also consider nonadditive combinations of predictors, which are called *interactions*.

17.1 MULTIPLE LINEAR REGRESSION

Figures 17.1 and 17.2 show examples of data generated by a model for multiple linear regression. The model specifies the dependency of y on x_1, x_2, but it does not specify the distribution of x_1, x_2. At any position, x_1, x_2, the values of y are normally distributed in a vertical direction, centered on the height of the plane at that position. The height of the plane is a linear combination of the x_1, x_2 values. Formally, $y \sim N(\mu, \tau)$, and $\mu = \beta_0 + \beta_1 x_1 + \beta_2 x_2$. For a review of how to interpret the coefficients, β_0, β_1, and β_2, see Figure 14.2, p. 365. The model assumes homogeneity of variance: At all values of x_1, x_2, the variance of y is the same.

17.1.1 The Perils of Correlated Predictors

Figures 17.1 and 17.2 show data generated from the same model. All that differs between them is the distribution of x_1, x_2, which is not specified by the model. In Figure 17.1, the x_1, x_2 values are distributed uniformly. In Figure 17.2, the x_1, x_2 values are negatively correlated: When x_1 is small, x_2 tends to be large, and when x_1 is large, x_2 tends to be small. The correlation of x_1 and x_2 can lead to misinterpretations of their individual influences on y. For instance, notice in Figure 17.2 that when x_1 is near zero, then the data y values are near 30, but when x_1 is near 10, then the data y values are near 20. This observation that y declines from 30 to 20 might leave the impression that an increase in x_1 predicts a *decrease* in y. But such an impression is wrong, because the data were generated by a function that *increases* y as x_1 increases (i.e., the coefficient β_1 on x_1 is +1). The reason that the y values appear to decline as x_1 increases is that x_2 decreases when x_1 decreases, and x_2 has an even bigger influence on y than x_1 does.

It is not unusual for predictors to be correlated in real data. For example, consider trying to predict a state's average high school SAT score on the basis of the amount of money the state spends per pupil. If you plot only mean SAT against money spent, there is actually a *decreasing* trend, as can be seen in the lower-left panel of Figure 17.3 (data from Guber, 1999). In other words, SAT scores tend to go down as spending goes up. Guber (1999) explains how some political commentators have used this sort of evidence to argue against funding public education.

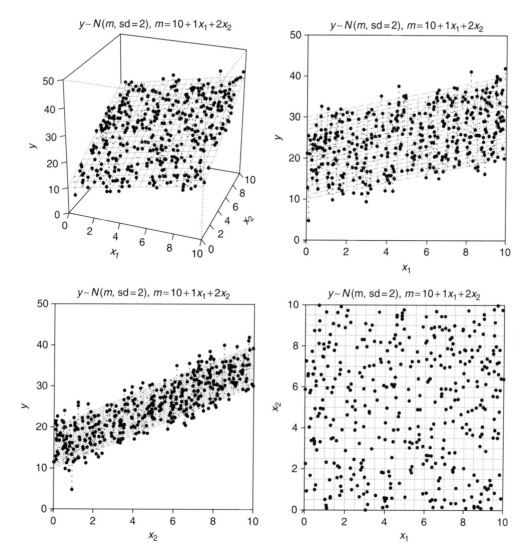

FIGURE 17.1
Data, y, that are normally distributed around the values in the plane. The x_1, x_2 values are sampled uniformly and independently of each other, as shown in the lower-right panel. The panels show different perspectives on the same plane and data. Notice that when the data are plotted against x_1 (marginalized across x_2), the points appear to rise along with the linear function that generated them. Compare with Figure 17.2.

This negative influence of spending on SAT scores seems counterintuitive. It turns out that the trend is an illusion caused by the influence of another factor, along with the correlation of spending with that other factor. The other factor is the proportion of students who take the SAT. Not all students at a high school take the SAT, because the test is used primarily for college entrance applications and therefore it is primarily students who intend to apply to college who take the SAT. Most of the top students at a high school will take the SAT, because most of the top students will apply to college. But students who are weaker academically may be less likely to take the SAT, because they are less likely to apply to college. Therefore, the more that a high school encourages mediocre students to take the SAT, the lower will be its average SAT score. It turns out that high schools that spend more money per pupil also have a much higher

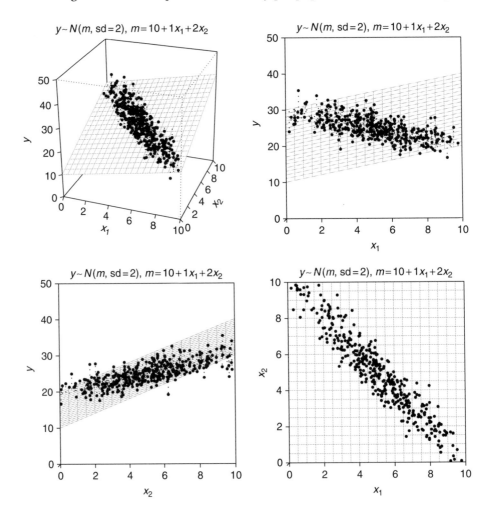

FIGURE 17.2

Data, y, that are normally distributed around the values in the plane. The x_1, x_2 values are correlated, as shown in the lower-right panel. The panels show different perspectives on the same plane and data. Notice that when the data are plotted against x_1 (marginalized across x_2), the points appear to drop, contrary to the linear function that generated them. Compare with Figure 17.1.

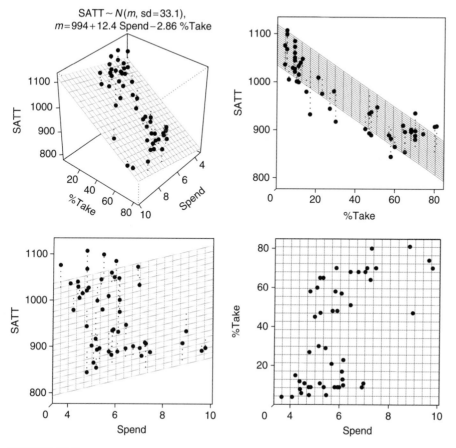

FIGURE 17.3
An example of multiple linear regression with real data. The data are plotted as dots, and the gridded plane shows the mean of the posterior slopes and intercept. The four panels show different perspectives on the same data and plane. "SATT" is the average total SAT score in a state. "%Take" is the percentage of students in the state who take the SAT. "Spend" is the spending per pupil, in thousands of dollars.

proportion of students who take the SAT. This correlation can be seen in the lower-right panel of Figure 17.3.

When both predictors (i.e., spending per pupil and percentage of students taking the SAT) are taken into account, the influence of spending on SAT score is seen to be positive, not negative. This positive influence of spending can be seen as the positive slope of the plane along the "Spend" direction in Figure 17.3. The negative influence, of percentage of students taking the SAT, is also clearly shown. To reiterate the main point of this example: It seems that the apparent drop in SAT due to spending is an artifact of spending being

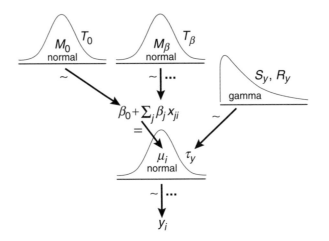

FIGURE 17.4

Hierarchical diagram for multiple linear regression. (In general, the slope parameters, β_j, may have different priors, instead of the same prior repeated for every j as is shown here for simplicity.)

correlated with the percentage of students taking the SAT, with the latter having a whoppingly negative influence on SAT scores.

The separate influences of the two predictors could be assessed in this example because the predictors had only mild correlation with each other. There was enough independent variation of the two predictors that their distinct relationships to the outcome variable could be detected. In some situations, however, the predictors are so tightly correlated that their distinct effects are difficult to tease apart. Correlation of predictors causes the estimates of their regression coefficients to trade off, as we will see when we examine the model and its posterior estimates.

17.1.2 The Model and BUGS Program

The hierarchical diagram for multiple linear regression is shown in Figure 17.4. It is merely a direct expansion of the one for simple linear regression (which appeared in Figure 16.3, p. 423). Instead of just one slope coefficient for a single predictor, there is another slope coefficient for every one of the multiple regressors. For every coefficient, the prior is normal, just as shown in Figure 16.3.

As usual, the BUGS model specification has a line of code for every arrow in the hierarchical diagram. The model specification looks like this (MultipleLinearRegressionBrugs.R):

```
11   model {
12       for( i in 1 : nData ) {
13           y[i] ~ dnorm( mu[i] , tau )
14           mu[i] <- b0 + inprod( b[] , x[i,] )
```

```
15          }
16          tau ~ dgamma(.01,.01)
17          b0 ~ dnorm(0,1.0E-12)
18          for ( j in 1:nPredictors ) {
19              b[j] ~ dnorm(0,1.0E-12)
20          }
21      }
```

Notice that what is written in mathematical notation as $\mu_i = \beta_0 + \sum_j \beta_j x_{ij}$ is expressed in BUGS on line 14 by using the BUGS `inprod` function. The `inprod` function is named for the mathematical "inner product" of two vectors. The function takes two vectors and returns the sum of their component-by-component products. Thus, for two vectors, `v` and `w`, the inner product `inprod` `(v[1:n],w[1:n])` is $\sum_{j=1}^{j=n} v_j w_j$. In BUGS, we do not need to specify the range of indices for the vector when it is on the right side of an assignment operator, and we intend to use all components of the vector. But in BUGS we do need to include the square brackets so that BUGS knows that the variable is a vector. If we wanted to be completely explicit, we could write the inner product in line 14 as `inprod( b[1:nPredictors] ,x[i,1:nPredictors] )`. The complete program is listed in Section 17.5.1 (`MultipleLinearRegression Brugs.R`).

The BUGS model specification happens to put the same prior on every slope parameter. This equivalence is applied as a generic convenience, but it is not required. Indeed, if there is prior information that suggests different priors on different predictors, then the prior knowledge should be respected.

17.1.2.1 MCMC Efficiency: Standardizing and Initializing

As described previously in Section 16.1.1.1 (p. 424), the MCMC sampling can be made much more efficient if the data are standardized. Standardizing each variable is straightforward. The MCMC sampling then finds regression coefficients that are appropriate for the standardized data. We would like to transform the parameters to the corresponding values that are appropriate to the original, nonstandardized scores. This can be done by generalizing Equation 16.2 to multiple predictors:

$$z_{\hat{y}} = \zeta_0 + \sum_j \zeta_j z_{x_j}$$

$$\frac{(\hat{y} - M_y)}{SD_y} = \zeta_0 + \sum_j \zeta_j \frac{(x_j - M_{x_j})}{SD_{x_j}}$$

$$\hat{y} = \underbrace{\zeta_0 SD_y + M_y - \sum_j \zeta_j SD_y M_{x_j}/SD_{x_j}}_{\beta_0} + \sum_j \underbrace{\zeta_j SD_y/SD_{x_j}}_{\beta_j} x_j \quad (17.1)$$

The estimate of σ_y is merely $\sigma_{z_y} SD_y$.

Even after standardizing, it can also help to start the chains near their posterior credible values. To do this, we use the built-in linear model function of R, called 1m. There is no need to delve into the inner workings of 1m, but suffice it to say that it returns the maximum-likelihood estimate (MLE) of the intercept and slope coefficients, which can be used to initialize the chains. In conclusion, the standardization of the data makes the chain efficient and less autocorrelated once it reaches the modal region of the posterior, and the initialization at the MLE implies that the burn-in period is minimal.

17.1.3 The Posterior: How Big Are the Slopes?

Figure 17.5 shows the posterior distribution of the results from the SAT data in Figure 17.3 and model in Figure 17.4. You can see that the slope on spending is credibly well above zero, with a mean slope of about 12.3, which suggests that SAT scores rise by about 12.3 points for every extra $1000 spent per pupil. The slope on percentage taking the SAT is also credibly nonzero, with a mean of −2.85, which suggests that SAT scores fall by about 2.85 points for every additional 1% of students who take the test.

The scatter plots in the bottom of Figure 17.5 show correlations among the credible parameter values. In particular, the lower-right scatter plots show that the coefficient for spending trades off with the coefficient on percentage taking the SAT. The correlation means that if we believe that the influence of spending is larger, then we must believe that the influence of percentage taking is smaller, to stay consistent with the data. Conversely, if we believe that the influence of spending is smaller, then we must believe that the influence of percentage taking is larger. This makes sense because those two predictors are correlated in the data, and therefore the two predictors are not differentially constraining the regression coefficients. Think of this simple example for two data points: Suppose $y = 1$ for $x_1 = 1$ and $x_2 = 1$, and $y = 2$ for $x_1 = 2$ and $x_2 = 2$. The linear model, $y = \beta_1 x_1 + \beta_2 x_2$, is supposed to satisfy both data points, which implies that $1 = \beta_1 + \beta_2$. Therefore, to satisfy the data, it could be that $\beta_1 = 2$ and $\beta_2 = -1$, or $\beta_1 = 0.5$ and $\beta_2 = 0.5$, or $\beta_1 = 0$ and $\beta_2 = 1$, and so on. In other words, the credible values of β_1 and β_2 are (anti-)correlated.

One of the benefits of Bayesian analysis is that correlations of credible parameter values are explicit in the posterior distribution. Traditional statistical methods provide only a single "best" (e.g., MLE) parameter value, without indicating the trade-offs among parameter values. The Bayesian posterior, however, naturally reveals trade-offs and redundancies among parameters. It is up to the user to actually look for and interpret the correlations of parameters, of course. Another benefit of Bayesian analysis is that the model doesn't "explode" when predictors are correlated. If predictors are correlated, the joint

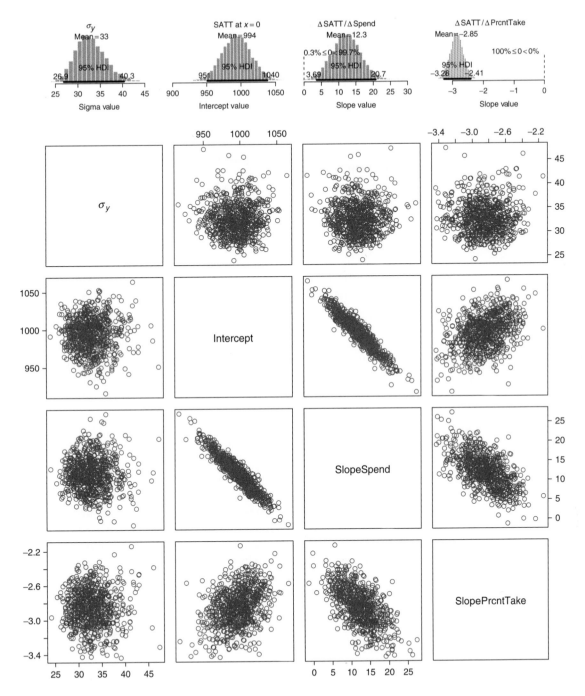

FIGURE 17.5

Posterior distribution for data in Figure 17.3 and model in Figure 17.4. Upper row indicates HDIs on regression coefficients. Scatter plots reveal correlations among credible parameter values; in particular, the coefficient on spending ("SlopeSpend") trades off with the coefficient on percentage taking the SAT ("SlopePrcntTake"), because those predictors are correlated in the data.

uncertainty in the regression coefficients is evident in the posterior, but the model happily generates a posterior regardless of correlations in the predictors. The classical, one-best-solution method is much less robust in the presence of strongly correlated predictors.

17.1.4 Posterior Prediction

Often we are interested in using the linear model to predict y values for various x values. It is straightforward to generate a large sample of credible y values for specified x values. From the distribution of y values, we can compute the mean and HDI to summarize the centrally predicted y value and the uncertainty of the prediction. As was the case for simple linear regression, illustrated back in Figure 16.6, the uncertainty in predicted y is greater for x values outside the bulk of the data. In other words, extrapolation is more uncertain than interpolation.

The last part of the code in Section 17.5.1 (MultipleLinearRegression Brugs.R) carries out these computations. At every step in the MCMC chain of posterior parameter values, the program randomly generates a y value based on the linear model. The x values, for which posterior predictions are desired, are stored in the matrix xPostPred, one row per point to be predicted. Thus, xPostPred has as many columns as there are predictors. The matrix yPostPred also has one row per point to be predicted, with each row containing randomly generated y values for the corresponding point. The program generates one y value per step in the MCMC chain, hence yPostPred has as many columns as there are steps in the chain. The vector b0Samp contains posterior values of the intercept, β_0, at each step in the MCMC chain. The vector bSamp[chainIdx,] contains the slope coefficients, β_j, for the predictors, at one step in the chain. The slope coefficients are forced to be a column vector by passing them through the cbind function in R. Then the slopes can be multiplied by the corresponding x values and summed together, simultaneously for all the points, in one matrix operation: xPostPred %*% cbind(bSamp[chainIdx,]). This matrix operation appears on line 235 in the following code (MultipleLinearRegressionBrugs.R):

```
232   for ( chainIdx in 1:chainLength ) {
233       yPostPred[,chainIdx] = rnorm( NROW(xPostPred) ,
234                       mean = b0Samp[chainIdx]
235                           + xPostPred %*% cbind(bSamp[chainIdx,]) ,
236                       sd = rep( sigmaSamp[chainIdx] , NROW(xPostPred) ) ) )
237   }
```

Notice that the code loops through every step in the MCMC chain, filling in the yPostPred matrix one column at a time. The values are generated randomly from a normal distribution, using the rnorm function, with a standard deviation of sigmaSamp[chainIdx].

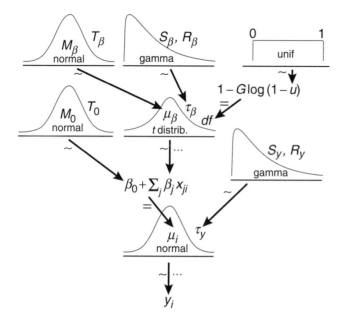

FIGURE 17.6
Hierarchical diagram for multiple linear regression, with hyperprior on slope coefficients across predictors. Compare with Figure 17.4.

17.2 HYPERPRIORS AND SHRINKAGE OF REGRESSION COEFFICIENTS

In some research, there are many candidate predictors which we suspect could possibly be informative about the predicted variable. For example, when predicting college GPA, we might include high school SAT, high school GPA, income of student, income of parents, years of education of the parents, spending per pupil at the student's high school, student IQ, student height, weight, shoe size, hours of sleep per night, distance from home to school, amount of caffeine consumed, hours spent studying, hours spend earning a wage, and so on. We can include all the candidate predictors in the model, with a regression coefficient for every predictor. Should all those regression coefficients be estimated *in isolation from the others*, as in the model of Figure 17.4? Probably not, because we probably have some prior knowledge that relates the influences of the predictors to each other. If nothing else, we can at least say that the candidate predictors all come from the set of remotely plausible predictors. What do we know about this set of remotely plausible predictors? Most candidate predictors probably have a very small relationship with the predicted variable, but a few candidate predictors may have sizable covariation with the predicted variable. In other words, the regression coefficients are probably distributed something like a t distribution, with lots near a mean of zero, but with a few off in the extended tails. We therefore put this prior knowledge into the model structure, as shown in Figure 17.6. (This method is mentioned in passing by Gelman et al., 2004, p. 405.)

Figure 17.6 indicates that the regression coefficients, β_j, are distributed as a t distribution. The parameters of the t distribution are estimated from the data. Presumably, many of the credible regression coefficients will be near zero, but a few will depart a lot from zero, and the t distribution will have credible τ and df values that reflect the distribution of regression coefficients in the data. Exercise 17.1 has you generate the prior from BUGS, so you get a better intuition for how the constants in the hyperdistribution affect the implied prior on the regression coefficients.

A desirable side-effect of incorporating this prior structure is that the estimates of the regression coefficients experience shrinkage. If many regression coefficients are near zero, then the t distribution will have a high precision (τ parameter), which in turn will shrink the estimates of the regression coefficients. The regression coefficients are mutually informing each other, via the prior knowledge that they should be distributed according to a t distribution.

The shrinkage is desirable not only because it expresses our prior knowledge but also because it rationally helps control for "false alarms" in declaring that a predictor has a nonzero regression coefficient. When there are many candidate predictors, some of them may spuriously appear to have credibly nonzero regression coefficients, even when the true coefficient is zero. This sort of false alarm is unavoidable because data are randomly sampled, and there will be occasional coincidences of data that are unrepresentative. By letting the regression coefficients be mutually informed by other predictors, and not only by the data of a single predictor, the coefficients are less likely to be spuriously distorted by a rogue sample.

As an example in which we can specify the true regression coefficients, let's randomly generate 100 data points from a linear regression model that has parameter values of $\sigma_y = 2$, $\beta_0 = 100$, $\beta_1 = 1$, $\beta_2 = 2$, and all other $\beta_j = 0$ for 21 other predictors. When using insulated regression coefficients for each predictor, as in the hierarchical diagram of Figure 17.4, the resulting posterior estimates are shown in the upper half of Figure 17.7. Notice that the estimate of β_{15} (denoted $\Delta Y/\Delta X15$) suggests that this predictor may have a nonzero regression coefficient. (If we could specify a ROPE of some nonzero width, we might not decide that β_{15} is nonzero, but ROPEs are best defined in meaningful contexts, not in a generic example like this.) In other words, this apparent nonzero value of β_{15} is a "false alarm," produced by quirks in the random sample of data.

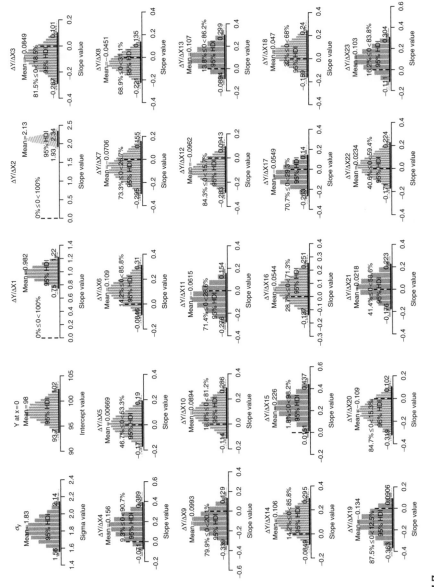

FIGURE 17.7
(*Continued*)

465

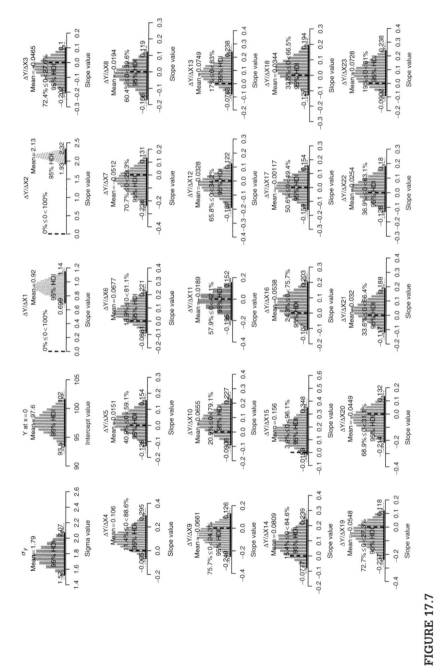

FIGURE 17.7

Posterior distribution for 100 data points randomly generated with true parameter values of $\sigma_y = 2$, $\beta_0 = 100$, $\beta_1 = 1$, $\beta_2 = 2$, and all other $\beta_j = 0$. *Upper half (previous page)*: The model is from Figure 17.4, wherein each slope coefficient is insulated from the other. Notice the "false alarm" for β_{15}. *Lower half (this page)*: The model is shown in Figure 17.6, wherein the slope coefficients mutually inform each other. Shrinkage prevents the "false alarm" for β_{15}.

The estimates in the upper half of Figure 17.7 (p. 465) did not use any prior knowledge about how the regression coefficients might be related. If we instead use the model of Figure 17.6, with the same data, the resulting posterior is shown in the lower half of Figure 17.7 (p. 466). Notice that the estimate of β_{15} now includes zero inside the HDI. All the estimates of the regression coefficients are reduced a little, relative to the upper half of the figure. The reason for this shrinkage is that the many regression coefficients near zero mutually inform each other via the overarching t distribution.

When using the model of Figure 17.6, the shrinkage of the regression coefficients is toward their mean, μ_β, not necessarily toward zero. If your prior knowledge suggests that the true mean of the regression coefficients is very close to zero, then that knowledge should be expressed in the model of Figure 17.6 by setting the mean M_β to zero and setting the precision T_β to a large value such as 10,000. This setting implies that μ_β is already fairly certain, and the real prior uncertainty is in the variability of regression coefficients across predictors.

The hyperprior over regression coefficients, shown in Figure 17.6, is meant to express a genuine belief, namely, that all the regression coefficients can be reasonably described by a t distribution. This assumption may be more or less tenable in different situations. For example, a region's agricultural crop yield might be predicted from rainfall amounts measured at 39 randomly selected locations. The predictabilities from each of these measuring locations are probably fairly similar, except for perhaps a few outliers, and therefore it seems reasonable to model the regression coefficients by a t distribution. In other situations, it may be less reasonable to treat all the regression coefficients as coming from a shared t distribution. At the least, the predictors were selected from some implicit set of reasonably likely predictors, and we can think of the overarching distribution as reflecting that set. We might still use the hyperprior model, but only as a convenience to impose some degree of shrinkage on the regression coefficients. It should be interpreted carefully. Beware of convenience priors that are used in routinized ways.

17.2.1 Informative Priors, Sparse Data, and Correlated Predictors

This book has emphasized the use of mildly informed priors, as opposed to conventionalized noninformed priors or strongly informed priors. On the other hand, this book has also mentioned that a benefit of Bayesian analysis is the potential for cumulative scientific progress through the use of priors that have been informed by previous research. Informed priors should be used whenever the skeptical scientific audience of the analysis deems it appropriate, especially if the analysis is accompanied by a check of posterior robustness. If the conclusion from the posterior is strong, even with a mildly informed prior, then the mildly informed prior should be used, because a broader audience may find the analysis to be convincing. But there are situations in which the use of a more strongly informed prior is appropriate.

In general, when the data are sparse (i.e., when the sample is small), the posterior will be imprecise if the prior is imprecise. But if the prior is more tightly constrained, then a small amount of data can lead to a more decisive posterior. An example of this phenomenon appeared in Exercise 5.7, p. 97, in which the prior allowed only two very different interpretations of the data. In this case, even a small amount of data shifted the posterior toward one or the other interpretation.

Sparse data can also lead to usefully precise posteriors in the context of multiple linear regression, if some of the regression coefficients use informed priors, and the predictors are correlated. To develop this idea, it is important to remember that when predictors are correlated, their regression coefficients are also (anti-)correlated. For example, recall the data from Figure 17.3, p. 457, in which the predictors are correlated in the data (i.e., spending per pupil and percent taking the exam are correlated). Consequently, the posterior estimates of the regression coefficients had a negative correlation, as shown in Figure 17.5, p. 461. The correlation of credible regression coefficients implies that a strong belief about the value of one regression coefficient constrains the value of the other coefficient. Look carefully at the scatter plot of the two slopes shown in Figure 17.5. It shows that if we believe that the slope on percent taking the exam is -3.2, then credible values of the slope on spending per pupil are around 18, with an HDI of roughly 13 to 22. Notice that this HDI is much smaller than the marginal HDI on spending per pupil, which goes from 3.7 to 20.7. Thus, constraining the beliefs about one slope also constrains beliefs about the other slope, because estimates of the slopes are correlated.

That influence of one slope estimate on another can be used for inferential advantage when we have prior knowledge about one of the slopes. If some previous or auxiliary research informs the prior of one regression coefficient, that constraint can propagate to the estimates of regression coefficients on other predictors that are correlated with the first. This is especially useful when the sample size is small, and a merely mildly informed prior would not yield a precise posterior. Of course, the informed prior on the first coefficient must be cogently justified for the skeptical audience. A robustness check also may be useful to show how strong the prior must be to draw strong conclusions. If the information used for the prior is compelling, then this technique can be very useful for leveraging novel implications from small samples. An accessible discussion and example from political science is provided by Western & Jackman (1994), and a mathematical discussion is provided by Learner (1978, p. 175ff).

17.3 MULTIPLICATIVE INTERACTION OF METRIC PREDICTORS

In some situations, the predicted value is not merely an additive combination of the predictors. For example, the effects of drugs are often nonadditive. As the dosage of one drug increases, there might be a moderate increasing effectiveness. And as the dosage of another drug increases (when administered by itself), there may be a moderate increase in effectiveness. But when the two drugs are administered together, when the dosages of both drugs are high, they might interact to greatly enhance or greatly decrease effectiveness, beyond a mere addition of the two separate effects. As another example, consider trying to predict subjective happiness from income and health. If either income or health is low, subjective happiness is probably also fairly low. But if both income and health are high, then happiness is more likely to be high. In other words, happiness might not increase additively with both income and health; instead, happiness may be better predicted by an interaction of income and health. In general, interaction means that the influence of one predictor varies, depending on the value of the other predictor.

Formally, interaction can be expressed different ways. We will consider *multiplicative* interaction. For two metric predictors, regression with multiplicative interaction has these algebraically equivalent expressions:

$$y = \beta_0 + \beta_1 x_1 + \beta_2 x_2 + \beta_{1\times2} x_1 x_2 \qquad (17.2)$$

$$= \beta_0 + \underbrace{\left(\beta_1 + \beta_{1\times2} x_2\right)}_{\text{slope of } x_1} x_1 + \beta_2 x_2 \qquad (17.3)$$

$$= \beta_0 + \beta_1 x_1 + \underbrace{\left(\beta_2 + \beta_{1\times2} x_1\right)}_{\text{slope of } x_2} x_2 \qquad (17.4)$$

These three expressions emphasize different interpretations of interaction, as illustrated in Figure 17.8.

The form of Equation 17.2 is illustrated in the left panel of Figure 17.8. The vertical arrows show that the curved-surface interaction is created by adding the product, $\beta_{1\times2} x_1 x_2$, to the planar linear combination.

The form of Equation 17.3 is illustrated in the middle panel of Figure 17.8. Its dark lines show that the slope in the x_1 direction depends on the value of x_2. In particular, when $x_2 = 0$, then the slope along x_1 is β_1, which in the graphed example is -1. But when $x_2 = 10$, then the slope along x_1 is $\beta_1 + \beta_{1\times2} x_2$, which in the graphed example is $+1$. Again, the slope in the x_1 direction changes when x_2 changes, and β_1 only indicates the slope along x_1 when $x_2 = 0$.

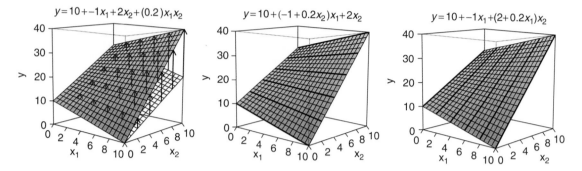

FIGURE 17.8
A multiplicative interaction of x_1 and x_2, parsed three ways. The left panel emphasizes that the interaction involves a multiplicative component that adds a vertical amount to the planar additive model, as indicated by the arrows. The middle panel shows the same function, but with the terms algebraically regrouped to emphasize that the slope in the x_1 direction depends on the value of x_2, as shown by the darkened lines. The right panel again shows the same function, but with the terms algebraically regrouped to emphasize that the slope in the x_2 direction depends on the value of x_1, as shown by the darkened lines.

The form of Equation 17.4 is illustrated in the right panel of Figure 17.8. It shows that the interaction can be expressed as the slope in the x_2 direction changing when x_1 changes. This is exactly analogous to the middle panel of Figure 17.8, but it is important to realize, and visualize, that the interaction can be expressed in terms of the slopes on either predictor.

Great care must be taken when interpreting the coefficients of a model that includes interaction terms (Braumoeller, 2004). In particular, low-order terms are especially difficult to interpret when higher-order interactions are present. In the simple two-predictor case, the coefficient β_1 describes the influence of predictor x_1 *only* at $x_2 = 0$, because the slope on x_1 is $\beta_1 + \beta_{1\times2}x_2$, as was shown in Equation 17.3 and graphed in the middle panel of Figure 17.8. In other words, it is not appropriate to say that β_1 indicates the *overall* influence of x_1 on y_1. Indeed, in many applications, the value of x_2 never realistically gets close to zero, and therefore β_1 has no realistic interpretation at all. For example, suppose we are predicting college GPA (y) from parental income (x_1) and high school GPA (x_2). If there is interaction, then the regression coefficient, β_1, on parental income, only indicates the slope on x_1 *when x_2 (GPA) is zero.* Of course, there are no GPAs of zero, and therefore β_1 by itself is not very informative.

17.3.1 The Hierarchical Model and BUGS Code

The model for regression with multiplicative interaction is the same as for linear regression but with an added term for the interaction. Because we are estimating the coefficient on the multiplication of the predictors, that coefficient must have a prior, analogous to the priors on the linear coefficients. For the case of two predictors, with an interaction, the BUGS model specification looks like this (`MultiLinRegressInterBrugs.R`):

```
11   model {
12       for( i in 1 : nData ) {
13           y[i] ~ dnorm( mu[i] , tau )
14           mu[i] <- b0  +  b1 * x[i,1]  +  b2 * x[i,2]  +  b12 * x[i,1] * x[i,2]
15       }
16       tau ~ dgamma(.001,.001)
17       b0  ~ dnorm(0,1.0E-12)
18       b1  ~ dnorm(0,1.0E-12)
19       b2  ~ dnorm(0,1.0E-12)
20       b12 ~ dnorm(0,1.0E-12)
21   }
```

Notice that the interaction coefficient, b12, has a normal prior analogous to the priors on the slope coefficients.

In models that involve many predictors, there can be many possible interaction terms. Just as we can put a hyperprior on the slopes, we can also put a hyperprior on the interaction coefficients. The idea is that most interaction terms are near zero, but a few might depart from zero. This distribution over interaction terms could be modeled as a t distribution. We will not pursue examples that involve numerous interactions.

17.3.1.1 Standardizing the Data and Initializing the Chains
We will, as before, standardize the data before entering them into the BUGS model. This helps reduce correlations in the parameters, but it does not eliminate correlations. When initializing the chains, it may suffice to set the interaction coefficient(s) to zero and start the slopes at their MLE values for a noninteractive model, especially when the interactions are small, as is often the case.

Transforming the standardized estimates back to the original scales is conceptually simple but algebraically much messier when interaction terms are involved. The expression, when there are merely two predictors, turns into this unwieldy form:

$$z(y) = \zeta_0 + \zeta_1 z(x_1) + \zeta_2 z(x_2) + \zeta_{1 \times 2} z(x_1) z(x_2)$$

$$\frac{y - m_y}{s_y} = \zeta_0 + \zeta_1 \frac{x_1 - m_1}{s_1} + \zeta_2 \frac{x_2 - m_2}{s_2} + \zeta_{1 \times 2} \frac{x_1 - m_1}{s_1} \frac{x_2 - m_2}{s_2}$$

$$y = \underbrace{\left(m_y + s_y \zeta_0 - \zeta_1 s_y \frac{m_1}{s_1} - \zeta_2 s_y \frac{m_2}{s_2} + \zeta_{1 \times 2} s_y \frac{m_1}{s_1} \frac{m_2}{s_2} \right)}_{\beta_0}$$

$$+ \underbrace{\left(\zeta_1 s_y \frac{1}{s_1} - \zeta_{1 \times 2} s_y \frac{m_1}{s_1} \frac{1}{s_2} \right)}_{\beta_1} x_1$$

$$+ \underbrace{\left(\zeta_2 s_y \frac{1}{s_2} - \zeta_{1 \times 2} s_y \frac{1}{s_1} \frac{m_2}{s_2} \right)}_{\beta_2} x_2$$

$$+ \underbrace{\left(\zeta_{1 \times 2} s_y \frac{1}{s_1} \frac{1}{s_2} \right)}_{\beta_{1 \times 2}} x_1 x_2 \tag{17.5}$$

When there are more predictors involved, with their interactions, the expression becomes quite protracted. In those situations, there is no avoiding matrix notation and matrix algebra, which greatly facilitates manipulating the forms. We will not venture into matrix expressions for the GLM in this book (although we did splurge a bit on matrices in Section 7.1.5, p. 126). Matrix operations are not easily expressed in BUGS, unfortunately.

17.3.2 Interpreting the Posterior

To illustrate some of the issues involved in interpreting the parameters of a model with interaction, consider again the SAT data from Figure 17.3. Recall that the mean SAT score in a state was predicted from the spending per pupil and the percentage of students who took the test. When no interaction term was included in the model, the posterior looked like Figure 17.5, which indicated a positive influence of spending and a negative influence of percentage of students.

Would we want to include an interaction term in the model? The meaning of interaction is that the effect of one predictor changes when the level of the other predictor changes. Does it make sense in this application that the effect of spending would depend on the percentage of students taking the test? Perhaps yes, because if very few students are taking the test, they are probably already at the top of the class and therefore might not have as much headroom for increasing their scores. In other words, it is plausible that the effect of spending is larger when the percentage of students taking the test is larger, and we would not be surprised if there were a positive interaction between those predictors. Therefore, it is theoretically meaningful to include an interaction term in the model.

When we incorporate a multiplicative interaction into the model, the posterior looks like Figure 17.9. The top-right histogram indicates that the mean of the believable interaction coefficients is indeed positive, as we anticipated it could be. The 95% HDI includes zero, however, which indicates that we do not have

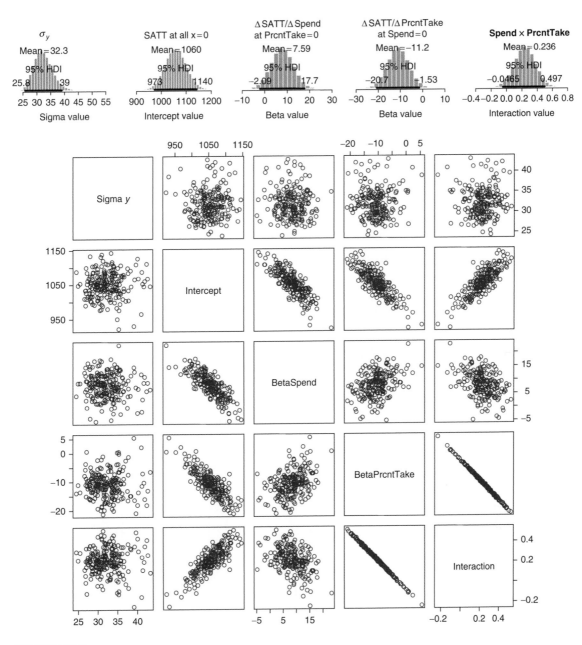

FIGURE 17.9
Posterior for SAT data in Figure 17.3. Compare this posterior, which includes an interaction term, with the posterior that excluded an interaction, shown in Figure 17.5.

strong precision in our estimate of the magnitude of the interaction. The scatter plots in the lower part of Figure 17.9 indicate that the interaction coefficient is strongly correlated with the beta coefficient on percentage of students taking the test. If the MCMC sampling were taking place in these original scales, it would be inefficient.

A cursory look at the middle histogram of Figure 17.9 might lead a person, inappropriately, to conclude that there is not a credible influence of spending on SAT scores, because zero is among the credible values of β_{spend}. This conclusion is inappropriate because β_{spend} only indicates the slope on spending *when the percentage of students taking the test is zero*. The slope changes when the percentage of students changes because of the interaction. Because the interaction tends to be positive, the slope of spending increases when the percentage of students taking the test increases.

Figure 17.10 shows the increase in slope along spending, as the percentage taking the test increases. Also plotted is the extent of the 95% HDI of the estimated slope. Notice that the HDI at PrcntTake $= 0$ matches the HDI shown in the middle histogram of Figure 17.9.

Interestingly and importantly, the extent of the HDI is not constant but also depends on the percentage of students taking the test. Mathematically, the change in extent of the HDI stems for two sources. First, the slope along

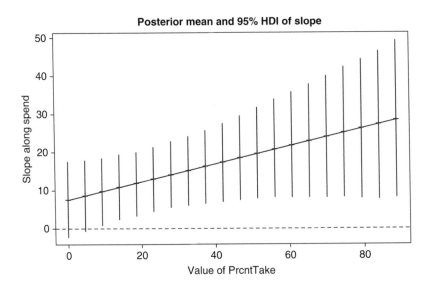

FIGURE 17.10

The slope in the x_{spend} direction is $\beta_{spend} + \beta_{spend \times \%take} x_{\%take}$. Shown here are the means and 95% HDIs for the slope in the x_{spend} direction, for various values of $x_{\%take}$.

x_1 is $\beta_1 + \beta_{1\times2}x_2$, which means that the uncertainty in $\beta_{1\times2}$ is being multiplied by x_2, and therefore the uncertainty in the slope depends on x_2. But the uncertainty in the slope does not necessarily always increase when x_2 increases because β_{spend} and $\beta_{\text{spend}\times\%\text{take}}$ are negatively correlated (see the scatter plot in Figure 17.9): When x_{spend} has a modest size, its negative correlation removes a bit of uncertainty in the slope. This relationship can be seen graphically in Figure 17.11, which shows an idealized subset of credible values for β_1 and $\beta_{1\times2}$. The left panel shows that when $x_2 = 0$, the credible parameters (i.e., the dots) span a range of $x1$ slopes from 0 to 15. The middle panel shows that when $x_2 = 25$, the parameters span a range of $x1$ slopes from 10 to 15 (i.e., a much smaller range). The right panel shows that when $x_2 = 50$, the parameters span a range of $x1$ slopes from 10 to 25, again a large range. Thus, because of the negative correlation of credible values of β_1 and $\beta_{1\times2}$, the narrowest range of $x1$ slopes is at an intermediate value of $x2$.

In summary, when there is interaction, then the influence of the individual predictors cannot be summarized by their individual regression coefficients alone, because those coefficients only describe the influence when the other variables are at zero. A careful analyst considers credible slopes across a variety of values for the other predictors, as in Figure 17.10. Notice that this is true even when the interaction coefficient did not exclude zero from its 95% HDI: Even though the estimate of the interaction was not precise, the interaction did have a considerable influence on the interpretation of the predictors.

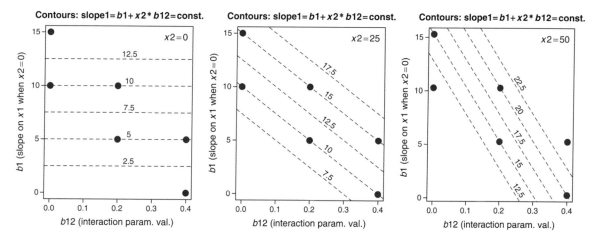

FIGURE 17.11

Dots show idealized parameter-value combinations, the same in all three panels. Contour lines mark parameter value combinations for slopes in the $x1$ direction when $x2$ has that value indicated in the panel.

17.4 WHICH PREDICTORS SHOULD BE INCLUDED?

In many research projects, there are specific theoretical motivations for measuring particular predictors and the predicted variable. In these situations, we know in advance which predictors and interactions we are interested in modeling. But in other situations, we come to a preexisting database that has many variables measured for other purposes, and we are curious to know which of the variables might be good predictors of a particular variable of interest. Or we might have some hybrid situation, in which we have particular predictors that we think may be relevant, but we measure several other variables on the off chance that they might be relevant, because they are needed for some other study, or because they are trivial to measure so we go ahead and measure them.

When there are many potential predictors, which ones should be included in our model? And which interactions? A reasonable answer is to include all the predictors and interactions that you think would have any chance of providing useful predictive information. If you do not include them, then you have essentially set the prior on those variables' regression coefficients to zero with complete certainty.

Whether or not to include candidate predictors depends on the purpose of the regression analysis. If the goal is to *predict* the outcome as accurately as possible, using any predictors regardless of how those predictors might be causally related to the predicted variable, then all reasonable predictors could be included. On the other hand, if the purpose of the analysis is to *explain* the outcome on the basis of the predictors, then only the predictors that can be meaningfully related to the outcome should be included (see, e.g., Keith, 2005, p. 70).

There are several costs of including a lot of candidate predictors. One cost is unwieldy interpretation of the results. With too many predictors and interactions, the complexity of the mathematical description may provide little meaningful insight into the data. Unless you really believe that the effect of income on happiness depends on shoe size and that such an interaction of income with shoe size would be theoretically meaningful, don't include shoe size among the predictors. But if you think there is some chance that a variable would be informative, include it.

Another cost of including numerous predictors is that "false alarms" may occur more often, such that predictors that really have no predictive value spuriously appear to have credibly nonzero regression coefficients. This problem can be addressed by using prior knowledge about relations among regression coefficients, as described in Section 17.2. By letting the predictors inform a hyperprior, they constrain each other's estimates, thereby producing shrinkage and attenuation of false alarms. If you have prior knowledge about the particular predictors that your skeptical audience would assent to, then try to express it in the mathematical form of the prior.

Another cost of including numerous candidate predictors is that the noise and parameter flexibility introduced by the extra predictors can produce a loss of precision in the estimation of the coefficients. Even when the true regression coefficients on additional predictors are all zero, the uncertainty in their values can introduce uncertainty in the estimates of the original predictors. This bleeding of uncertainty is not always large, especially when the predictors are uncorrelated and there are many data points, but it does occur. You can experiment with this issue by using the random data generator in the program of Section 17.5.1 (`MultipleLinearRegressionBrugs.R`) and including different numbers of predictors with zero coefficients.

All of these issues are distinct from the peril of correlated predictors discussed at the beginning of the chapter. To reiterate, a predictor may appear to have a particular relation with the predicted variable, when other candidate predictors are left out of the analysis. But the apparent relation between the predictor and the outcome might be an illusory artifact, produced instead by the predictor being correlated with some other efficacious factor, whereas the putative predictor itself has zero or opposite influence on the outcome. Therefore, it is important to include all reasonable predictors in the analysis, so that each has a better opportunity for being correctly interpreted, while also keeping in mind the costs mentioned in preceding paragraphs.

When considering the inclusion of *interaction* terms, and the goal of the analysis is explanation, then the main criterion is whether it is theoretically meaningful that the effect of one predictor should depend on the level of another predictor. Inclusion of an interaction term can also cause loss of precision in the estimates of the lower-order terms. Moreover, interpretation of interactions and their lower-order terms can be subtle, as we saw, for example, in Figure 17.10. If the goal of the analysis is prediction, without emphasis on explanation, then interaction terms may be included to the extent that they enhance predictability without loss of parsimony. Bayesian model comparison can be useful in this case (see references cited at the end of this section).

Whenever an interaction term is included in a model, it is important also to include all lower-order terms. For example, if an $x_i \times x_j$ interaction is included, then both x_i and x_j should also be included in the model. Although we did not discuss them, it is also possible to include three-way interactions such as $x_i \times x_j \times x_k$, if it is theoretically meaningful to do so. When this is done, it is important to include all the lower-order interactions and single predictors, including $x_i \times x_j$, $x_i \times x_k$, $x_j \times x_k$, x_i, x_j, and x_k. When the lower-order terms are omitted, this is artificially setting their regression coefficients to zero and distorting the posterior estimates on the other terms. For clear discussion and examples of this issue, see Braumoeller (2004), and Brambor, Clark, & Golder (2006).

In some situations, it may be theoretically tenable to suppose that the candidate predictors have exactly zero influence on the variable to be predicted, and our goal is to identify which predictors have exactly zero influence. In this case, we can establish a model-comparison framework in which different models have different regression coefficients fixed at zero. A large-scale Bayesian model comparison then reveals which combinations of zero coefficients are most credible. There are numerous variations on this approach, and it is an active area of research (e.g., Casella & Moreno, 2006; Clyde & George, 2004; George, 2000; Greenland, 2008; Liang et al., 2008; Scott & Berger, 2006).

17.5 R CODE

17.5.1 Multiple Linear Regression

This program was used to produce Figure 17.5, among others. Its data section includes three different data sets. The rest of the program is designed to be generically applicable, so that the user can substitute other data sets without modifying the remainder of the program.

The use of the BUGS function inprod() allows the model specification to remain unchanged when the number of predictors changes. Unfortunately, inprod() is processed slowly by BUGS. If you adapt this program for a large application and it is running too slowly, try changing inprod(b[],x[i,]) to an explicit sum of products: b[1]*x[i,1] + b[2]*x[i,2] + ...

(MultipleLinearRegressionBrugs.R)

```
1   graphics.off()
2   rm(list=ls(all=TRUE))
3   fname = "MultipleLinearRegressionBrugs"
4   library(BRugs)              # Kruschke, J. K. (2010). Doing Bayesian data analysis:
5                               # A Tutorial with R and BUGS. Academic Press / Elsevier.
6   #------------------------------------------------------------------------------------
7   # THE MODEL.
8
9   modelstring = "
10  # BUGS model specification begins here...
11  model {
12      for( i in 1 : nData ) {
13          y[i] ~ dnorm( mu[i] , tau )
14          mu[i] <- b0 + inprod( b[] , x[i,] )
15      }
16      tau ~ dgamma(.01,.01)
17      b0 ~ dnorm(0,1.0E-12)
18      for ( j in 1:nPredictors ) {
19          b[j] ~ dnorm(0,1.0E-12)
20      }
21  }
22  # ... end BUGS model specification
23  " # close quote for modelstring
```

```
24   writeLines(modelstring,con="model.txt")
25   modelCheck( "model.txt" )
26
27   #------------------------------------------------------------------------------
28   # THE DATA.
29
30   dataSource = c("Guber1999","McIntyre1994","random")[1]
31
32   if ( dataSource=="Guber1999" ) {
33     fname = "Guber1999" # file name for saved graphs
34     dataMat = read.table( file="Guber1999data.txt" ,
35                           col.names = c( "State","Spend","StuTchRat","Salary",
36                                          "PrcntTake","SATV","SATM","SATT") )
37     # Specify variables to be used in BUGS analysis:
38     predictedName = "SATT"
39     predictorNames = c( "Spend" , "PrcntTake" )
40     #predictorNames = c( "Spend" , "PrcntTake" , "Salary" , "StuTchRat" )
41     nData = NROW( dataMat )
42     y = as.matrix( dataMat[,predictedName] )
43     x = as.matrix( dataMat[,predictorNames] )
44     nPredictors = NCOL( x )
45   }
46
47   if ( dataSource=="McIntyre1994" ) {
48     fname = "McIntyre1994" # file name for saved graphs
49     dataMat = read.csv(file="McIntyre1994data.csv")
50     predictedName = "CO"
51     predictorNames = c("Tar","Nic","Wt")
52     nData = NROW( dataMat )
53     y = as.matrix( dataMat[,predictedName] )
54     x = as.matrix( dataMat[,predictorNames] )
55     nPredictors = NCOL( x )
56   }
57
58   if ( dataSource=="random" ) {
59     fname = "Random"  # file name for saved graphs
60     # Generate random data.
61     # True parameter values:
62     betaTrue = c( 100 , 1 , 2 , rep(0,21) ) # beta0 is first component
63     nPredictors = length( betaTrue ) - 1
64     sdTrue = 2
65     tauTrue = 1/sdTrue^2
66     # Random X values:
67     set.seed(47405)
68     xM = 5 ; xSD = 2
69     nData = 100
70     x = matrix( rnorm( nPredictors*nData , xM , xSD ) , nrow=nData )
71     predictorNames = colnames(x) = paste("X",1:nPredictors,sep="")
72     # Random Y values generated from linear model with true parameter values:
73     y = x %*% matrix(betaTrue[-1],ncol=1) + betaTrue[1] + rnorm(nData,0,sdTrue)
74     predictedName = "Y"
75     # Select which predictors to include
76     includeOnly = 1:nPredictors # default is to include all
77     #includeOnly = 1:10 # subset of predictors overwrites default
```

```
78      x = x[,includeOnly]
79      predictorNames = predictorNames[includeOnly]
80      nPredictors = NCOL(x)
81    }
82
83    # Prepare data for BUGS:
84    # Re-center data at mean, to reduce autocorrelation in MCMC sampling.
85    # Standardize (divide by SD) to make prior specification easier.
86    standardizeCols = function( dataMat ) {
87       zDataMat = dataMat
88       for ( colIdx in 1:NCOL( dataMat ) ) {
89          mCol = mean( dataMat[,colIdx] )
90          sdCol = sd( dataMat[,colIdx] )
91          zDataMat[,colIdx] = ( dataMat[,colIdx] - mCol ) / sdCol
92       }
93       return( zDataMat )
94    }
95    zx = standardizeCols( x )
96    zy = standardizeCols( y )
97
98    # Get the data into BUGS:
99    datalist = list(
100             x = zx ,
101             y = as.vector( zy ) , # BUGS does not treat 1-column mat as vector
102             nPredictors = nPredictors ,
103             nData = nData
104    )
105   modelData( bugsData( datalist ) )
106
107   #------------------------------------------------------------------------------
108   # INTIALIZE THE CHAINS.
109
110   nChain = 3
111   modelCompile( numChains = nChain )
112
113   genInitList <- function(nPred=nPredictors) {
114       lmInfo = lm( datalist$y ~ datalist$x ) # R function returns MLE
115       bInit = lmInfo$coef[-1]
116       tauInit = length(datalist$y) / sum(lmInfo$res^2)
117       list(
118           b0 = 0 ,
119           b = bInit ,
120           tau = tauInit
121       )
122   }
123   for ( chainIdx in 1 : nChain ) {
124       modelInits( bugsInits( genInitList ) )
125   }
126
127   #------------------------------------------------------------------------------
128   # RUN THE CHAINS
129
130   # burn in
131   BurnInSteps = 100
```

```
132    modelUpdate( BurnInSteps )
133    # actual samples
134    samplesSet( c( "b0" , "b" , "tau" ) )
135    stepsPerChain = ceiling(10000/nChain)
136    thinStep = 2
137    modelUpdate( stepsPerChain , thin=thinStep )
138
139    #------------------------------------------------------------------------------
140    # EXAMINE THE RESULTS
141
142    source("plotChains.R")
143    source("plotPost.R")
144
145    checkConvergence = F
146    if ( checkConvergence ) {
147      b0Sum  = plotChains( "b0"  , saveplots=F , filenameroot=fname )
148      bSum   = plotChains( "b"   , saveplots=F , filenameroot=fname )
149      tauSum = plotChains( "tau" , saveplots=F , filenameroot=fname )
150    }
151
152    # Extract chain values:
153    zb0Samp = matrix( samplesSample( "b0" ) )
154    zbSamp = NULL
155    for ( j in 1:nPredictors ) {
156        zbSamp = cbind( zbSamp , samplesSample( paste("b[",j,"]",sep="") ) )
157    }
158    zTauSamp = matrix( samplesSample( "tau" ) )
159    zSigmaSamp = 1 / sqrt( zTauSamp ) # Convert precision to SD
160    chainLength = length(zTauSamp)
161
162    # Convert to original scale:
163    bSamp = zbSamp * matrix( sd(y)/apply(x,2,sd) , byrow=TRUE ,
164                        ncol=nPredictors , nrow=NROW(zbSamp) )
165    b0Samp = ( zb0Samp * sd(y)
166            + mean(y)
167            - rowSums( zbSamp
168            * matrix( sd(y)/apply(x,2,sd) , byrow=TRUE ,
169                        ncol=nPredictors , nrow=NROW(zbSamp) )
170            * matrix( apply(x,2,mean) , byrow=TRUE ,
171                        ncol=nPredictors , nrow=NROW(zbSamp) ) ) )
172    sigmaSamp = zSigmaSamp * sd(y)
173
174    # Save MCMC sample:
175    save( b0Samp , bSamp , sigmaSamp ,
176        file="MultipleLinearRegressionBrugsGuber1999.Rdata" )
177
178    # Scatter plots of parameter values, pairwise:
179    if ( nPredictors <= 6 ) { # don't display if too many predictors
180        windows()
181        thinIdx = round(seq(1,length(zb0Samp),length=200))
182        pairs( cbind( zSigmaSamp[thinIdx] , zb0Samp[thinIdx] , zbSamp[thinIdx,] ) ,
183            labels=c("Sigma zy","zIntercept",paste("zSlope",predictorNames,sep="")))
184        windows()
185        thinIdx = seq(1,length(b0Samp),length=700)
```

```
186     pairs( cbind( sigmaSamp[thinIdx] , b0Samp[thinIdx] , bSamp[thinIdx,] ) ,
187        labels=c( "Sigma y" , "Intercept", paste("Slope",predictorNames,sep="")))
188     dev.copy2eps(file=paste(fname,"PostPairs.eps",sep=""))
189  }
190  # Show correlation matrix on console:
191  cat("\nCorrlations of posterior sigma, b0, and bs:\n")
192  show( cor( cbind( sigmaSamp , b0Samp , bSamp ) ) )
193
194  # Display the posterior:
195  nPlotPerRow = 5
196  nPlotRow = ceiling((2+nPredictors)/nPlotPerRow)
197  nPlotCol = ceiling((2+nPredictors)/nPlotRow)
198  windows(3.5*nPlotCol,2.25*nPlotRow)
199  layout( matrix(1:(nPlotRow*nPlotCol),nrow=nPlotRow,ncol=nPlotCol,byrow=T) )
200  par( mar=c(4,3,2.5,0) , mgp=c(2,0.7,0) )
201  histInfo = plotPost( sigmaSamp , xlab="Sigma Value" , compVal=NULL ,
202                       breaks=30 , main=bquote(sigma[y]) ,
203                       cex.main=1.67 , cex.lab=1.33 )
204  histInfo = plotPost( b0Samp , xlab="Intercept Value" , compVal=NULL ,
205                       breaks=30 , main=bquote(.(predictedName) *" at "* x==0) ,
206                       cex.main=1.67 , cex.lab=1.33 )
207  for ( sIdx in 1:nPredictors ) {
208  histInfo = plotPost( bSamp[,sIdx] , xlab="Slope Value" , compVal=0.0 ,
209                       breaks=30 ,
210                       main=bquote( Delta * .(predictedName) /
211                                    Delta * .(predictorNames[sIdx]) ) ,
212                       cex.main=1.67 , cex.lab=1.33 )
213  }
214  dev.copy2eps(file=paste(fname,"PostHist.eps",sep=""))
215
216  # Posterior prediction:
217  # Specify x values for which predicted y's are needed.
218  # xPostPred is a matrix such that ncol=nPredictors and nrow=nPostPredPts.
219  xPostPred = rbind(
220     apply(x,2,mean)-3*apply(x,2,sd) , # mean of data x minus thrice SD of data x
221     apply(x,2,mean)                 , # mean of data x
222     apply(x,2,mean)+3*apply(x,2,sd)   # mean of data x plus thrice SD of data x
223  )
224  # Define matrix for recording posterior predicted y values for each xPostPred.
225  # One row per xPostPred value, with each row holding random predicted y values.
226  postSampSize = chainLength
227  yPostPred = matrix( 0 , nrow=NROW(xPostPred) , ncol=postSampSize )
228  # Define matrix for recording HDI limits of posterior predicted y values:
229  yHDIlim = matrix( 0 , nrow=NROW(xPostPred) , ncol=2 )
230  # Generate posterior predicted y values.
231  # This gets only one y value, at each x, for each step in the chain.
232  for ( chainIdx in 1:chainLength ) {
233     yPostPred[,chainIdx] = rnorm( NROW(xPostPred) ,
234                           mean = b0Samp[chainIdx]
235                                  + xPostPred %*% cbind(bSamp[chainIdx,]) ,
236                           sd = rep( sigmaSamp[chainIdx] , NROW(xPostPred) ) )
237  }
238  source("HDIofMCMC.R")
239  for ( xIdx in 1:NROW(xPostPred) ) {
```

```
240      yHDIlim[xIdx,] = HDIofMCMC( yPostPred[xIdx,] )
241    }
242    cat( "\nPosterior predicted y for selected x:\n" )
243    show( cbind( xPostPred , yPostPredMean=rowMeans(yPostPred) , yHDIlim ) )
244
245    #-----------------------------------------------------------------------
```

17.5.2 Multiple Linear Regression with Hyperprior on Coefficients

This program was used to create Figure 17.7, among others. See the comments regarding inprod() before the previous program.

(MultiLinRegressHyperBrugs.R)
```
1    graphics.off()
2    rm(list=ls(all=TRUE))
3    fname = "MultiLinRegressHyper"
4    library(BRugs)              # Kruschke, J. K. (2010). Doing Bayesian data analysis:
5                                # A Tutorial with R and BUGS. Academic Press / Elsevier.
6    #-----------------------------------------------------------------------
7    # THE MODEL.
8
9    modelstring = "
10   # BUGS model specification begins here...
11   model {
12       for( i in 1 : nData ) {
13           y[i] ~ dnorm( mu[i] , tau )
14           mu[i] <- b0 + inprod( b[] , x[i,] )
15       }
16       tau ~ dgamma(.01,.01)
17       b0 ~ dnorm(0,1.0E-12)
18       for ( j in 1:nPredictors ) {
19           b[j] ~ dt( muB , tauB , tdfB )
20       }
21       muB ~ dnorm( 0 , .100 )
22       tauB ~ dgamma(.01,.01)
23       udfB ~ dunif(0,1)
24       tdfB <- 1 + tdfBgain * ( -log( 1 - udfB ) )
25   }
26   # ... end BUGS model specification
27   " # close quote for modelstring
28   writeLines(modelstring,con="model.txt")
29   modelCheck( "model.txt" )
30
31   #-----------------------------------------------------------------------
32   # THE DATA.
33
34   tdfBgain = 1
35
36   dataSource = c("Guber1999","McIntyre1994","random")[3]
37
38   if ( dataSource=="Guber1999" ) {
39       fname = paste("Guber1999Brugs","tdf",tdfBgain,sep="")
```

```
40        dataMat = read.table( file="Guber1999data.txt" ,
41                          col.names = c( "State","Spend","StuTchRat","Salary",
42                                      "PrcntTake","SATV","SATM","SATT") )
43      # Specify variables to be used in BUGS analysis:
44      predictedName = "SATT"
45      predictorNames = c( "Spend" , "PrcntTake" )
46      #predictorNames = c( "Spend" , "PrcntTake" , "Salary" , "StuTchRat" )
47      nData = NROW( dataMat )
48      y = as.matrix( dataMat[,predictedName] )
49      x = as.matrix( dataMat[,predictorNames] )
50      nPredictors = NCOL( x )
51    }
52
53    if ( dataSource=="McIntyre1994Hyper" ) {
54      fname = paste("McIntyre1994Brugs","tdf",tdfBgain,sep="")
55      dataMat = read.csv(file="McIntyre1994data.csv")
56      predictedName = "CO"
57      predictorNames = c("Tar","Nic","Wt")
58      nData = NROW( dataMat )
59      y = as.matrix( dataMat[,predictedName] )
60      x = as.matrix( dataMat[,predictorNames] )
61      nPredictors = NCOL( x )
62    }
63
64    if ( dataSource=="random" ) {
65      fname = paste("RandomHyper","tdf",tdfBgain,sep="")
66      # Generate random data.
67      # True parameter values:
68      betaTrue = c( 100 , 1 , 2 , rep(0,21) ) # beta0 is first component
69      nPredictors = length( betaTrue ) - 1
70      sdTrue = 2
71      tauTrue = 1/sdTrue^2
72      # Random X values:
73      set.seed(47405)
74      xM = 5 ; xSD = 2
75      nData = 100
76      x = matrix( rnorm( nPredictors*nData , xM , xSD ) , nrow=nData )
77      predictorNames = colnames(x) = paste("X",1:nPredictors,sep="")
78      # Random Y values generated from linear model with true parameter values:
79      y = x %*% matrix(betaTrue[-1],ncol=1) + betaTrue[1] + rnorm(nData,0,sdTrue)
80      predictedName = "Y"
81      # Select which predictors to include
82      includeOnly = 1:nPredictors # default is to include all
83      #includeOnly = 1:6 # subset of predictors overwrites default
84      x = x[,includeOnly]
85      predictorNames = predictorNames[includeOnly]
86      nPredictors = NCOL(x)
87    }
88
89    # Prepare data for BUGS:
90    # Re-center data at mean, to reduce autocorrelation in MCMC sampling.
91    # Standardize (divide by SD) to make initialization easier.
92    standardizeCols = function( dataMat ) {
93      zDataMat = dataMat
```

```
94      for ( colIdx in 1:NCOL( dataMat ) ) {
95          mCol = mean( dataMat[,colIdx] )
96          sdCol = sd( dataMat[,colIdx] )
97          zDataMat[,colIdx] = ( dataMat[,colIdx] - mCol ) / sdCol
98      }
99      return( zDataMat )
100 }
101 zx = standardizeCols( x )
102 zy = standardizeCols( y )
103
104 # Get the data into BUGS:
105 datalist = list(
106          x = zx ,
107          y = as.vector( zy ) , # BUGS does not treat 1-column mat as vector
108          nPredictors = nPredictors ,
109          nData = nData ,
110          tdfBgain = tdfBgain
111 )
112 modelData( bugsData( datalist ) )
113
114 #------------------------------------------------------------------------
115 # INTIALIZE THE CHAINS.
116
117 nChain = 3
118 modelCompile( numChains = nChain )
119
120 genInitList <- function(nPred=nPredictors) {
121      lmInfo = lm( y ~ x ) # R function returns least-squares (normal MLE) fit.
122      bInit = lmInfo$coef[-1] * apply(x,2,sd) / sd(y)
123      tauInit = (length(y)*sd(y)^2)/sum(lmInfo$res^2)
124      list(
125          b0 = 0 ,
126          b = bInit ,
127          tau = tauInit ,
128          muB =  mean( bInit ) ,
129          tauB = 1 / sd( bInit )^2 ,
130          udfB = 0.95 # tdfB = 4
131      )
132 }
133 for ( chainIdx in 1 : nChain ) {
134      modelInits( bugsInits( genInitList ) )
135 }
136
137 #------------------------------------------------------------------------
138 # RUN THE CHAINS
139
140 # burn in
141 BurnInSteps = 100
142 modelUpdate( BurnInSteps )
143 # actual samples
144 samplesSet( c( "b0" , "b" , "tau" , "muB" , "tauB" , "tdfB" ) )
145 stepsPerChain = ceiling(10000/nChain)
146 thinStep = 2
147 modelUpdate( stepsPerChain , thin=thinStep )
```

```
148   #----------------------------------------------------------------
149   # EXAMINE THE RESULTS
150
151
152   source("plotChains.R")
153   source("plotPost.R")
154
155   checkConvergence = F
156   if ( checkConvergence ) {
157     b0Sum  = plotChains( "b0"   , saveplots=F , filenameroot=fname )
158     bSum   = plotChains( "b"    , saveplots=F , filenameroot=fname )
159     tauSum = plotChains( "tau"  , saveplots=F , filenameroot=fname )
160     muBSum = plotChains( "muB"  , saveplots=F , filenameroot=fname )
161     tauBSum = plotChains( "tauB" , saveplots=F , filenameroot=fname )
162     tdfBSum = plotChains( "tdfB" , saveplots=F , filenameroot=fname )
163   }
164
165   # Extract chain values:
166   zb0Samp = matrix( samplesSample( "b0" ) )
167   zbSamp = NULL
168   for ( j in 1:nPredictors ) {
169      zbSamp = cbind( zbSamp , samplesSample( paste("b[",j,"]",sep="") ) )
170   }
171   zTauSamp = matrix( samplesSample( "tau" ) )
172   zSigmaSamp = 1 / sqrt( zTauSamp ) # Convert precision to SD
173   chainLength = length(zTauSamp)
174
175   # Convert to original scale:
176   bSamp = zbSamp * matrix( sd(y)/apply(x,2,sd) , byrow=TRUE ,
177                      ncol=nPredictors , nrow=NROW(zbSamp) )
178   b0Samp = ( zb0Samp * sd(y)
179            + mean(y)
180            - rowSums( zbSamp
181            * matrix( sd(y)/apply(x,2,sd) , byrow=TRUE ,
182                      ncol=nPredictors , nrow=NROW(zbSamp) )
183            * matrix( apply(x,2,mean) , byrow=TRUE ,
184                      ncol=nPredictors , nrow=NROW(zbSamp) ) ) )
185   sigmaSamp = zSigmaSamp * sd(y)
186
187   # Scatter plots of parameter values, pairwise:
188   if ( nPredictors <= 6 ) { # don't display if too many predictors
189      windows()
190      thinIdx = round(seq(1,length(zb0Samp),length=200))
191      pairs( cbind( zSigmaSamp[thinIdx] , zb0Samp[thinIdx] , zbSamp[thinIdx,] )  ,
192        labels=c("Sigma zy","zIntercept",paste("zSlope",predictorNames,sep="")))
193      windows()
194      thinIdx = seq(1,length(b0Samp),length=700)
195      pairs( cbind( sigmaSamp[thinIdx] , b0Samp[thinIdx] , bSamp[thinIdx,] ) ,
196        labels=c( "Sigma y" , "Intercept", paste("Slope",predictorNames,sep="")))
197      dev.copy2eps(file=paste(fname,"PostPairs.eps",sep=""))
198   }
199   # Show correlation matrix on console:
200   cat("\nCorrlations of posterior sigma, b0, and bs:\n")
```

```
201   show( cor( cbind( sigmaSamp , b0Samp , bSamp ) ) )
202
203   # Display the posterior:
204   nPlotPerRow = 5
205   nPlotRow = ceiling((2+nPredictors)/nPlotPerRow)
206   nPlotCol = ceiling((2+nPredictors)/nPlotRow)
207   windows(3.5*nPlotCol,2.25*nPlotRow)
208   layout( matrix(1:(nPlotRow*nPlotCol),nrow=nPlotRow,ncol=nPlotCol,byrow=T) )
209   par( mar=c(4,3,2.5,0) , mgp=c(2,0.7,0) )
210   histInfo = plotPost( sigmaSamp , xlab="Sigma Value" , compVal=NULL ,
211                        breaks=30 , main=bquote(sigma[y]) ,
212                        cex.main=1.67 , cex.lab=1.33 )
213   histInfo = plotPost( b0Samp , xlab="Intercept Value" , compVal=NULL ,
214                        breaks=30 , main=bquote(.(predictedName) *" at "* x==0) ,
215                        cex.main=1.67 , cex.lab=1.33 )
216   for ( sIdx in 1:nPredictors ) {
217   histInfo = plotPost( bSamp[,sIdx] , xlab="Slope Value" , compVal=0.0 ,
218                        breaks=30 ,
219                        main=bquote( Delta * .(predictedName) /
220                                     Delta * .(predictorNames[sIdx]) ) ,
221                        cex.main=1.67 , cex.lab=1.33 )
222   }
223   dev.copy2eps(file=paste(fname,"PostHist.eps",sep=""))
224
225   # Posterior prediction:
226   # Specify x values for which predicted y's are needed.
227   # xPostPred is a matrix such that ncol=nPredictors and nrow=nPostPredPts.
228   xPostPred = rbind(
229       apply(x,2,mean)-3*apply(x,2,sd) , # mean of data x minus thrice SD of data x
230       apply(x,2,mean)                 , # mean of data x
231       apply(x,2,mean)+3*apply(x,2,sd)   # mean of data x plus thrice SD of data x
232   )
233   # Define matrix for recording posterior predicted y values for each xPostPred.
234   # One row per xPostPred value, with each row holding random predicted y values.
235   postSampSize = chainLength
236   yPostPred = matrix( 0 , nrow=NROW(xPostPred) , ncol=postSampSize )
237   # Define matrix for recording HDI limits of posterior predicted y values:
238   yHDIlim = matrix( 0 , nrow=NROW(xPostPred) , ncol=2 )
239   # Generate posterior predicted y values.
240   # This gets only one y value, at each x, for each step in the chain.
241   for ( chainIdx in 1:chainLength ) {
242       yPostPred[,chainIdx] = rnorm( NROW(xPostPred) ,
243                             mean = b0Samp[chainIdx]
244                                    + xPostPred %*% cbind(bSamp[chainIdx,]) ,
245                             sd = rep( sigmaSamp[chainIdx] , NROW(xPostPred) ) )
246   }
247   source("HDIofMCMC.R")
248   for ( xIdx in 1:NROW(xPostPred) ) {
249       yHDIlim[xIdx,] = HDIofMCMC( yPostPred[xIdx,] )
250   }
251   cat( "\nPosterior predicted y for selected x:\n" )
252   show( cbind( xPostPred , yPostPredMean=rowMeans(yPostPred) , yHDIlim ) )
253
254   #- - - - - - - - - - - - - - - - - - - - - - - - - - - - - - - - - - - - - - - - - - -
```

17.6 EXERCISES

Exercise 17.1. [Purpose: To view the prior on the regression coefficients, when there is a hyperprior.] The hyperprior on regression coefficients in Figure 17.6 may be difficult to intuit. Therefore, generate graphs of the *prior distribution* on the regression coefficients for the program in Section 17.5.2 (`MultiLinRegress HyperBrugs.R`), when `tdf` is set at different values. To accomplish this, do the following:

1. Because BUGS balks at extreme values of tau, and extreme values can be sampled under the vague prior, the gamma distributions for tau need to be censored. Therefore, change the tau specifications to
   ```
   tau  ~ dgamma(.01,.01)I(0.0001,10000)
   tauB ~ dgamma(.01,.01)I(0.0001,10000)
   ```
2. In the `datalist`, comment out only the single line that specifies the y values. The other lines, that specify the x values and the number of predictors, and so on, must remain, because they specify the structure of the model. The model only predicts y as a function of x; the model does not predict x.
3. In the initialization of the chains, comment out the data-based initialization, and instead use `modelGenInits()`.
4. Do not use the `plotChains()` command, because some of the MCMC values may be too extreme for some of the BUGS graphics routines that are called by `plotChains()`, and BUGS will crash.
5. When plotting the slope values in the last part of the code, manually specify the axis limits so that you can see the central parts of the distributions, perhaps like this:

```
histInfo = plotPost( bSamp[,sIdx] , xlab="Slope Value" , compVal=0.0 ,
        breaks=c(-1000000,seq(-400,400,21),1000000),xlim=c(-400,400) , ...)
```

Run the program and display the prior for tdfBgain=1 and for tdfBgain=100. Point out and discuss any differences in the priors for those different hyperprior constants.

Exercise 17.2. [Purpose: To conduct a power analysis for multiple regression.] Consider the SAT data shown in Figure 17.3, p. 457, and the posterior for a linear regression model (with no interaction) shown in Figure 17.5, p. 461. The marginal posterior distribution for the slope on spending per pupil has a 95% HDI that excludes zero, and that has a width of very nearly 17. Consider two different goals for the research. One goal is to show that the 95% HDI on spending per pupil excludes a ROPE of $[-1, +1]$, and the second goal is to show that the width of the 95% HDI is less than 10. How can we assess the probability of achieving these goals?

To address the question, we think of the y values as being randomly generated according to the linear regression model, at the x values that are provided by the world. The model only describes the dependency of y on x; the model does not describe the distribution of the x values. In different applications the x values have different actual generators. In experiments, the x values are selected by the experimenter, and power analysis can explicitly manipulate the values of the x's and the frequencies of each. In observational studies, the x values are generated by the world, not by the experimenter. In some observational studies, we can think of the x values as randomly drawn from some underlying distribution (e.g., x could be a person's height) and we can easily randomly sample another person to get another x value. In other observational studies, it is a conceptual stretch to think of x as being a random value that is sampled from some underlying population (e.g., spending per pupil in each of the 50 states). But even in that last case, we can think of each state as being representative of some universe of possible states, from which 50 were drawn.

In the present nonexperimental, observational study (i.e., the SAT data for 50 states), the x values are not selected by researcher. Moreover, we have no model of the x values. Therefore, we use the actual x values themselves as the best available description of the underlying distribution of x values. When we simulate a new datum, we first randomly select one of the points from the actual x values, and then we randomly generate a y value according the credible parameter values in the model.

In our specific situation, there are only 50 states altogether; therefore we can do a retrospective power analysis using a sample size of 50. But we can also imagine increasing the sample size if, instead of using state-average data, we imagine that the data are from individual school districts within states. Under this interpretation, the posterior from the state-average data, and the range of x values, might not be fully representative of district data, but at least it's better than nothing.

Answer these questions: What is the retrospective power for each of the two goals? (Hint: It's about 0.67 and 0.00, when the x points are randomly sampled with replacement from the 50 actual points.) What is the power for each of the two goals when $N = 130$? (Hint: It's about 0.85 and 0.30.) What's the minimal N needed to achieve a power of 0.80 for the second goal?

Programming hints: Use the power-analysis program of Section 13.6.2 (FilconBrugsPower.R) as a template for adapting the linear regression program in Section 17.5.1 (MultipleLinearRegressionBrugs.R). You'll want to pass the raw data into the GoalAchievedForSample() function because that way the data can be standardized for BUGS but the resulting parameter values can be converted back to the original scale (using information from the original data).

CHAPTER 18

Metric Predicted Variable with One Nominal Predictor

Familywise error rates breed rumors of incest,
Hounding for quarry in multiple t tests.
Barking at research, poor dog got run over;
Should have done Bayesian oneway ANOVA.

In this chapter we consider situations with a metric predicted variable and a nominally scaled predictor variable. These cases occur frequently in real-world research. For example, we may want to predict weight loss (a metric variable) as a function of which diet the person follows (e.g., low-carb, vegetarian, or low-fat). As another example, we may want to predict severity of psychosis (measured on a metric scale) as a function of which antipsychotic drug the person takes. Or we may want to predict income as a function of political party affiliation. This combination of predicted and predictor scale types occurs in the first row, fourth cell, of Table 14.1 (p. 385).

Doing Bayesian Data Analysis: A Tutorial with R and BUGS. DOI: 10.1016/B978-0-12-381485-2.00018-3
© 2011, Elsevier Inc. All rights reserved.

In traditional NHST analyses, these situations are addressed by "oneway analysis of variance" (ANOVA). The term *oneway* refers to the fact that a single nominal variable is being used as the predictor. The phrase *analysis of variance* refers to the fact that the overall variance across all the data is decomposed (i.e., analyzed) into two parts: variance within the levels of the nominal predictors and variance between the levels of the nominal predictors. The variance within levels of the nominal predictor is called *noise* or *error* (i.e., variability that cannot be predicted by the predictor). The complementary variance between the levels of the nominal predictor is called the *effect* of the predictor. Usually we do the research with the goal of detecting an effect, which means that we would like the magnitude of the variance between levels to be large compared to the noise within levels. The ratio of variance between to variance within is called the *F-ratio*. In the Bayesian approach, we rarely if ever refer to the *F-ratio*. But because the model we use is based on the model of traditional ANOVA, we will refer to our analysis as Bayesian ANOVA or sometimes BANOVA.

18.1 BAYESIAN ONEWAY ANOVA

The basic idea of oneway ANOVA was introduced in Section 14.1.6.1, p. 368. The predictor is a variable measured on a nominal scale. For example, if income is predicted as a function of political party affiliation, notice that the predictor has nominal levels such as libertarian, green, democratic, republican, and so on. We denote the predictor variable as $\vec{x}$, which is a vector with one component per nominal level. For example, suppose that the predictor is political party affiliation, with Green as level 1, Democrat as level 2, Republican as level 3, Libertarian as level 4, and Other as level 5. Then Democrat is represented as $\vec{x} = \langle 0, 1, 0, 0, 0 \rangle$, and Libertarian is represented as $\vec{x} = \langle 0, 0, 0, 1, 0 \rangle$. Political party affiliation is being treated here as a categorical label only, with no ordering along a liberal-conservative scale.

The formal model indicates how to derive the predicted value from the predictor. The idea is that there is a baseline quantity of the predicted variable, and each level of the predictor indicates a deflection above or below that baseline. We will denote the baseline value of the prediction as β_0. The deflection for the jth level of the predictor is denoted β_j. When the predictor has value $\vec{x}_i = \langle \ldots, x_{ji}, \ldots \rangle$, then the predicted value is

$$\mu_i = \beta_0 + \sum_j \beta_j x_{ji}$$

$$= \beta_0 + \vec{\beta} \cdot \vec{x}_i \qquad (18.1)$$

where the notation $\vec{\beta} \cdot \vec{x}$ denotes the *dot product* of the vectors. In Equation 18.1, the coefficient β_j indicates how much μ changes when $\vec{x}$ changes from neutral to level j. In other words, β_j indicates how much μ changes when

$\vec{x}$ changes from all $x_j = 0$ to $x_j = 1$. The baseline is constrained such that the deflections sum to zero across the levels of $\vec{x}$:

$$\sum_{j=1} \beta_j = 0 \qquad (18.2)$$

The expression of the model in Equation 18.1 is not complete without the constraint in Equation 18.2. Examples were shown in Figure 14.4, p. 370, and it is worth the effort to go now to that Figure for a quick review.

The predicted value, μ_i, in Equation 18.1 is for the central tendency in the data. The data themselves are assumed to be randomly generated around that central tendency. As usual, we will assume a normal distribution, $y_i \sim N(\mu_i, \tau)$, where τ is the precision of the normal distribution. As discussed in previous chapters, if the data have outliers, a t distribution may be used instead.

18.1.1 The Hierarchical Prior

Our primary interest is in estimating the deflection parameters, β_j, for each level of $\vec{x}$. We could just put a separate prior on each parameter and estimate them separately from each other. It is typical, however, that the levels of $\vec{x}$ are not utterly unrelated to each other, and therefore data from one level may inform estimates in another level. For example, the deflections for republicans, libertarians, and greens can inform an estimate of the deflection for democrats. Thus, if the deflection for libertarians is $+1.0$, for republicans is $+0.5$, and for greens is -1.0, then the deflection for democrats should be somewhere in that general range, and not out at, say, -12.0. At the least, we might have prior beliefs that the deflections for most levels of $\vec{x}$ may be small, with only a few deflections being large, and therefore we can let the various levels mutually inform each other's estimates based on this structural assumption.

The form of the hierarchical model for oneway BANOVA is displayed in Figure 18.1 (Gelman, 2005, 2006). In the upper middle of the diagram is the normal distribution that describes the distribution of deflections, β_j, across levels of $\vec{x}$. This normal distribution has a mean at zero, reflecting the fact that the deflections are constrained to fall both above and below the baseline, because they must sum to zero. Importantly, the precision of this normal distribution, τ_β, is estimated, not preset at a constant. Thus, if many of the levels of $\vec{x}$ have a small deflection in the data, then the precision τ_β is estimated to be high, and this in turn shrinks the estimates of other β_j.

The prior for τ_β derives from the recommendation of Gelman (2006). First, the precision is converted to standard deviation: $\tau_\beta = 1/\sigma_\beta^2$. Then a folded-$t$ distribution is used as a prior on σ_β. The folded-t is just the positive side of the usual t distribution. Notice that it is defined only over nonzero values, as is required for σ_β, and it extends to positive infinity. Unlike

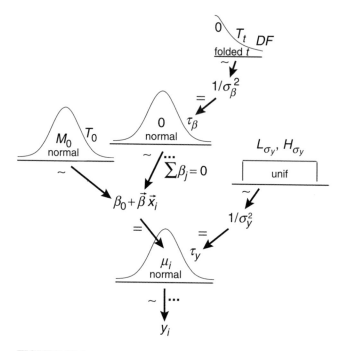

FIGURE 18.1

Hierarchical dependencies for model of oneway Bayesian ANOVA. The baseline is β_0, and the deflection away from that baseline for the jth level of x is β_j. The standard deviation of the β_j's has a folded-t prior. The variance within levels of x is estimated by the precision τ_y, which is here assumed to be homogeneous across groups, although it need not be in general.

the gamma(ϵ, ϵ) distribution that is often used for precisions, however, the folded-t does not have infinite density near zero. Because noisy data can never rule out deflections of all zero, there can be unintended distortions in the estimates if the prior places extreme densities at either end of the scale (see Gelman, 2006, for details).

A folded-t prior could also be used for the noise σ_y, but we will use a uniform, again as recommended by Gelman (2006). One motivation is that a uniform may have a more intuitive form than a folded-t when expressing a prior belief. A second reason is that the infelicities of estimation that affect σ_β are not present so prominently at this level in the model, because the within-level noise is typically not near zero, and there are enough data points to overwhelm any mildly informed prior.

18.1.1.1 Homogeneity of Variance

The model described here assumes equal variances across all levels of $\vec{x}$. As a concrete example, the model assumes that the variance of income is the same for republicans as for democrats as for libertarians as for greens. This assumption of homogeneous variances is vestigial from two precursors. First, the analogous assumption is made in linear regression, and ANOVA may be construed, mathematically, as a special case of linear regression. Second, homogeneity of variance is assumed in NHST ANOVA to simplify derivation of F distributions. Neither of these precursors actually demands that we make the assumption of equal variances in BANOVA.

The model described here assumes homogeneity of variance merely for simplicity in presentation. By assuming equal variances for all levels of x, the focus could be on the estimation of the group deflection parameters, β_j. Also, by assuming equal variances, the results of BANOVA can be more directly compared to the results of NHST ANOVA, if such a comparison desired.

In principle, the BANOVA model can (and should) estimate difference variances for each level of $\vec{x}$. The model in Figure 18.1 can be expanded analogously to the model in Figure 16.11, p. 436. Instead of a single precision, τ_y, used for all levels of $\vec{x}$, a separate precision τ_j is estimated for each level of $\vec{x}$, as in the lower right of Figure 16.11. A higher-level distribution describes the spread of the τ_j across levels of $\vec{x}$. This structure provides shrinkage of the estimates of the τ_j, to the extent that the data suggest homogeneity of variance. Exercise 18.3 has you give this scheme a test drive.

18.1.2 Doing It with R and BUGS

As usual, every arrow in the hierarchical diagram of Figure 18.1 has a corresponding line in the BUGS model specification. The parameters that appear as "β_j" in Figure 18.1 are denoted by "a[j]" in the model specification.

To understand the way that the model is specified in the BUGS code, it is important to understand how the data are formatted. The $\vec{x}$ values in the program are coded as integer indices 1, 2, 3,..., and *not* as vectors $\langle 1, 0, 0, \ldots \rangle$, $\langle 0, 1, 0, \ldots \rangle$, $\langle 0, 0, 1, \ldots \rangle$,.... By coding $\vec{x}$ as integers, then nested indexing can be used instead of dot products of vectors. Thus, $\vec{\beta} \cdot \vec{x}$ becomes coded as a[x], not inprod(a[],x[]). For the ith observation, the value of x is coded as x[i]. Thus, x[i] $\in \{1, 2, 3, \ldots, \text{NxLvl}\}$ for $i \in \{1, \ldots, \text{Ntotal}\}$, where NxLvl is the number of levels of $\vec{x}$ and Ntotal is the total number of observations.

Here is the BUGS model specification (ANOVAonewayBRugs.R):

```
11    model {
12      for ( i in 1:Ntotal ) {
13        y[i] ~ dnorm( mu[i] , tau )
14        mu[i] <- a0 + a[x[i]]
15      }
16      #
17      tau <- pow( sigma , -2 )
18      sigma ~ dunif(0,10) # y values are assumed to be standardized
```

```
19    #
20    a0 ~ dnorm(0,0.001) # y values are assumed to be standardized
21    #
22    for ( j in 1:NxLvl ) { a[j] ~ dnorm( 0.0 , atau ) }
23    atau <- 1 / pow( aSD , 2 )
24    aSD <- abs( aSDunabs ) + .1
25    aSDunabs ~ dt( 0 , 0.001 , 2 )
26    }
```

The constraint, that the deflections sum to zero, does not appear in the model specification. The BUGS code estimates the baseline and deflections without the constraint, but the MCMC estimates are recentered at zero by subsequent R code. The noncentered baseline is denoted in the BUGS model as a0, and the noncentered deflections are denoted a[j]. Those noncentered estimates are transformed to respect the sum-to-zero constraint merely by subtracting the mean of the a[j]'s from each a[j], and adding the mean to the baseline. Thus, b[j] = a[j] - mean(a) and b0 = a0 + mean(a).

The constants for the top-level priors are set with the assumption that the data values, y, have been standardized according to Equation 16.1, p. 425. (Of course, the x values cannot be standardized because they are nominal.) This standardization makes it easier to establish reasonable default priors for a range of applications, without having to change the prior constants when the application changes, for example, from income, on the order of 10^5 dollars, to width of hairs, on the order of 10^{-1} millimeters. Nevertheless, when there is strong prior information, it should be incorporated. Exercise 18.2 has you explore robustness of the results when you use different priors.

There is one other trick in the BUGS model specification that is not in the hierarchical diagram of Figure 18.1. One line of the BUGS model specifies that the standard deviation of the group effects, denoted aSDunabs, comes from a t distribution: aSDunabs ~ dt(0,0.001,2). Another line takes the absolute value to "fold" the t distribution onto the non-negative numbers: aSD <- abs(aSDunabs) + .1. But that line also mysteriously adds a small constant, namely 0.1. This constant keeps aSD from venturing extremely close to zero. The reason for keeping aSD away from zero is that shrinkage can become overwhelmingly strong when there are many groups with few data points per group. This becomes especially problematic in the next chapter when we consider interaction of factors.

It turns out that MCMC sampling for this model can be extremely inefficient. One important way to reduce burn-in time is to start the chain at reasonable positions. We start the overall baseline at the grand mean of the data, and start the deflections at the level means minus the grand mean. The variances are also initialized near the corresponding data variances. The full code, including initialization of chains, is presented in Section 18.4.1 (ANOVAonewayBRugs.R).

Because the chains can be highly autocorrelated, extensive thinning is needed, keeping a step only once out of several hundred. Running such long chains can take a long time and become boring for your computer, which would rather be searching the web for exciting software updates. In the examples presented here, we simply tolerate the modest waiting times. But there are various methods for reparameterizing the models so that the chains are sampled with less autocorrelation (e.g., Gelman, 2006; Gelman & Hill, 2007, Ch. 19).

One tempting but inappropriate approach is to impose the sum-to-zero constraint in the BUGS model specification like this:

```
a[1] <- -sum( a[2:NxLvl] )
for ( j in 2:NxLvl ) { a[j] ~ dnorm( 0.0 , atau ) }
```

Notice that the first deflection is forced to equal the negative sum of the remaining deflections; therefore the first deflection is not an estimated parameter. Only deflections indexed 2 and higher have a prior specification. This approach works fine when the prior on the deflections has no hyperprior, that is, when `atau` is a constant (Ntzoufras, 2009). But when `atau` is itself being estimated, it must be informed by all the deflections, not only by deflections 2 and higher. For example, it might be that group 1 is very different from groups 2 through NxLvl, whereas groups 2 through NxLvl are nearly equal. This situation would cause the estimate of the precision `atau` to be artificially high, because it would not be affected by the group 1. Therefore, despite the fact that this approach to model specification reduces autocorrelation dramatically, the approach is not appropriate when we are using a hyperprior to estimate the deflections.

18.1.3 A Worked Example

With all the emphasis these days on physical fitness and muscle building, it's only appropriate to consider an example about muscles. In particular, we'd like to know if geographical location influences muscle size, which might be affected by the weather or amount of daylight. Consider some data regarding muscles from five geographic locations: (1) Tillamook, Oregon; (2) Newport, Oregon; (3) Petersburg, Alaska; (4) Magadan, Russia (Pacific coast); and (5) Tvarminne, Finland. The values in the data set are the length of the anterior adductor muscle scar divided by total muscle length, in the mussel species Mytilus trossulus. These ratios of scar length to total length tend to be between 5% and 15% (McDonald, 2009; McDonald, Seed, & Koehn, 1991).

Results of the BRugs program listed in Section 18.4.1 (`ANOVAonewayBRugs.R`) are shown in Figure 18.2. The histograms in the upper row show the (marginal) posterior distributions of the β_j values for the five geographical locations. These β_j values are deflections away from the baseline β_0, which is not shown. Some things to keep in mind when interpreting the results: First, the estimates of deflection are subject to shrinkage, because the model

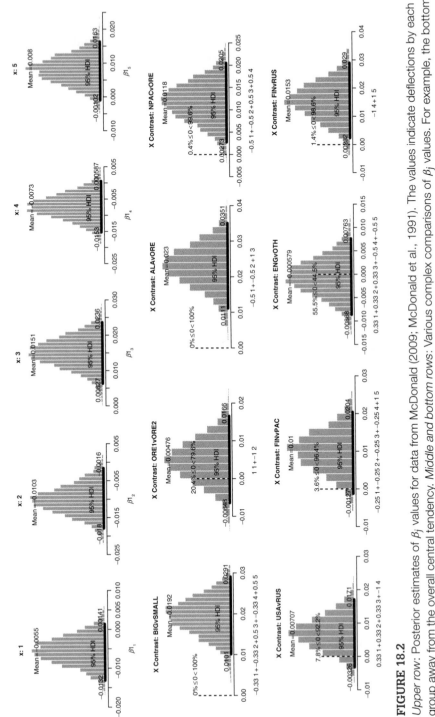

FIGURE 18.2

Upper row: Posterior estimates of β_j values for data from McDonald (2009; McDonald et al., 1991). The values indicate deflections by each group away from the overall central tendency. *Middle and bottom rows*: Various complex comparisons of β_j values. For example, the bottom row, third panel, compares the three English-speaking sites against the two non-English-speaking sites (where they might say "midiya myshtsy" or "simpukka lihas" instead of "mussel muscle").

incorporates the prior structural assumption that all the deflections come from the same overarching distribution. The mean deflections shown in Figure 18.2 are, in fact, a little smaller than the deflections of the actual sample means. Second, the model assumes that the precision is the same for all groups (i.e., there is homogeneity of variance). The posterior β_j values are the ones that are believable when also assuming homogeneous variances. If the groups actually have wildly different variances, then the estimates for β_j may be distorted. Third, the marginal distributions on the β_j cannot be used to directly infer differences between groups, because the parameters might be correlated. Indeed, the deflections tend to be negatively correlated, because increasing the estimated deflection for one group suggests decreasing the estimated deflection for another, if they are to remain symmetric around the baseline. To judge differences between groups, the differences must be computed directly.

18.1.3.1 Contrasts and Complex Comparisons

The middle and bottom rows of Figure 18.2 shows several comparisons for the mussel muscle results. A comparison amounts to a difference between an average of some groups and an average of other groups. For example, to compare the four Pacific Ocean mussels against the one non-Pacific (Baltic Sea) mussel, we multiply the deflections (β_j's) of the first four groups by 1/4 to get their average, and subtract it from the deflection (β_5) of the fifth group to get the difference. The difference is called a *contrast*, and when the comparison involves a contrast of averages, instead of a contrast of two specific groups, it is sometimes called a *complex* comparison. The contrast is fully specified by the coefficients on the groups, which can be placed into a vector of *contrast coefficients*. For example, the contrast coefficients for comparing Pacific Ocean mussels against Baltic Sea mussels are $-1/4, -1/4, -1/4, -1/4, +1$. Notice that the coefficients sum to zero. We compute the difference at every step in the MCMC chain, and examine the resulting distribution of believable differences. The distribution for this particular example is shown in the bottom row, second panel, of Figure 18.2, where it can be seen that just over 96% of the believable differences lie on one side of zero, and the 95% HDI just spans zero. From these results we may not want to declare categorically that there is a credible difference between Finland and the other sites; the decision depends on how you set your HDI and ROPE. Regardless of your decision rule, the posterior does tells us the most believable difference and the uncertainty in that difference.

Figure 18.2 shows a variety of comparisons that might be of interest. For example, the first panel of the middle row compares the two sites with the biggest muscles against the three other sites. This sort of comparison would be labeled "post hoc" by traditional analyses, because we might not have specified which sites would be biggest before collecting the data. The second panel in the middle row contrasts the two sites in Oregon. The third panel in the middle row compares the Alaska site against the average of the two Oregon sites.

We can make all the comparisons shown in Figure 18.2, and as many others as we like, without worrying about inflated false alarm rates, because the posterior distribution does not change when we consider additional comparisons. The posterior distribution is the best inference we can make based on the data we have and the prior beliefs we started with. It is possible that the random data in our sample are spuriously unrepresentative of the underlying population, but we cannot know. Fortunately, because of the incorporation of our prior knowledge about how estimates in the different locations can mutually inform each other, the estimates undergo shrinkage, which helps to mitigate the effect of rogue data. In many applications, the shrinkage yields decisions similar to those that would result from NHST "corrections" for multiple comparisons. But unlike NHST corrections, the shrinkage in the Bayesian approach is based on explicit structural prior knowledge, and is not affected by which or how many comparisons are intended. (For previous discussion of these issues, see Section 17.2, regarding decisions about multiple regression coefficients, and Section 11.4, regarding multiple comparisons of groups.)

18.1.3.2 Is There a Difference?

The contrasts and complex comparisons in Figure 18.2 were judged to be credibly nonzero if the 95% HDI excluded (a ROPE around) zero. A difference would be deemed to be practically equivalent to zero if its HDI fell entirely within a ROPE. This decision procedure is attractive because all the group β_j's are simultaneously estimated, with mutually informed shrinkage, and from priors that are also appropriately informed (which entails also being agreeable to a skeptical audience).

Some researchers prefer to pose the question "Is there a difference?" as a model comparison on two priors. One prior expresses the null hypothesis that the contrast has zero magnitude; the other prior expresses a complementary hypothesis that any magnitude contrast is possible. This approach was discussed extensively in Section 12.2.

There are two attractions to the two-prior, model-comparison approach. One attraction is that the model comparison can yield posterior odds in favor of the null, unlike NHST, which can only reject a null hypothesis but never accept it. Another attraction is that the complementary hypothesis is usually intended to be an "automatic" uninformed prior that is chosen for mathematical felicity. The hope is that an automatic prior obviates debate about how prior information should be expressed.

As was argued in Section 12.2, the two-prior approach should be applied cautiously. First, it is important to emphasize that the two-prior approach only indicates which prior is relatively less unbelievable. If either prior is theoretically untenable in the first place, then the "automatic" model comparison is automatically uninformative. Thus, the two-prior approach should only be applied to situations in which (1) it is theoretically appropriate to posit that a particular contrast really can be exactly zero, and (2) the alternative prior incorporates prior knowledge about the plausible magnitude of the difference.

As an example, consider a situation presented by Solari, Liseo, & Sun (2008, Table 3, p. 495). There were nine groups, with a metric dependent variable. The dependent variable was the acetic acid content of tomatoes, and the nine groups were different types of manuring during growth of the tomatoes. The mean of group 3 appeared to be different than other groups. To test whether group 3 was different, the authors conducted a Bayesian model comparison of two priors: The null-hypothesis prior had all nine groups with identical means. The alternative prior had group 3 with a separately estimated mean, while the other eight groups had identical means. The resulting Bayes factor (BF) strongly favored the alternative prior. Does this result suggest that the alternative prior is what we should believe? Unfortunately, no. The BF tells us that the prior with eight equal means and one different mean for group 3 is more believable than the prior with nine equal means (assuming that the priors on the two hypotheses were 50-50). But the prior with eight equal means, on groups other than group 3, is already untenable because we do not believe that the eight groups have identical means. Moreover, the estimate of difference, between group 3 and the other groups, is not what we want, because the estimate does not take into account variation among the eight other groups.

When instead we conduct a Bayesian analysis using the BANOVA model, we obtain a posterior that simultaneously estimates all the separate group deflections, with shrinkage, from a plausibly informed prior. The complex comparison of group 3 against the other eight groups is shown in Figure 18.3, where it can be seen that the magnitude of the contrast is credibly greater than zero. In this application, there is no need to pursue a BF approach to group comparisons.

It is also worth reiterating that the two-prior, model-comparison approach can arrive at a conclusion opposite that of the one-prior, estimation approach. Recall Figure 12.5, p. 308, which showed that a model comparison preferred the null hypothesis of identical groups to the alternative hypothesis of all different groups, even though an estimation of effects in the alternative hypothesis showed a credible difference among groups. The point in that case was that the null model, even though it was a poor model, was less bad than the alternative model. Follow-up model comparisons would be required to narrow down which combination of group equivalences was least implausible. Even after that, we would not necessarily want to believe that any of the groups are truly equivalent, because we know in advance that they were treated differently. Instead, we desire an estimate of the differences and the precision of the estimate. That situation involved a dichotomous dependent variable, but the analogous situation can arise for metric dependent variables.

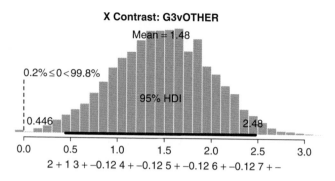

FIGURE 18.3

A comparison of group 3 versus the average of other groups, for the data in Solari et al. (2008, Table 3, p. 495). The specification of contrast coefficients on the x axis overflows the margins of the figure because there are so many groups. The contrast coefficients on the nine groups are $-1/8, -1/8, +1, -1/8, \ldots$, which, when rounded to two decimal places, appear as $-0.12, -0.12, +1, -0.12, \ldots$. (Reprinted with permission from Figure 6 of Kruschke, 2010a,b).

The two-prior, model-comparison approach can be appropriate in situations where actual equivalence is tenable and the goal is to identify which conditions are plausibly equivalent, or situations in which zero-magnitude effects are tenable and the goal is to identify which conditions have zero effect. In those situations, it behooves the researcher to pursue the model-comparison or related approaches (see, e.g., Berry & Hochberg, 1999; Gopalan & Berry, 1998; Mueller, Parmigiani, & Rice, 2007; Scott & Berger, 2006). Moreover, Bayesian model comparison is highly advisable when the two models are genuinely viable competitors that express different explanations of the data. In these situations, it is important that the priors in the two models are equivalently informed so that neither model is at a disadvantage because of an infelicity in an arbitrary, automatic prior.

18.2 MULTIPLE COMPARISONS

In 20th-century null-hypothesis significance testing (NHST), there is immense literature regarding how to compute the "true" significance (i.e., probability of false alarm) of an apparent difference between groups, when the analyst is conducting comparisons of multiple groups. The problem is that when more comparisons are conducted, there are more opportunities for a spuriously large difference to appear by accident. In other words, there are more opportunities for false alarms. Notice that this problem of inflated false alarm rates arises because NHST is based on the intentions of the analyst. If the analyst intends to make lots of comparisons between various combinations of groups, then there is greater opportunity for false alarms. If the analyst intends to make only a few comparisons between groups, then there is less opportunity for false alarms.

For example, consider again the sea mussel data. Group 4 (Pacific coast Russia) and group 5 (Finland) seem to be different, and it is meaningful to plan a comparison between them because of their geographical difference. If we run a two-group t test, we get $t = 2.53, p = 0.028$, which denotes a significant difference. On the other hand, if we run a post hoc test of all pairwise comparisons, using Tukey's "Honest Significant Difference" correction, then we find that $p = 0.093$, and the difference is *not* significant. So do Russia and Finland really differ? According to NHST, the answer depends on your intentions: If you intended to compare only those two locales, then they are significantly different, but if you intended to make all pairwise comparisons, then they are not significantly different.

Section 11.4, p. 281, discussed multiple comparisons in NHST, in the context of a dichotomous dependent variable. Here we reiterate those ideas in the context of a metric dependent variable.

Suppose that we have two groups: One group is patients treated with a placebo and a second group is patients treated with a totally ineffective drug. We measure a metric variable (e.g., body temperature). Because there is no actual difference between the treatments, the underlying distributions of body temperatures are identical for the two groups; we will suppose that they are normally distributed with equal means and equal variances. When we run an experiment, we are collecting a random sample of data from each of the groups. The random samples might show a spuriously large difference between their means, just by chance, despite the fact that on average, in the long run, the groups are identical.

To determine how often the spuriously large differences occur, we can simulate conducting the experiment over and over. For every simulated experiment, we compute the difference of means between the samples from the groups. The difference of sample means is in units of the original measurement scale (e.g., degrees Fahrenheit or degrees Celsius). To get rid of the arbitrary influence of the measurement scale, we standardize the difference of means and call the result the t statistic. Because the true difference between groups is zero, the t value typically will be near zero. Occasionally, by chance, the t value will be far above or far below zero. The lowest curve in Figure 18.4 shows the probability that the sampled t value falls above the critical t value on the abscissa. For example, the probability that the sampled t value falls above $t_{crit} = 2.23$ is $p(FA) = 0.05$; this is marked by an arrow. In NHST, the decision rule is to reject the null hypothesis if the sample t exceeds a critical value that is selected to keep false alarms to only 5%. Thus, when comparing group 1 with

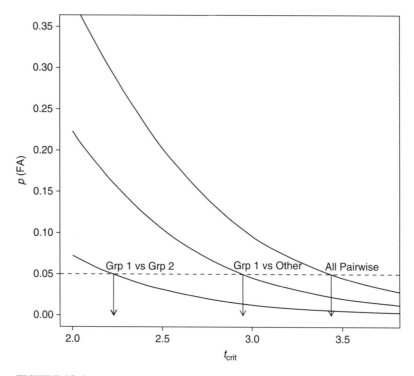

FIGURE 18.4

The probability of false alarm as a function of critical t value, with separate curves for different sets of comparisons. All groups have $N = 6$ fixed by intention. The curve labeled "Grp 1 vs Grp 2" is for a single comparison of two groups, and corresponds with the usual two-group t distribution. The curve labeled "Grp 1 vs Other" refers to four paired comparisons, of Group 1 versus each of the other four groups. The last curve is for the set of all 10 paired comparisons. (Reprinted with permission from Figure 2 of Kruschke, 2010a.)

group 2, we would reject the null hypothesis if $t > 2.23$, because that would happen only 5% of the time by chance alone.

Now consider an expanded experiment, in which there is a placebo treatment and four distinct drugs, for a total of five treatment groups. According to the null hypothesis, the five treatment groups have identical distributions of body temperatures (normally distributed with equal means and variances). However, because of random sampling in any particular experiment, some treatment samples will have higher or lower mean temperatures than other treatment samples. Suppose that before we collect any real data, we plan to compare the placebo group (group 1) with each of the four drug groups (i.e., we plan four pairwise comparisons). Each of these comparisons might yield a fairly large difference merely by chance, even when there is truly no difference in the underlying distributions. We can determine how often these chance extremes happen by running a Monte Carlo simulation. For a simulated experiment, we randomly sample six scores from each of the five groups, and compute the t values of each of the four comparisons. The simulated experiment is repeated many times. For each candidate t_{crit}, we see what proportion of simulated experiments had a comparison that exceeded that critical value. The middle curve of Figure 18.4 shows the result. Notice that at any given value of t_{crit}, there is now a much higher probability that the simulated experiment will have at least one comparison with larger t. In particular, to keep the false alarm rate down to 5%, t_{crit} must be about 2.95 instead of 2.33.

If we did not plan only four tests but instead decided to compare every group with every other group, then we would have even more opportunity for false alarms. With five groups, there are 10 pairwise comparisons. If we simulate experiments from equal distributions as before, but this time consider all 10 t values, the probability of false alarm is higher yet, as shown in the right curve of Figure 18.4. The critical value has risen even higher, to approximately 3.43.[1]

Now, suppose we actually run the experiment. We randomly assign 30 people to the five groups, six people per group. The first group gets the placebo, and the other four groups get the corresponding four drugs. *We are careful to make this a double-blind experiment: Neither the subjects nor experimenters know who is getting which treatment. Moreover, no one knows whether any other person is even in the experiment.* We collect the data. Our first question is to compare the placebo and the first drug (i.e., group 1 versus group 2). We compute the t statistic for the data from the two groups and find that $t = 2.95$. Do we decide that the two treatments had significantly different effects?

The answer, bizarrely, depends on the intentions of the person we ask. Suppose, for instance, that we handed the data from the first two groups to a research assistant, who is asked to test for a difference between groups. The assistant runs a t test and finds $t = 2.95$, declaring it to be *highly significant* because it greatly exceeds the critical value of 2.23 for a two-group t test. Suppose, on the other hand, that we handed the data from all five groups to a different research assistant, who is asked to compare the first group against each of the other four. This assistant runs a t test of group 1 versus group 2 and finds $t = 2.95$, declaring it to be *marginally significant* because it just squeezes past the critical value of 2.95 for these four planned comparisons. Suppose, on yet another hand, that we handed the data from all five groups to a different research assistant, who is told to conduct all pairwise comparisons post hoc because we have no strong hypotheses about which treatments will have beneficial or detrimental or neutral effects. This assistant runs a t test of group 1

---

[1] For a discussion of various correction procedures and when to use them, see Figure 5.1 of Maxwell & Delaney (2004). If you must learn NHST methods, this is an excellent resource.

versus group 2 and finds $t = 2.95$, declaring it to be *not significant* because it fails to exceed the critical value of 3.43 that is used for post hoc pairwise comparisons. Notice that regardless of which assistant analyzed the data, the t value for the two groups stayed the same because the data of the two groups stayed the same. Indeed, the data were completely uninfluenced by the intentions of the analyst. So why should the interpretation of the data be influenced by the intentions of the analyst? It shouldn't.

If you believe that the interpretation should be influenced by the intentions of the analyst, how do you determine the intentions of the analyst? Did the analyst truly plan only those particular comparisons, or did the analyst really plan others but jettison them once the data were in? Or did the analyst actually plan fewer comparisons but realize later that additional comparisons should be made to address other theoretical issues? Or did the analyst actually plan to include two other treatment groups in the study but then not actually include those groups in the analysis because of administrative errors committed during the data collection? Or what if the experiment was planned by a team of people, some of whom planned some comparisons and others of whom planned other comparisons? Conclusion: Establishing the true intentions of the analyst is not only pointless, it is also impossible.

Multiple comparisons are not a problem in a Bayesian analysis (e.g., Gelman, Hill, & Yajima, 2009). The posterior distribution is a fixed entity in high-dimensional parameter space, and making comparisons between groups is simply examining that posterior distribution from different perspectives or margins. The posterior does not change when new comparisons come to mind.

The posterior is not immune to spurious coincidences of rogue data, of course. False alarms are mitigated, however, by incorporating prior knowledge into the structure of the model. The estimates of the groups are mutually informative via estimation of higher-level structure, and shrinkage of estimates across groups attenuates false alarms. The attenuation of false alarms is governed by the data, not by unknowable intentions.

18.3 TWO-GROUP BAYESIAN ANOVA AND THE NHST t TEST

The idea behind an NSHT t test is simple: We have two groups, each with a mean. We compute the difference of the means and standardize that difference relative to the standard deviation of the scores within the groups. The resulting standardized difference is called the t value. We want to know whether the observed t value is significantly different from zero, so we compare the t value to a sampling distribution of t values (Gosset, 1908). The sampling distribution assumes that the intention of the researcher was to stop when there were exactly $N1$ values observed for the first group, and exactly $N2$ values observed for the second group.

The t test is a special case of NHST ANOVA when there are only two groups. More specifically, when the two groups are assumed to have equal variances in the underlying population, then the t value squared equals the F value in two-group ANOVA. (And what's an F value, you may ask? The F value is the summary statistic used in NHST ANOVA to express how much the groups differ from each other. It's the ratio of the variance between group means, relative to the variance within groups.)

In typical applications of BANOVA, the prior on the between-group variance is only mildly informed. In this case, a BANOVA on two groups imposes little shrinkage on the group estimates because there are so few groups. It is only when several groups "gang up" that they strongly inform the estimate of the variation between groups and therefore constrain the estimates of other groups. When the prior on the variance within groups is also vague, the results of a two-group BANOVA closely agree with the results of an NHST t test. Exercise 18.1 has you explore this correspondence.

18.4 R CODE

18.4.1 Bayesian Oneway ANOVA

(ANOVAonewayBRugs.R)

```
1   graphics.off()
2   rm(list=ls(all=TRUE))
3   fnroot = "ANOVAonewayBrugs"
4   library(BRugs)              # Kruschke, J. K. (2010). Doing Bayesian data analysis:
5                              # A Tutorial with R and BUGS. Academic Press / Elsevier.
6   #------------------------------------------------------------------------
7   # THE MODEL.
8
9   modelstring = "
10  # BUGS model specification begins here...
11  model {
12    for ( i in 1:Ntotal ) {
13      y[i] ~ dnorm( mu[i] , tau )
14      mu[i] <- a0 + a[x[i]]
15    }
16    #
17    tau <- pow( sigma , -2 )
18    sigma ~ dunif(0,10) # y values are assumed to be standardized
19    #
20    a0 ~ dnorm(0,0.001) # y values are assumed to be standardized
21    #
22    for ( j in 1:NxLvl ) { a[j] ~ dnorm( 0.0 , atau ) }
23    atau <- 1 / pow( aSD , 2 )
24    aSD <- abs( aSDunabs ) + .1
```

```
25    aSDunabs ~ dt( 0 , 0.001 , 2 )
26  }
27  # ... end BUGS model specification
28  " # close quote for modelstring
29  # Write model to a file, and send to BUGS:
30  writeLines(modelstring,con="model.txt")
31  modelCheck( "model.txt" )
32
33  #------------------------------------------------------------------
34  # THE DATA.
35
36  # Specify data source:
37  dataSource = c( "McDonaldSK1991" , "SolariLS2008" , "Random" )[1]
38  # Load the data:
39
40  if ( dataSource == "McDonaldSK1991" ) {
41    fnroot = paste( fnroot , dataSource , sep="" )
42    datarecord = read.table( "McDonaldSK1991data.txt", header=T ,
43                        colClasses=c("factor","numeric") )
44    y = as.numeric(datarecord$Size)
45    Ntotal = length(datarecord$Size)
46    x = as.numeric(datarecord$Group)
47    xnames = levels(datarecord$Group)
48    NxLvl = length(unique(datarecord$Group))
49    contrastList = list( BIGvSMALL = c(-1/3,-1/3,1/2,-1/3,1/2) ,
50                        ORE1vORE2 = c(1,-1,0,0,0) ,
51                        ALAvORE = c(-1/2,-1/2,1,0,0) ,
52                        NPACvORE = c(-1/2,-1/2,1/2,1/2,0) ,
53                        USAvRUS = c(1/3,1/3,1/3,-1,0) ,
54                        FINvPAC = c(-1/4,-1/4,-1/4,-1/4,1) ,
55                        ENGvOTH = c(1/3,1/3,1/3,-1/2,-1/2) ,
56                        FINvRUS = c(0,0,0,-1,1) )
57  }
58
59  if ( dataSource == "SolariLS2008" ) {
60    fnroot = paste( fnroot , dataSource , sep="" )
61    datarecord = read.table("SolariLS2008data.txt", header=T ,
62                        colClasses=c("factor","numeric") )
63    y = as.numeric(datarecord$Acid)
64    Ntotal = length(datarecord$Acid)
65    x = as.numeric(datarecord$Type)
66    xnames = levels(datarecord$Type)
67    NxLvl = length(unique(datarecord$Type))
68    contrastList = list( G3vOTHER = c(-1/8,-1/8,1,-1/8,-1/8,-1/8,-1/8,-1/8,-1/8) )
69  }
70
71  if ( dataSource == "Random" ) {
72    fnroot = paste( fnroot , dataSource , sep="" )
73    #set.seed(47405)
74    ysdtrue = 4.0
75    a0true = 100
76    atrue = c( 2 , -2 ) # sum to zero
77    npercell = 8
78    datarecord = matrix( 0, ncol=2 , nrow=length(atrue)*npercell )
```

```
79    colnames(datarecord) = c("y","x")
80    rowidx = 0
81    for ( xidx in 1:length(atrue) ) {
82      for ( subjidx in 1:npercell ) {
83        rowidx = rowidx + 1
84        datarecord[rowidx,"x"] = xidx
85        datarecord[rowidx,"y"] = ( a0true + atrue[xidx] + rnorm(1,0,ysdtrue) )
86      }
87    }
88    datarecord = data.frame( y=datarecord[,"y"] , x=as.factor(datarecord[,"x"]) )
89    y = as.numeric(datarecord$y)
90    Ntotal = length(y)
91    x = as.numeric(datarecord$x)
92    xnames = levels(datarecord$x)
93    NxLvl = length(unique(x))
94    # Construct list of all pairwise comparisons, to compare with NHST TukeyHSD:
95    contrastList = NULL
96    for ( g1idx in 1:(NxLvl-1) ) {
97      for ( g2idx in (g1idx+1):NxLvl ) {
98        cmpVec = rep(0,NxLvl)
99        cmpVec[g1idx] = -1
100       cmpVec[g2idx] = 1
101       contrastList = c( contrastList , list( cmpVec ) )
102     }
103   }
104 }
105
106 # Specify the data in a form that is compatible with BRugs model, as a list:
107 ySDorig = sd(y)
108 yMorig = mean(y)
109 z = ( y - yMorig ) / ySDorig
110 datalist = list(
111   y = z ,
112   x = x ,
113   Ntotal = Ntotal ,
114   NxLvl = NxLvl
115 )
116 # Get the data into BRugs:
117 modelData( bugsData( datalist ) )
118
119 #------------------------------------------------------------------------------
120 # INTIALIZE THE CHAINS.
121
122 # Autocorrelation within chains is large, so use several chains to reduce
123 # degree of thinning. But we still have to burn-in all the chains, which takes
124 # more time with more chains (on serial CPUs).
125 nchain = 5
126 modelCompile( numChains = nchain )
127
128 if ( F ) {
129   modelGenInits() # often won't work for diffuse prior
130 } else {
131   # initialization based on data
132   theData = data.frame( y=datalist$y , x=factor(x,labels=xnames) )
```

```
133    a0 = mean( theData$y )
134    a = aggregate( theData$y , list( theData$x ) , mean )[,2] - a0
135    ssw = aggregate( theData$y , list( theData$x ) ,
136                    function(x){var(x)*(length(x)-1)} )[,2]
137    sp = sqrt( sum( ssw ) / length( theData$y ) )
138    genInitList <- function() {
139      return(
140          list(
141              a0 = a0 ,
142              a = a ,
143              sigma = sp ,
144              aSDunabs = sd(a)
145          )
146      )
147    }
148    for ( chainIdx in 1 : nchain ) {
149      modelInits( bugsInits( genInitList ) )
150    }
151  }
152
153  #-------------------------------------------------------------------------------
154  # RUN THE CHAINS
155
156  # burn in
157  BurnInSteps = 10000
158  modelUpdate( BurnInSteps )
159  # actual samples
160  samplesSet( c( "a0" , "a" , "sigma" , "aSD" ) )
161  stepsPerChain = ceiling(5000/nchain)
162  thinStep = 750
163  modelUpdate( stepsPerChain , thin=thinStep )
164
165  #-------------------------------------------------------------------------------
166  # EXAMINE THE RESULTS
167
168  source("plotChains.R")
169  source("plotPost.R")
170
171  checkConvergence = T
172  if ( checkConvergence ) {
173    sumInfo = plotChains( "a0" , saveplots=T , filenameroot=fnroot )
174    sumInfo = plotChains( "a" , saveplots=T , filenameroot=fnroot )
175    sumInfo = plotChains( "sigma" , saveplots=T , filenameroot=fnroot )
176    sumInfo = plotChains( "aSD" , saveplots=T , filenameroot=fnroot )
177  }
178
179  # Extract and plot the SDs:
180  sigmaSample = samplesSample("sigma")
181  aSDSample = samplesSample("aSD")
182  windows()
183  layout( matrix(1:2,nrow=2) )
184  par( mar=c(3,1,2.5,0) , mgp=c(2,0.7,0) )
185  plotPost( sigmaSample , xlab="sigma" , main="Cell SD" , breaks=30 )
186  plotPost( aSDSample , xlab="aSD" , main="a SD" , breaks=30 )
```

```
187    dev.copy2eps(file=paste(fnroot,"SD.eps",sep=""))
188
189    # Extract a values:
190    a0Sample = samplesSample( "a0" )
191    chainLength = length(a0Sample)
192    aSample = array( 0 , dim=c( datalist$NxLvl , chainLength ) )
193    for ( xidx in 1:datalist$NxLvl ) {
194       aSample[xidx,] = samplesSample( paste("a[",xidx,"]",sep="") )
195    }
196
197    # Convert to zero-centered b values:
198    mSample = array( 0, dim=c( datalist$NxLvl , chainLength ) )
199    for ( stepIdx in 1:chainLength ) {
200       mSample[,stepIdx ] = ( a0Sample[stepIdx] + aSample[,stepIdx] )
201    }
202    b0Sample = apply( mSample , 2 , mean )
203    bSample = mSample - matrix(rep( b0Sample ,NxLvl),nrow=NxLvl,byrow=T)
204    # Convert from standardized b values to original scale b values:
205    b0Sample = b0Sample * ySDorig + yMorig
206    bSample = bSample * ySDorig
207
208    # Plot b values:
209    windows(datalist$NxLvl*2.75,2.5)
210    layout( matrix( 1:datalist$NxLvl , nrow=1 ) )
211    par( mar=c(3,1,2.5,0) , mgp=c(2,0.7,0) )
212    for ( xidx in 1:datalist$NxLvl ) {
213       plotPost( bSample[xidx,] , breaks=30 ,
214                 xlab=bquote(beta*1[.(xidx)]) ,
215                 main=paste("x:",xnames[xidx])   )
216    }
217    dev.copy2eps(file=paste(fnroot,"b.eps",sep=""))
218
219    # Display contrast analyses
220    nContrasts = length( contrastList )
221    if ( nContrasts > 0 ) {
222       nPlotPerRow = 5
223       nPlotRow = ceiling(nContrasts/nPlotPerRow)
224       nPlotCol = ceiling(nContrasts/nPlotRow)
225       windows(3.75*nPlotCol,2.5*nPlotRow)
226       layout( matrix(1:(nPlotRow*nPlotCol),nrow=nPlotRow,ncol=nPlotCol,byrow=T) )
227       par( mar=c(4,0.5,2.5,0.5) , mgp=c(2,0.7,0) )
228       for ( cIdx in 1:nContrasts ) {
229          contrast = matrix( contrastList[[cIdx]],nrow=1) # make it a row matrix
230          incIdx = contrast!=0
231          histInfo = plotPost( contrast %*% bSample , compVal=0 , breaks=30 ,
232                    xlab=paste( round(contrast[incIdx],2) , xnames[incIdx] ,
233                               c(rep("+",sum(incIdx)-1),"") , collapse=" " ) ,
234                    cex.lab = 1.0 ,
235                    main=paste( "X Contrast:", names(contrastList)[cIdx] ) )
236       }
237       dev.copy2eps(file=paste(fnroot,"xContrasts.eps",sep=""))
238    }
239
240    #==============================================================================
```

```
241    # Do NHST ANOVA and t tests:
242
243    theData = data.frame( y=y , x=factor(x,labels=xnames) )
244    aovresult = aov( y ~ x , data = theData ) # NHST ANOVA
245    cat("\n----------------------------------------------------------------\n\n")
246    print( summary( aovresult ) )
247    cat("\n----------------------------------------------------------------\n\n")
248    print( model.tables( aovresult , "means" ) , digits=4 )
249    windows()
250    boxplot( y ~ x , data = theData )
251    cat("\n----------------------------------------------------------------\n\n")
252    print( TukeyHSD( aovresult , "x" , ordered = FALSE ) )
253    windows()
254    plot( TukeyHSD( aovresult , "x" ) )
255    if ( T ) {
256      for ( xIdx1 in 1:(NxLvl-1) ) {
257        for ( xIdx2 in (xIdx1+1):NxLvl ) {
258          cat("\n----------------------------------------------------------------\n\n")
259          cat( "xIdx1 = " , xIdx1 , ", xIdx2 = " , xIdx2 ,
260               ", M2-M1 = " , mean(y[x==xIdx2])-mean(y[x==xIdx1]) , "\n" )
261          print( t.test( y[x==xIdx2] , y[x==xIdx1] , var.equal=T ) ) # t test
262        }
263      }
264    }
265    cat("\n----------------------------------------------------------------\n\n")
266
267    #================================================================
```

18.5 EXERCISES

Exercise 18.1. [Purpose: To notice that Bayesian ANOVA with two groups tends to agree with an NHST *t* test.] The BRugs program of Section 18.4.1 (ANOVAonewayBRugs.R) allows you to specify random data. It executes a Bayesian ANOVA, and at the end of the program it also conducts an NHST ANOVA and *t* tests (using R's aov and t.test functions). Run the program ten times with different random data by commenting out the set.seed command. Specify ysdtrue = 4.0, atrue = c(2,-2) (which implies two groups because there are two deflections) and npercell = 8. For each run, record, by hand, (1) how much of the posterior difference between means falls on one side of zero (see the posterior histogram with the main title "X Contrast" and *x* axis labeled "−1 1 + 1 2"), (2) whether the 95% HDI excludes zero, and (3) the confidence interval and *p* value of the NHST *t* test. Do the *t* test and the BANOVA usually agree in their decisions about whether the group means are different?

Exercise 18.2. [Purpose: To understand the influence of the prior in Bayesian ANOVA.] In the model section of the BRugs program of Section 18.4.1 (ANOVAonewayBRugs.R), and correspondingly in the diagram of Figure 18.1, there are several constants that determine the prior. These constants include the mean value of the baseline (M_0 in the diagram), the precision on the baseline (T_0 in the diagram), the precision of the folded-t distribution (T_t in the diagram), and the upper value of the uniform distribution on σ_y (H_{σ_y} in the diagram). Because the data are standardized, M_0 should be set at zero, and T_0 can be modest (not terribly small). H_{σ_y} also can be set to a modest value because the data are standardized. But what about the precision of the folded-t distribution, T_t? This constant modulates the degree of shrinkage: A large value of T_t indicates prior knowledge that the groups do not differ much, and it imposes a high degree of shrinkage that must be overcome by the data.

Run the program on the mussel data using a small value of T_t, such as 1.0E-6, and a large value of T_t, such as 1000. Are the results very different? Discuss which prior value might be appropriate.

Exercise 18.3. [Purpose: To understand Bayesian ANOVA without assuming equal variances.] Modify the program in Section 18.4.1 (`ANOVAonewayBRugs.R`) so that it allows a different variance for each group, with the different variances coming from a hyperdistribution that has its precision informed by the data. In other words, instead of assuming the same τ_y ($= 1/\sigma_y^2$) for all the levels of x, we allow each group to have its own variance. Denote the precision of the jth group as τ_j, analogous to the deflection β_j. Just as the group deflections are assumed to come from a higher-level distribution, we will assume that the group SDs come from a higher-level distribution. Because SDs must be non-negative, use a gamma density for the higher-level distribution. The gamma distribution has two parameters for which you need to establish a prior. *See the right side of Figure 16.11 for guidance.* Corresponding code is offered in a hint, below. Run the program on the mussel muscle data. Are the conclusions about the group means any different than when assuming equal variances across groups?

Hint regarding the conclusion: The posteriors on the group means are only a little different in this case, because the group variances are roughly the same. But because the group variances are less constrained when they are all allowed to be different, they are less certain. Therefore, the group means are a little less certain, and thus the differences of means are a little less certain.

Programming hints: Here are some code snippets, showing the model specification and chain initialization.

(ANOVAonewayNonhomogvarBrugs.R)

```
11   model {
12     for ( i in 1:Ntotal ) {
13       y[i] ~ dnorm( mu[i] , tau[x[i]] )
14       mu[i] <- a0 + a[x[i]]
15     }
16     a0 ~ dnorm(0,0.001)
17     for ( j in 1:NxLvl ) {
18       a[j] ~ dnorm( 0.0 , atau )
19       tau[j] ~ dgamma( sG , rG )
```

```
20        }
21        sG <- pow(m,2)/pow(d,2)
22        rG <- m/pow(d,2)
23        m ~ dgamma(1,1)
24        d ~ dgamma(1,1)
25        atau <- 1 / pow( aSD , 2 )
26        aSD <- abs( aSDunabs ) + .1
27        aSDunabs ~ dt( 0 , 0.001 , 2 )
28      }
```

(ANOVAonewayNonhomogvarBrugs.R)

```
133     #  initialization based on data
134     theData = data.frame( y=datalist$y , x=factor(x,labels=xnames) )
135     a0 = mean( theData$y )
136     a = aggregate( theData$y , list( theData$x ) , mean )[,2] - a0
137     tau = 1/(aggregate( theData$y , list( theData$x ) , sd )[,2])^2
138     genInitList <- function() {
139       return(
140         list(
141             a0 = a0 ,
142             a = a ,
143             tau = tau ,
144             m = mean( tau ) ,
145             d = sd( tau ) ,
146             aSDunabs = sd(a)
147         )
148       )
149     }
150     for ( chainIdx in 1 : nchain ) {
151       modelInits( bugsInits( genInitList ) )
152     }
```

CHAPTER 19

Metric Predicted Variable with Multiple Nominal Predictors

CONTENTS

Sometimes I wonder just how it could be, that
Factors aligned so you'd end up with me.
All of the priors made everyone think, that
Our interaction was destined to shrink.

In this chapter we consider situations with a metric predicted variable and multiple nominal predictor variables. For example, we might want to predict income (a metric variable) on the basis of political party affiliation (a nominal variable) and ethnicity (another nominal variable). Or we may want to predict response time (a metric variable) on the basis of hand used for the response (a nominal value: dominant hand or nondominant hand) and modality of

Doing Bayesian Data Analysis: A Tutorial with R and BUGS. DOI: 10.1016/B978-0-12-381485-2.00019-5
© 2011, Elsevier Inc. All rights reserved.

stimulus (another nominal value: visual, auditory, or tactile). These situations are modeled by the cell in the first row and last column of Table 14.1, p. 385.

In traditional NHST, this situation is known as multifactor ANOVA. We use the same underlying model, but without reference to F sampling distributions; instead we use hierarchical priors that provide additional structural constraints. Multifactor ANOVA is a straightforward extension of the model presented in the previous chapter, but with a new concept of interaction between nominal variables. Just as multiple regression considered interaction of metric predictors, multifactor ANOVA considers interaction of nominal predictors.

19.1 BAYESIAN MULTIFACTOR ANOVA

Recall from the previous chapter that in oneway ANOVA, we describe the effect of each level of the predictor as a deflection away from an overall baseline, where the baseline is the central tendency across all levels of the predictor. In multifactor ANOVA, the same idea applies to two or more predictors, and the deflections resulting from each predictor are *added*. We'll use notation analogous to the previous chapter, but with extra subscripts to indicate the different predictors, just as we used in multiple regression on continuous predictors.

The mathematical notation was introduced as a case of the generalized linear model in Section 14.1.6.2, p. 370. Suppose we have two nominal predictors, cleverly denoted $\vec{x}_1$ and $\vec{x}_2$. These predictor vectors can only take on values of $\langle 1, 0, 0, \ldots \rangle$, $\langle 0, 1, 0, \ldots \rangle$, and so on, with the jth component having the value 1 when the predictor has its jth nominal level.

When the effects of the two predictors are additive, the predicted tendency is as follows:

$$y = \beta_0 + \vec{\beta}_1 \vec{x}_1 + \vec{\beta}_2 \vec{x}_2$$

$$= \beta_0 + \sum_{j=1}^{J_1} \beta_{1,j} x_{1,j} + \sum_{k=1}^{J_2} \beta_{2,k} x_{2,k}$$

To make the parameter values unique, we include the constraints

$$\sum_{j=1}^{J_1} \beta_{1,j} = 0 \quad \text{and} \quad \sum_{j=k}^{J_2} \beta_{2,k} = 0$$

Those equations repeat Equations 14.7 and 14.8. In words, the value β_0 establishes the overall baseline from which the predictors indicate deflections. When predictor x_1 has value $x_{1,j}$, a deflection of $\beta_{1,j}$ is added to the baseline, and when predictor x_2 has value $x_{2,k}$, a deflection of $\beta_{2,k}$ is also added to the baseline. The deflections may be negative. Indeed, across all levels of the predictors, the constraints demand as much negative deflection as positive deflection, so that the deflections sum to zero for each predictor.

19.1.1 Interaction of Nominal Predictors

The effect of two predictors may be nonadditive, in which case we say that there is an "interaction" of the predictors. For example, if a flame is put under a hot-air balloon, its levity will increase. And if hydrogen is added to a balloon, its levity will increase. But if hydrogen and flame are added to a balloon, there is a nonadditive interaction, such that levity is not increased.

Figure 19.1 displays a simple interaction. Both predictors have only two levels. The abscissa groups the two levels of predictor $\vec{x}_1$, and the shading of the bars indicates the two levels of predictor $\vec{x}_2$. All three panels of Figure 19.1 show the same data, but the nature of the interaction is highlighted differently in each panel.

In the left panel of Figure 19.1, the dashed parallelogram indicates the best *additive* model for the data. The dashed lines indicate the average change when the levels of the predictors change. *The vertical arrows highlight the nonadditive deflections, away from the additive average, that constitute the interaction.* Notice that the arrows sum to zero across each edge of the parallelogram. Thus, the interaction components do not change the average deflections of each predictor.

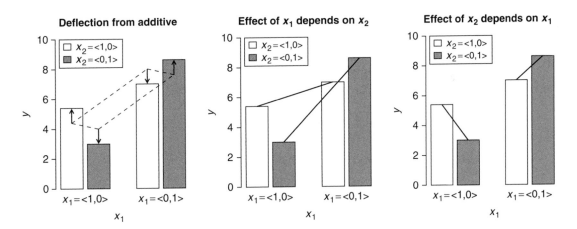

FIGURE 19.1

An interaction of nominal variables $\vec{x}_1$ and $\vec{x}_2$, parsed three ways. The left panel emphasizes that the interaction involves a nonadditive, torsion-like deflection away from the additive model, as indicated by the arrows. The middle panel shows the same data, with lines that emphasize that the effect of $\vec{x}_1$ depends on the value of $\vec{x}_2$. The right panel again shows the same data, but with lines that emphasize that the effect of $\vec{x}_2$ depends on the value of $\vec{x}_1$.

The middle and right panels of Figure 19.1 highlight different interpretations of the interaction. The middle panel shows that the effect of $\vec{x}_1$, that is, the amount that y changes when $\vec{x}_1$ changes, depends on the level of $\vec{x}_2$: When $\vec{x}_2 = \langle 1, 0 \rangle$, there is only a small change in y when $\vec{x}_1$ changes, but when $\vec{x}_2 = \langle 0, 1 \rangle$, there is a larger change in y when $\vec{x}_1$ changes. The right panel makes the same point but with the roles of the predictors reversed: When $\vec{x}_1 = \langle 1, 0 \rangle$, the effect of $\vec{x}_2$ is to decrease y, but when $\vec{x}_1 = \langle 0, 1 \rangle$, the effect of $\vec{x}_2$ is to increase y.

The average deflection from baseline due to a predictor is called the *main effect* of the predictor. The main effects of the predictors correspond to the dashed lines in the left panel of Figure 19.1. When there is nonadditive interaction between predictors, the effect of one predictor depends on the level of the other predictor. The deflection from baseline for a predictor, at a fixed level of the other predictor, is called the *simple effect* of the predictor at the level of the other predictor. When there is interaction, the simple effects do not equal the main effect.

It may be edifying to compare Figure 19.1, which shows interaction of *nominal* predictors, with Figure 17.8, p. 470, which shows interaction of *metric* predictors. The essential notion of interaction is the same in both cases: Interaction is the nonadditive portion of the prediction, and interaction means that the effect of one predictor depends on the level of the other predictor.

The mathematical formalism for nonadditive interactions was introduced in Section 14.1.6.3, p. 371, and is repeated here. The nonadditive components, indicated by the vertical arrows in Figure 19.1, are denoted $\beta_{1\times2,j,k}$, which means the interaction of predictors 1 and 2 (denoted 1×2) at level j of predictor 1 and level k of predictor 2. The formal expression merely expands the additive model by including the interaction. Recall from Equations 14.9 and 14.10 that the model with interaction term can be written as

$$y = \beta_0 + \vec{\beta}_1 \vec{x}_1 + \vec{\beta}_2 \vec{x}_2 + \vec{\beta}_{1\times2} \vec{x}_{1\times2}$$

$$= \beta_0 + \sum_{j=1}^{J_1} \beta_{1,j} x_{1,j} + \sum_{k=1}^{J_2} \beta_{2,k} x_{2,k} + \sum_{j=1}^{J_1} \sum_{k=1}^{J_2} \beta_{1\times2,j,k} x_{1\times2,j,k}$$

with the constraints

$$\sum_{j=1}^{J_1} \beta_{1,j} = 0 \quad \text{and} \quad \sum_{k=1}^{J_2} \beta_{2,k} = 0 \quad \text{and}$$

$$\sum_{j=1}^{J_1} \beta_{1\times2,j,k} = 0 \ \forall k \quad \text{and} \quad \sum_{k=1}^{J_2} \beta_{1\times2,j,k} = 0 \ \forall j$$

In those last equations, the symbol "∀" means "for all." In words, the last two equations simply mean that the interaction deflections sum to zero along every level of the two predictors. A graphic example of this was presented in the left panel of Figure 19.1, which shows that the heights of the arrows sum to zero along every edge of the parallelogram.

Our goal is to estimate the additive and interactive deflections, based on the observed data. It is important to understand that the observed data are *not* the bars in Figure 19.1; instead, the data are swarms of points at various heights near the heights of the bars. The bars represent the central tendency of the data at each combination of the predictors. Thus, what the equations above actually predict is the central tendency μ at each combination of predictors, and the data are typically modeled as being normally distributed around μ.

19.1.2 The Hierarchical Prior

The complete generative model of the data is shown in Figure 19.2. It might look daunting, but it really is merely the diagram for oneway ANOVA, in Figure 18.1, with the hyperprior replicated for each predictor and interaction.

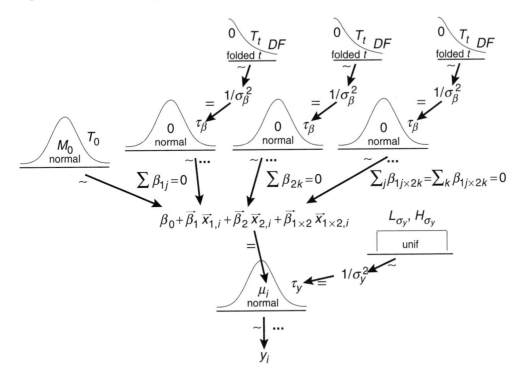

FIGURE 19.2
Hierarchical dependencies for model of two-way Bayesian ANOVA. Compare with Figure 18.1.

The lowest level of Figure 19.2 indicates that the observed data points, y_i, are distributed normally around the predicted value μ_i. Moving upward in the diagram, the arrow impinging on μ_i indicates that the predicted value is baseline plus additive deflection due to each predictor plus interactive deflection due to the combination of predictors. The upper levels of the diagram indicate prior structural assumptions about the deflections. We assume that the deflections produced by a predictor are centered at zero, and we allow the variance (i.e., precision) of the deflections to be estimated from the data. Thus, if most of the deflections are small, the estimated variance is small, and the hyperdistribution creates shrinkage in the estimates of other deflections.

A key conceptual aspect of the hyperdistributions is that they apply separately to the different predictors and interactions. In other words, there is not just one hyperdistribution that governs all deflections for all predictors and interactions. This division of generative structure reflects a prior assumption that the magnitude of the effect of one predictor might not be informative regarding the magnitude of the effect of a different predictor. But within a predictor, the magnitude of deflection produced by one level may inform the magnitude of deflection produced by other levels of that same predictor.[1]

As was assumed in the case of oneway ANOVA, we will assume homogeneity of variance: The variability of the observed data is the same within each combination of predictors. This is indicated in Figure 19.2 by the *single* parameter σ_y that is used in the likelihood function, regardless of the values of the predictors. As before, there are two reasons for this assumption. First, the assumption is a natural simplification in multiple regression on metric predictors, and ANOVA can be construed as a special case of multiple regression. Second, the assumption of equal variances is made in NHST ANOVA, and we will also make it here in BANOVA to facilitate comparing across the techniques. But there is no requirement in BANOVA to assume equal variances. If the the situation suggests that different levels of the predictors produce radically different variances in the data, then the hierarchical prior can allow different variances.

19.1.3 An Example in R and BUGS

Figure 19.3 shows the mean annual salaries of faculty in four departments at three levels of seniority. The four departments are business finance, counseling and educational psychology, chemistry, and theater. These departments

---

[1] By analogy to multiple regression, if there are many predictors included in a model, it is reasonable in principle to include a higher-level distribution *across predictors* such that the estimated variance of one predictor informs the estimated variance of another predictor. This would be especially useful if the application includes many nominal predictors, each with many levels. Such applications are rare.

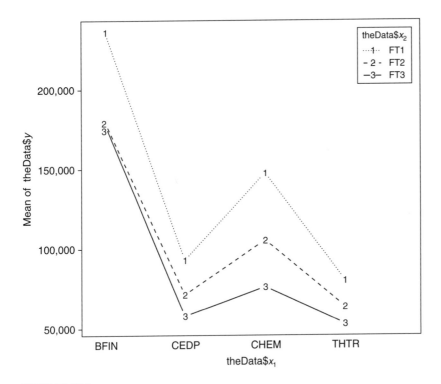

FIGURE 19.3
Mean annual salaries of faculty in four departments at three levels of seniority.

are the nominal levels of a predictor denoted $\vec{x}_1$. The three levels of seniority are full professor, associate professor, and assistant professor. Assistant professors are usually within 7 years after completing their doctoral or postdoctoral studies. Associate professors are usually within about 10 years of their doctoral or postdoctoral studies. Full professors are anywhere from 10 to 40 years post graduate school. Although seniority could, and perhaps should, be treated as an ordinal variable, we will treat it as a nominal predictor, denoted $\vec{x}_2$. A glance at the means suggests that there are effects of department and of seniority. There appears also to be an interaction, meaning that the change in salary due to seniority depends on the department. Our goal is to estimate the baseline salary, the main effect of department membership, the main effect of seniority, and the interaction of department and seniority.

The display of the means in Figure 19.3 obscures the fact that different combinations of department and seniority had different numbers of data points. In other words, the number of associate professors in business finance was not necessarily equal to the number of full professors in theater. In traditional

NHST ANOVA, this sort of "unbalanced" design can cause serious computational difficulties (e.g., Maxwell & Delaney, 2004, pp. 320–343). But Bayesian ANOVA has no problem with unbalanced designs.

The model of Figure 19.2 was implemented in R and BRugs and is listed in Section 19.3.1 (`ANOVAtwowayBRugs.R`). Several tricks for running the model in BUGS are described in that section, before the program listing. The essentials, however, are much like the oneway ANOVA model of the previous chapter.

The results are shown in Figure 19.4. (The means and HDI limits are displayed with only three significant digits, but more precise values can be obtained directly from the program.) The top-left histogram shows that the baseline for these four departments is 111,381. Notice, however, that most of the data fall below this baseline because the overall data are skewed by the much higher salaries in one department. For salaries in the department of chemistry, the fourth histogram in the top row indicates that 2164 should be subtracted from the baseline. For salaries of assistant professors, the first histogram in the bottom row indicates that 20,100 should be subtracted from the baseline. Thus, for assistant professors in the department of chemistry, the *linearly* predicted salary is $111,381 − 2164 − 20,100 = 89,117$. But there is a notable nonlinear interaction component for that combination: The fourth histogram of the bottom row shows that 10,938 must be subtracted from the linear combination to get the mean estimate for that combination, namely, 78,179.

19.1.4 Interpreting the Posterior

In most applications, we are interested not only in estimation of effects for each group, but we are also interested in deciding whether two groups are credibly different. Just as we compared groups in oneway ANOVA in the previous chapter, we can compare groups in multifactor ANOVA.

The top and middle rows of Figure 19.5 show selected contrasts of levels of the main effects. We may ask whether there is a credible difference in salaries, on average, between business finance (BFIN) and counseling and educational psychology (CEDP). The top-left histogram indicates that the average difference is about $122,000, and the 95% HDI falls far from zero. We may also ask whether there is a credible difference in salaries, on average, between CEDP and theater (THTR). The top-right histogram indicates that the average difference is about $7780, but the 95% HDI spans zero, which indicates that the uncertainty in the estimated difference is fairly large relative to the estimated difference itself. The middle row of Figure 19.5 shows contrasts regarding levels of seniority: There is a credible difference between full professors (FT1) and associate professors (FT2), and between FT2 and assistant professors (FT3).

It is important to understand that the main effects of department and seniority are *average* effects, when the other factors are collapsed. For example, the

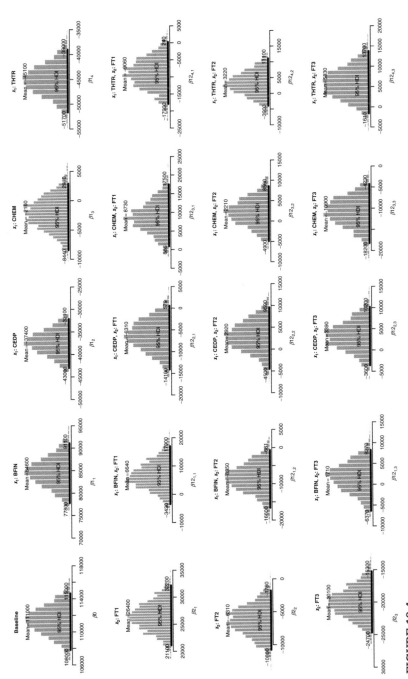

FIGURE 19.4

Posterior distribution for data in Figure 19.3. Baseline (β_0) is shown in upper left. Remainder of top row is main effect of $\vec{x}_1$ (department). Remainder of left column is main effect of $\vec{x}_2$ (seniority). Remaining cells show the interaction effects.

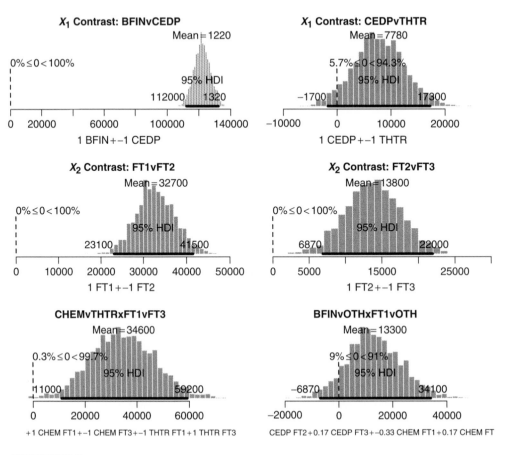

FIGURE 19.5
Selected contrasts for posterior in Figure 19.4. The values for the means and HDI limits are
rounded to three leading digits. The x-axis labels of the bottom row are obscured because
they exceed the boundaries of the plot.

contrast between FT2 and FT3 (middle row, right panel of Figure 19.5) is the
average difference between FT2 and FT3, collapsed across all departments. But
if you look at the data in Figure 19.3, you can see that the difference between
FT2 and FT3 is not the same in every department: There is a fairly large difference in CHEM, but a very small difference in BFIN. The effect of changing from
FT2 to FT3 depends on the department, which means that there is interaction.

Main effects must be interpreted and described cautiously when there are interactions. It would be a mistake to say that "the" difference between FT2 and FT3
is 13,800. Instead, that is the average difference across departments. The actual
difference within any particular department might be quite different. Similarly,

it would be a mistake to say that "the" effect of FT3 is to subtract 20,100 from the baseline, because the effect of seniority interacts with department.

19.1.4.1 Combining Metric and Nominal Predictors: ANCOVA

Consider again the salary data in Figure 19.3. You can see that the mean salary for FT1's in chemistry is much higher than in theater. This difference might be attributable solely to being in one department or the other. But the difference might also be attributable to some other factor, such as years on the job. In other words, the FT1's in chemistry might happen to have been employed for decades, whereas the FT1's in theater might happen to be relatively young. If we had the age of each employee, or, better yet, the number of years that the employee had been at the current level of seniority, then we could include that information as an additional predictor of salary. We could then assess whether department membership contributed any predictiveness beyond number of years on the job.

When a nominal predictor, such as department membership, is combined with a metric predictor, such as years on the job, the model is sometimes referred to as analysis of *co*variance, or ANCOVA. The metric predictor is the "covariate."

Programming ANCOVA in BUGS is a trivial combination of the models we've used for linear regression and ANOVA. Denote the nominal group member- ship for individual i as xNom[i], and denote the metric covariate value as xMet[i]. Then the core of the BUGS model specification is

```
mu[i] <- a0 + a[ xNom[i] ] + bMet * xMet[i]
y[i] ~ dnorm( mu[i] , tau )
```

where a[] is the deflection of each group from baseline, and bMet is the regression coefficient on the covariate. As in standard ANOVA, the deflec- tions a[] and intercept a0 should be transformed so that the deflections sum to zero.

When initializing the chains for ANCOVA in BUGS, it can help to start at the maximum likelihood estimate (MLE). The lm() function in R provides the MLE. If we type

```
xNom = factor( xNom ) # makes xNom a "factor"
lmInfo = lm( y ~ xNom + xMet )
```

then lmInfo is a list of information about the linear model that "best" fits the data. The best fitting coefficients are stored in lmInfo$coef. The first compo- nents of lmInfo$coef are the deflections for the levels of xNom and the last component of lmInfo$coef is the slope coefficient for xMet (because of the ordering of the variables in the call to lm). The deflections are parameterized relative to the first component, however. To convert to the parameterization that we use, in which the deflections sum to zero around a baseline, we can use the following code:

```
NxNomLevels = length( levels( xNom ) ) # number of levels of xNom
# Next line adds first xNom component to other xNom components:
a = c( lmInfo$coef[1] , lmInfo$coef[1] + lmInfo$coef[2:NxNomLevels] )
a0Init = mean( a )
aInit = a - mean( a )
```

ANCOVA models can also involve a separate slope for every level of the nominal predictor. Then the core of the BUGS model specification is

```
mu[i] <- a0 + a[ xNom[i] ] + ( bMet + bMetI[ xNom[i] ] ) * xMet[i]
y[i] ~ dnorm( mu[i] , tau )
```

where bMet+bMetI[xNom[i]] is the group-specific regression coefficient on the covariate for the xNom[i]th group. The coefficient bMet is the overall slope, and deflection bMetI[xNom[i]] adjusts steeper or shallower for each group. To make the slopes identifiable, the group-specific deflections are constrained to sum to zero: $\sum_{xNom}$ bMetI[xNom] = 0. Just as in multiple regression, a hyperprior can be put on the group-specific slopes, whereby the group-specific slopes come from a normal (or t) distribution, and the precision of that distribution is itself estimated.

The ANCOVA model, with a distinct intercept and slope for each group, closely resembles the model for repeated-measures simple linear regression in Section 16.3, p. 433. The model in that section had a distinct intercept and slope for each *subject*. If the subject variable in that model is considered to be a nominal predictor analogous to xNom here, then that model is essentially equivalent to one used here. The two model expressions are different, however, in how naturally they generalize to situations with more predictors. The formulation in the present section uses the general ANOVA formulation for the group-specific coefficients (i.e., deflections that sum to zero) and therefore generalizes naturally to situations with multiple nominal predictors.

Additional information about non-Bayesian ANCOVA can be found in a variety of other sources. A brief Bayesian treatment can be found in the book by Ntzoufras (2009), but beware that the formulation there uses no hyperprior on the nominal or metric coefficients, and the method used there to implement the sum-to-zero constraint cannot be used with hyperpriors, as was discussed previously on p. 497.

19.1.4.2 Interaction Contrasts
Just as we can ask whether differences among particular levels of predictors are credible, we can ask whether interactions among particular combinations of predictors are credible. Consider again the data in Figure 19.3. The difference

between full professors (FT1) and assistant professors (FT3) appears to be large in the chemistry department (CHEM) but smaller in the theater department (THTR). Is the simple effect of seniority bigger in chemistry than it is in theater? In other words, is (CHEM.FT1−CHEM.FT3)−(THTR.FT1−THTR.FT3) credibly nonzero?

This sort of difference of differences is called an *interaction contrast*. In general, an interaction contrast is constructed by taking any set of contrast coefficients on $\vec{x}_1$, and any set of contrast coefficients on $\vec{x}_2$, and computing their outer product. The outer product was described in Section 8.8.1 (BernTwoGrid.R), p. 178. Formally, the outer product of two vectors is denoted by the symbol "⊗." To provide an example of an interaction contrast as an outer product of main-effect contrasts, we will recast the one we are presently considering, namely, (CHEM.FT1−CHEM.FT3)−(THTR.FT1−THTR.FT3), in generic notation. Notice that CHEM is level 3 of predictor 1, and hence can be written as $\vec{x}_{1,3}$. Writing the other components in the same fashion, the interaction contrast is $(\vec{x}_{1,3}.\vec{x}_{2,1} - \vec{x}_{1,3}.\vec{x}_{2,3}) - (\vec{x}_{1,4}.\vec{x}_{2,1} - \vec{x}_{1,4}.\vec{x}_{2,3})$. That can be algebraically rearranged to highlight the coefficients on the particular combinations:

$$(+1)\,\vec{x}_{1,3}.\vec{x}_{2,1} + (-1)\,\vec{x}_{1,3}.\vec{x}_{2,3} + (-1)\,\vec{x}_{1,4}.\vec{x}_{2,1} + (+1)\,\vec{x}_{1,4}.\vec{x}_{2,3}$$

Those highlighted coefficients can be obtained as the outer product of main-effect contrasts, namely, the contrast $\vec{c}_1 = \langle 0, 0, +1, -1 \rangle$, which expresses CHEM minus THTR, and the contrast $\vec{c}_2 = \langle +1, 0, -1 \rangle$, which expresses FT1 minus FT3:

$$\vec{c}_1 \otimes \vec{c}_2 = \quad \langle \begin{matrix} \vec{x}_{1,1} & \vec{x}_{1,2} & \vec{x}_{1,3} & \vec{x}_{1,4} \\ 0 & 0 & +1 & -1 \end{matrix} \rangle \otimes \langle \begin{matrix} \vec{x}_{2,1} & \vec{x}_{2,2} & \vec{x}_{2,3} \\ +1 & 0 & -1 \end{matrix} \rangle$$

$$= \begin{matrix} & \begin{matrix} \vec{x}_{2,1} & \vec{x}_{2,2} & \vec{x}_{2,3} \end{matrix} \\ \begin{matrix} \vec{x}_{1,1} \\ \vec{x}_{1,2} \\ \vec{x}_{1,3} \\ \vec{x}_{1,4} \end{matrix} & \begin{bmatrix} 0 & 0 & 0 \\ 0 & 0 & 0 \\ +1 & 0 & -1 \\ -1 & 0 & +1 \end{bmatrix} \end{matrix}$$

Notice that the coefficients in the matrix match the highlighted coefficients in the difference of differences that was expressed a few sentences previously. The posterior of this interaction contrast is shown in the bottom-left histogram of Figure 19.5. The mean of 34,600 indicates that the difference between FT1 and FT2 is about 34,600 greater for CHEM than for THTR. The 95% HDI clearly excludes zero, indicating that this interaction contrast is credibly nonzero.

Interaction contrasts can involve "complex" comparisons just as simply as pairwise comparisons. For example, suppose we are interested in comparing BFIN against the average of the other nonbusiness departments, specifically for a contrast between FT1 and lesser ranks. This interaction contrast is expressed as $\langle +1, -1/3, -1/3, -1/3 \rangle \otimes \langle +1, -1/2, -1/2 \rangle$. The posterior of this contrast is shown in the bottom-right histogram of Figure 19.5. (The label on the x axis exceeds the margins of the figure because there are 12 combinations of levels involved in the contrast specification.) The result suggests that there is considerable uncertainty in the larger difference, between FT1 and other ranks, in BFIN than in other departments. Therefore, we would not want to conclude that the interaction contrast is credibly nonzero. Exercise 19.2 gives you hands-on practice with specification of interaction contrasts.

19.1.5 Noncrossover Interactions, Rescaling, and Homogeneous Variances

When interpreting interactions, it can be important to consider the scale on which the data are measured. This is because an interaction means nonadditive effects when measured in the current scale. If the data are nonlinearly transformed to a different scale, then the nonadditivity can also change.

Consider an example, using utterly fictional numbers merely for illustration. Suppose the average salary of Democratic women is 10 monetary units, for Democratic men it's 12 units, for Republican women it's 15 units, and for Republican men it's 18 units. These data indicate that there is a nonadditive interaction of political party and gender, because the change in pay from women to men is 2 units for Democrats, but 3 units for Republicans. Another way of describing the interaction is to notice that the change in pay from Democrats to Republicans is 5 units for women but 6 units for men. A researcher might be tempted to interpret the interaction as indicating some extra advantage attained by Republican men, or some special disadvantage suffered by Democratic women. But such an interpretation may be inappropriate, because a mere rescaling of the data makes the interaction disappear, as will be described next.

Salary raises and comparisons are often measured by percentages and ratios, not by additive or subtractive differences. Consider the salary data in percentage terms. Among Democrats, men make 20% more than women. Among Republicans, the men again make 20% more than the women. Among women, Republicans make 50% more than Democrats. Among men, Republicans again make 50% more than Democrats. In these ratio terms, there is no interaction of gender and political party: Change from female to male predicts a 20% increase in salary regardless of party, and change from Democrat to Republican predicts a 50% increase in salary regardless of gender.

Equal ratios are transformed to equal distances by a logarithmic transformation. If we measure salary in terms of the logarithm of monetary units, then the salary of Democratic women is $\log_{10}(10) = 1.000$, the salary of Democratic men is $\log_{10}(12) = 1.079$, the salary of Republican women is $\log_{10}(15) = 1.176$, and the salary of Republican men is $\log_{10}(18) = 1.255$. With this logarithmic scaling, the increase in salary from women to men is 0.079 for both parties, and the increase from Democrat to Republican is 0.176 for both genders. In other words, when salary is measured on a logarithmic scale, there is no interaction of gender and political party.

It may seem strange to measure salary on a logarithmic scale, but there are many situations for which the scale is arbitrary. The pitch of a sound can be measured in terms of frequency (i.e., cycles per second) or in terms of perceived pitch, which is essentially the logarithm of the frequency. The magnitude of an earthquake can be measured by its energy or by its value on the Richter scale, which is the logarithm of energy. The pace of a dragster on a racetrack can be measured by the average speed during the run or by the duration from start to finish (which is the reciprocal of average speed). Thus, measurement scales are not unique and are instead determined by convention.

The general issue is illustrated in Figure 19.6. Suppose that predictor x_1 has two levels, as does predictor x_2. Suppose we have three data points at each combination of levels, yielding 12 data points altogether. The means at each combination of levels are shown in the top-left graph of Figure 19.6. You can see that there is an interaction, with the effect of x_1 being bigger when $x_2 = 2$ than when $x_2 = 1$. But this interaction goes away when the data are transformed by taking the logarithm, as shown in the lower-left graph. Each individual data point was transformed, and then the means in each cell were computed. Of course, the transformation can go the other way: Data with no interaction, as in the lower-left plot, can be made to have an interaction when they are rescaled as in the upper-left plot, via an exponential transformation.

The transformability from interaction to noninteraction is only possible for *noncrossover* interactions. This terminology is merely a description of the graph: The lines do not cross over each other (and they have the same sign slope). In this situation, the y axis can have different portions stretched or shrunken so that the lines become parallel. If, however, the lines cross, as in the middle column of Figure 19.6, then there is no way to uncross the lines merely by stretching or shrinking intervals of the y axis. The right column of Figure 19.6 shows the same data as the middle column, but it is plotted with the roles of x_1 and x_2 exchanged. When plotted this way, the lines do not cross, but they do have *opposite-sign slopes* (i.e., one slope is positive and the other slope is negative). There is no way that stretching or shrinking the y axis can change the signs of the slopes, hence the interaction cannot be removed merely by

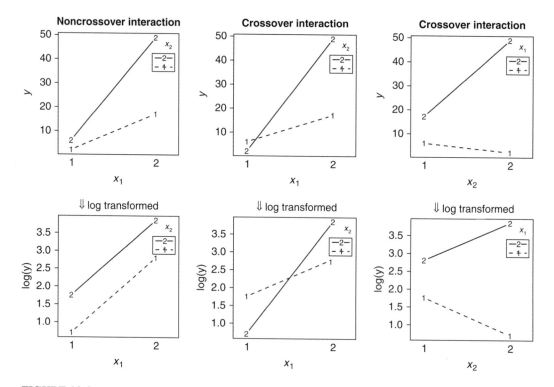

FIGURE 19.6
Top row shows means of original data; bottom row shows means of logarithmically transformed data. Left column shows a noncrossover interaction; middle and right columns show the same crossover interaction plotted against x_1 or x_2.

transforming the data. Because these data have crossing lines when plotted as in the middle column, they are said to have a crossover interaction even when they are plotted as in the right column. (Test your understanding: Is the interaction in Figure 19.1 a crossover interaction?)

It is important to note that the transformation applies to individual raw data values, not to the means of the conditions. A consequence of transforming the data, therefore, is changes in the variances of the data within each condition. For example, suppose one condition has data values of 100, 110, and 120, whereas a second condition has data values of 1100, 1110, and 1120. For both conditions, the variance is 66.7 (i.e., there is homogeneity of variance). When the data are logarithmically transformed, the variance of the first group becomes 1.05e−3, but the variance of the second group becomes two orders of magnitude smaller, namely, 1.02e−5 (i.e., there is *not* homogeneity of variance).

Therefore, when applying the hierarchical model of Figure 19.2, we must be aware that it assumes homogeneity of variance. If we transform the data, we are changing the variances within the levels of the predictors. The transformed variances might or might not be fairly homogeneous. If they are not, then either the data should be transformed in such a way as to respect homogeneity of variance or the model should be changed to allow unequal variances.

The models we have been using also assume a normal likelihood function, which means that the data at any level of the predictors should be normally distributed. When the data are transformed to a different scale, the shape of their distribution also changes. If the distributions become radically non-normal, it may be misleading to use a model with a normal likelihood function. For a discussion of these issues, review Section 15.1.4, p. 399.

In summary, this section has made two main points. First, if you have a noncrossover interaction, be careful what you claim about it. A noncrossover interaction merely means nonadditivity *in the scale you are using*. If this scale is the only meaningful scale, or if the this scale is the overwhelmingly dominant scale used in that field of research, then you can cautiously interpret the nonadditive interaction with respect to that scale. But if transformed scales are reasonable, then keep in mind that nonadditivity is scale specific, and there might be no interaction in a different scale. With a crossover interaction, however, no rescaling can undo the interaction. Second, nonlinear transformations change the within-cell variances and the shapes of the within-cell distributions. Be sure that the model you are using is appropriate to the homogeneity or non-homogeneity of variances in the data and to the shapes of the distributions, on whatever scale you are using. Exercise 19.1 has you consider these issues "hands on."

19.2 REPEATED MEASURES, A.K.A. WITHIN-SUBJECT DESIGNS

In many situations, a single "subject" contributes data to multiple levels of the predictors. For example, suppose we are interested in how quickly people can press a button in response to a stimulus onset. The stimulus could appear in the visual modality as a light, or in the auditory modality as a tone. The subject could respond with his or her dominant hand or with the nondominant hand. Thus, there are two nominal predictors, namely modality and hand. The new aspect is that a single subject contributes data to all combinations of the predictors. On many successive trials, the subject gets either a tone or light and is instructed to respond with either the dominant or nondominant hand. Because every subject is measured many times, this situation is sometimes called a *repeated measures* design. Because the levels of the predictors change within subjects, this situation is also called a *within-subject* design. I favor the latter terminology because it more explicitly connotes the essential aspect of the design, that the same subject contributes data in more than one condition. Within-subject designs are contrasted with *between-subject* designs, in which different subjects contribute data to different levels of the predictors.

When every subject contributes many data points to every combination of predictors, then the model of the situation is a straightforward extension of the models we've already considered. We merely add "subject" as another predictor in the model, with each individual subject being a level of the predictor. If there is one predictor other than subject, the model becomes

$$y = \beta_0 + \vec{\beta}_1 \vec{x}_1 + \vec{\beta}_S \vec{x}_S + \vec{\beta}_{1 \times S} \vec{x}_{1 \times S}$$

This is exactly the two-predictor model we have already considered, with the second predictor being subject. When there are two predictors other than subject, the model becomes

$$y = \beta_0 + \vec{\beta}_1 \vec{x}_1 + \vec{\beta}_2 \vec{x}_2 + \vec{\beta}_S \vec{x}_S + \vec{\beta}_{1 \times 2} \vec{x}_{1 \times 2} + \vec{\beta}_{1 \times S} \vec{x}_{1 \times S}$$
$$+ \vec{\beta}_{2 \times S} \vec{x}_{2 \times S} + \vec{\beta}_{1 \times 2 \times S} \vec{x}_{1 \times 2 \times S}$$

This model includes all the two-way interactions of the factors, plus the three-way interaction. Again, subject merely plays the role of the third predictor.

The preceding model, which includes all the high-order interactions with subject, is fine in principle but may be overkill in practice. Unless you have specific theoretical motivations to seek out and interpret high-order interactions of subject with other predictors, there is little reason to model them, and there is difficulty making sense of them even if you did model them. Instead, if you have many data points from each subject in every cell, an alternative approach is to apply a Bayesian ANOVA model to each subject's data, and then put a higher-order prior across the subject parameter estimates, so that different subjects mutually inform each other's estimates and provide a stable group-level estimate. Thus, every subject has a baseline, β_{0s}, and there is a higher-order, group-level prior on the distribution of β_{0s} across subjects. Each predictor also has subject-specific estimates, with the effect of the jth level of predictor 1 denoted $\beta_{1s,j}$. Each of these effect parameters has a higher, group-level prior across subjects. (This was the modeling approach taken for repeated measures in simple linear regression in Section 16.3, p. 433.) Finally, the group-level effects have a hyperprior that provides shrinkage on the effects of a predictor. In other words, the shrinkage prior, on effects of a predictor, is set at the group level, not at the subject level.

There are other situations, however, in which each subject contributes only one datum to a combination of the other predictors. For example, in the case of the response-time study described earlier, perhaps we have only the median response time of the subject in each combination of hand and modality. As another example, suppose the value to be predicted is IQ, as measured by a lengthy exam, with one predictor being noisy versus quiet exam environment and the other predictor being paper versus computerized exam format. Although it is conceivable that subjects could be repeatedly tested in each condition, it would be challenging enough to get people to sit through all four combinations even once. Thus, each subject would contribute one value to each condition.

In the situation when each subject contributes only one datum per condition, the models described earlier, with all the interaction terms, are not *identifiable*, meaning that there are more parameters than data points. The simplest case of this situation is trying to estimate the mean and variance of a normal distribution from a single data point. A Bayesian analysis can still be conducted, but there will be high uncertainty in the parameter estimates, governed largely by the priors. Therefore, instead of attempting to estimate all the interactions of subjects with other predictors, we assume a simpler model in which the only influence of subjects is on the baseline:

$$ y = \beta_0 + \vec{\beta}_S \vec{x}_S + \vec{\beta}_1 \vec{x}_1 + \vec{\beta}_2 \vec{x}_2 + \vec{\beta}_{1\times2} \vec{x}_{1\times2} $$

In other words, we assume a main effect of subject but no interaction of subject with other predictors. In this model, the subject effect (deflection) is constant across treatments, and the treatment effects (deflections) are constant across subjects. Notice that the model makes no requirement that every subject contributes a datum to every condition. Indeed, the model allows zero or more than one datum per subject per condition. As mentioned earlier, the computations in Bayesian ANOVA make no assumptions or requirements that the design is "balanced." If you do have many observations per subject in every combination of predictors, then one of the previously described models may be considered.

19.2.1 Why Use a Within-Subject Design? And Why Not?

The primary reason to use a within-subject design is that you can achieve much greater precision in the estimates of the effects than in a between-subject design. For example, suppose you are interested in measuring the effect on response time of using the dominant versus nondominant hand. Suppose there is a population of four subjects from whom you could measure data. If we could measure every subject in every condition, we would know that for the first subject, his or her response times for dominant and nondominant hands are 300 and 320 msec. For the second subject, the response times are 350 and 370. For the third subject, the response times are 400 and 420, and for the fourth subject, the response times are 450 and 470. Thus, for every subject, the difference between dominant and nondominant hands is 20 msec. Suppose we have the resources to measure only two data points in each condition. We measure response times from the dominant hands of two subjects. Should we measure response times from the nondominant hands of the *same* two subjects or the nondominant hands of two *other* subjects? If we measure from the same two subjects, then the estimated effect for each subject is 20 msec, and we have high certainty in the magnitude of the effect. If we measure from two other subjects, then the estimated effect of dominant versus nondominant hand is the average of the first two subjects versus the average of the second two subjects, and the difference is badly affected by random sampling. The between-subject design yields lower precision in the estimate of the effect. Exercise 19.3 has you examine, hands on, a case of this improvement in precision.

Because of the gain in precision, it is desirable to use within-subject designs. But there are many dangers of within-subject designs that need to be considered before they are applied in any particular situation. The key problem is that, in most situations, when you measure the subject you change the subject, and therefore subsequent measurements are not measuring the same subject. The simplest examples of this are mere fatigue or generic practice effects. In measures of response time, if you measure repeatedly from the same subject, you will find improvement over the first several trials because of the subject gaining practice with the task, but after a while, as the subject tires, there will be a decline in performance. The problem is that if you measure the dominant hand in the early trials and the nondominant hand in the later trials, then the effect of practice or fatigue will contaminate the effect of handedness. The repeated measurement process affects and contaminates the measure that is supposed to be a signature of the predictor.

Practice and fatigue effects can be overcome by randomly distributing and repeating the conditions throughout the repeated measures, *if* the practice and fatigue effects influence all conditions equally. Thus, if practice improves both the dominant and nondominant hand by 50 msec, then the difference between dominant and nondominant hands is unaffected by practice. But practice might affect the nondominant hand much more than the dominant hand. You can imagine that in complex designs with many predictors, each with many levels, it can become difficult to justify an assumption that repeated measures have comparable effects on all conditions.

Worse yet, in some situations there can be *differential carryover effects* from one condition to the next. For example, having just experienced practice in the visual modality with the nondominant hand might improve subsequent performance in the auditory modality with the nondominant hand, but it might not improve subsequent performance in the visual modality with the dominant hand. Thus, the carryover effect is different for different subsequent conditions.

When you suspect strong differential carryover effects, you may be able to explicitly manipulate the ordering of the conditions and measure the carry-over effects, but this might be impossible mathematically and impractical, depending on the specifics of your situation. In this case, you must revert to a between-subject design and simply include many subjects to attenuate between-subject noise.

In general, all the models we have been using assume independence of observations. The probability of the combined data is the product of the prob-abilities of the individual data points. When we use repeated measures, this assumption is much less easy to justify. On the one hand, when we repeatedly flip a coin, we might be safe to assume that its underlying bias does not change much from one flip to the next. But, on the other hand, when we repeatedly test the response time of a human subject, it is less easy to justify an assump-tion that the underlying response time remains unaffected by the previous trial. Researchers will often make the assumption of independence merely as an approximation of convenience, hoping that by arranging conditions ran-domly across many repeated measures, the differential carryover effects will be minimized.

19.3 R CODE

19.3.1 Bayesian Two-Factor ANOVA

Several implementation details of the program are the same as the oneway ANOVA program of the previous chapter:

- Data are normalized so that prior constants can be more generic.
- Initialization of chains is based on the data. It is important to do this, otherwise burn-in can take forever.
- Because there is nasty autocorrelation, we use a large thinning constant and we also use multiple chains. For a reminder of the issues of burn-in and thinning, see Section 23.2, p. 623.

A new detail arises in how the uncentered parameter estimates are recentered to respect the sum-to-zero constraints. The uncentered estimates from BUGS are a0, a1[], a2[], and a1a2[,]. By definition of the ANOVA model, the predicted mean of cell i, j is

```
m[i,j] = a0 + a1[i] + a2[j] + a1a2[i,j]
```

We use these predicted means to construct the zero-centered parameters. First, b0 is the mean across all the predicted means:

```
b0 = mean( m[,] )
```

Then the main effect deflections are the marginal means minus the overall mean:

```
b1[i] = mean( m[i,] ) - b0
b2[j] = mean( m[,j] ) - b0
```

It is easy (honest!) to check that those deflections do indeed sum to zero (i.e., `sum( b1[] ) = 0` and `sum( b2[] ) = 0`). Finally, the interaction deflections are the residuals after the additive effect of b1 and b2 is taken into account:

```
b1b2[i,j] = m[i,j] - ( b0 + b1[i] + b2[j] )
```

Again, it is easy to check that the rows and columns of `b1b2[,]` all sum to zero.

In the data section of the program, one option is to load data from the article of Qian & Shen (2007). The program here uses a hierarchical structure similar to that used by Qian & Shen (2007), but their program did not recenter the parameters as is done here. It may be instructive to compare the results of the program here with the results reported by Qian & Shen (2007).

BUGS for many factors. The program that follows applies only for cases of two nominal predictors. If you have many nominal predictors, along with their two-way, three-way, and higher-order interactions, it becomes unwieldy to explicitly and separately name all the deflection parameters. Instead, it can be more elegant to use a technique of *dummy coding*, whereby we essentially revert back to using vectors for coding the values of the predictors instead of integer indices. That is, $\vec{x}_1 = level\ 2$ is coded by the "dummy" vector $\langle 0, 1, 0, \ldots \rangle$ instead of by the integer index 2. Interactions are represented by matrices of dummy codes that have been flattened into vectors. For an example of programming this technique in BUGS, see Ntzoufras (2009, Ch. 6). Unfortunately, those examples do not incorporate the higher-level prior structure emphasized in Figure 19.2.

(ANOVAtwowayBRugs.R)

```
1   graphics.off()
2   rm(list=ls(all=TRUE))
3   fnroot = "ANOVAtwowayBrugs"
4   library(BRugs)          # Kruschke, J. K. (2010). Doing Bayesian data analysis:
5                           # A Tutorial with R and BUGS. Academic Press / Elsevier.
6   #------------------------------------------------------------------------------
7   # THE MODEL.
8
9   modelstring = "
10  # BUGS model specification begins here...
11  model {
```

```
12   for ( i in 1:Ntotal  ) {
13     y[i] ~ dnorm( mu[i] , tau )
14     mu[i] <- a0 + a1[x1[i]] + a2[x2[i]] + a1a2[x1[i],x2[i]]
15   }
16   #
17   tau <- pow( sigma , -2 )
18   sigma ~ dunif(0,10) # y values are assumed to be standardized
19   #
20   a0 ~ dnorm(0,0.001) # y values are assumed to be standardized
21   #
22   for ( j1 in 1:Nx1Lvl ) { a1[j1] ~ dnorm( 0.0 , a1tau ) }
23   a1tau <- 1 / pow( a1SD , 2 )
24   a1SD <- abs( a1SDunabs ) + .1
25   a1SDunabs ~ dt( 0 , 0.001 , 2 )
26   #
27   for ( j2 in 1:Nx2Lvl ) { a2[j2] ~ dnorm( 0.0 , a2tau ) }
28   a2tau <- 1 / pow( a2SD , 2 )
29   a2SD <- abs( a2SDunabs ) + .1
30   a2SDunabs ~ dt( 0 , 0.001 , 2 )
31   #
32   for ( j1 in 1:Nx1Lvl ) { for ( j2 in 1:Nx2Lvl ) {
33     a1a2[j1,j2] ~ dnorm( 0.0 , a1a2tau )
34   } }
35   a1a2tau <- 1 / pow( a1a2SD , 2 )
36   a1a2SD <- abs( a1a2SDunabs ) + .1
37   a1a2SDunabs ~ dt( 0 , 0.001 , 2 )
38   }
39   # ... end BUGS model specification
40   " # close quote for modelstring
41   # Write model to a file, and send to BUGS:
42   writeLines(modelstring,con="model.txt")
43   modelCheck( "model.txt" )
44
45   #-------------------------------------------------------------------------
46   # THE DATA.
47   # Specify data source:
48   dataSource = c( "QianS2007" , "Salary" , "Random" , "Ex19.3" )[4]
49
50   # Load the data:
51   if ( dataSource == "QianS2007" ) {
52     fnroot = paste( fnroot , dataSource , sep="" )
53     datarecord = read.table( "QianS2007SeaweedData.txt" , header=TRUE , sep="," )
54     # Logistic transform the COVER value:
55     # Used by Appendix 3 of QianS2007 to replicate Ramsey and Schafer (2002).
56     datarecord$COVER = -log( ( 100 / datarecord$COVER ) - 1 )
57     y = as.numeric(datarecord$COVER)
58     x1 = as.numeric(datarecord$TREAT)
59     x1names = levels(datarecord$TREAT)
60     x2 = as.numeric(datarecord$BLOCK)
61     x2names = levels(datarecord$BLOCK)
62     Ntotal = length(y)
63     Nx1Lvl = length(unique(x1))
64     Nx2Lvl = length(unique(x2))
```

```
65      x1contrastList = list( f_Effect=c( 1/2 , -1/2 , 0 , 1/2 , -1/2 , 0 ) ,
66                              F_Effect=c( 0 , 1/2 , -1/2 , 0 , 1/2 , -1/2 ) ,
67                              L_Effect=c( 1/3 , 1/3 , 1/3 , -1/3 , -1/3 , -1/3 ) ) )
68      x2contrastList = NULL # list( vector(length=Nx2Lvl) )
69      x1x2contrastList = NULL # list( matrix( 1:(Nx1Lvl*Nx2Lvl) , nrow=Nx1Lvl ) )
70    }
71
72    if ( dataSource == "Salary" ) {
73      fnroot = paste( fnroot , dataSource , sep="" )
74      datarecord = read.table( "Salary.csv" , header=TRUE , sep="," )
75      y = as.numeric(datarecord$Salary)
76      if ( F ) { # take log10 of salary
77        y = log10( y )
78        fnroot = paste( fnroot , "Log10" , sep="" )
79      }
80      x1 = as.numeric(datarecord$Org)
81      x1names = levels(datarecord$Org)
82      x2 = as.numeric(datarecord$Post)
83      x2names = levels(datarecord$Post)
84      Ntotal = length(y)
85      Nx1Lvl = length(unique(x1))
86      Nx2Lvl = length(unique(x2))
87      x1contrastList = list( BFINvCEDP = c( 1 , -1 , 0 , 0 ) ,
88                             CEDPvTHTR = c( 0 , 1 , 0 , -1 ) )
89      x2contrastList = list( FT1vFT2 = c( 1 , -1 , 0 ) , FT2vFT3 = c(0,1,-1) )
90      x1x2contrastList = list(
91            CHEMvTHTRxFT1vFT3 = outer( c(0,0,+1,-1) , c(+1,0,-1) ) ,
92            BFINvOTHxFT1vOTH = outer( c(+1,-1/3,-1/3,-1/3) , c(+1,-1/2,-1/2) ) ) )
93    }
94
95    if ( dataSource == "Random" ) {
96      fnroot = paste( fnroot , dataSource , sep="" )
97      set.seed(47405)
98      ysdtrue = 3.0
99      a0true = 100
100     a1true = c( 2 , 0 , -2 ) # sum to zero
101     a2true = c( 3 , 1 , -1 , -3 ) # sum to zero
102     a1a2true = matrix( c( 1,-1,0, -1,1,0, 0,0,0, 0,0,0 ),# row and col sum to zero
103                        nrow=length(a1true) , ncol=length(a2true) , byrow=F )
104     npercell = 8
105     datarecord = matrix( 0, ncol=3 , nrow=length(a1true)*length(a2true)*npercell )
106     colnames(datarecord) = c("y","x1","x2")
107     rowidx = 0
108     for ( x1idx in 1:length(a1true) ) {
109       for ( x2idx in 1:length(a2true) ) {
110         for ( subjidx in 1:npercell ) {
111           rowidx = rowidx + 1
112           datarecord[rowidx,"x1"] = x1idx
113           datarecord[rowidx,"x2"] = x2idx
114           datarecord[rowidx,"y"] = ( a0true + a1true[x1idx] + a2true[x2idx]
115                                      + a1a2true[x1idx,x2idx] + rnorm(1,0,ysdtrue) )
116         }
117       }
118     }
```

```
119    datarecord = data.frame( y=datarecord[,"y"] ,
120                             x1=as.factor(datarecord[,"x1"]) ,
121                             x2=as.factor(datarecord[,"x2"]) )
122    y = as.numeric(datarecord$y)
123    x1 = as.numeric(datarecord$x1)
124    x1names = levels(datarecord$x1)
125    x2 = as.numeric(datarecord$x2)
126    x2names = levels(datarecord$x2)
127    Ntotal = length(y)
128    Nx1Lvl = length(unique(x1))
129    Nx2Lvl = length(unique(x2))
130    x1contrastList = list( X1_1v3 = c( 1 , 0 , -1 ) ) #
131    x2contrastList = list( X2_12v34 = c( 1/2 , 1/2 , -1/2 , -1/2 ) ) #
132    x1x2contrastList = list(
133      IC_11v22 = outer( c(1,-1,0) , c(1,-1,0,0) ) ,
134      IC_23v34 = outer( c(0,1,-1) , c(0,0,1,-1) )
135    )
136  }
137
138  # Load the data:
139  if ( dataSource == "Ex19.3" ) {
140    fnroot = paste( fnroot , dataSource , sep="" )
141    y = c( 101,102,103,105,104, 104,105,107,106,108, 105,107,106,108,109, 109,
142    108,110,111,112 )
143    x1 = c( 1,1,1,1,1, 1,1,1,1,1, 2,2,2,2,2, 2,2,2,2,2 )
144    x2 = c( 1,1,1,1,1, 2,2,2,2,2, 1,1,1,1,1, 2,2,2,2,2 )
145    # S = c( 1,2,3,4,5, 1,2,3,4,5, 1,2,3,4,5, 1,2,3,4,5 )
146    x1names = c("x1.1","x1.2")
147    x2names = c("x2.1","x2.2")
148    # Snames = c("S1","S2","S3","S4","S5")
149    Ntotal = length(y)
150    Nx1Lvl = length(unique(x1))
151    Nx2Lvl = length(unique(x2))
152    # NSLvl = length(unique(S))
153    x1contrastList = list( X1.2vX1.1 = c( -1 , 1 ) )
154    x2contrastList = list( X2.2vX2.1 = c( -1 , 1 ) )
155    x1x2contrastList = NULL # list( matrix( 1:(Nx1Lvl*Nx2Lvl) , nrow=Nx1Lvl ) )
156  }
157
158  # Specify the data in a form that is compatible with BRugs model, as a list:
159  ySDorig = sd(y)
160  yMorig = mean(y)
161  z = ( y - yMorig ) / ySDorig
162  datalist = list(
163    y = z ,
164    x1 = x1 ,
165    x2 = x2 ,
166    Ntotal = Ntotal ,
167    Nx1Lvl = Nx1Lvl ,
168    Nx2Lvl = Nx2Lvl
169  )
170  # Get the data into BRugs:
171  modelData( bugsData( datalist ) )
172
```

```
173  #-------------------------------------------------------------------------
174  # INTIALIZE THE CHAINS.
175
176  # Autocorrelation within chains is large, so use several chains to reduce
177  # degree of thinning. But we still have to burn-in all the chains, which takes
178  # more time with more chains.
179  nchain = 10
180  modelCompile( numChains = nchain )
181
182  if ( F ) {
183    modelGenInits() # often won't work for diffuse prior
184  } else {
185    #  initialization based on data
186    theData = data.frame( y=datalist$y , x1=factor(x1,labels=x1names) ,
187                          x2=factor(x2,labels=x2names) )
188    a0 = mean( theData$y )
189    a1 = aggregate( theData$y , list( theData$x1 ) , mean )[,2] - a0
190    a2 = aggregate( theData$y , list( theData$x2 ) , mean )[,2] - a0
191    linpred = as.vector( outer( a1 , a2 , "+" ) + a0 )
192    a1a2 = aggregate( theData$y, list(theData$x1,theData$x2), mean)[,3] - linpred
193    genInitList <- function() {
194      return(
195        list(
196          a0 = a0 ,
197          a1 = a1 ,
198          a2 = a2 ,
199          a1a2 = matrix( a1a2 , nrow=Nx1Lvl , ncol=Nx2Lvl ) ,
200          sigma = sd(theData$y)/2 , # lazy
201          a1SDunabs = sd(a1) ,
202          a2SDunabs = sd(a2) ,
203          a1a2SDunabs = sd(a1a2)
204        )
205      )
206    }
207    for ( chainIdx in 1 : nchain ) {
208      modelInits( bugsInits( genInitList ) )
209    }
210  }
211
212  #-------------------------------------------------------------------------
213  # RUN THE CHAINS
214
215  # burn in
216  BurnInSteps = 10000
217  modelUpdate( BurnInSteps )
218  # actual samples
219  samplesSet( c( "a0" , "a1" , "a2" , "a1a2" ,
220                 "sigma" , "a1SD" , "a2SD" , "a1a2SD" ) )
221  stepsPerChain = ceiling(2000/nchain)
222  thinStep = 500 # 750
223  modelUpdate( stepsPerChain , thin=thinStep )
224
225  #-------------------------------------------------------------------------
226  # EXAMINE THE RESULTS
```

```
227
228   source("plotChains.R")
229   source("plotPost.R")
230
231   checkConvergence = F
232   if ( checkConvergence ) {
233     sumInfo = plotChains( "a0" , saveplots=F , filenameroot=fnroot )
234     sumInfo = plotChains( "a1" , saveplots=F , filenameroot=fnroot )
235     sumInfo = plotChains( "a2" , saveplots=F , filenameroot=fnroot )
236     sumInfo = plotChains( "a1a2" , saveplots=F , filenameroot=fnroot )
237     readline("Press any key to clear graphics and continue")
238     graphics.off()
239     sumInfo = plotChains( "sigma" , saveplots=F , filenameroot=fnroot )
240     sumInfo = plotChains( "a1SD" , saveplots=F , filenameroot=fnroot )
241     sumInfo = plotChains( "a2SD" , saveplots=F , filenameroot=fnroot )
242     sumInfo = plotChains( "a1a2SD" , saveplots=F , filenameroot=fnroot )
243     readline("Press any key to clear graphics and continue")
244     graphics.off()
245   }
246
247   # Extract and plot the SDs:
248   sigmaSample = samplesSample("sigma")
249   a1SDSample = samplesSample("a1SD")
250   a2SDSample = samplesSample("a2SD")
251   a1a2SDSample = samplesSample("a1a2SD")
252   windows()
253   layout( matrix(1:4,nrow=2) )
254   par( mar=c(3,1,2.5,0) , mgp=c(2,0.7,0) )
255   plotPost( sigmaSample , xlab="sigma" , main="Cell SD" , breaks=30 )
256   plotPost( a1SDSample , xlab="a1SD" , main="a1 SD" , breaks=30 )
257   plotPost( a2SDSample , xlab="a2SD" , main="a2 SD" , breaks=30 )
258   plotPost( a1a2SDSample , xlab="a1a2SD" , main="Interaction SD" , breaks=30 )
259   dev.copy2eps(file=paste(fnroot,"SD.eps",sep=""))
260
261   # Extract a values:
262   a0Sample = samplesSample( "a0" )
263   chainLength = length(a0Sample)
264   a1Sample = array( 0 , dim=c( datalist$Nx1Lvl , chainLength ) )
265   for ( x1idx in 1:datalist$Nx1Lvl ) {
266     a1Sample[x1idx,] = samplesSample( paste("a1[",x1idx,"]",sep="") )
267   }
268   a2Sample = array( 0 , dim=c( datalist$Nx2Lvl , chainLength ) )
269   for ( x2idx in 1:datalist$Nx2Lvl ) {
270     a2Sample[x2idx,] = samplesSample( paste("a2[",x2idx,"]",sep="") )
271   }
272   a1a2Sample = array(0, dim=c( datalist$Nx1Lvl , datalist$Nx2Lvl , chainLength ) )
273   for ( x1idx in 1:datalist$Nx1Lvl ) {
274     for ( x2idx in 1:datalist$Nx2Lvl ) {
275       a1a2Sample[x1idx,x2idx,] = samplesSample( paste( "a1a2[",x1idx,",",x2idx,"]",
276                                                     sep="" ) )
277     }
278   }
279
```

```
280    # Convert to zero-centered b values:
281    m12Sample = array( 0, dim=c( datalist$Nx1Lvl , datalist$Nx2Lvl , chainLength ) )
282    for ( stepIdx in 1:chainLength ) {
283        m12Sample[,,stepIdx ] = ( a0Sample[stepIdx]
284                                    + outer( a1Sample[,stepIdx] ,
285                                             a2Sample[,stepIdx] , "+" )
286                                    + a1a2Sample[,,stepIdx] )
287    }
288    b0Sample = apply( m12Sample , 3 , mean )
289    b1Sample = ( apply( m12Sample , c(1,3) , mean )
290                 - matrix(rep( b0Sample ,Nx1Lvl),nrow=Nx1Lvl,byrow=T) )
291    b2Sample = ( apply( m12Sample , c(2,3) , mean )
292                 - matrix(rep( b0Sample ,Nx2Lvl),nrow=Nx2Lvl,byrow=T) )
293    linpredSample = array(0,dim=c(datalist$Nx1Lvl,datalist$Nx2Lvl,chainLength))
294    for ( stepIdx in 1:chainLength ) {
295        linpredSample[,,stepIdx ] = ( b0Sample[stepIdx]
296                                      + outer( b1Sample[,stepIdx] ,
297                                               b2Sample[,stepIdx] , "+" ) )
298    }
299    b1b2Sample = m12Sample - linpredSample
300    # Convert from standardized b values to original scale b values:
301    b0Sample = b0Sample * ySDorig + yMorig
302    b1Sample = b1Sample * ySDorig
303    b2Sample = b2Sample * ySDorig
304    b1b2Sample = b1b2Sample * ySDorig
305
306    # Plot b values:
307    windows((datalist$Nx1Lvl+1)*2.75,(datalist$Nx2Lvl+1)*2.0)
308    layoutMat = matrix( 0 , nrow=(datalist$Nx2Lvl+1) , ncol=(datalist$Nx1Lvl+1) )
309    layoutMat[1,1] = 1
310    layoutMat[1,2:(datalist$Nx1Lvl+1)] = 1:datalist$Nx1Lvl + 1
311    layoutMat[2:(datalist$Nx2Lvl+1),1] = 1:datalist$Nx2Lvl + (datalist$Nx1Lvl + 1)
312    layoutMat[2:(datalist$Nx2Lvl+1),2:(datalist$Nx1Lvl+1)] = matrix(
313        1:(datalist$Nx1Lvl*datalist$Nx2Lvl) + (datalist$Nx2Lvl+datalist$Nx1Lvl+1) ,
314        ncol=datalist$Nx1Lvl , byrow=T )
315    layout( layoutMat )
316    par( mar=c(4,0.5,2.5,0.5) , mgp=c(2,0.7,0) )
317    histinfo = plotPost( b0Sample , xlab=expression(beta * 0) , main="Baseline" ,
318                         breaks=30  )
319    for ( x1idx in 1:datalist$Nx1Lvl ) {
320      histinfo = plotPost( b1Sample[x1idx,] , xlab=bquote(beta*1[.(x1idx)]) ,
321                           main=paste("x1:",x1names[x1idx]) , breaks=30 )
322    }
323    for ( x2idx in 1:datalist$Nx2Lvl ) {
324      histinfo = plotPost( b2Sample[x2idx,] , xlab=bquote(beta*2[.(x2idx)]) ,
325                           main=paste("x2:",x2names[x2idx]) , breaks=30 )
326    }
327    for ( x2idx in 1:datalist$Nx2Lvl ) {
328      for ( x1idx in 1:datalist$Nx1Lvl ) {
329        histinfo = plotPost( b1b2Sample[x1idx,x2idx,] , breaks=30 ,
330                  xlab=bquote(beta*12[.(x1idx)*","*.(x2idx)]) ,
331                  main=paste("x1:",x1names[x1idx],", x2:",x2names[x2idx])  )
332      }
333    }
```

```
334    dev.copy2eps(file=paste(fnroot,"b.eps",sep=""))
335
336    # Display contrast analyses
337    nContrasts = length( x1contrastList )
338    if ( nContrasts > 0 ) {
339       nPlotPerRow = 5
340       nPlotRow = ceiling(nContrasts/nPlotPerRow)
341       nPlotCol = ceiling(nContrasts/nPlotRow)
342       windows(3.75*nPlotCol,2.5*nPlotRow)
343       layout( matrix(1:(nPlotRow*nPlotCol),nrow=nPlotRow,ncol=nPlotCol,byrow=T) )
344       par( mar=c(4,0.5,2.5,0.5) , mgp=c(2,0.7,0) )
345       for ( cIdx in 1:nContrasts ) {
346          contrast = matrix( x1contrastList[[cIdx]],nrow=1) # make it a row matrix
347          incIdx = contrast!=0
348          histInfo = plotPost( contrast %*% b1Sample , compVal=0 , breaks=30 ,
349                      xlab=paste( round(contrast[incIdx],2) , x1names[incIdx] ,
350                             c(rep("+",sum(incIdx)-1),"") , collapse=" " ) ,
351                      cex.lab = 1.0 ,
352                      main=paste( "X1 Contrast:", names(x1contrastList)[cIdx] ) )
353       }
354       dev.copy2eps(file=paste(fnroot,"x1Contrasts.eps",sep=""))
355    }
356    #
357    nContrasts = length( x2contrastList )
358    if ( nContrasts > 0 ) {
359       nPlotPerRow = 5
360       nPlotRow = ceiling(nContrasts/nPlotPerRow)
361       nPlotCol = ceiling(nContrasts/nPlotRow)
362       windows(3.75*nPlotCol,2.5*nPlotRow)
363       layout( matrix(1:(nPlotRow*nPlotCol),nrow=nPlotRow,ncol=nPlotCol,byrow=T) )
364       par( mar=c(4,0.5,2.5,0.5) , mgp=c(2,0.7,0) )
365       for ( cIdx in 1:nContrasts ) {
366          contrast = matrix( x2contrastList[[cIdx]],nrow=1) # make it a row matrix
367          incIdx = contrast!=0
368          histInfo = plotPost( contrast %*% b2Sample , compVal=0 , breaks=30 ,
369                      xlab=paste( round(contrast[incIdx],2) , x2names[incIdx] ,
370                             c(rep("+",sum(incIdx)-1),"") , collapse=" " ) ,
371                      cex.lab = 1.0 ,
372                      main=paste( "X2 Contrast:", names(x2contrastList)[cIdx] ) )
373       }
374       dev.copy2eps(file=paste(fnroot,"x2Contrasts.eps",sep=""))
375    }
376    #
377    nContrasts = length( x1x2contrastList )
378    if ( nContrasts > 0 ) {
379       nPlotPerRow = 5
380       nPlotRow = ceiling(nContrasts/nPlotPerRow)
381       nPlotCol = ceiling(nContrasts/nPlotRow)
382       windows(3.75*nPlotCol,2.5*nPlotRow)
383       layout( matrix(1:(nPlotRow*nPlotCol),nrow=nPlotRow,ncol=nPlotCol,byrow=T) )
384       par( mar=c(4,0.5,2.5,0.5) , mgp=c(2,0.7,0) )
385       for ( cIdx in 1:nContrasts ) {
386          contrast = x1x2contrastList[[cIdx]]
387          contrastArr = array( rep(contrast,chainLength) ,
```

```
388                              dim=c(NROW(contrast),NCOL(contrast),chainLength) )
389          contrastLab = ""
390          for ( x1idx in 1:Nx1Lvl ) {
391            for ( x2idx in 1:Nx2Lvl ) {
392              if ( contrast[x1idx,x2idx] != 0 ) {
393                contrastLab = paste( contrastLab , "+" ,
394                              signif(contrast[x1idx,x2idx],2) ,
395                              x1names[x1idx] , x2names[x2idx] )
396              }
397            }
398          }
399          histInfo = plotPost( apply( contrastArr * b1b2Sample , 3 , sum ) ,
400                  compVal=0 , breaks=30 , xlab=contrastLab , cex.lab = 0.75 ,
401                  main=paste( names(x1x2contrastList)[cIdx] ) )
402        }
403      dev.copy2eps(file=paste(fnroot,"x1x2Contrasts.eps",sep=""))
404    }
405
406    #===============================================================================
407    # Do NHST ANOVA:
408
409    theData = data.frame( y=y , x1=factor(x1,labels=x1names) ,
410                          x2=factor(x2,labels=x2names) )
411    windows()
412    interaction.plot( theData$x1 , theData$x2 , theData$y , type="b" )
413    dev.copy2eps(file=paste(fnroot,"DataPlot.eps",sep=""))
414    aovresult = aov( y ~ x1 * x2 , data = theData )
415    cat("\n---------------------------------------------------------------\n\n")
416    print( summary( aovresult ) )
417    cat("\n---------------------------------------------------------------\n\n")
418    print( model.tables( aovresult , type = "effects", se = TRUE ) , digits=3 )
419    cat("\n---------------------------------------------------------------\n\n")
420
421    #===============================================================================
```

19.4 EXERCISES

Exercise 19.1. [Purpose: To inspect an interaction for transformed data.] Consider the data plotted in Figure 19.3, p. 521.

(A)   Is the interaction a crossover interaction or not? Briefly explain your answer.

(B)   Suppose we are interested in salaries thought of in terms of *percentage* (i.e., ratio) differences rather than additive differences. Therefore we take the logarithm, base 10, of the individual salaries (the R code has this option built into to the data section, where the salary data are loaded). Run the analysis on the transformed data, producing the results and contrasts analogous to those in Figures 19.4 and 19.5. Do any of the conclusions change?

(C)   Examine the within-cell variances in the original and in the transformed data. (Hint: Try using the aggregate function on the data. As a guide, see how the function is used to initialize a1a2. Instead of applying the `mean` to the aggregated data, apply the standard deviation. The result is the within-cell standard deviations. Are they all roughly the same?) Do the original or the transformed data better respect the assumption of homogeneous variances?

Exercise 19.2. [Purpose: To investigate a case of two-factor Bayesian ANOVA.] In the data specification of the program in Section 19.3.1 (`ANOVAtwowayBRugs`), you can load data from Qian & Shen (2007), regarding how quickly seaweed regenerates when in the presence of different types of grazers. Data were collected from eight different tidal areas of the Oregon coast; this predictor ($\vec{x}_2$) is referred to as the *Block*. In each of the eight Blocks, there were six different combinations of seaweed grazers established by the researchers. This predictor ($\vec{x}_1$) had six levels: control, with no grazers allowed; only small fish allowed; small and large fish allowed; only limpets allowed; limpets and small fish allowed; and all three grazers allowed. The predicted variable was the percentage of the plot covered by regenerated seaweed, logarithmically transformed.

(A) Load the data and run the program. You will find that there are too many levels of the two predictors to fit all the posterior histograms into a single multipanel display. Therefore, modify the plotPost.R program so that it produces only the mean and HDI limits, marked by a horizontal bar with a circle at the mean (without a histogram) and perhaps without a `main` title. Name your program something other than plotPost.R, and use it in the plotting section at the end of the program instead of plotPost.R. Show your results. (A secondary goal of this part of the exercise is to give you experience modifying graphics in R to suit your own purposes.) Hints: There are many ways to do this, but here are some options. To suppress plotting of the histogram, just put this argument in the `hist` function: `plot=F`. To suppress a title on a plot, just use the argument `main=""`. To adjust the font size, specify the "character expansion": `cex` for text, `cex.lab` for axis labels, and so forth. To reduce the margins around a plot, so there is more room for the plot itself, try variations of these margin specifications: `par( mar=c(2,0.5,1,0.5), mgp=c(0.5,0,0) )`. The `par` command needs to be called before the plots are made.

(B) The program already includes contrasts that consider whether there is an effect of small fish, an effect of large fish, and an effect of limpets. What conclusions do you reach from the posteriors of these contrasts?

(C) Construct a contrast of the average of Blocks 3 and 4 versus the average of Blocks 1 and 2. Show your specification, the graph of the posterior on the contrast, and state your conclusion.

(D) Is the effect of limpets different in Block 6 than in Block 7? To answer this question, construct an interaction contrast using an outer product (Hint: refer to the already-coded `L_effect` for the contrast that specifies the effect of limpets). Is the effect of small fish different in Blocks 1 and 7 than in Blocks 3 and 5? For both questions, show the contrast vectors that you constructed and show the posterior of the contrast, and state your conclusion.

Exercise 19.3. [Purpose: To notice that within-subject designs can be more sensitive (hence more powerful) than between-subject designs.] Consider these data:

$\vec{x}_1$	$\vec{x}_2$	y	S	$\vec{x}_1$	$\vec{x}_2$	y	S
1	1	101	1	2	1	105	1
1	1	102	2	2	1	107	2
1	1	103	3	2	1	106	3
1	1	105	4	2	1	108	4
1	1	104	5	2	1	109	5
1	2	104	1	2	2	109	1
1	2	105	2	2	2	108	2
1	2	107	3	2	2	110	3
1	2	106	4	2	2	111	4
1	2	108	5	2	2	112	5

Note: The table is split into two halves so it fits the page more compactly. The continuation of the first column appears in the fifth column. The continuation of the second column appears in the sixth column, and so forth.

(A) Ignoring the last column, which indicates the subject who generated the data, conduct a Bayesian ANOVA using $\vec{x}_1$ and $\vec{x}_2$ as predictors of y. Show the code you used to load the data, and show the resulting posterior histograms of β_0, $\beta_{1,j}$, $\beta_{2,k}$, and $\beta_{1\times 2,jk}$. Also show the posterior of the contrast $\beta_{1,2} - \beta_{1,1}$ (i.e., the marginal difference between levels 1 and 2 of factor 1, also called the *main effect of factor 1*) and the posterior of the contrast $\beta_{2,2} - \beta_{2,1}$ (i.e., the marginal difference between levels 1 and 2 of factor 2, also called the *main effect of factor 2*).

(B) Now include the subject as a predictor by expanding the model to include a deflection from baseline due to subject. (Do not include any subject interaction terms.) Again show the posteriors of the β's requested in the previous part. Are the certainties on the estimates and contrasts different than in the previous part? In what way, and why?
(Hint regarding the answer: Figure 19.7 shows posterior histograms for the main effect of factor 2, when the data are considered to be between subject or within subject. Notice that the means are essentially the same in both histograms, but the uncertainties are very different!

Programming hints: The model specification without a subject factor is

```
mu[i] <- a0 + a1[x1[i]] + a2[x2[i]] + a1a2[x1[i],x2[i]]
```

but with a subject factor becomes

```
mu[i] <- a0 + a1[x1[i]] + a2[x2[i]] + a1a2[x1[i],x2[i]] + aS[S[i]]
```

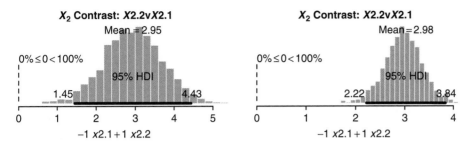

FIGURE 19.7

For Exercise 19.3. *Left panel:* Posterior for difference between levels of factor 2 when data are considered to be between subject. *Right panel:* Posterior for difference between levels of factor 2 when data are considered to be within subject. Notice that the means are (essentially) the same in both histograms, but the uncertainties are very different!

where $S[i]$ is the subject number for the ith datum, and $aS[]$ are the deflections from baseline for each subject. You must, of course, specify a prior on $aS[]$ analogous to the prior on $a1[]$.

The conversion of the $a[]$ values to zero-centered $b[]$ values proceeds analogously to what was explained at the beginning of Section 19.3.1 (ANOVA twowayBRugs.R). The code merely needs to be expanded to include the additional subject factor. Here is a guide (ANOVAtwowayBRugsWithinSubj.R):

```
209   # Convert the a values to zero-centered b values.
210   # m12Sample is predicted cell means at every step in MCMC chain:
211   m12Sample = array( 0, dim=c( datalist$Nx1Lvl , datalist$Nx2Lvl ,
212                               datalist$NSLvl , chainLength ) )
213   for ( stepIdx in 1:chainLength ) {
214     for ( a1idx in 1:Nx1Lvl ) {
215       for ( a2idx in 1:Nx2Lvl ) {
216         for ( aSidx in 1:NSLvl ) {
217           m12Sample[ a1idx , a2idx , aSidx , stepIdx ] = (
218             a0Sample[stepIdx]
219             + a1Sample[a1idx,stepIdx]
220             + a2Sample[a2idx,stepIdx]
221             + a1a2Sample[a1idx,a2idx,stepIdx]
222             + aSSample[aSidx,stepIdx] )
223         }
224       }
225     }
226   }
227
228   # b0Sample is mean of the cell means at every step in chain:
229   b0Sample = apply( m12Sample , 4 , mean )
230   # b1Sample is deflections of factor 1 marginal means from b0Sample:
231   b1Sample = ( apply( m12Sample , c(1,4) , mean )
232               - matrix(rep( b0Sample ,Nx1Lvl),nrow=Nx1Lvl,byrow=T) )
233   # b2Sample is deflections of factor 2 marginal means from b0Sample:
234   b2Sample = ( apply( m12Sample , c(2,4) , mean )
```

```
235                    - matrix(rep( b0Sample ,Nx2Lvl),nrow=Nx2Lvl,byrow=T) )
236     # bSSample is deflections of factor S marginal means from b0Sample:
237     bSSample = ( apply( m12Sample , c(3,4) , mean )
238                    - matrix(rep( b0Sample ,NSLvl),nrow=NSLvl,byrow=T) )
239     # linpredSample is linear combination of the marginal effects:
240     linpredSample = 0*m12Sample
241     for ( stepIdx in 1:chainLength ) {
242       for ( a1idx in 1:Nx1Lvl ) {
243         for ( a2idx in 1:Nx2Lvl ) {
244           for ( aSidx in 1:NSLvl ) {
245             linpredSample[a1idx,a2idx,aSidx,stepIdx ] = (
246               b0Sample[stepIdx]
247               + b1Sample[a1idx,stepIdx]
248               + b2Sample[a2idx,stepIdx]
249               + bSSample[aSidx,stepIdx] )
250           }
251         }
252       }
253     }
254     # b1b2Sample is the interaction deflection, i.e., the difference
255     # between the cell means and the linear combination:
256     b1b2Sample = apply( m12Sample - linpredSample , c(1,2,4) , mean )
257
258     # Convert from standardized b values to original scale b values:
259     b0Sample = b0Sample * ySDorig + yMorig
260     b1Sample = b1Sample * ySDorig
261     b2Sample = b2Sample * ySDorig
262     bSSample = bSSample * ySDorig
263     b1b2Sample = b1b2Sample * ySDorig
```

Exercise 19.4. [Purpose: To conduct a power analysis for Bayesian ANOVA, for within-subject versus between-subject designs.] Conduct power analyses for the between-subject and within-subject versions of the previous exercise. Specifically, suppose the goal is for the 95% HDI of the contrast on factor 2 to have a width of 2.0 or less. Conduct a retrospective power analysis for this goal, for the within-subject version and the between-subject version. Caution: This exercise demands a lot of programming and could be time consuming, but the results drive home the point that within-subject designs can be more powerful than between-subject designs.

Dichotomous Predicted Variable

Fortune and Favor make fickle decrees, it's
Heads or it's tails with no middle degrees.
Flippant commandments decreed by law gods, have
Reasons so rare they have minus log odds.

In many situations the value to be predicted is dichotomous (instead of metric). For example, we might want to predict whether a patient is cured or not (a dichotomous outcome) on the basis of the dosage of drug administered and the age of the patient (two metric predictors). What the model does, in this

Doing Bayesian Data Analysis: A Tutorial with R and BUGS. DOI: 10.1016/B978-0-12-381485-2.00020-1
© 2011, Elsevier Inc. All rights reserved.

case, is generate a prediction of the *probability* that a patient is cured, given the specified dosage and age. In other situations, the predictors might be nominal. For example, we could predict cured versus not cured on the basis of type of drug and gender of patient. The model specifies the probability that a patient is cured, given the drug and the gender. We will consider such situations in this chapter.

The formal framework for this situation was presented in the second row of Table 14.1 on p. 385. As you may recall, the link function, which maps a linear combination of predictors to an outcome tendency, is a logistic function. Therefore, the models we use in this chapter are referred to as cases of logistic regression or logistic ANOVA.

20.1 LOGISTIC REGRESSION

Suppose we have metric predictors and a dichotomous predicted variable. As a concrete example, suppose we want to predict the gender of a person (coded by male $= 1$ and female $= 0$) on the basis of the person's height and weight. Figure 20.1 shows realistic data from 70 people. Males are plotted as

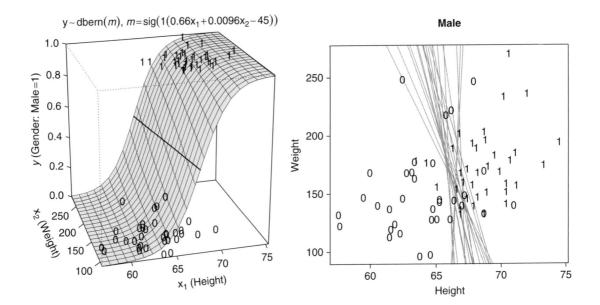

FIGURE 20.1
Gender (male $= 1$, female $= 0$), height, and weight. The left panel shows data in perspective, with the MLE logistic surface, and the $p(y = 1) = 0.5$ level contour marked as a dark line. The data points are all at $y = 1$ or $y = 0$, not on the logistic surface. The right panel shows the data "from above" with several blue lines that indicate a smattering of credible $p(y = 1) = 0.5$ level contours derived from Bayesian logistic regression.

numeral 1's, and females are plotted as numeral 0's. You can see that there is notable variation of heights and weights across people, and, importantly, there is considerable overlap in the distributions of males and females. Thus, at any specific combination of height and weight, the best we can do is predict the *probability* that a person with that height and weight is male.

The model we will use to formulate the probability that $y = 1$, as a function of the two predictors, is the logistic transform of a linear combination of the predictors. Formal details are reviewed later, but a graphical preview is provided in the left panel of Figure 20.1. The smoothly ascending surface plots the probability that $y = 1$ as a function of x_1 and x_2. The parameters of the model control the "cliff face," specifically its orientation, location, and steepness.

The maximum likelihood estimate (MLE) of the model parameters provides a single logistic surface that summarizes the distribution of the data. The MLE surface is shown in the left panel of Figure 20.1. Also plotted on the surface, as a dark line, is the level contour at which $p(y=1) = 0.5$. This level contour marks the predictions that are halfway up the logistic cliff. The level contour is a convenient summary of the orientation and location of the logistic surface, but the level contour loses information about the steepness of the cliff.

The right panel of Figure 20.1 shows the same data viewed from above. Also plotted are several credible $p(y=1) = 0.5$ level contours, from a Bayesian analysis. Unlike the single MLE contour in the left panel, the distribution of credible contours reveals our uncertainty in the parameter estimates of the logistic-regression model. The mechanics of the Bayesian analysis are described in the next sections.

20.1.1 The Model

Recall that for multiple *linear* regression, the central tendency of the predicted value is a linear combination of the predictors, as in Equation 14.3 (p. 364), and as explained at length in Chapter 17. In multiple *logistic* regression, the linear combination is transformed by a logistic squashing function, and the resulting value, between zero and one, is used as the probability that the predicted value is one. Formally, as in the second row of Table 14.1 on p. 385, logistic regression can be written as

$$\mu = \text{sig}\left(\beta_0 + \beta_1 x_1 + \beta_2 x_2 + \cdots\right)$$

$$y \sim \text{dbern}(\mu)$$

where sig refers to the sigmoid function, which is merely another name for the logistic function:

$$\text{sig}(x) = 1/(1 + \exp(-x)) \tag{20.1}$$

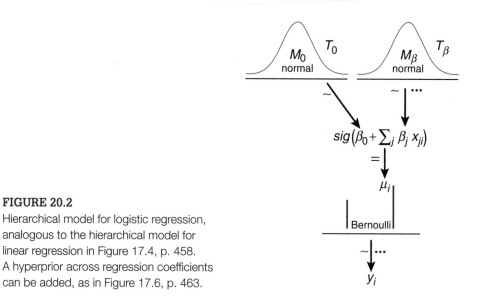

FIGURE 20.2
Hierarchical model for logistic regression,
analogous to the hierarchical model for
linear regression in Figure 17.4, p. 458.
A hyperprior across regression coefficients
can be added, as in Figure 17.6, p. 463.

Graphs of the logistic (a.k.a. sigmoid) function were shown in Figure 14.6
(p. 375) for a single predictor, Figure 14.7 (p. 376) for two predictors, and
Figure 14.10 (p. 381) for two predictors with dichotomous predicted values
superimposed. Please do review those figures now!

The logit function is the inverse of the logistic function. Formally, for $0 <
p < 1$, $\mathrm{logit}(p) = \log\left(p/(1 - p)\right)$. In applications to logistic regression, the "p"
is the probability that $y = 1$, and therefore we can write $\mathrm{logit}\left(p(y{=}1)\right) =
\log\left(p(y{=}1)/p(y{=}0)\right)$, where the logarithm is the natural logarithm (i.e., the
inverse of the exponential function). The ratio, $p(y{=}1)/p(y{=}0)$, is called the
odds of outcome 1 to outcome 0. The logistic regression model can be written
in terms of the logit function, also called the *log-odds form*, like this:

$$\mathrm{logit}(\mu) = \beta_0 + \beta_1 x_1 + \beta_2 x_2 + \cdots \tag{20.2}$$
$$y \sim \mathrm{dbern}(\mu)$$

I prefer the expression in terms of the logistic (i.e., sigmoid) in Equation 20.1
because it provides an explicit specification of μ that is natural to depict in hierarchical diagrams. But the logit form in Equation 20.2 is useful for interpreting
the regression coefficients, as will be described in Section 20.1.3.

The hierarchical diagram for the model is shown in Figure 20.2. It is just like
the one for multiple linear regression in Figure 17.4, p. 458, except for inclusion of a sigmoid and a change in the likelihood function at the bottom of the
diagram, from a normal distribution to a Bernoulli distribution.

To reiterate, what the model does is take the ith individual's predictor values, x_{ji}, and generate the probability μ_i that $y = 1$ for that individual. The Bayesian estimation yields values of the parameters, β_0 and β_j, that respect the data and the prior.

20.1.2 Doing It in R and BUGS

The implementation of logistic regression in R, BRugs, and BUGS is straight-forward, with only a few modifications of the linear regression program presented earlier in the book. The full logistic regression program is listed in Section 20.5.1 (`MultipleLogisticRegressionBrugs.R`). The only notable changes, other than the model, are (1) the use of R's `glm` function to initialize the chains at the MLE and (2) a simpler formula for conversion of standardized data to original-scale data. A more thorough explanation is provided before the listing in Section 20.5.1 (`MultipleLogisticRegressionBrugs.R`).

20.1.3 Interpreting the Posterior

Consider again the height, weight, and gender data in Figure 20.1. Aspects of the posterior distribution are shown in Figure 20.3. Often we are primarily interested in knowing the magnitude of a predictor's influence on the predicted value and, specifically, whether that influence is credibly nonzero.

The middle panel of Figure 20.3 indicates that height provides a credibly nonzero predictiveness for gender. The mean value of the coefficient on height is 0.721, which means that the log odds, $\log\left(p(\text{male})/p(\text{female})\right)$, increases by 0.721 when height increases by 1 inch.

This notion of constant increase in log odds is not instantly intuitive. Recall from Equation 20.2 that the regression equation can be expressed in log-odds form as

$$\text{logit}(\mu) = \log\left(\frac{p(y=1)}{p(y=0)}\right) = \beta_0 + \beta_1 x_1 + \beta_2 x_2 + \cdots$$

Therefore, when x_i increases one unit, the log odds increase β_i units. Some numerical examples might help. Consider a hypothetical person who weighs 160 pounds. We will compute what happens to the log odds when the person's height increases by 1 inch. First, consider an increase from 63 inches tall to 64 inches tall. According to the logistic function with parameters set at the mean of the posterior, if the 160-pound person were 63 inches tall, then the probability of being male is 0.0728, and if the person were 64 inches tall, the probability of being male is 0.1390. That increase of 6.62 percentage points corresponds to a change in log odds of $\log(0.1390/(1-0.1390)) - \log(0.0728/(1-0.0728))$, which equals 0.721. Next, consider an increase from 66 inches tall to 67 inches tall. If the person were 66 inches tall, the probability of being male is 0.4056, and if the

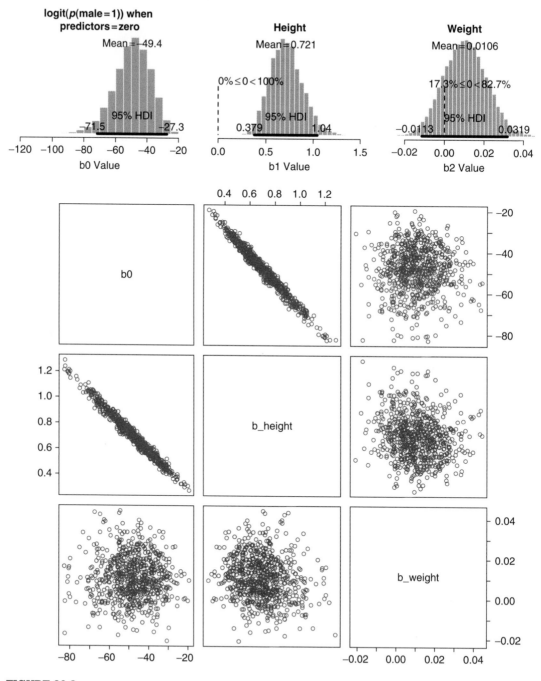

FIGURE 20.3

Posterior distribution of Bayesian logistic regression of gender on height and weight.

person were 1 inch taller, at 67 inches, then the probability of being male is 0.5839. The increase of 17.83 percentage points corresponds to a change in log odds of $\log(0.5839/(1 - 0.5839)) - \log(0.4056/(1 - 0.4056))$, which *again* equals 0.721. Thus, for any increase of one inch in height, the increase in predicted log odds of being male is the same, namely, 0.721.

The right panel of Figure 20.3 indicates that weight does not provide a credibly nonzero predictiveness for gender, because zero is well among the credible values in the posterior. The mean value of the coefficient on weight is 0.0106, which means that the log odds, $\log(p(\text{male})/p(\text{female}))$, increases by only 0.0106 when weight increases by 1 pound.

The lower part of Figure 20.3 shows pairwise scatter plots of the three-dimensional posterior. Notice that the intercept and the coefficient on height are strongly negatively correlated. This correlation results from two properties of the data. First, height is predictive of gender, and therefore the intercept is constrained by height. Second, these original-scale data are not standardized at heights of zero, so if the coefficient on height is increased, the intercept must be decreased to keep the logistic cliff at the appropriate position.

20.1.4 Perils of Correlated Predictors

When predictors are correlated, care must be taken in interpreting their regression coefficients. This issue was emphasized in Section 17.1.1 in the context of multiple *linear* regression, but analogous perils arise in *logistic* regression.

Consider a situation with two predictors. When the predictors are *un*correlated, the posterior will have relatively high certainty regarding the regression coefficients. But when the predictors are strongly correlated, the posterior will have relatively low certainty about the regression coefficients, because many different logistic surfaces can fit the data fairly well. Figure 20.7 in Exercise 20.1 provides an example.

Another peril arising from correlated predictors arises when one of the correlated predictors is not included in the model. Suppose that one predictor is included in the model, but another predictor is not included, perhaps because the data for the excluded predictor were lost or unobtainable or not even considered in the first place. Suppose further that the true regression coefficient on the included predictor is *zero*, but the true regression coefficient on the excluded predictor is *large positive*. The posterior estimate of the included regression coefficient will be nonzero if the included predictor is correlated with the excluded predictor. The reason is intuitively straightforward: As the included predictor increases, the excluded predictor changes (because it is correlated with the included predictor), and therefore the outcome changes, even though the included predictor has no direct predictive value for the outcome. Exercise 20.1 provides an example of this situation.

20.1.5 When There Are Few 1's in the Data

In linear regression, the predicted variable is on a metric scale. Usually the data are distributed over a range of values, without being severely clumped over a single value. In logistic regression, however, the predicted variable is dichotomous, and the data can sometimes consist of mostly 1's or mostly 0's. For example, if the predicted variable is the occurrence of a rare disease, then, by definition, there are few 1's. As another example, the predicted variable might be the occurrence of a defect on a factory assembly line, which is expected to be rare.

The problem with data that have only a few 1's or a few 0's is that the estimate of the regression coefficients is usually relatively uncertain. This makes intuitive sense, because it is only the transition from 0's to 1's that constrains the regression coefficients. Exercise 20.2 provides an example.

20.1.6 Hyperprior Across Regression Coefficients

When there are many candidate predictors, we can use prior knowledge to mutually constrain the estimates of the coefficients. Because all the predictors come from a pool of factors that might have some remotely plausible predictive relation with the dichotomous value, we could reasonably put a hyperprior on the regression coefficients that expresses the assumption that most coefficients are near zero, but some may be notably farther from zero. A distribution that captures this prior knowledge is a t distribution, with a precision that is estimated from the data. This scheme was described in the context of multiple linear regression in Section 17.2, p. 463. A diagram of the hierarchical prior was displayed in Figure 17.6, p. 463. The same scheme can be applied to logistic regression, merely by substituting the Bernoulli likelihood function, as in Figure 20.2.

The knowledge expressed in the hyperprior provides shrinkage on the coefficients in logistic regression. When several predictors have estimated coefficients near zero, they drive plausible values of the precision of the overarching t toward higher values, thereby shrinking estimates of other predictors' coefficients. The shrinkage attenuates false alarms, as discussed at length in Section 17.2.

20.2 INTERACTION OF PREDICTORS IN LOGISTIC REGRESSION

Logistic regression can include a multiplicative interaction, just as when the predicted variable is metric (recall Figures 14.3 and 17.8). The interpretation of such an interaction requires care.

Suppose that the predicted variable is a two-valued self-report of happiness: The respondent says she or he is happy or not happy. The two predictors are annual income and healthiness assayed on a metric scale. Intuitively, we can imagine that happiness depends on the *conjunction* of good health and high income; either one alone may be insufficient for happiness. To put it another way, a person can be *un*happy either because of poor health or because of poverty. Thus, the probability of being happy is a conjunctive interaction of health and wealth.

As another example, consider predicting the success of a lighter-than-air vehicle. One predictor is the intensity of heat under the balloon. A second predictor is the amount of hydrogen in the balloon. Intuitively, the balloon will rise only if there is sufficient heat or sufficient hydrogen but not both. This is called an *exclusive-OR* interaction between the predictors.

Figure 20.4 shows graphs of multiplicative interactions of predictors in a logistic function. The top-right panel shows an exclusive-OR: The predicted probability, m, is high only if x_1 is large or x_2 is large but not both. The top-left panel shows absence of all effects as a comparison and reminder that the baseline for a logistic is 0.5.

The lower-right panel of Figure 20.4 shows a conjunctive interaction: The predicted probability is high only if x_1 and x_2 are large. The lower-left panel shows a comparison that has zero interaction. The two lower panels are subtly different: Level contours (not shown) on the zero-interaction cliff are straight lines, but level contours (not shown) on the interactive cliff are curved, resembling hyperbolas.

Multiplicative interactions can be incorporated into logistic regression models in the same way they are incorporated into metric-predictor models. As was emphasized in the chapter on multiple linear regression, the coefficients on the predictors must be interpreted with care, especially when interactions are involved.

20.3 LOGISTIC ANOVA

In many situations the variable to be predicted is dichotomous and the predictors are nominal. As an example, recall the filtration/condensation experiment introduced in Section 9.3.1, p. 219, and used many times thereafter. The predicted variable was accuracy on each trial (i.e., correct or incorrect). The predictor was a nominal variable that indicated which of four category structures the learner experienced. As another example, recall the relevance-shift experiment of Exercise 13.4, p. 347. Again the predicted variable was accuracy on each trial, and the predictor was a nominal variable indicating which of four relevance shifts the learner experienced. If the predicted variable was metric instead of dichotomous, then these cases could be analyzed with oneway ANOVA as in Chapter 18.

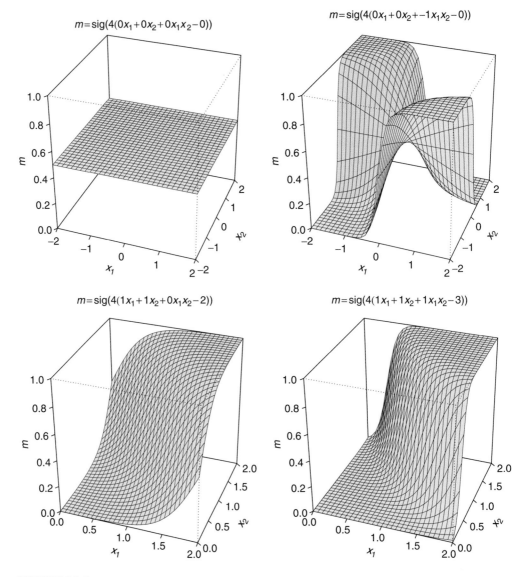

FIGURE 20.4

Right panels: Logistic regression *with interaction*. *Left panels*: Corresponding function with*out* any interaction component. Notice that the upper panels include negative values of the predictors, whereas the lower panels do not. In the lower-right graph, when both x_1 and x_2 are sufficiently large negative values (not shown), then the surface rises up toward $m = 1$.

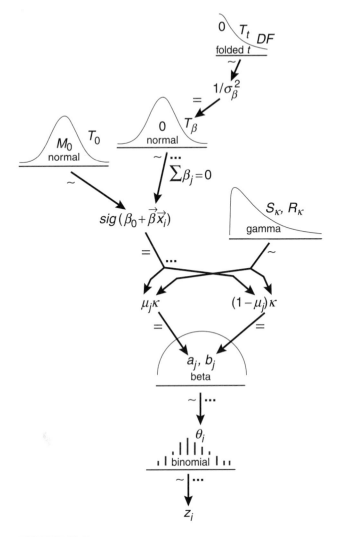

FIGURE 20.5
Bayesian logistic ANOVA. The ANOVA model, in the upper parts, generates the μ_j for the groups through the sigmoidal transform. The beta distribution, in the lower part of the diagram, describes the within-group variability across subjects. The same κ is assumed for all groups, which is analogous to homogeneity of variance in metric ANOVA.

The logistic function can be used to adapt ANOVA models to dichotomous outcomes. We start with the ANOVA model to generate a predicted tendency for each condition, then pass that value through a sigmoid so it becomes the predicted probability that $y = 1$ for that condition. We also assume that subjects can vary around the condition's predicted probability, just as in the

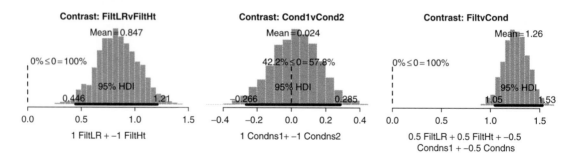

FIGURE 20.6
Results of logistic ANOVA applied to filtration/condensation experiment. Compare with Figure 9.16, p. 224, which did not use an ANOVA model for these data. The scale on these histograms is difference of ANOVA-model β_j values, not difference of μ_c values as in Figure 9.16.

ANOVA model. Figure 20.5 shows the logistic ANOVA model for a single factor (i.e., "oneway" logistic ANOVA). The basic ANOVA model comes directly from the top part of Figure 18.1, p. 494. The beta distribution for subjects within conditions comes directly from the left side of Figure 9.17, p. 226. Notice that this model assumes equal variances for all conditions, like traditional ANOVA, insofar as the same κ value is used for all groups simultaneously.

Figure 20.6 shows the results of running logistic oneway ANOVA on the filtration/condensation data. The comparisons are differences of the β parameters in the ANOVA model. The β parameters are not probabilities, but the β parameters can be mapped to probabilities via the logistic function. The comparisons of the β parameters from different conditions yield the same conclusions as the analysis from earlier chapters, as in Figure 9.16, p. 224, which did not use an ANOVA model for these data.

If the ANOVA model leads to the same conclusions as the non-ANOVA approach, what are the advantages or disadvantages of the two models? A disadvantage of the logistic ANOVA model is that it tends to have larger autocorrelation in its parameters and therefore takes longer to run because more thinning is needed. Another disadvantage of the logistic ANOVA model is that the parameters are not instantly interpretable as probabilities but must be converted through the logistic. But the ANOVA approach has great advantages when the situation to be modeled becomes more complex. In particular, when there are two or more predictors, the ANOVA model provides a natural representation for the main effects and interactions of the predictors. This generalizability of the logistic ANOVA model is its trump card.

Some analysts of dichotomous data will treat z_i/N_i as if it were a metric value and apply standard ANOVA. Because standard ANOVA assumes that the data come from normal distributions, the analyst will transform the values to $2 \arcsin\sqrt{z_i/N_i}$ (e.g., Winer, Brown, & Michels, 1991, p. 356), to approximate normality. Unfortunately, the approximation is not good for small N or for values of z_i/N_i that are close to 0 or 1. Most important, the ratio z_i/N_i loses information about the magnitude of N_i: If all we know is that $z_i/N_i = 0.75$, we don't know whether $z_i = 3$ and $N_i = 4$, or $z_i = 3000$ and $N_i = 4000$. This matters because a proportion based on a large sample should influence our beliefs more than a proportion based on a small sample. Therefore, converting to proportions fails in unbalanced designs (i.e., when there are different sample sizes for different subjects). In the Bayesian approach used here, every individual observation has an equal influence, and the Bayesian analysis automatically handles unbalanced designs appropriately.

20.3.1 Within-Subject Designs

Fortunately, there is no news when it comes to repeated measures in within-subject designs. All the concepts of Section 19.2, p. 531, apply the same way to logistic ANOVA. There were two main messages of that section. First, in terms of model structure, if every subject contributes many observations to every combination of conditions, then it makes sense to apply a logistic ANOVA model to every individual subject, with hyperdistributions on the ANOVA coefficients to allow mutually informed estimation across subjects. If every subject contributes only one observation to any combination of conditions, then it makes sense to include a subject factor in the model, with no interactions of that subject factor with the other factors. Second, in terms of whether to use a repeated measures design at all, there must be a good rationale for assuming that repeated measures from the same subject are reasonably independent of each other, so that any of these models can be applied.

20.4 SUMMARY

The extension of linear regression and ANOVA to logistic regression and logistic ANOVA is straightforward. The hierarchical models of Figures 20.2 and 20.5 illustrated how the core linear model of regression or ANOVA is passed through a logistic function, and the resulting value is used as the mean for a Bernoulli (or binomial) likelihood function. All the basic structural concepts of linear regression and ANOVA are applied to logistic regression and ANOVA.

Only a few new technical concepts were introduced in this chapter. One was the notion of level contour on the logistic surface. The contour at level 0.5 was useful for marking the boundary between $\vec{x}$ values for which $p(y=1) < 0.5$ and $\vec{x}$ values for which $p(y=1) > 0.5$.

Another new technical concept was interpreting a coefficient in logistic regression in terms of the increase in log odds of the outcome when the predictor increases by one unit. Although it was not previously mentioned, the same idea applies to logistic ANOVA: The β_j value indicates the increase in log odds of the outcome relative to the baseline, β_0. This interpretation must be qualified, however, when the model includes interaction terms, whether it is regression or ANOVA.

The remaining technical issues arise only in the mechanics of computing the analyses. The next section, which includes the R code, explains these details. There are two main issues, one regarding standardizing the predictors (for regression only) and transforming the parameters back to the original scales and the other issue regarding how to initialize the chains by starting with the maximum likelihood estimate (MLE).

20.5 R CODE

20.5.1 Logistic Regression Code

The program for logistic regression is much like the analogous program for linear regression, with only a few changes. One obvious change is that the likelihood function is Bernoulli, not normal. Beyond the model specification, it is important to understand the following details.

The predictor values are standardized, to reduce autocorrelation in MCMC sampling. Notice, for example, the highly correlated original-scale parameter values in Figure 20.3, which would be difficult to navigate by most MCMC sampling algorithms. Unlike linear regression, however, the predicted values are not standardized, because the predicted values must be zero or one. Because the predicted values are not standardized, the conversion of the standardized parameter values to original-scale parameter values is less complicated:

$$
\begin{aligned}
\operatorname{logit}(\mu_i) &= \zeta_0 + \sum_j \zeta_j z_{ji} \\
&= \zeta_0 + \sum_j \zeta_j \frac{x_{ji} - \bar{x}_j}{s_{x_j}} \\
&= \underbrace{\zeta_0 - \sum_j \frac{\zeta_j \bar{x}_j}{s_{x_j}}}_{\beta_0} + \sum_j \underbrace{\frac{\zeta_j}{s_{x_j}} x_{ji}}_{\beta_j}
\end{aligned}
$$

where $\bar{x}_j = \frac{1}{N} \sum_i^N x_{ji}$ is the arithmetic mean of the x_j values and s_{x_j} is the standard deviation of the x_j values. This conversion is executed at every step in the MCMC chain and is performed at the end of the program.

Initialization of the chains is based on parameter values that are plausible given the data. Specifically, the values of the MLE are used, because R has a built-in function for determining the MLE of logistic regression, namely, the glm function with the family argument specified as binomial(logit).

The final lines of the program perform an MLE of the logistic regression, just for comparison with the Bayesian results. The MLE is typically among the most credible values in the Bayesian posterior. But the posterior provides more information about correlations of parameters and the range of uncertainty in each parameter.

(MultipleLogisticRegressionBrugs.R)

```
1    graphics.off()
2    rm(list=ls(all=TRUE))
3    fname = "MultipleLogisticRegressionBrugs"
4    library(BRugs)              # Kruschke, J. K. (2010). Doing Bayesian data analysis:
5                               # A Tutorial with R and BUGS. Academic Press / Elsevier.
6    #-------------------------------------------------------------------------------
7    # THE MODEL.
8    modelstring = "
9    # BUGS model specification begins here...
10   model {
11     for( i in 1 : nData ) {
12       y[i] ~ dbern( mu[i] )
13       mu[i] <- 1/(1+exp(-( b0 + inprod( b[] , x[i,] ))))
14     }
15     b0 ~ dnorm( 0 , 1.0E-12 )
16     for ( j in 1 : nPredictors ) {
17       b[j] ~ dnorm( 0 , 1.0E-12 )
18     }
19   }
20   # ... end BUGS model specification
21   " # close quote for modelstring
22   # Write model to a file:
23   writeLines(modelstring,con="model.txt")
24   modelCheck( "model.txt" )
25
26   #-------------------------------------------------------------------------------
27   # THE DATA.
28
29   dataSource = c( "HtWt" , "Cars" , "HeartAttack" )[3]
30
31   if ( dataSource == "HtWt" ) {
32     fname = paste( fname , dataSource , sep="" )
33     # Generate random but realistic data:
34     source( "HtWtDataGenerator.R" )
35     dataMat = HtWtDataGenerator( nSubj = 70 , rndsd=474 )
36     predictedName = "male"
37     predictorNames = c( "height" , "weight" )
38     nData = NROW( dataMat )
39     y = as.matrix( dataMat[,predictedName] )
40     x = as.matrix( dataMat[,predictorNames] )
41     nPredictors = NCOL( x )
42   }
```

```
43
44   if ( dataSource == "Cars" ) {
45     fname = paste( fname , dataSource , sep="" )
46     dataMat = read.table(file="Lock1993data.txt",header=T,sep=" ")
47     predictedName = "AirBag"
48     predictorNames = c( "MidPrice" , "RPM" , "Uturn" )
49     nData = NROW( dataMat )
50     y = as.matrix( as.numeric( dataMat[,predictedName] > 0 ) ) # 0,1,2 to 0,1
51     x = as.matrix( dataMat[,predictorNames] )
52     nPredictors = NCOL( x )
53   }
54
55   if ( dataSource == "HeartAttack" ) {
56     fname = paste( fname , dataSource , sep="" )
57     dataMat = read.table(file="BloodDataGeneratorOutput.txt",header=T,sep=" ")
58     predictedName = "HeartAttack"
59     predictorNames = c( "Systolic", "Diastolic", "Weight", "Cholesterol",
60                         "Height", "Age" )
61   # predictorNames = c( "Systolic", "Diastolic" )
62     nData = NROW( dataMat )
63     y = as.matrix( dataMat[,predictedName] )
64     x = as.matrix( dataMat[,predictorNames] )
65     nPredictors = NCOL( x )
66   }
67
68   # Prepare data for BUGS:
69   # Re-center data at mean, to reduce autocorrelation in MCMC sampling.
70   # Standardize (divide by SD) to make initialization easier.
71   standardizeCols = function( dataMat ) {
72       zDataMat = dataMat
73       for ( colIdx in 1:NCOL( dataMat ) ) {
74           mCol = mean( dataMat[,colIdx] )
75           sdCol = sd( dataMat[,colIdx] )
76           zDataMat[,colIdx] = ( dataMat[,colIdx] - mCol ) / sdCol
77       }
78       return( zDataMat )
79   }
80   zx = standardizeCols( x )
81   zy = y   # y is not standardized; must be 0,1
82
83   # Get the data into BUGS:
84   datalist = list(
85               x = zx ,
86               y = as.vector( zy ) , # BUGS does not treat 1-column mat as vector
87               nPredictors = nPredictors ,
88               nData = nData
89   )
90   modelData( bugsData( datalist ) )
91
92
93   #-------------------------------------------------------------------------
94   # INTIALIZE THE CHAINS.
95
96   nchain = 3
```

```
97    modelCompile( numChains = nchain )
98
99    genInitList <- function() {
100       glmInfo = glm( datalist$y ~ datalist$x , family=binomial(logit) ) # R func.
101       show( glmInfo ) ; flush.console() # display in case glm() has troubles
102       b0Init = glmInfo$coef[1]
103       bInit = glmInfo$coef[-1]
104       return( list(
105           b0 = b0Init ,
106           b = bInit
107       ) )
108    }
109    for ( chainIdx in 1 : nchain ) {
110       modelInits( bugsInits( genInitList ) )
111    }
112
113    #------------------------------------------------------------------------------
114    # RUN THE CHAINS
115
116    # burn in
117    BurnInSteps = 1000
118    modelUpdate( BurnInSteps )
119    # actual samples
120    samplesSet( c( "b0" , "b" ) )
121    stepsPerChain = ceiling(5000/nchain)
122    thinStep = 50  # check autocorrelation and increase as needed
123    modelUpdate( stepsPerChain , thin=thinStep )
124
125    #------------------------------------------------------------------------------
126    # EXAMINE THE RESULTS
127
128    source("plotChains.R")
129    source("plotPost.R")
130
131    # Check chains for mixing
132    checkConvergence = T
133    if ( checkConvergence ) {
134      b0Sum = plotChains( "b0" , saveplots=F , filenameroot=fname )
135      bSum  = plotChains( "b"  , saveplots=F , filenameroot=fname )
136    }
137
138    # Extract chain values:
139    zb0Sample = matrix( samplesSample( "b0" ) )
140    chainLength = length(zb0Sample)
141    zbSample = NULL
142    for ( j in 1:nPredictors ) {
143       zbSample = cbind( zbSample , samplesSample( paste("b[",j,"]",sep="") ) )
144    }
145
146    # Convert to original scale:
147    x = dataMat[,predictorNames]
148    y = dataMat[,predictedName]
149    My = mean(y)
150    SDy = sd(y)
```

```
151    Mx = apply(x,2,mean)
152    SDx = apply(x,2,sd)
153    b0Sample = 0 * zb0Sample
154    bSample = 0 * zbSample
155    for ( stepIdx in 1:chainLength ) {
156        b0Sample[stepIdx] = ( zb0Sample[stepIdx]
157                            - sum( Mx / SDx * zbSample[stepIdx,] ) )
158        for ( j in 1:nPredictors ) {
159          bSample[stepIdx,j] = zbSample[stepIdx,j] / SDx[j]
160        }
161    }
162
163    # Examine sampled values, z scale:
164    windows()
165    thinIdx = ceiling(seq(1,chainLength,length=700))
166    pairs(   cbind( zb0Sample[thinIdx] , zbSample[thinIdx,] )  ,
167          labels=c( "zb0", paste("zb",predictorNames,sep="") ) )
168    # Examine sampled values, original scale:
169    windows()
170    pairs( cbind( b0Sample[thinIdx] , bSample[thinIdx,] ) ,
171          labels=c( "b0", paste("b_",predictorNames,sep="") ) )
172    dev.copy2eps(file=paste(fname,"PostPairs.eps",sep=""))
173
174    # Display the posterior :
175    windows(3.5*(1+nPredictors),2.75)
176    layout( matrix(1:(1+nPredictors),nrow=1) )
177    histInfo = plotPost( b0Sample , xlab="b0 Value" , compVal=NULL , breaks=30 ,
178                       main=paste( "logit(p(", predictedName ,
179                                      "=1)) when predictors = zero" , sep="" ) )
180    for ( bIdx in 1:nPredictors ) {
181    histInfo = plotPost( bSample[,bIdx] , xlab=paste("b",bIdx," Value",sep="") ,
182                       compVal=0.0 , breaks=30 ,
183                       main=paste(predictorNames[bIdx]) )
184    }
185    dev.copy2eps(file=paste(fname,"PostHist.eps",sep=""))
186
187    # Plot data with .5 level contours of believable logistic surfaces.
188    # The contour lines are best interpreted when there are only two predictors.
189    for ( p1idx in 1:(nPredictors-1) ) {
190      for ( p2idx in (p1idx+1):nPredictors ) {
191        windows()
192        xRange = range(x[,p1idx])
193        yRange = range(x[,p2idx])
194        # make empty plot
195        plot( NULL , NULL , main=predictedName , xlim=xRange , ylim=yRange ,
196              xlab=predictorNames[p1idx] , ylab=predictorNames[p2idx] )
197        # Some of the 50% level contours from the posterior sample.
198        for ( chainIdx in ceiling(seq( 1 , chainLength , length=20 )) ) {
199          abline( -( b0Sample[chainIdx]
200                    + if (nPredictors>2) {
201                        bSample[chainIdx,c(-p1idx,-p2idx)]*Mx[c(-p1idx,-p2idx)]
202                      } else { 0 } )
203                   / bSample[chainIdx,p2idx] ,
204                -bSample[chainIdx,p1idx]/bSample[chainIdx,p2idx] ,
```

```
205              col="grey" , lwd = 2 )
206        }
207      # The data points:
208      for ( yVal in 0:1 ) {
209        rowIdx = ( y == yVal )
210        points( x[rowIdx,p1idx] , x[rowIdx,p2idx] , pch=as.character(yVal) ,
211               cex=1.75 )
212      }
213      dev.copy2eps(file=paste(fname,"PostContours",p1idx,p2idx,".eps",sep=""))
214    }
215  }
216
217  #-------------------------------------------------------------------------
218
219  # MLE logistic regression:
220  glmRes = glm( datalist$y ~ as.matrix(x) , family=binomial(logit) )
221  show( glmRes )
```

20.5.2 Logistic ANOVA Code

This program implements the model shown in Figure 20.5.

(LogisticOnewayAnovaBrugs.R)

```
1   graphics.off()
2   rm(list=ls(all=TRUE))
3   fnroot = "LogisticOnewayAnovaBrugs"
4   library(BRugs)          # Kruschke, J. K. (2010). Doing Bayesian data analysis:
5                           # A Tutorial with R and BUGS. Academic Press / Elsevier.
6   #-------------------------------------------------------------------------
7   # THE MODEL.
8
9   modelstring = "
10  # BUGS model specification begins here...
11  model {
12    for ( i in 1:Ntotal ) {
13      z[i] ~ dbin( theta[i] , n[i] )
14      theta[i] ~ dbeta( aBeta[x[i]] , bBeta[x[i]] )I(0.001,0.999)
15    }
16    for ( j in 1:NxLvl ) {
17      aBeta[j] <- mu[j] * k
18      bBeta[j] <- (1-mu[j]) * k
19      mu[j] <- 1 / ( 1 + exp( -( a0 + a[j] ) ) )
20      a[j] ~ dnorm( 0.0 , atau )
21    }
22    k ~ dgamma( 1.0 , 0.01 )
23    a0 ~ dnorm( 0 , 0.001 )
24    atau <- 1 / pow( aSD , 2 )
25    aSD <- abs( aSDunabs ) + .1
26    aSDunabs ~ dt( 0 , 0.001 , 2 )
27  }
28  # ... end BUGS model specification
29  " # close quote for modelstring
30  # Write model to a file, and send to BUGS:
```

```
31    writeLines(modelstring,con="model.txt")
32    modelCheck( "model.txt" )
33
34    #------------------------------------------------------------------------
35    # THE DATA.
36
37    # Specify data source:
38    dataSource = c( "Filcon" , "Relshift" , "Random" )[1]
39    # Load the data:
40
41    sigmoid = function( x ) { return( 1 / ( 1 + exp( -x ) ) ) }
42    logit = function( y ) { return( log( y / (1-y) ) ) }
43
44    if ( dataSource == "Filcon" ) {
45      fnroot = paste( fnroot , dataSource , sep="" )
46      x = c(1,1,1,1,1,1,1,1,1,1,1,1,1,1,1,1,1,1,1,1,1,1,1,1,1,1,1,1,1,1,1,1,1,1,1,1,1,1,1,
47          2,2,2,2,2,2,2,2,2,2,2,2,2,2,2,2,2,2,2,2,2,2,2,2,2,2,2,2,2,2,2,2,2,2,2,2,2,2,2,3,3,
48          3,3,3,3,3,3,3,3,3,3,3,3,3,3,3,3,3,3,3,3,3,3,3,3,3,3,3,3,3,3,3,3,3,3,3,3,4,4,4,4,
49          4,4,4,4,4,4,4,4,4,4,4,4,4,4,4,4,4,4,4,4,4,4,4,4,4,4,4,4,4,4,4,4,4)
50      n = c(64,64,64,64,64,64,64,64,64,64,64,64,64,64,64,64,64,64,64,64,64,64,64,64,64,
51          64,64,64,64,64,64,64,64,64,64,64,64,64,64,64,64,64,64,64,64,64,64,64,64,64,64,
52          64,64,64,64,64,64,64,64,64,64,64,64,64,64,64,64,64,64,64,64,64,64,64,64,64,64,
53          64,64,64,64,64,64,64,64,64,64,64,64,64,64,64,64,64,64,64,64,64,64,64,64,64,64,
54          64,64,64,64,64,64,64,64,64,64,64,64,64,64,64,64,64,64,64,64,64,64,64,64,64,64,
55          64,64,64,64,64,64,64,64,64,64,64,64,64,64,64,64,64,64,64,64)
56      z = c(45,63,58,64,58,63,51,60,59,47,63,61,60,51,59,45,61,59,60,58,63,56,63,64,64,60,
57          64,62,49,64,64,58,64,52,64,64,64,62,64,61,59,59,55,62,51,58,55,54,59,57,58,60,54,42,
58          59,57,59,53,53,42,59,57,29,36,51,64,60,54,54,38,61,60,61,60,62,55,38,43,58,60,44,44,
59          32,56,43,36,38,48,32,40,40,34,45,42,41,32,48,36,29,37,53,55,50,47,46,44,50,56,58,42,
60          58,54,57,54,51,49,52,51,49,51,46,46,42,49,46,56,42,53,55,51,55,49,53,55,40,46,56,47,
61          54,54,42,34,35,41,48,46,39,55,30,49,27,51,41,36,45,41,53,32,43,33)
62      Ntotal = length(z)
63      xnames = c("FiltLR","FiltHt","Condns1","Condns2")
64      NxLvl = length(unique(x))
65      contrastList = list( FiltLRvFiltHt = c(1,-1,0,0) ,
66                           Cond1vCond2 = c(0,0,1,-1) ,
67                           FiltvCond = c(1/2,1/2,-1/2,-1/2) )
68    }
69
70    if ( dataSource == "Relshift" ) {
71      fnroot = paste( fnroot , dataSource , sep="" )
72      #source( "Kruschke1996CSdata.R" ) # if it has not yet been run
73      load("Kruschke1996CSdatsum.Rdata") # loads CondOfSubj, nCorrOfSubj, nTrlOfSubj
74      x = CondOfSubj
75      n = nTrlOfSubj
76      z = nCorrOfSubj
77      Ntotal = length(z)
78      xnames = c("Rev","Rel","Irr","Cmp")
79      NxLvl = length(unique(x))
80      contrastList = list( REVvREL = c(1,-1,0,0) , RELvIRR = c(0,1,-1,0) ,
81                           IRRvCMP = c(0,0,1,-1) , CMPvOneRel = c(0,-1/2,-1/2,1) ,
82                           FourExvEightEx = c(-1,1/3,1/3,1/3) ,
83                           OneRelvTwoRel = c(-1/2,1/2,1/2,-1/2) )
84    }
```

```
85
86   if ( dataSource == "Random" ) {
87     fnroot = paste( fnroot , dataSource , sep="" )
88     #set.seed(47405)
89     a0true = -0.5
90     atrue = c( 0.8 , -0.3 , -0.5 ) # sum to zero
91     ktrue = 100
92     subjPerCell = 50
93     nPerSubj = 100
94     datarecord = matrix( 0, ncol=3 , nrow=length(atrue)*subjPerCell )
95     colnames(datarecord) = c("x","z","n")
96     rowidx = 0
97     for ( xidx in 1:length(atrue) ) {
98       for ( subjidx in 1:subjPerCell ) {
99         rowidx = rowidx + 1
100        datarecord[rowidx,"x"] = xidx
101        mu = sigmoid(a0true+atrue[xidx])
102        theta = rbeta( 1 , mu*ktrue , (1-mu)*ktrue )
103        datarecord[rowidx,"z"] = rbinom( 1 , prob=theta , size=nPerSubj )
104        datarecord[rowidx,"n"] = nPerSubj
105      }
106    }
107    datarecord = data.frame( x=as.factor(datarecord[,"x"]) , z=datarecord[,"z"] ,
108                             n=datarecord[,"n"] )
109    z = as.numeric(datarecord$z)
110    Ntotal = length(z)
111    n = as.numeric(datarecord$n)
112    x = as.numeric(datarecord$x)
113    xnames = levels(datarecord$x)
114    NxLvl = length(unique(x))
115    # Construct list of all pairwise comparisons, to compare with NHST TukeyHSD:
116    contrastList = NULL
117    for ( g1idx in 1:(NxLvl-1) ) {
118      for ( g2idx in (g1idx+1):NxLvl ) {
119        cmpVec = rep(0,NxLvl)
120        cmpVec[g1idx] = -1
121        cmpVec[g2idx] = 1
122        contrastList = c( contrastList , list( cmpVec ) )
123      }
124    }
125  }
126
127  # Specify the data in a form that is compatible with BRugs model, as a list:
128  datalist = list(
129    z = z ,
130    n = n ,
131    x = x ,
132    Ntotal = Ntotal ,
133    NxLvl = NxLvl
134  )
135  # Get the data into BRugs:
136  modelData( bugsData( datalist ) )
137
138  #------------------------------------------------------------------------------
```

```
139   # INTIALIZE THE CHAINS.
140
141   # Autocorrelation within chains is large, so use several chains to reduce
142   # degree of thinning. But we still have to burn-in all the chains, which takes
143   # more time with more chains.
144   nchain = 10
145   modelCompile( numChains = nchain )
146
147   if ( F ) {
148       modelGenInits() # often won't work for diffuse prior
149   } else {
150     #  initialization based on data
151     theData = data.frame( pr=.01+.98*datalist$z/datalist$n ,
152                           x=factor(x,labels=xnames) )
153     a0 = mean( logit(theData$pr) )
154     a = aggregate( logit(theData$pr) , list( theData$x ) , mean )[,2] - a0
155     mGrp = aggregate( theData$pr , list( theData$x ) , mean )[,2]
156     sdGrp = aggregate( theData$pr , list( theData$x ) , sd )[,2]
157     kGrp = mGrp*(1-mGrp)/sdGrp^2 - 1
158     k = mean(kGrp)
159     genInitList <- function() {
160       return(
161           list(
162               a0 = a0 ,
163               a = a ,
164               aSDunabs = sd(a) ,
165               theta = theData$pr ,
166               k = k
167           )
168       )
169     }
170     for ( chainIdx in 1 : nchain ) {
171       modelInits( bugsInits( genInitList ) )
172     }
173   }
174
175   #---------------------------------------------------------------------------
176   # RUN THE CHAINS
177
178   # burn in
179   BurnInSteps = 10000
180   modelUpdate( BurnInSteps )
181   # actual samples
182   samplesSet( c( "a0" , "a" , "aSD" , "k" ) )
183   stepsPerChain = ceiling(2000/nchain)
184   thinStep = 750
185   modelUpdate( stepsPerChain , thin=thinStep )
186
187   #---------------------------------------------------------------------------
188   # EXAMINE THE RESULTS
189
190   source("plotChains.R")
191   source("plotPost.R")
192
```

```
193    checkConvergence = T
194    if ( checkConvergence ) {
195        sumInfo = plotChains( "a0" , saveplots=F , filenameroot=fnroot )
196        sumInfo = plotChains( "a" , saveplots=F , filenameroot=fnroot )
197        sumInfo = plotChains( "aSD" , saveplots=F , filenameroot=fnroot )
198        sumInfo = plotChains( "k" , saveplots=F , filenameroot=fnroot )
199    }
200
201    # Extract and plot the SDs:
202    aSDSample = samplesSample("aSD")
203    windows()
204    par( mar=c(3,1,2.5,0) , mgp=c(2,0.7,0) )
205    histInfo = plotPost( aSDSample , xlab="aSD" , main="a SD" , breaks=30 )
206    dev.copy2eps(file=paste(fnroot,"SD.eps",sep=""))
207
208    # Extract a values:
209    a0Sample = samplesSample( "a0" )
210    chainLength = length(a0Sample)
211    aSample = array( 0 , dim=c( datalist$NxLvl , chainLength ) )
212    for ( xidx in 1:datalist$NxLvl ) {
213        aSample[xidx,] = samplesSample( paste("a[",xidx,"]",sep="") )
214    }
215
216    # Convert to zero-centered b values:
217    mSample = array( 0, dim=c( datalist$NxLvl , chainLength ) )
218    for ( stepIdx in 1:chainLength ) {
219        mSample[,stepIdx ] = ( a0Sample[stepIdx] + aSample[,stepIdx] )
220    }
221    b0Sample = apply( mSample , 2 , mean )
222    bSample = mSample - matrix(rep( b0Sample ,NxLvl),nrow=NxLvl,byrow=T)
223
224    # Plot b values:
225    windows(datalist$NxLvl*2.75,2.5)
226    layout( matrix( 1:datalist$NxLvl , nrow=1 ) )
227    par( mar=c(3,1,2.5,0) , mgp=c(2,0.7,0) )
228    for ( xidx in 1:datalist$NxLvl ) {
229        plotPost( bSample[xidx,] , breaks=30 ,
230                  xlab=bquote(beta[.(xidx)]) ,
231                  main=paste(xnames[xidx])  )
232    }
233    dev.copy2eps(file=paste(fnroot,"b.eps",sep=""))
234
235    # Consider parameter correlations:
236    kSample = samplesSample("k")
237    windows()
238    pairs( cbind( b0Sample , t(bSample) , kSample ) , labels=c("b0",xnames,"k") )
239
240    # Display contrast analyses
241    nContrasts = length( contrastList )
242    if ( nContrasts > 0 ) {
243        nPlotPerRow = 5
244        nPlotRow = ceiling(nContrasts/nPlotPerRow)
245        nPlotCol = ceiling(nContrasts/nPlotRow)
246        windows(3.75*nPlotCol,2.5*nPlotRow)
```

```
247    layout( matrix(1:(nPlotRow*nPlotCol),nrow=nPlotRow,ncol=nPlotCol,byrow=T) )
248    par( mar=c(4,0.5,2.5,0.5) , mgp=c(2,0.7,0) )
249    for ( cIdx in 1:nContrasts ) {
250        contrast = matrix( contrastList[[cIdx]],nrow=1) # make it a row matrix
251        incIdx = contrast!=0
252        histInfo = plotPost( contrast %*% bSample , compVal=0 , breaks=30 ,
253                    xlab=paste( round(contrast[incIdx],2) , xnames[incIdx] ,
254                            c(rep("+",sum(incIdx)-1),"") , collapse=" " ) ,
255                    cex.lab = 1.5 ,
256                    main=paste( "Contrast:", names(contrastList)[cIdx] ) )
257    }
258    dev.copy2eps(file=paste(fnroot,"xContrasts.eps",sep=""))
259  }
260
261  #==============================================================================
262  # Do NHST ANOVA:
263
264  theData = data.frame( y=z/n , x=factor(x,labels=xnames) )
265  aovresult = aov( y ~ x , data = theData )
266  cat("\n----------------------------------------------------------------\n\n")
267  print( summary( aovresult ) )
268  cat("\n----------------------------------------------------------------\n\n")
269  print( model.tables( aovresult , "means" ) , digits=4 )
270  windows()
271  boxplot( y ~ x , data = theData )
272  cat("\n----------------------------------------------------------------\n\n")
273  print( TukeyHSD( aovresult , "x" , ordered = FALSE ) )
274  windows()
275  plot( TukeyHSD( aovresult , "x" ) )
276  if ( F ) {
277    for ( xIdx1 in 1:(NxLvls-1) ) {
278      for ( xIdx2 in (xIdx1+1):NxLvls ) {
279        cat("\n----------------------------------------------------\n\n")
280        cat( "xIdx1 = " , xIdx1 , ", xIdx2 = " , xIdx2 ,
281             ", M2-M1 = " , mean(score[x==xIdx2])-mean(score[x==xIdx1]) ,
282             "\n" )
283        print( t.test( score[ x == xIdx2 ] , score[ x == xIdx1 ] ) )
284      }
285    }
286  }
287  cat("\n----------------------------------------------------------------\n\n")
288
289  #==============================================================================
```

20.6 EXERCISES

Exercise 20.1. [Purpose: To observe correlated predictors in logistic regression.] For this exercise, we'll use some fictitious data regarding whether or not a patient suffered a heart attack within a year after a check-up that included measurements of systolic and diastolic blood pressures, cholesterol, weight, height, and age. The data are completely fabricated and fictional, for illustrative purposes only. The data are generated by the program BloodDataGenerator.R, which is available at the author's website.

(A) For this part, we will generate two data sets. In one set, the six predictors have correlations of zero with each other. In the second set, the first two predictors have a strong positive correlation, but still zero correlation with all the other predictors. For both sets, the true regression coefficient on the first predictor is zero, but the true regression coefficient on the second predictor is positive. Use the program `BloodDataGenerator.R` to generate these two data sets by selecting the relevant correlation matrix at the top of the program. For each data set, conduct a Bayesian analysis using the program in Section 20.5.1 (`MultipleLogisticRegressionBrugs.R`). Be sure to specify that the only predictors are the first two, namely, systolic and diastolic pressure. Include the posterior contour plots, which should look similar to those in Figure 20.7. Be sure also to include the histograms of the marginal posterior distributions. In which case are the HDIs narrower?

(B) For this part, we generate a different fictitious data set, in which the first two predictors (systolic and diastolic) have zero correlation with each other, but they are correlated with another predictor (weight). The first predictor (systolic) has a true regression coefficient of zero, but the second (diastolic) and third (weight) predictors have positive regression coefficients. Run the Bayesian logistic regression analysis *using only the first two predictors*. What is the estimate of the regression coefficient on the first predictor? Include the histograms of the marginals of the posterior. Now rerun the Bayesian logistic regression analysis *including all six predictors*. What is the estimate of the regression coefficient on the first predictor? Include the histograms of the marginals of the posterior (for which you might need to manually stretch the graph window or change

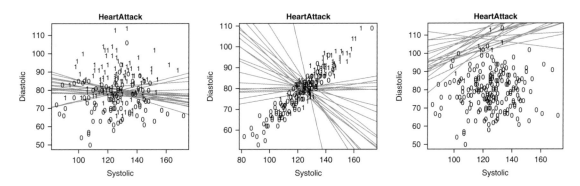

FIGURE 20.7

In all panels, the true regression coefficient on systolic is zero, and the true regression coefficient on diastolic is positive (nonzero). In the left panel, the predictors are uncorrelated, and the credible $p=0.5$ level contours are fairly certain. In the middle panel, the predictors are highly correlated, and the credible $p=0.5$ level contours are very *un*certain. In the right panel, only a small percentage of the data has value 1, and again the level contours are relatively *un*certain. See Exercises 20.1 and 20.2.

the character sizes in the plot). Why are the estimates of the regression coefficient for the first predictor different in the two analyses?

Exercise 20.2. [Purpose: To observe the consequences of a small proportion of 1's in logistic regression.] For this exercise, we again use fictitious data generated by the program BloodDataGenerator.R. Set all the inter-predictor correlations to zero by selecting the appropriate correlation matrix at the top of the program. *Change the* proportionOnes *constant to 0.02.* This causes the proportion of 1's in the data to be very small. In other words, the threshold for the sigmoidal function is set very high, so that the weighted sum of the predictors must be very large before $p(y=1)$ gets very large. This sort of situation, in which there are few 1's, is not unusual in real data. In heart attack data, for example, there will be few people who have a heart attack in the year following a routine check-up. Run a Bayesian logistic regression on the data, using only the first two predictors. What are the estimates of the regression coefficients? Include the histograms of the marginals of the posterior, and include the posterior contour plot, which should look similar to the right panel of Figure 20.7. Compare the widths of the HDIs to the results of the first part of the previous exercise.

Exercise 20.3. [Purpose: To observe logistic ANOVA *without* assuming homogeneity of variance.] Extend the program of Section 20.5.2 (LogisticOneway AnovaBrugs.R) so that it allows different certainty (κ) parameters for each condition, with a hyperdistribution across conditions, just as in the model on the *right* side of Figure 9.17, p. 226. Apply the model to the filtration-condensation data, and compare the results with the results of Exercise 9.2, p. 236. If you already did Exercise 18.3, you can simply modify your program from that exercise. Programming hint: The model specification might look like this:

(LogisticOnewayAnovaHeteroVarBrugs.R)

```
11    model {
12      for ( i in 1:Ntotal ) {
13        z[i] ~ dbin( theta[i] , n[i] )
14        theta[i] ~ dbeta( aBeta[x[i]] , bBeta[x[i]] )I(0.001,0.999)
15      }
16      for ( j in 1:NxLvl ) {
17        aBeta[j] <- mu[j] * k[j]
18        bBeta[j] <- (1-mu[j]) * k[j]
19        mu[j] <- 1 / ( 1 + exp( -( a0 + a[j] ) ) )
20        a[j] ~ dnorm( 0.0 , atau )
21        k[j] ~ dgamma( skappa , rkappa )
22      }
23      a0 ~ dnorm( 0 , 0.001 )
24      atau <- 1 / pow( aSD , 2 )
25      aSD <- abs( aSDunabs ) + .1
26      aSDunabs ~ dt( 0 , 0.001 , 2 )
27      skappa <- pow(mg,2)/pow(sg,2)
28      rkappa <- mg/pow(sg,2)
29      mg ~ dunif(0,50)
30      sg ~ dunif(0,30)
31    }
```

Ordinal Predicted Variable

CONTENTS

The winner is first, and that's all that he knows, whether
Won by a mile or won by a nose. But
Second recalls every inch of that distance, in
Vivid detail and with haunting persistence.

Very often the predicted variable is ordinal, such as a rating on a scale from
1 to 5. Rate how much you agree with this statement: "Bayesian methods
are the most informative, useful, and rational way to analyze data," where
5 = strongly agree, 4 = moderately agree, 3 = neither agree nor disagree,
2 = moderately disagree, and 1 = strongly disagree. This sort of response scale
is often called a *Likert* (pronounced LICK-ert) scale (Likert, 1932), and it typically has five, seven, or more levels (Dawes, 2008). Vast amounts of survey data
use Likert rating scales.

Doing Bayesian Data Analysis: A Tutorial with R and BUGS. DOI: 10.1016/B978-0-12-381485-2.00021-3
© 2011, Elsevier Inc. All rights reserved.

We often want to predict the ordinal measure on the basis of other predictors. For example, predictors might include age of respondent, number of years of education, CPU speed of the computer used by respondent, and so forth. To predict ordinal values from metric predictors, we need a way of mapping a weighted linear combination of predictors onto ordinal scale values. The linking function we will use is a probit function with thresholds. This type of analysis is therefore called *ordinal probit regression*. It appeared in the third row of Table 14.1, p. 385.

21.1 ORDINAL PROBIT REGRESSION

21.1.1 What the Data Look Like

Some stereotypical (fictitious) data that could be modeled by ordinal probit regression are shown in Figure 21.1. The predicted variable, y, is from a 5-point ordinal scale. The predictor variable, x, is metric. If you like concrete examples, imagine that the y value is a rating of overall happiness and the x value is annual income in dollars. The key aspect of y that makes it ordinal, not metric, is that we don't know whether the distance from 1 to 2 is the same as the distance from 2 to 3, or from 3 to 4, and so on. Is an increase in happiness from 1 to 2 the same amount of differential happiness as an increase in happiness from 2 to 3? It may be, but we don't know. All we know is the order of the ratings, not the distances between them.

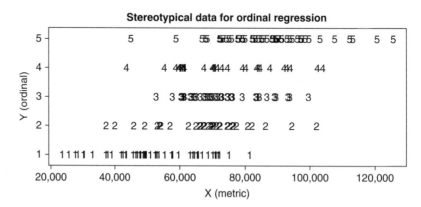

FIGURE 21.1

Some stereotypical (fictitious) data of the type modeled by ordinal regression. Points are labeled with their y value instead of plotted merely as dots. Although the y values are plotted at equal intervals on the y axis, we really only know their order, not their separation. The conceptual emphasis is on *vertical* slices through the scatter plot: At any given value of x, we are interested in the probability distribution across the discrete y values.

Regarding Figure 21.1, notice that when x is relatively small, around 40,000, then the y values tend to be 1's or 2's. When x is relatively large, around 100,000, then the y values tend to include more 4's and 5's. But notice also that there is lots of random variability, in the sense that there is a range of y values at any given x value.

21.1.2 The Mapping from Metric x to Ordinal y

Our goal is to construct a model for this sort of data. To accomplish the goal, we start with linear regression on x and then link the continuous prediction to ordinal y values via a thresholded cumulative-normal function. Figure 21.2 shows the underlying mappings from x to y. The right side of the diagram shows simple linear regression: The metric predictor value, x, is mapped to an underlying metric tendency μ via the simple linear relationship, $\mu = \beta_0 + \beta x$. The underlying metric value is noisy, distributed around μ as a normal distribution with standard devition σ, which is drawn sideways in the diagram. So far, this is merely simple linear regression.

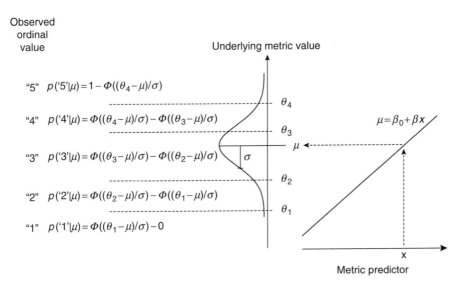

FIGURE 21.2

The underlying mappings in ordinal probit regression. The lower right shows the predictor value, x, which gets mapped to an underlying metric value via a linear relation. There is normally distributed noise in the underlying metric value, which has mean μ and standard deviation σ. The underlying metric value is mapped to an ordinal value depending on which thresholds it falls between. The thresholds are not evenly spaced. The probability of falling between thresholds is given by the cumulative normal function, Φ.

The new part, on the left of Figure 21.2, gets us from the underlying metric value to an observed ordinal value. To achieve this mapping, the underlying metric scale is carved into intervals by thresholds, located at scale values θ_1, θ_2, and so on, as shown. The threshold values are estimated from the data. When the underlying metric value falls between θ_{n-1} and θ_n, ordinal value n is generated. Because the underlying metric value is normally distributed, the probability that it falls in any interval is given by the cumulative normal distribution.

As you may recall from Section 14.1.7.2 and Figure 14.8, p. 378, the cumulative normal function is denoted Φ. The value of $\Phi(z)$ indicates how much of the normal distribution falls to the left of z, where z is a standardized value. For example, $\Phi(0) = 0.5$, because half of the standardized normal distribution falls to the left of 0. In R, the cumulative normal function is pnorm(z). In BUGS, the cumulative normal function is phi(z).

To determine how much of the normal distribution falls between two values, z_n and z_{n-1}, we simply compute the difference, $\Phi(z_n) - \Phi(z_{n-1})$. We apply this result to the thresholds in Figure 21.2, but we first standardize the thresholds relative to the mean and standard deviation of the noise distribution. Thus, the mass of the normal distribution between θ_n and θ_{n-1} is $\Phi((\theta_n - \mu)/\sigma) - \Phi((\theta_{n-1} - \mu)/\sigma)$. This mass is the probability of ordinal value n. In summary, and as written in Figure 21.2 for each interval,

$$p(\text{"}n\text{"} \mid \mu) = \Phi((\theta_n - \mu)/\sigma) - \Phi((\theta_{n-1} - \mu)/\sigma) \qquad (21.1)$$

where "n," in quotes, means that the value n is emitted by the subject. A more traditional, but awkward, way of writing $p(\text{"}n\text{"} \mid \mu) = \ldots$ is $p(y=n \mid \mu) = \ldots$. For the lowest and highest ordinal values, the outside thresholds are at negative and positive infinity. The area under the normal distribution to the left of positive infinity is 1, hence Equation 21.1 becomes $1 - \Phi((\theta_{n-1} - \mu)/\sigma)$ at the highest ordinal value.

Because the mapping from underlying metric value to ordinal value uses a cumulative normal function, this type of regression might be called *cumulative normal regression* or *phi regression*, by analogy to "logistic regression". But it's not. Instead, it's called *probit regression* to indicate that a cumulative normal is used as the link function, and it's called *ordinal probit regression* to indicate that the predicted value is ordinal.

21.1.3 The Parameters and Their Priors

Inspection of Figure 21.2 reveals the parameters that are estimated in ordinal probit regression. There are the usual parameters for linear regression, including β_0, β_j for each predictor, and σ. Also, there are the threshold parameters, θ_1 through θ_{L-1}, where L is the number of levels in the ordinal scale. It turns out that the intercept trades off with the thresholds, because we can add any constant to the intercept and subtract that constant from the thresholds and have an equivalent model. Therefore, we set the intercept to an arbitrary convenient constant. Also, the noise standard deviation, σ, trades off with the slopes and thresholds, because if we multiply the noise by a constant, we can divide all the slopes and thresholds by that constant and end up with an equivalent model (also translating the thresholds if the intercept is not zero). In summary, we estimate the slopes on the predictors, β_j, and the thresholds, θ_n, but we fix the intercept β_0 and the noise σ at convenient constants. (Which constants are convenient? This is answered in the description of the R code in Section 21.5 (OrdinalProbitRegressionBrugs.R).)

To estimate these parameters, we must define our prior uncertainty. The prior on the regression coefficients is the same as for linear or logistic regression. For example, each regression coefficient could have a normal prior. If there are several predictors, then a hyperprior could be posited to express the mutual informativeness of the regression coefficients, as in Figure 17.6.

The prior on the thresholds can be (mildly) informed by the scale of the predicted value and by the fact that the thresholds need to be in order (i.e., it must be that $\theta_n > \theta_{n-1}$). Because the y values are 1 through L, and we will set up the underlying metric value to match that scale, it is reasonable to believe that the thresholds should be approximately $1.5, 2.5, \ldots, L - 0.5$. Our uncertainty in those values is moderate, not extreme, and therefore we may opt to put a normal prior on each threshold with a standard deviation of about 1 unit. This sort of prior, in which each threshold has a normal distribution that is independent of the others, makes it possible to have thresholds that are out of order. There are various ways to express prior knowledge that the thresholds must be ordered. One simple way is to put the knowledge into the likelihood function itself, by changing Equation 21.1 to this:

$$p(\text{“}n\text{”} \mid \mu) = \max\left(0, \Phi\left((\theta_n - \mu)/\sigma\right) - \Phi\left((\theta_{n-1} - \mu)/\sigma\right)\right)$$

Notice that if $\theta_n < \theta_{n-1}$, then the probability of "n" is zero. That would be fine if the value n never occurs in the data, but if it does, then threshold inversions will never be found to be credible.

The prior on the regression coefficients is explained in Exercise 21.2. In that exercise you will also explore the robustness of the posterior to changes in the prior (and you'll find that the posterior is robust).

21.1.4 Standardizing for MCMC Efficiency

To reduce correlation among regression coefficients and thereby make the MCMC sampling more efficient, the predictor values are standardized before the parameters are estimated. In other words, the x_j values are transformed according to Equation 16.1: $x_j^{(z)} = (x_j - M_j)/SD_j$. For the formulas in this section, the standardized parameters for the standardized predictors all have a z superscript, like this: $x^{(z)}$, $\beta_j^{(z)}$, and so on. After estimation, the parameters must be transformed back to the original scales of the predictors. To do this back-transformation, consider the following sequence of equalities:

$$p(\text{"}n\text{"} \mid \vec{x}) = \Phi\left((\theta_n^{(z)} - \mu^{(z)})/\sigma^{(z)}\right) - \Phi\left((\theta_{n-1}^{(z)} - \mu^{(z)})/\sigma^{(z)}\right)$$

that's Equation 21.1 with z superscripts

$$p(\text{"}\leq n\text{"} \mid \vec{x}) = \Phi\left((\theta_n^{(z)} - \mu^{(z)})/\sigma^{(z)}\right)$$

notice "$\leq n$" on the left-hand side!

$$\text{probit}\left(p(\text{"}\leq n\text{"} \mid \vec{x})\right) = (\theta_n^{(z)} - \mu^{(z)})/\sigma^{(z)}$$

$$= \frac{1}{\sigma^{(z)}}\left[\theta_n^{(z)} - \left(\beta_0^{(z)} + \sum_j \beta_j^{(z)} x_j^{(z)}\right)\right]$$

$$= \frac{1}{\sigma^{(z)}}\left[\theta_n^{(z)} - \left(\beta_0^{(z)} + \sum_j \beta_j^{(z)} \left(\frac{x_j - M_j}{SD_j}\right)\right)\right]$$

$$= \underbrace{\frac{1}{\sigma^{(z)}}\left(\theta_n^{(z)} - \beta_0^{(z)} + \sum_j \beta_j^{(z)} M_j/SD_j\right)}_{\theta_n} - \sum_j \underbrace{\frac{\beta_j^{(z)}}{\sigma^{(z)} SD_j}}_{\beta_j} x_j$$

(21.2)

with $\beta_0 \equiv 0$ and $\sigma \equiv 1$

where all the parameters without z superscripts are on the original scale. Why is $\beta_0 \equiv 0$? Because it's arbitrary. We could set it to any value, adding it into the sum on the far right and subtracting it out of the thresholds. Similarly, the value of σ is arbitrary; we could set it to any value if we multiplied all the thresholds and regression coefficients by that value, and then divided by σ before passing into the Φ function.

21.1.5 Posterior Prediction

The slope and threshold values from Equation 21.2 are arbitrary, determined only relative to $\beta_0 \equiv 0$ and $\sigma \equiv 1$. Their usefulness is being able to easily compute predicted probabilities from the original scale values, as follows:

$$p(\text{"}n\text{"} \mid \vec{x}) = \Phi\left(\theta_n - \sum_j \beta_j x_j\right) - \Phi\left(\theta_{n-1} - \sum_j \beta_j x_j\right)$$ (21.3)

This expression (in Equation 21.3) is merely Equation 21.2 rewritten with the original-scale parameters set where the underbraces indicate they belong. Exercise 21.3 has you use this equation to make predictions.

21.2 SOME EXAMPLES

Consider the data shown in the top panel of Figure 21.3. There are two metric predictors, labeled X1 and X2 in the graph, and the data to be predicted are ordinal, ranging from 1 to 7. Each data point is labeled with its y value. These data were randomly generated, but with little noise in the y value, so that the transitions from one y level to the next would be clear. Notice that as X1 increases, the y value tends to decrease. Hence, the regression coefficient on X1 should be negative. Notice also that as X2 increases, the y values tend to increase. Hence, the regression coefficient on X2 should be positive.

The program in Section 21.5 (`OrdinalProbitRegressionBrugs.R`), which does Bayesian ordinal probit regression, was run with these data. Aspects of the resulting posterior are shown in Figure 21.3. The first histogram indicates credible values of β_1, which are all negative, as inspection of the trends in the data suggested. The second histogram indicates credible values of β_2, which are all positive, again as inspection of the trends in the data suggested.

The next six histograms show credible values of the thresholds. One thing to notice about the distributions of the thresholds is that they appear to overlap tremendously. The histogram for θ_1 covers much of the same range as the histogram for θ_2, and so on. This overlap would seem to suggest a violation of the necessary ordering of the thresholds, because, it seems, θ_1 could easily be larger than θ_2. What's wrong? Nothing, because credible values of the thresholds are strongly correlated. The continuation of Figure 21.3 on p. 583 shows pairwise scatter plots of the parameter values and reveals the extreme correlation. In fact, if you plot histograms of the posterior $\theta_n - \theta_{n-1}$, you will find that the differences do not overlap zero at all (i.e., there is no violation of ordering in the thresholds). Exercise 21.1 provides more details regarding how to do this.

The plot of the data, in the top panel of Figure 21.3 on p. 582, also includes examples of level contours from the posterior. A set of six (i.e., $L - 1$) level contours comes from a single step in the MCMC chain. The six level contours correspond to the six thresholds for the β_1 and β_2 values at that step in the chain. The level contours within a set have the same slope because they all correspond to the same β_1 and β_2 values. Different sets of contours have different slopes because they have different β_1 and β_2 values. Because these data were generated with little noise in the y values, there is little uncertainty in

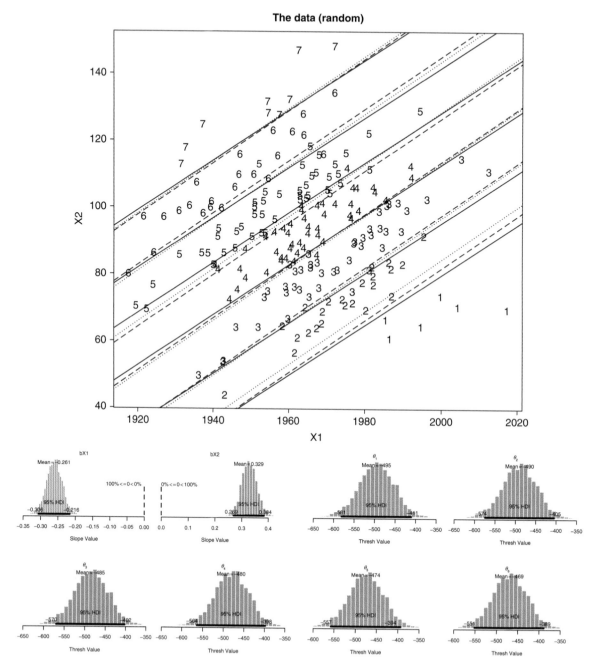

FIGURE 21.3

Upper panel: Ordinal data plotted against two predictors, superimposed with a smattering of threshold lines from the posterior. Histograms and scatter plots (next page) show posterior on the regression coefficients and thresholds.

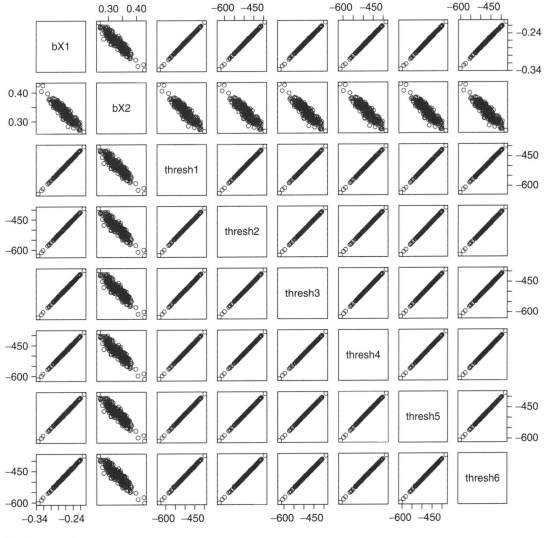

FIGURE 21.3

(Continued)

the slopes or thresholds. One indicator of the lack of noise is that fact that between level contours in the plot there is mostly just one value of y.

The level contours in the top panel of Figure 21.3 were derived as follows. Consider the line that marks the transition from the "$n-1$" region to the "n" region. The transition occurs when $p("\leq n" \mid \vec{x})$ is 0.5, which means that $\text{probit}\left(p("\leq n" \mid \vec{x})\right) = 0$. From Equation 21.3 (and/or Eqn. 21.2), we know

$$\text{probit}\left(p("\leq n" \mid \vec{x})\right) = \theta_n - \sum_j \beta_j x_j$$

It follows that the transition line occurs for $\vec{x}$ values that satisfy

$$0 = \theta_n - \sum_j \beta_j x_j$$

$$0 = \theta_n - \beta_1 x_1 - \beta_2 x_2 \quad \text{when there are two predictors}$$

$$x_2 = \frac{\theta_n}{\beta_2} + \frac{-\beta_1}{\beta_2} x_1$$

The level contours in the top panel of Figure 21.3 used that formula; they are applied in the last few lines of the code in Section 21.5 (OrdinalProbitRegressionBrugs.R).

The previous example used fictitious data so that the concepts could be clearly illustrated. Realistic data are typically *much* messier. Consider, for example, ratings of movies (Moore, 2006), as shown in the top panel of Figure 21.4. The predictor on the abscissa is the year the movie was made, and the predictor on the ordinate is length (i.e., duration) of the movie in minutes. Notice that ratings are rather noisy; between any pair of contour lines there occur many different rating values.

Despite the noise in the data, the regression coefficients on the two predictors are credibly nonzero, as can be seen in the first two histograms of Figure 21.4. Ratings clearly tend to decrease as year of production increases, and ratings clearly tend to increase as length increases. Nevertheless, the degree of uncertainty in the regression coefficients is much larger than the previous example. This uncertainty is apparent in the smattering of posterior level contours superimposed on the data. The slants of the level contours vary greatly from one set to another.

The posterior distribution of thresholds again shows the strong correlations, in the scatter plots of the continuation of Figure 21.4 on p. 586. The thresholds never violate the proper order, although exactly where they should be is uncertain.

21.2.1 Why Are Some Thresholds Outside the Data?

In Figure 21.4, the upper panel shows a scatter plot of data, along with a few of the posterior contour lines superimposed. Notice that some of the extreme threshold contours fall outside all the data points. For example, the lower right of the plot shows that the first and second threshold contours have almost no data below them. How can these thresholds be among the most credible?

The answer lies in gaining a better intuition for the predictions of the ordinal probit model. Figure 21.5 shows the predictions of the model for a single x variable and thresholds set at 1 through 5. The upper panel shows the predictions when the noise is very small compared to the separation between thresholds, namely, $\sigma = 0.1$. In this case, the predictions match intuition: When

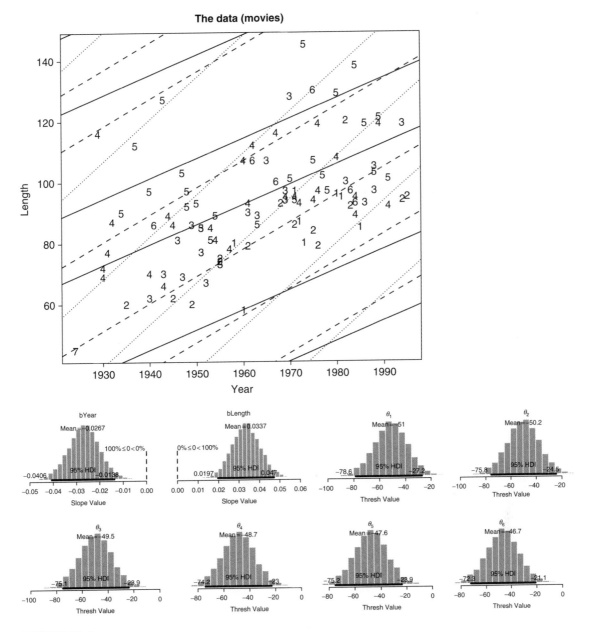

FIGURE 21.4
Upper panel: Movie ratings from Moore (2006) plotted against Length and Year, superimposed with a smattering of threshold lines from the posterior. Histograms and scatter plots (next page) show posterior on the regression coefficients and thresholds.

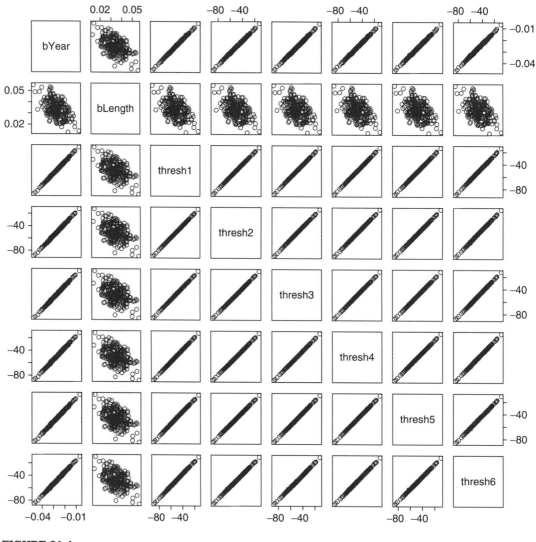

FIGURE 21.4
(*Continued*)

x is between thresholds *n* and *n* + 1, the most probable response is *n* + 1. It's only when *x* is very close to a threshold that an adjacent response gains probability.

Moving down Figure 21.5, the successive panels show greater amounts of noise (i.e., larger values of σ). Consider what happens to the predicted response for $x = 2.2$. When σ is small, the highest probability response is 3. As the noise increases, the dominance of this response becomes weaker. Finally, in the bottom panel, with high noise, 3 is no longer the dominant response;

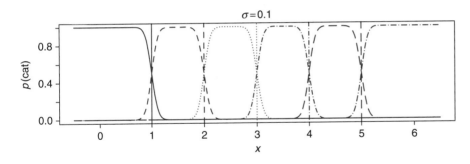

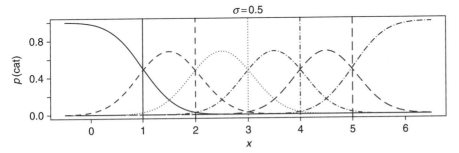

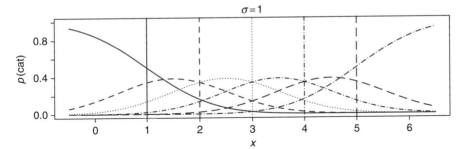

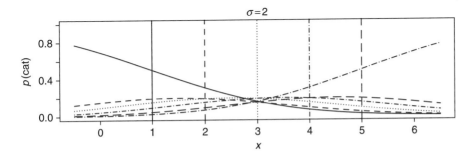

FIGURE 21.5

Each panel shows the probability of an ordinal response as a function of the stimulus value, x. The upper panel is for small σ, and the lower panel is for large σ. Response thresholds are arbitrarily placed at 1, 2, 3, 4, and 5.

instead 1 is! Thus, even though $x = 2.2$, which falls between the second and third thresholds, the highest probability response is 1.

Suppose that the data consist of x values scattered between $x = 2$ and $x = 4$. Suppose also that responses are noisy, with $\sigma = 2$, as in the lowest panel of Figure 21.5. For a point near $x = 2$, the most probable response is 1, even though the point is far above the threshold for 1. For a point near $x = 4$, the most probable response is 6, even though the point is far below the threshold for 6. It is this sort of pattern that we see in the movie data of Figure 21.4, in which the responses are noisy.

21.3 INTERACTION

Just as in multiple linear regression and multiple logistic regression, there can be multiplicative interaction of the predictors in ordinal probit regression. Much as the logistic function squashes the predictions, the cumulative normal function squashes the predictions. Hence, the prediction surface for ordinal probit regression with interaction is much like the prediction surface for logistic regression, as was shown in Figure 20.4, p. 558. Ordinal probit regression is analogous to logistic regression when there are only two levels of the ordinal variable y, because logistic regression has a predicted variable with two levels. In fact, in ordinal regression, the levels can be grouped into "$\leq n$" and "$> n$," and therefore every transition between adjacent levels can be thought of as analogous to logistic regression on two levels. Thus, when there is multiplicative interaction of the predictors, the interaction can express ideas such as conjuction: The y value is "$> n$" when x_1 is large *and* x_2 is large, but not otherwise. A multiplicative interaction can also express an exclusive-OR: The y value is "$> n$" when x_1 is large *or* x_2 is large, but *not both*.

21.4 RELATION TO LINEAR AND LOGISTIC REGRESSION

This chapter took the core model of multiple linear regression and extended it to ordinal predicted values by mapping the linear prediction through a thresholded cumulative normal function. The only real news regarded the mathematical details of dealing with the cumulative normal function, and implementing the model in R and BUGS. Otherwise, the basic concepts of multiple regression apply to ordinal probit regression.

When dealing with ordinal predicted values, some analysts will assume, for convenience, that the ordinal scale values are equally spaced on an interval scale. With this assumption, the problem reverts to ordinary linear regression. Often the results of such an analysis are similar to ordinal regression, in terms of the regression coefficients on the predictors. But if the emphasis is prediction of the probabilities of each ordinal level, the ordinal regression gives a more direct answer (via Equation 21.3).

When the predicted variable has only two levels, it is tantamount to an ordinal scale with two levels (because order can be assigned arbitrarily, with "1">"0" or "0">"1," without changing the results). In this case, ordinal probit regression will yield results essentially equivalent to logistic regression. The only difference will be subtle because of the slight difference in shape between the logistic function and the cumulative normal function. In fact, when there are only two levels of the predicted variable, some analysts prefer probit (cumulative normal) regression to logistic regression. Because the results tend to be so similar, the choice between link function usually is governed by meaningfulness for the domain of application. If the link function is to be thought of as describing normally distributed noise on an underlying linear relation, then the probit is more meaningful. If the analyst finds it meaningful to discuss regression coefficients in terms of log odds (see Section 20.1.3, p. 553), then the logistic link function is more useful.

21.5 R CODE

Initialization of the chains uses the maximum likelihood estimate (MLE) from linear regression as a heuristic approximation to start the slope estimates for ordinal regression. The R function $lm()$ is used for this purpose. The MLE also returns an estimate of the intercept (which should be close to the mean of the y values, because the x values are standardized), which is used as the arbitrary fixed value for β_0 in the model. The MLE has a residual error, which is used to compute a corresponding standard deviation that is used as the arbitrary fixed value for σ.

(OrdinalProbitRegressionBrugs.R)

```
1   graphics.off()
2   rm(list=ls(all=TRUE))
3   fname = "OrdinalProbitRegressionBrugs"
4   library(BRugs)        # Kruschke, J. K. (2010). Doing Bayesian data analysis:
5                         # A Tutorial with R and BUGS. Academic Press / Elsevier.
6   #------------------------------------------------------------------------------
7   # THE MODEL.
8
9   modelstring = "
10  # BUGS model specification begins here...
11  model {
12      for( i in 1 : nData ) {
13          y[i] ~ dcat( pr[i,1:nYlevels] )
14          pr[i,1] <- phi( ( thresh[1] - mu[i] ) / sigma )
15          for ( k in 2:(nYlevels-1) ) {
16              pr[i,k] <- max( 0 ,  phi( ( thresh[ k ] - mu[i] ) / sigma )
17                              - phi( ( thresh[k-1] - mu[i] ) / sigma ) )
```

```
18              }
19              pr[i,nYlevels] <- 1 - phi( ( thresh[nYlevels-1] - mu[i] ) / sigma )
20              mu[i] <- b0 + inprod( b[1:nPredictors] , x[i,1:nPredictors] )
21          }
22          bPrec <- pow( nYlevels/4 , -2 ) # max plausible slope is 1SD
23          for ( j in 1:nPredictors ) {
24              b[j] ~ dnorm(0,bPrec) # modest precision because of normalized x,y values
25          }
26          threshPriorPrec <- 1
27          for ( k in 1:(nYlevels-1) ) {
28              threshPriorMean[k] <- k+0.5
29              thresh[k] ~ dnorm( threshPriorMean[k] , threshPriorPrec )
30          }
31      }
32      # ... end BUGS model specification
33      " # close quote for modelstring
34      writeLines(modelstring,con="model.txt")
35      modelCheck( "model.txt" )
36
37      #------------------------------------------------------------------------------
38      # THE DATA.
39
40      dataSource = c("Random","Movies")[2]
41
42      # The loading of data must produce a matrix called dataMat that has
43      # one row per datum, where the first column is the ordinal predicted value
44      # and the 2nd - last columns are the predictor values. The columns should
45      # be named.
46
47      if ( dataSource=="Random" ) {
48          fname = paste( fname , dataSource , sep="" )
49          # Generate some random toy data.
50          source( "OrdinalProbitDataGenerator.R" )
51          nYlevels = 7
52          dataMat = OrdinalProbitDataGenerator( nData = 200 ,
53                  normPrec=200 , slope=c(-1.0,1.26) , # c(-1.0,1.26) matches Movies
54                  thresh=c(-Inf,seq(-1.2,1.2,length=nYlevels-1),Inf) ,
55                  nYlevels=nYlevels , makePlots=F , rndSeed=47405 )
56          # Change x values to arbitrary non-standardized scales:
57          dataMat[,2] =  1963.64 + 18.13 * dataMat[,2]
58          dataMat[,3] =   92.87 + 18.26 * dataMat[,3]
59      }
60
61      if ( dataSource=="Movies" ) {
62          fname = paste( fname , dataSource , sep="" )
63          dataFram = read.table( "Moore2006data.txt" , header=T )
64          rateVals = sort( unique( dataFram[,"Rating"] ) )
65          rankVals = match( dataFram[,"Rating"] , rateVals ) # convert to ranks
66          dataMat = cbind( rankVals , dataFram[,"Year"] , dataFram[,"Length"] )
67          colnames(dataMat) = c("Rating","Year","Length")
68      }
69
70      # Rename for use by generic processing later:
```

```
71   nData = NROW(dataMat)
72   x = dataMat[,-1]
73   predictorNames = colnames(dataMat)[-1]
74   nPredictors = NCOL(x)
75   y = as.matrix(dataMat[,1])
76   predictedName = colnames(dataMat)[1]
77   nYlevels = max(y)
78
79   # Re-center x values at mean, to reduce autocorrelation in MCMC sampling.
80   # Standardize (divide by SD) to make prior-setting easier.
81   standardizeCols = function( dataMat ) {
82       zDataMat = dataMat
83       for ( colIdx in 1:NCOL( dataMat ) ) {
84           mCol = mean( dataMat[,colIdx] )
85           sdCol = sd( dataMat[,colIdx] )
86           zDataMat[,colIdx] = ( dataMat[,colIdx] - mCol ) / sdCol
87       }
88       return( zDataMat )
89   }
90   zx = standardizeCols( x )
91   # Don't standarize y because they must be integers, 1 to nYlevels
92
93   lmInfo = lm( y ~ zx ) # R function returns MLE
94   b0Init = lmInfo$coef[1]
95   bInit = lmInfo$coef[-1]
96   sigmaInit = sqrt(sum(lmInfo$res^2)/nData)
97
98   # Get the data into BUGS:
99   datalist = list(
100              x = zx ,
101              y = as.vector( y ) , # BUGS does not treat 1-column mat as vector
102              nPredictors = nPredictors ,
103              nData = nData ,
104              nYlevels = nYlevels ,
105              sigma = sigmaInit , # fixed, not estimated
106              b0 = b0Init  # fixed, not estimated
107   )
108  modelData( bugsData( datalist ) )
109
110  #------------------------------------------------------------------------------
111  # INTIALIZE THE CHAINS.
112
113  nChain = 3
114  modelCompile( numChains = nChain )
115
116  genInitList <- function() {
117      list(
118          b = bInit , # from lm(y~zx), above
119          thresh = 1:(nYlevels-1)+.5
120      )
121  }
122  for ( chainIdx in 1 : nChain ) {
123      modelInits( bugsInits( genInitList ) )
124  }
```

```
125
126   #-------------------------------------------------------------------------
127   # RUN THE CHAINS
128
129   # burn in
130   BurnInSteps = 2000
131   modelUpdate( BurnInSteps )
132   # actual samples
133   samplesSet( c( "b" , "thresh" ) )
134   stepsPerChain = ceiling(5000/nChain)
135   thinStep = 20
136   modelUpdate( stepsPerChain , thin=thinStep )
137
138   #-------------------------------------------------------------------------
139   # EXAMINE THE RESULTS
140
141   source("plotChains.R")
142   source("plotPost.R")
143
144   checkConvergence = T
145   if ( checkConvergence ) {
146     bSum      = plotChains( "b"      , saveplots=F , filenameroot=fname )
147     threshSum = plotChains( "thresh" , saveplots=F , filenameroot=fname )
148   }
149
150   # Extract chain values:
151   zbSamp = NULL
152   for ( j in 1:nPredictors ) {
153      zbSamp = cbind( zbSamp , samplesSample( paste("b[",j,"]",sep="") ) )
154   }
155   chainLength = NROW(zbSamp)
156   zthreshSamp = NULL
157   for ( j in 1:(nYlevels-1) ) {
158      zthreshSamp = cbind( zthreshSamp ,
159                          samplesSample(paste("thresh[",j,"]",sep="")) )
160   }
161
162
163   # Convert to original scale:
164   bSamp = zbSamp * matrix( 1/(sigmaInit*apply(x,2,sd)) , byrow=TRUE ,
165                           ncol=nPredictors , nrow=chainLength )
166   threshSamp = (1/sigmaInit) * ( zthreshSamp - b0Init +
167                   rowSums( zbSamp * matrix( apply(x,2,mean)/apply(x,2,sd) ,
168                                             byrow=TRUE , ncol=nPredictors ,
169                                             nrow=chainLength ) ) )
170   b0 = 0
171   sigma = 1
172
173   # Scatter plots of parameter values, pairwise:
174   if ( ( nPredictors+nYlevels) <= 10 ) { # don't display if too many
175      windows()
176      thinIdx = ceiling(seq(1,chainLength,length=200))
177      pairs( cbind( zbSamp[thinIdx,] , zthreshSamp[thinIdx,] )  ,
178             labels=c( paste("zb",predictorNames,sep="") ,
```

```
179                             paste("zthresh",1:nYlevels,sep="")) )
180      windows()
181      pairs( cbind( bSamp[thinIdx,] , threshSamp[thinIdx,] )  ,
182              labels=c( paste("b",predictorNames,sep="") ,
183                        paste("thresh",1:nYlevels,sep="")) )
184      dev.copy2eps(file=paste(fname,"PostPairs.eps",sep=""))
185    }
186
187    # Display the posterior:
188    nPlotPerRow = 5
189    nPlotRow = ceiling((nPredictors+nYlevels-1)/nPlotPerRow)
190    nPlotCol = ceiling((nPredictors+nYlevels-1)/nPlotRow)
191    windows(3.5*nPlotCol,2.25*nPlotRow)
192    layout( matrix(1:(nPlotRow*nPlotCol),nrow=nPlotRow,ncol=nPlotCol,byrow=T) )
193    par( mar=c(4,3,2.5,0) , mgp=c(2,0.7,0) )
194    for ( sIdx in 1:nPredictors ) {
195    histInfo = plotPost( bSamp[,sIdx] , xlab="Slope Value" , compVal=0.0 ,
196                          breaks=30 ,
197                          main=bquote( b *.(predictorNames[sIdx]) ) ,
198                          cex.main=1.67 , cex.lab=1.33 )
199    }
200    for ( sIdx in 1:(nYlevels-1) ) {
201    histInfo = plotPost( threshSamp[,sIdx] , xlab="Thresh Value" , compVal=NULL ,
202                          breaks=30 ,
203                          main=bquote( theta * .(sIdx) ) ,
204                          cex.main=1.67 , cex.lab=1.33 )
205    }
206    dev.copy2eps(file=paste(fname,"PostHist.eps",sep=""))
207
208    # Plot the data
209    if ( nPredictors == 2 ) {
210    windows()
211    plot( x[,1] , x[,2] , xlab=colnames(x)[1] , ylab=colnames(x)[2] ,
212          main=paste( "The Data (" , dataSource , ")" , sep="") ,
213          pch=as.character(y) )
214    for ( chainIdx in round(seq(1,chainLength,len=3)) ) {
215      for ( threshIdx in 1:(nYlevels-1) ) {
216        abline( threshSamp[chainIdx,threshIdx]/bSamp[chainIdx,2] ,
217                -bSamp[chainIdx,1]/bSamp[chainIdx,2] ,
218                lwd = 2 , lty=chainIdx , col="grey" )
219      }
220    }
221    dev.copy2eps(file=paste(fname,"Data.eps",sep=""))
222
223    } # end if nPredictors == 2
224
225    # Posterior prediction.
226    xProbe = c( 1991 , 94 ) # Note order of values: x1 is year and x2 is duration.
227    # Set up a matrix for storing the values of p(y|xProbe) at each step in chain.
228    py = matrix( 0 , nrow=chainLength , ncol=nYlevels )
229    # Step through chain and compute p(y|xProbe) and each step:
230    for ( chainIdx in 1:chainLength ) {
231        yValue = 1
232        py[chainIdx,yValue] = (
```

```
233              pnorm( threshSamp[chainIdx,yValue]
234                     - sum( bSamp[chainIdx,] * xProbe ) ) )
235         for ( yValue in 2:(nYlevels-1) ) {
236             py[chainIdx,yValue] = (
237                 pnorm( threshSamp[chainIdx,yValue]
238                         - sum( bSamp[chainIdx,] * xProbe ) )
239                 - pnorm( threshSamp[chainIdx,yValue-1]
240                         - sum( bSamp[chainIdx,] * xProbe ) ) )
241         }
242         yValue = nYlevels
243         py[chainIdx,yValue] = ( 1 -
244             pnorm( threshSamp[chainIdx,yValue-1]
245                     - sum( bSamp[chainIdx,] * xProbe ) ) )
246    }
247    # Now average across the chain:
248    pyAve = colMeans( py )
```

21.6 EXERCISES

Exercise 21.1. [Purpose: To investigate posterior differences of thresholds.] Run the program in Section 21.5 (OrdinalProbitRegressionBrugs.R) on the Movies data (Moore, 2006). At the end of the program, notice that the posterior thresholds are stored in a matrix named threshSamp, one column per threshold.

(A)   Make histograms of the differences between every pair of adjacent thresholds (i.e., threshSamp[,n]-threshSamp[,n-1], for n in 2:nYlevels). (Hint: Use plotPost with compVal=0.) Are any differences between adjacent thresholds close to zero?

(B)   Make a scatter plot of θ_4 against θ_3. (Hint: plot(threshSamp[,4], threshSamp[,3]).) Superimpose a line that marks $\theta_4 = \theta_3$. (Hint: abline(0,1).) What is the relation of this scatter plot to the histogram you made in the previous part? In particular, what is the relation of the line in the scatter plot to the histogram?

Exercise 21.2. [Purpose: To examine robustness when the prior is changed.] In the program in Section 21.5 (OrdinalProbitRegressionBrugs.R), the priors on the regression coefficients and thresholds are mildly informed. Consider the precision of the normal prior for the regression coefficients. Because the predictors are standardized, their range is approximately -2 to $+2$. If we have a prior assumption that the steepest possible relationship between x and y maps the lowest value of x to the lowest value of y and the highest value of x to the highest value of y, then the steepest regression line would rise nYlevels as x goes from -2 to $+2$. Hence the steepest plausible slope is nYlevels/4, and the program makes that value the standard deviation of the prior on each regression coefficient. Consider now the prior on the thresholds. Because the thresholds are separated in the prior by 1 unit, and the thresholds should not violate consecutive ordering, it is reasonable to set their standard deviations no larger than 1.

(A) Set the prior on the regression coefficients so that the precision is very small, say, 1.0E-6. Run the program on the Movies data. Is the posterior much different?

(B) Set the prior on the thresholds so that the precision is very small, say, 1.0E-6. (Set the prior on the regression coefficients back to the original setting.) Run the program on the Movies data. Is the posterior much different?

Exercise 21.3. [Purpose: To observe predicted values for novel predictor values.] Run the program in Section 21.5 (`OrdinalProbitRegressionBrugs.R`) on the movies data (Moore, 2006). Consider a movie that was not included in the rated movies. Suppose it was made in 1991, and had a length of 94 minutes. What is the probability of each rating for the movie? In other words, what is the probability that it would be rated 7, what is the probability that it would be rated 6, and so on. Answer the same question for a movie that had a length of 94 minutes but was made in 1931. (Hint: Use Equation 21.3, averaging the result across all the steps in the MCMC chain. Extra special bonus hint: Code for doing this is already included at the end of the program; your job is to understand what it's doing.)

Contingency Table Analysis

CONTENTS

Count me the hours that we've been together, I'll
Count you the hours I'm light as a feather, but
'Cause every hour you're all that I see, there's
No telling if there's a contingency.

Consider a situation in which we observe two nominal values about every item measured. For example, suppose there is an election, and we poll randomly selected people regarding their political party affiliation (which is a nominal variable) and their candidate preference (which is also a nominal variable). Presumably the two variables are not independent, which is to say that the proportion of people who prefer candidate A should differ from one political party to another. This is the type of situation addressed in this chapter.

Traditionally, this situation comes under the rubric of chi-square tests of independence. But we will, of course, take a Bayesian approach, and consequently

Doing Bayesian Data Analysis: A Tutorial with R and BUGS. DOI: 10.1016/B978-0-12-381485-2.00022-5
© 2011, Elsevier Inc. All rights reserved.

have no need to compute chi-square values (except for comparison with the Bayesian conclusions). There are many advantages of a Bayesian approach. As usual, a significant advantage is never having to compute a p value. Better yet is that the Bayesian analysis provides credible intervals on the conjoint probabilities and on any desired comparison of conditions.

I will call our modeling framework *Poisson exponential ANOVA* because it uses a Poisson likelihood distribution with an exponential link function from an underlying ANOVA model. This terminology is not conventional, and it might even be misleading if readers mistakenly infer from the term "ANOVA" that there is a metric predicted variable involved. Nevertheless, the terminology is highly descriptive of the structural elements of the model. The model appears in the lower right cell of Table 14.1, p. 385, where its relation to other cases of the generalized linear model is evident.

22.1 POISSON EXPONENTIAL ANOVA

22.1.1 What the Data Look Like

To motivate the model, we need first to understand the structure of the data. An example of the sort of data we'll be dealing with is shown in Table 22.1. The data come from a classroom poll of students at the University of Delaware (Snee, 1974). Respondents reported their hair color and eye color, with each variable split into four nominal levels as indicated in Table 22.1. The cells of the table indicate the frequency with which each combination occurred in the sample. Each respondent falls in one and only one cell of the table.

The data to be predicted are the cell frequencies. The predictors are the nominal variables. This situation is analogous to two-way ANOVA, which also had two nominal predictors but had several metric values in each cell instead of a single frequency.

For data like these, we can ask a number of questions. We could wonder about one variable at a time and ask questions such as "Are there more brown-eyed

Table 22.1 Frequencies of Different Combinations of Hair Color and Eye Color (Data from Snee, 1974.)

Hair Color	Eye Color			
	Blue	*Brown*	*Green*	*Hazel*
Black	20	68	5	15
Blond	94	7	16	10
Brunette	84	119	29	54
Red	17	26	14	14

people than non-brown-eyed people?" Or "Are there more brunette-haired people than non-brunette-haired people?" Those questions are analogous to main effects in ANOVA. But usually we display the conjoint frequencies specifically because we're interested in the relationship between the variables. We would like to know if the distribution of frequencies across one variable depends on, or is contingent on, the level of the other variable. For example, does the distribution of hair colors depend on eye color, and, specifically, is the proportion of blond-haired people the same for brown-eyed people as for blue-eyed people? These questions are analogous to interaction contrasts in ANOVA.

22.1.2 The Exponential Link Function

The model can be motivated two different ways. One way is simply to start with the two-way ANOVA model and find a way to map the predicted value to frequency data. The predicted μ value from the ANOVA model could be any value from negative to positive infinity, but frequencies are non-negative. Therefore, we must transform the ANOVA predictions to non-negative values, while preserving order. The natural way to do this, mathematically, is via the exponential transformation. But this transformation only gets us to an underlying continuous predicted value, not to the probability of any discrete frequency. A natural candidate for the needed likelihood distribution is the Poisson (described below), which takes a non-negative value λ and gives a probability for each integer from zero to infinity. But this motivation may seem a bit arbitrary, even if there's nothing wrong with it in principle.

A different motivation was introduced in Section 14.1.10, p. 381. We start by treating the cell frequencies as representative of underlying cell probabilities and then asking whether the two nominal variables are independent of each other. (Recall the definition of independence way back in Section 3.4.3, p. 46.) For example, in Table 22.1, there's a particular marginal probability that hair color is black and a particular marginal probability that eye color is brown. If hair color and eye color are independent, then the conjoint probability of black hair and brown eyes is the product of the marginal probabilities. The attributes of hair color and eye color are independent if that relationship holds for every cell in the table. Independence of hair color and eye color means that the proportion of black hair among brown-eyed people is the same as the proportion of black hair among blue-eyed people, and so on for all hair and eye colors.

To check for independence of attributes, we need to estimate the marginal probabilities of the attributes. Denote the marginal (i.e., total) frequency of the rth row as $f(r)$, and denote the marginal frequency of the cth column as $f(c)$. Then the estimated marginal probabilities are $f(r)/N$ and $f(c)/N$, where N is the total of the entire table. If the attributes are independent, then the *predicted* conjoint probability, $\hat{p}(r,c)$, should equal the product of the marginal probabilities, which means $\hat{p}(r,c) = p(r) \cdot p(c)$, hence $\hat{f}(r,c)/N = f(r)/N \cdot f(c)/N$. Because the models we've been dealing with involve additive combinations, not multiplicative combinations, we convert the multiplicative expression of independence into an additive expression by using the fact that $\log(a \cdot b) = \log(a) + \log(b)$, as follows:

$$\hat{f}(r,c)/N = f(r)/N \cdot f(c)/N$$

$$\hat{f}(r,c) = f(r) \cdot f(c) \cdot 1/N$$

$$\underbrace{\hat{f}(r,c)}_{\lambda_{rc}} = \exp\left(\underbrace{\log(f(r))}_{\beta_r} + \underbrace{\log(f(c))}_{\beta_c} + \underbrace{\log(1/N)}_{\beta_0} \right) \qquad (22.1)$$

If we abstract the form of Equation 22.1 away from the specific frequencies, we get the equation $\lambda_{rc} = \exp(\beta_r + \beta_c + \beta_0)$. The idea is that whatever are the values of the β's, the resulting λ's will obey multiplicative independence. An example is shown in Table 22.2. The choice of β's is shown in the margins of the table, and the resulting λ's are shown in the cells of the table. Notice that every row has the same relative probabilities, namely, 10, 100, and 1. In other words, the row and column attributes are independent.

We have dealt before with additive combinations of row and column influences, in the context of ANOVA. In ANOVA, β_0 is a baseline representing the overall central tendency, and β_r is a deflection away from baseline due to being in the rth row, and β_c is a deflection away from baseline as a result of being in the cth column. The deflections are constrained to sum to zero, and the example in Table 22.2 respects this constraint.

In ANOVA, when the cell data are not captured by an additive combination of row and column effects, we include an interaction term, denoted here as β_{rc}. The interaction terms are constrained so that every row and every column sums to zero: $\sum_r \beta_{rc} = 0 \; \forall c$ and $\sum_c \beta_{rc} = 0 \; \forall r$. The key idea is that the interaction

---

Table 22.2 Example of Exponentiated Linear Model with Zero Interaction

$\beta_0 = 4.60517$	$\beta_{c1} = 0$	$\beta_{c2} = 2.30259$	$\beta_{c2} = -2.30259$
$\beta_{r1} = 0$	100	1000	10
$\beta_{r2} = 2.30259$	1000	10000	100
$\beta_{r3} = -2.30259$	10	100	1

Margins show the values of the β's, and cells show $\lambda_{rc} = \exp(\beta_0 + \beta_r + \beta_c)$. Notice that every row has the same relative probabilities, namely, 10, 100, 1. In other words, the row and column attributes are independent.

term in the model, which indicates violations of additivity in standard ANOVA, indicates violations of multiplicative independence in exponentiated ANOVA. To estimate the magnitude of the violation of multiplicative independence, we estimate the magnitude of the interaction coefficients in exponentiated ANOVA. To summarize, the model of the cell tendencies is

$$\lambda_{rc} = \exp(\beta_0 + \beta_r + \beta_c + \beta_{rc}) \quad \text{with the constraints}$$

$$\sum_r \beta_r = 0 \ \text{ and } \ \sum_c \beta_c = 0 \ \text{ and } \ \sum_r \beta_{rc} = 0 \ \forall c \ \text{ and } \ \sum_c \beta_{rc} = 0 \ \forall r \quad (22.2)$$

If the researcher is interested in violations of independence, then the interest is on the magnitudes of the β_{rc} interaction terms. The model is especially convenient for this purpose, because arbitrary interaction contrasts can be investigated to determine where the nonindependence is arising.

22.1.3 The Poisson Likelihood

The value of λ_{rc} in Equation 22.2 is a cell tendency, not a predicted frequency per se. In particular, the value of λ_{rc} can be any non-negative real value, but frequencies can only be integers. What we need to complete the model is a likelihood function that maps the parameter value λ_{rc} to a probability of possible frequencies. The Poisson distribution is a natural choice. The Poisson distribution is named after the French mathematician Simon-Denis Poisson (1781–1840), and is defined as

$$p(y \mid \lambda) = \lambda^y \exp(-\lambda)/y! \quad (22.3)$$

where y is a non-negative integer and λ is a non-negative real number. The mean of the Poisson distribution is λ. Importantly, the variance of the Poisson distribution is also λ. Thus, in the Poisson distribution, the variance is completely yoked to the mean. Examples of the Poisson distribution are shown in Figure 22.1. Notice that the distribution is discrete, having masses only at non-negative integer values. Notice that the visual central tendency of the distribution does indeed correspond with λ. And notice that as λ increases, the width of the distribution also increases. The examples in Figure 22.1 use noninteger vales of λ to emphasize that λ is not necessarily an integer, even though y is an integer.

The Poisson distribution is often used to model discrete occurrences in time (or across space) when the probability of occurrence is the same at any moment in time (e.g., Sadiku & Tofighi, 1999). For example, suppose that customers arrive at a retail store at an *average* rate of 35 people per hour. Then the Poisson distribution, with $\lambda = 35$, is a model of the probability that any particular number of people will arrive in an hour. As another example,

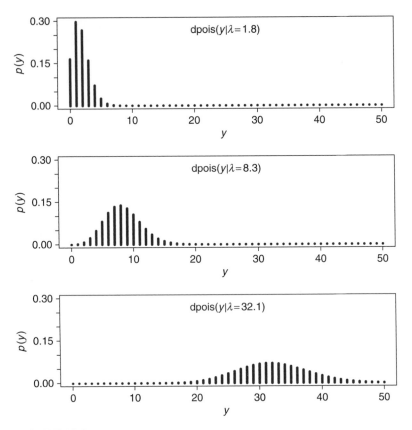

FIGURE 22.1
Examples of the Poisson distribution; y includes integers from zero to positive infinity.

suppose that 11.2% of the students at the University of Delaware in the early 1970s had black hair and brown eyes, and suppose that an average of 600 students per term will fill out a survey. That means an average of 67.2 ($= 11.2\% \cdot 600$) students per term will indicate they have black hair and brown eyes. The Poisson distribution, $p(y \mid \lambda = 67.2)$, gives the probability that any particular number of people will give that response.

We will use the Poisson distribution as the likelihood function for modeling the probability of the observed frequency, $f(r, c)$, given the mean, λ_{rc}, from Equation 22.2. The idea is that each particular r, c combination has an underlying average rate of occurrence, λ_{rc}. We collect data for a certain period of time, during which we happen to observe particular frequencies, $f(r, c)$, of each combination. From the observed frequencies, we infer the underlying average rates.

22.1.4 The Parameters and the Hierarchical Prior

The parameters of the Poisson exponential ANOVA model include all the β parameters from standard ANOVA. The prior on those parameters is the same as standard ANOVA. Figure 22.2 shows the Poisson-exponential model, with its hierarchical prior. Above the linear core, the hierarchy is identical to the two-way ANOVA model in Figure 19.2, p. 519. The diagram retains the original notation from standard ANOVA (e.g., showing $\vec{\beta}_1 \vec{x}_{1,i}$ instead of $\beta_{r,i}$), but I hope that the correspondence of notation is easy to make.

Below the linear core in Figure 22.2, the diagram simply shows the exponentiation of the ANOVA summation to specify λ and the use of the Poisson distribution to specify $p(y|\lambda)$. This lower portion of the diagram replaces both the normal likelihood of standard ANOVA and the prior on the normal distribution's precision. The Poisson, of course, has no separate precision parameter.

The model is easily implemented in R, BRugs, and BUGS, as listed in Section 22.4 (PoissonExponentialBrugs.R). The code is a straightforward modification of the program for two-factor ANOVA (that was listed

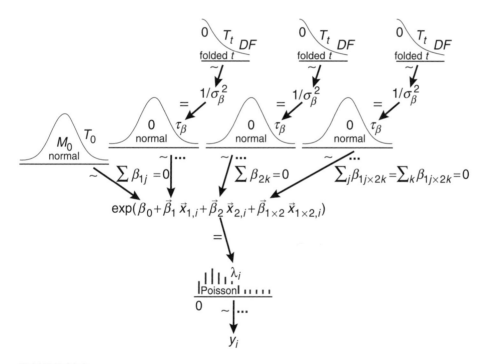

FIGURE 22.2
Hierarchical dependencies for Poisson-exponential model of a two-way frequency table. Compare with Figure 19.2, p. 519, for two-way ANOVA.

in Section 19.3.1 (ANOVAtwowayBRugs.R)). The modifications are described immediately before the listing.

22.2 EXAMPLES

Consider again the hair color and eye color data in Table 22.1. When the program in Section 22.4 (PoissonExponentialBrugs.R) is run with these data, the posterior has marginal histograms as shown in Figure 22.3. Inspection of the histograms of the individual cell interaction parameters reveals robustly nonzero interactions. For example, the interaction parameter for blue eyes and black hair is credibly below zero, which means simply that the combination of blue eyes with black hair happens less frequently than

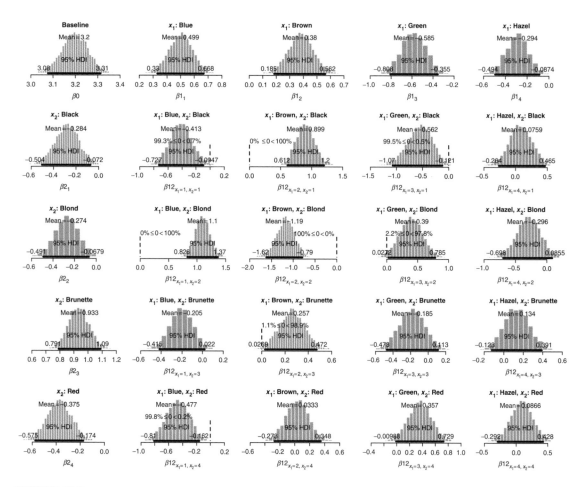

FIGURE 22.3
Posterior of Poisson exponential model applied to data of Table 22.1.

would be expected if eye color and hair color were independent. As another example, the interaction parameter for brown eyes and black hair is credibly above zero, which means simply that this combination happens more frequently than would be expected if eye color and hair color were independent. Exercise 22.2 has you consider some other data and how to interpret main effects and interaction contrasts.

Another way to investigate interaction is by examining specific interaction contrasts. Interaction contrasts can be more sensitive and revealing than single-cell interaction terms. To illustrate this point, consider the toy data in Table 22.3. The upper-left and lower-right cells have higher frequencies than the upper-right and lower-left cells. When a Bayesian analysis is conducted on these data, the resulting posterior looks like Figure 22.4. Notice that not one of the single-cell interaction coefficients excludes zero from its HDI. The bottom histogram in Figure 22.4 shows the posterior distribution of an interaction contrast, namely, $\langle 1, 1, -1, -1 \rangle / 2 \otimes \langle 1, 1, -1, -1 \rangle / 2$. This contrast takes the average of the interaction coefficients in the top-left and bottom-right cells and subtracts the average of the interaction coefficients in the top-right and bottom-left cells. The histogram of the contrast magnitude clearly excludes zero from the credible values. Therefore, we would conclude that the attributes in Table 22.3 are not independent. Exercise 22.3 has you investigate the pattern of this toy example with different total sample sizes and compare the Bayesian results with classical chi-square tests.

22.2.1 Credible Intervals on Cell Probabilities

Although the posterior distributions of the beta coefficients are useful for making inferences about differences between cells, the magnitudes of the beta coefficients do not explicitly tell us about the credible cell proportions. To get those cell proportions, we need to use Equation 22.2 to compute the

Table 22.3 Some Toy Data for Illustrating That Interaction Contrasts Can Be More Sensitive to Interaction Than Individual Cells

	Column Attribute			
Row Attribute	C1	C2	C3	C4
R1	20	20	10	10
R2	20	20	10	10
R3	10	10	20	20
R4	10	10	20	20

Posterior appears in Figure 22.4.

predicted cell mean frequencies, and then we divide by their total to get the predicted cell proportions. Thus, at every step in the MCMC chain, we compute $\lambda_{rc} = \exp(\beta_0 + \beta_r + \beta_c + \beta_{rc})$, and then we divide, again at each step in the chain, by $\sum_r \sum_c \lambda_{rc}$. The resulting posterior distribution of normalized λ_{rc}'s

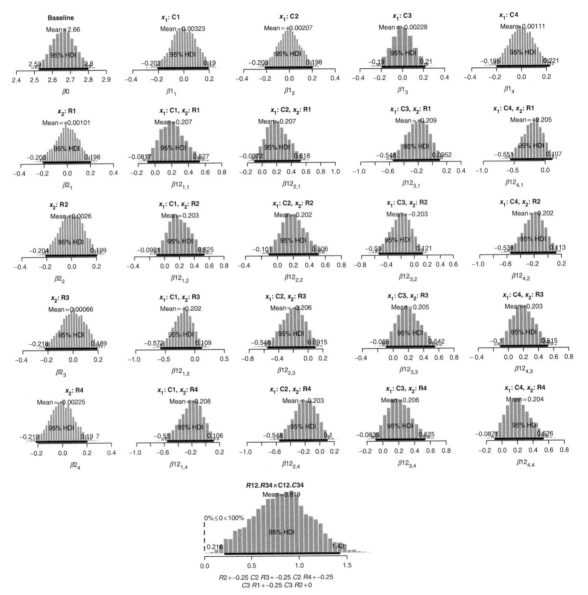

FIGURE 22.4

Posterior of Poisson exponential model applied to data of Table 22.3. The interaction contrast excludes zero, but no single interaction coefficient does. (Label of x-axis on lowest histogram is truncated due to lack of space.)

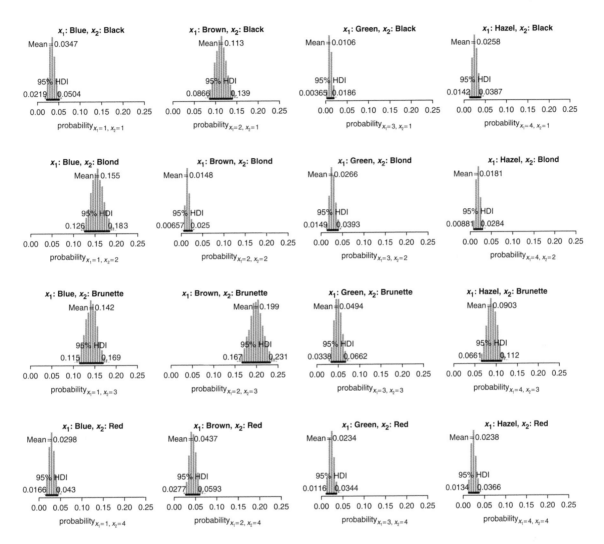

FIGURE 22.5
Posterior distribution on estimated cell probabilities for the data of Table 22.1, p. 598. See Section 22.2.1 for details.

shows the credible proportions of each combination of attributes. Figure 22.5 shows the resulting estimated cell probabilities for the hair/eye color data from Table 22.1.

22.3 LOG LINEAR MODELS FOR CONTINGENCY TABLES

This chapter only scratches the surface of methods for analyzing count data from nominal predictors. As mentioned earlier, these sorts of data are often displayed as tables, and the counts in each cell are thought of as contingent upon the level of the nominal predictor. Therefore, the data are referred to as *contingency tables*. There can be more than two predictors, and models are generalized in the same way as ANOVA is generalized to more than two predictors. The formulation presented here has emphasized linkage expressed as $\lambda = \exp(\beta_0 + \beta_r + \beta_c + \beta_{rc})$, but this equation can also be written $\log(\lambda) = \beta_0 + \beta_r + \beta_c + \beta_{rc}$. This latter form lends the usual name for these models: *log linear models for contingency tables*. This is the terminology to use when you want to explore these models more deeply. Agresti & Hitchcock (2005) provide a brief review of Bayesian log-linear models for contingency tables, but the method used in this chapter is not included, because the hierarchical ANOVA model was only popularized later (Gelman, 2005, 2006). For a description of Bayesian inference regarding contingency tables with a Poisson ANOVA model, without the Gelman-style hyperprior, see Marin & Robert (2007, pp. 109–118), and Congdon (2005, p. 134 and p. 202).

22.4 R CODE FOR THE POISSON EXPONENTIAL MODEL

This program constitutes a small adaptation of the two-way ANOVA program listed in Section 19.3.1 (ANOVAtwowayBRugs.R). The only notable changes are that (1) the normal likelihood is changed to a Poisson, (2) all references to the precision parameter of the normal likelihood are removed, (3) the *y* values are not standardized, because they must be non-negative integers, and (4) the initialization of the parameters is based on $\log(y)$, not *y*, because the parameters are exponentiated to map to *y*.

At the end of the program, some code for conducting a chi-square test is included. This is merely for comparison of Bayesian results with NHST.

(PoissonExponentialBrugs.R)

```
1   graphics.off()
2   rm(list=ls(all=TRUE))
3   fnroot = "PoissonExponentialBrugs"
4   library(BRugs)               # Kruschke, J. K. (2010). Doing Bayesian data analysis:
5                                # A Tutorial with R and BUGS. Academic Press / Elsevier.
6   #-----------------------------------------------------------------------------
7   # THE MODEL.
8
9   modelstring = "
10  # BUGS model specification begins here...
11  model {
12    for ( i in 1:Ncells ) {
13      y[i] ~ dpois( lambda[i] )
14      lambda[i] <- exp( a0 + a1[x1[i]] + a2[x2[i]] + a1a2[x1[i],x2[i]] )
15    }
```

```
16    #
17    a0 ~ dnorm(10,1.0E-6)
18    #
19    for ( j1 in 1:Nx1Lvl ) { a1[j1] ~ dnorm( 0.0 , a1tau ) }
20    a1tau <- 1 / pow( a1SD , 2 )
21    a1SD <- abs( a1SDunabs ) + .1
22    a1SDunabs ~ dt( 0 , 0.001 , 2 )
23    #
24    for ( j2 in 1:Nx2Lvl ) { a2[j2] ~ dnorm( 0.0 , a2tau ) }
25    a2tau <- 1 / pow( a2SD , 2 )
26    a2SD <- abs( a2SDunabs ) + .1
27    a2SDunabs ~ dt( 0 , 0.001 , 2 )
28    #
29    for ( j1 in 1:Nx1Lvl ) { for ( j2 in 1:Nx2Lvl ) {
30      a1a2[j1,j2] ~ dnorm( 0.0 , a1a2tau )
31    } }
32    a1a2tau <- 1 / pow( a1a2SD , 2 )
33    a1a2SD <- abs( a1a2SDunabs ) + .1
34    a1a2SDunabs ~ dt( 0 , 0.001 , 2 )
35    }
36    # ... end BUGS model specification
37    " # close quote for modelstring
38    # Write model to a file, and send to BUGS:
39    writeLines(modelstring,con="model.txt")
40    modelCheck( "model.txt" )
41
42    #-------------------------------------------------------------------
43    # THE DATA.
44    # Specify data source:
45    dataSource = c( "HairEye" , "CrimeDrink" , "Toy" )[1]
46
47    # Load the data:
48    if ( dataSource == "HairEye" ) {
49      fnroot = paste( fnroot , dataSource , sep="" )
50      dataFrame = data.frame( # from Snee (1974)
51        Freq = c(68,119,26,7,20,84,17,94,15,54,14,10,5,29,14,16) ,
52        Eye  = c("Brown","Brown","Brown","Brown","Blue","Blue","Blue","Blue","Hazel",
53          "Hazel","Hazel","Hazel","Green","Green","Green","Green"),
54        Hair = c("Black","Brunette","Red","Blond","Black","Brunette","Red","Blond","Black",
55          "Brunette","Red","Blond","Black","Brunette","Red","Blond") )
56      y = as.numeric(dataFrame$Freq)
57      x1 = as.numeric(dataFrame$Eye)
58      x1names = levels(dataFrame$Eye)
59      x2 = as.numeric(dataFrame$Hair)
60      x2names = levels(dataFrame$Hair)
61      Ncells = length(y)
62      Nx1Lvl = length(unique(x1))
63      Nx2Lvl = length(unique(x2))
64      x1contrastList = list( GREENvHAZEL = c(0,0,1,-1) )
65      x2contrastList = list( BLONDvRED = c(0,1,0,-1) )
66      x1x2contrastList = list( BLUE.BROWNxBLACK.BLOND
67                          = outer(c(-1,1,0,0),c(-1,1,0,0)) )
68    }
69
```

```
70    if ( dataSource == "CrimeDrink" ) {
71      fnroot = paste( fnroot , dataSource , sep="" )
72      dataFrame = data.frame( # from Kendall (1943) via Snee (1974)
73        Freq = c(50,88,155,379,18,63,43,62,110,300,14,144) ,
74        Crime = c("Arson","Rape","Violence","Theft","Coining","Fraud","Arson","Rape",
75          "Violence","Theft","Coining","Fraud"),
76        Drink = c("Drinker","Drinker","Drinker","Drinker","Drinker","Drinker","Nondrink",
77          "Nondrink","Nondrink","Nondrink","Nondrink","Nondrink") )
78      y = as.numeric(dataFrame$Freq)
79      x1 = as.numeric(dataFrame$Crime)
80      x1names = levels(dataFrame$Crime)
81      x2 = as.numeric(dataFrame$Drink)
82      x2names = levels(dataFrame$Drink)
83      Ncells = length(y)
84      Nx1Lvl = length(unique(x1))
85      Nx2Lvl = length(unique(x2))
86      x1contrastList = list( FRAUDvOTHER = c(-1/5,-1/5,1,-1/5,-1/5,-1/5) ,
87                             FRAUDvRAPE = c(0,0,1,-1,0,0) )
88      x2contrastList = list( DRINKERvNON = c(1,-1) )
89      x1x2contrastList = list( FRAUD.OTHERxDRINKER.NON
90                               = outer(c(-1/5,-1/5,1,-1/5,-1/5,-1/5),c(-1,1)) ,
91                               FRAUD.RAPExDRINKER.NON
92                               = outer(c(0,0,1,-1,0,0),c(-1,1)) )
93    }
94
95    if ( dataSource == "Toy" ) {
96      dataMultiplier = 2 # Try 2 (chi-sq warns) , 6 (p>.05) , 7 (p<.05) , 10
97      fnroot = paste( fnroot , dataSource , dataMultiplier , sep="" )
98      dataFrame = data.frame(
99        Freq = c( 2,2,1,1, 2,2,1,1, 1,1,2,2, 1,1,2,2 ) * dataMultiplier ,
100       Col = c("C1","C2","C3","C4", "C1","C2","C3","C4", "C1","C2","C3","C4", "C1","C2",
101         "C3","C4"),
102       Row = c("R1","R1","R1","R1", "R2","R2","R2","R2", "R3","R3","R3","R3", "R4","R4",
103         "R4","R4") )
104     y = as.numeric(dataFrame$Freq)
105     x1 = as.numeric(dataFrame$Col)
106     x1names = levels(dataFrame$Col)
107     x2 = as.numeric(dataFrame$Row)
108     x2names = levels(dataFrame$Row)
109     Ncells = length(y)
110     Nx1Lvl = length(unique(x1))
111     Nx2Lvl = length(unique(x2))
112     x1contrastList = NULL
113     x2contrastList = NULL
114     x1x2contrastList = list( R12.R34xC12.C34 = outer(c(1,1,-1,-1)/2,c(1,1,-1,-1)/2) )
115   }
116
117   # Specify the data in a form that is compatible with BRugs model, as a list:
118   datalist = list(
119     y = y ,
120     x1 = x1 ,
121     x2 = x2 ,
122     Ncells = Ncells ,
123     Nx1Lvl = Nx1Lvl ,
```

```
124      Nx2Lvl = Nx2Lvl
125    )
126    # Get the data into BRugs:
127    modelData( bugsData( datalist ) )
128
129    #-----------------------------------------------------------------
130    # INTIALIZE THE CHAINS.
131
132    nchain = 5
133    modelCompile( numChains = nchain )
134
135    if ( F ) {
136      modelGenInits() # often won't work for diffuse prior
137    } else {
138      #  initialization based on data
139      theData = data.frame( y=log(y) , x1=factor(x1,labels=x1names) ,
140                                       x2=factor(x2,labels=x2names) )
141      a0 = mean( theData$y )
142      a1 = aggregate( theData$y , list( theData$x1 ) , mean )[,2] - a0
143      a2 = aggregate( theData$y , list( theData$x2 ) , mean )[,2] - a0
144      linpred = as.vector( outer( a1 , a2 , "+" ) + a0 )
145      a1a2 = aggregate( theData$y, list(theData$x1,theData$x2), mean)[,3] - linpred
146      genInitList <- function() {
147        return(
148            list(
149                a0 = a0 ,
150                a1 = a1 ,
151                a2 = a2 ,
152                a1a2 = matrix( a1a2 , nrow=Nx1Lvl , ncol=Nx2Lvl ) ,
153                a1SDunabs = sd(a1) ,
154                a2SDunabs = sd(a2) ,
155                a1a2SDunabs = sd(a1a2)
156            )
157        )
158      }
159      for ( chainIdx in 1 : nchain ) {
160        modelInits( bugsInits( genInitList ) )
161      }
162    }
163
164    #-----------------------------------------------------------------
165    # RUN THE CHAINS
166
167    # burn in
168    BurnInSteps = 1000
169    modelUpdate( BurnInSteps )
170    # actual samples
171    samplesSet( c( "a0" , "a1" , "a2" , "a1a2" , "a1SD" , "a2SD" , "a1a2SD" ) )
172    stepsPerChain = ceiling(5000/nchain)
173    thinStep = 500
174    modelUpdate( stepsPerChain , thin=thinStep )
175
176    #-----------------------------------------------------------------
177    # EXAMINE THE RESULTS
```

```
178
179   source("plotChains.R")
180   source("plotPost.R")
181
182   checkConvergence = F
183   if ( checkConvergence ) {
184      sumInfo = plotChains( "a0" , saveplots=F , filenameroot=fnroot )
185      sumInfo = plotChains( "a1" , saveplots=F , filenameroot=fnroot )
186      sumInfo = plotChains( "a2" , saveplots=F , filenameroot=fnroot )
187      sumInfo = plotChains( "a1a2" , saveplots=F , filenameroot=fnroot )
188      readline("Press any key to clear graphics and continue")
189      graphics.off()
190      sumInfo = plotChains( "a1SD" , saveplots=F , filenameroot=fnroot )
191      sumInfo = plotChains( "a2SD" , saveplots=F , filenameroot=fnroot )
192      sumInfo = plotChains( "a1a2SD" , saveplots=F , filenameroot=fnroot )
193      readline("Press any key to clear graphics and continue")
194      graphics.off()
195   }
196
197   # Extract and plot the SDs:
198   a1SDSample = samplesSample("a1SD")
199   a2SDSample = samplesSample("a2SD")
200   a1a2SDSample = samplesSample("a1a2SD")
201   windows(10,3)
202   layout( matrix(1:3,nrow=1) )
203   par( mar=c(3,1,2.5,0) , mgp=c(2,0.7,0) )
204   histInfo = plotPost( a1SDSample , xlab="a1SD" , main="a1 SD" , breaks=30 )
205   histInfo = plotPost( a2SDSample , xlab="a2SD" , main="a2 SD" , breaks=30 )
206   histInfo = plotPost( a1a2SDSample , xlab="a1a2SD" , main="Interaction SD" ,
207                        breaks=30 )
208   dev.copy2eps(file=paste(fnroot,"SD.eps",sep=""))
209
210   # Extract a values:
211   a0Sample = samplesSample( "a0" )
212   chainLength = length(a0Sample)
213   a1Sample = array( 0 , dim=c( datalist$Nx1Lvl , chainLength ) )
214   for ( x1idx in 1:datalist$Nx1Lvl ) {
215      a1Sample[x1idx,] = samplesSample( paste("a1[",x1idx,"]",sep="") )
216   }
217   a2Sample = array( 0 , dim=c( datalist$Nx2Lvl , chainLength ) )
218   for ( x2idx in 1:datalist$Nx2Lvl ) {
219      a2Sample[x2idx,] = samplesSample( paste("a2[",x2idx,"]",sep="") )
220   }
221   a1a2Sample = array(0, dim=c( datalist$Nx1Lvl , datalist$Nx2Lvl , chainLength ) )
222   for ( x1idx in 1:datalist$Nx1Lvl ) {
223     for ( x2idx in 1:datalist$Nx2Lvl ) {
224       a1a2Sample[x1idx,x2idx,] = samplesSample( paste( "a1a2[",x1idx,",",x2idx,"]",
225                                                         sep="" ) )
226     }
227   }
228
229   # Convert to zero-centered b values:
230   m12Sample = array( 0, dim=c( datalist$Nx1Lvl , datalist$Nx2Lvl , chainLength ) )
231   for ( stepIdx in 1:chainLength ) {
```

```
232    m12Sample[,,stepIdx ] = ( a0Sample[stepIdx]
233                        + outer( a1Sample[,stepIdx] ,
234                                 a2Sample[,stepIdx] , "+" )
235                        + a1a2Sample[,,stepIdx] )
236    }
237    b0Sample = apply( m12Sample , 3 , mean )
238    b1Sample = ( apply( m12Sample , c(1,3) , mean )
239               - matrix(rep( b0Sample ,Nx1Lvl),nrow=Nx1Lvl,byrow=T) )
240    b2Sample = ( apply( m12Sample , c(2,3) , mean )
241               - matrix(rep( b0Sample ,Nx2Lvl),nrow=Nx2Lvl,byrow=T) )
242    linpredSample = array(0,dim=c(datalist$Nx1Lvl,datalist$Nx2Lvl,chainLength))
243    for ( stepIdx in 1:chainLength ) {
244        linpredSample[,,stepIdx ] = ( b0Sample[stepIdx]
245                          + outer( b1Sample[,stepIdx] ,
246                                   b2Sample[,stepIdx] , "+" ) )
247    }
248    b1b2Sample = m12Sample - linpredSample
249
250    # Plot b values:
251    windows((datalist$Nx1Lvl+1)*2.75,(datalist$Nx2Lvl+1)*2.25)
252    layoutMat = matrix( 0 , nrow=(datalist$Nx2Lvl+1) , ncol=(datalist$Nx1Lvl+1) )
253    layoutMat[1,1] = 1
254    layoutMat[1,2:(datalist$Nx1Lvl+1)] = 1:datalist$Nx1Lvl + 1
255    layoutMat[2:(datalist$Nx2Lvl+1),1] = 1:datalist$Nx2Lvl + (datalist$Nx1Lvl + 1)
256    layoutMat[2:(datalist$Nx2Lvl+1),2:(datalist$Nx1Lvl+1)] = matrix(
257        1:(datalist$Nx1Lvl*datalist$Nx2Lvl) + (datalist$Nx2Lvl+datalist$Nx1Lvl+1) ,
258        ncol=datalist$Nx1Lvl , byrow=T )
259    layout( layoutMat )
260    par( mar=c(4,0.5,2.5,0.5) , mgp=c(2,0.7,0) )
261    histinfo = plotPost( b0Sample , xlab=expression(beta * 0) , main="Baseline" ,
262                         breaks=30  )
263    for ( x1idx in 1:datalist$Nx1Lvl ) {
264      histinfo = plotPost( b1Sample[x1idx,] , xlab=bquote(beta*1[.(x1idx)]) ,
265                           main=paste("x1:",x1names[x1idx]) , breaks=30 )
266    }
267    for ( x2idx in 1:datalist$Nx2Lvl ) {
268      histinfo = plotPost( b2Sample[x2idx,] , xlab=bquote(beta*2[.(x2idx)]) ,
269                           main=paste("x2:",x2names[x2idx]) , breaks=30 )
270    }
271    for ( x2idx in 1:datalist$Nx2Lvl ) {
272      for ( x1idx in 1:datalist$Nx1Lvl ) {
273        hdiLim = HDIofMCMC(b1b2Sample[x1idx,x2idx,])
274        if ( hdiLim[1]>0 | hdiLim[2]<0 ) { compVal=0 } else { compVal=NULL }
275        histinfo = plotPost( b1b2Sample[x1idx,x2idx,] , breaks=30 , compVal=compVal ,
276                   xlab=bquote(beta*12[list(x1==.(x1idx),x2==.(x2idx))]) ,
277                   main=paste("x1:",x1names[x1idx],", x2:",x2names[x2idx])  )
278      }
279    }
280    dev.copy2eps(file=paste(fnroot,"b.eps",sep=""))
281
282    # Display contrast analyses
283    nContrasts = length( x1contrastList )
284    if ( nContrasts > 0 ) {
285      nPlotPerRow = 5
```

```
286      nPlotRow = ceiling(nContrasts/nPlotPerRow)
287      nPlotCol = ceiling(nContrasts/nPlotRow)
288      windows(3.75*nPlotCol,2.5*nPlotRow)
289      layout( matrix(1:(nPlotRow*nPlotCol),nrow=nPlotRow,ncol=nPlotCol,byrow=T) )
290      par( mar=c(4,0.5,2.5,0.5) , mgp=c(2,0.7,0) )
291      for ( cIdx in 1:nContrasts ) {
292          contrast = matrix( x1contrastList[[cIdx]],nrow=1) # make it a row matrix
293          incIdx = contrast!=0
294          histInfo = plotPost( contrast %*% b1Sample , compVal=0 , breaks=30 ,
295                      xlab=paste( round(contrast[incIdx],2) , x1names[incIdx] ,
296                          c(rep("+",sum(incIdx)-1),"") , collapse=" " ) ,
297                      cex.lab = 1.0 ,
298                      main=paste( "X1 Contrast:", names(x1contrastList)[cIdx] ) )
299      }
300      dev.copy2eps(file=paste(fnroot,"x1Contrasts.eps",sep=""))
301  }
302  #
303  nContrasts = length( x2contrastList )
304  if ( nContrasts > 0 ) {
305      nPlotPerRow = 5
306      nPlotRow = ceiling(nContrasts/nPlotPerRow)
307      nPlotCol = ceiling(nContrasts/nPlotRow)
308      windows(3.75*nPlotCol,2.5*nPlotRow)
309      layout( matrix(1:(nPlotRow*nPlotCol),nrow=nPlotRow,ncol=nPlotCol,byrow=T) )
310      par( mar=c(4,0.5,2.5,0.5) , mgp=c(2,0.7,0) )
311      for ( cIdx in 1:nContrasts ) {
312          contrast = matrix( x2contrastList[[cIdx]],nrow=1) # make it a row matrix
313          incIdx = contrast!=0
314          histInfo = plotPost( contrast %*% b2Sample , compVal=0 , breaks=30 ,
315                      xlab=paste( round(contrast[incIdx],2) , x2names[incIdx] ,
316                          c(rep("+",sum(incIdx)-1),"") , collapse=" " ) ,
317                      cex.lab = 1.0 ,
318                      main=paste( "X2 Contrast:", names(x2contrastList)[cIdx] ) )
319      }
320      dev.copy2eps(file=paste(fnroot,"x2Contrasts.eps",sep=""))
321  }
322  #
323  nContrasts = length( x1x2contrastList )
324  if ( nContrasts > 0 ) {
325      nPlotPerRow = 5
326      nPlotRow = ceiling(nContrasts/nPlotPerRow)
327      nPlotCol = ceiling(nContrasts/nPlotRow)
328      windows(3.75*nPlotCol,2.5*nPlotRow)
329      layout( matrix(1:(nPlotRow*nPlotCol),nrow=nPlotRow,ncol=nPlotCol,byrow=T) )
330      par( mar=c(4,0.5,2.5,0.5) , mgp=c(2,0.7,0) )
331      for ( cIdx in 1:nContrasts ) {
332          contrast = x1x2contrastList[[cIdx]]
333          contrastArr = array( rep(contrast,chainLength) ,
334                          dim=c(NROW(contrast),NCOL(contrast),chainLength) )
335          contrastLab = ""
336          for ( x1idx in 1:Nx1Lvl ) {
337            for ( x2idx in 1:Nx2Lvl ) {
338              if ( contrast[x1idx,x2idx] != 0 ) {
339                contrastLab = paste( contrastLab , "+" ,
```

```
340                                        signif(contrast[x1idx,x2idx],2) ,
341                          x1names[x1idx] , x2names[x2idx] )
342              }
343            }
344          }
345        histInfo = plotPost( apply( contrastArr * b1b2Sample , 3 , sum ) ,
346                   compVal=0 , breaks=30 , xlab=contrastLab , cex.lab = 0.75 ,
347                   main=paste( names(x1x2contrastList)[cIdx] ) )
348      }
349      dev.copy2eps(file=paste(fnroot,"x1x2Contrasts.eps",sep=""))
350  }
351
352  # Compute credible cell probability at each step in the MCMC chain
353  lambda12Sample = 0 * b1b2Sample
354  for ( chainIdx in 1:chainLength ) {
355      lambda12Sample[,,chainIdx] = exp(
356          b0Sample[chainIdx]
357          + outer( b1Sample[,chainIdx] , b2Sample[,chainIdx] , "+" )
358          + b1b2Sample[,,chainIdx] )
359  }
360  cellp = 0 * lambda12Sample
361  for ( chainIdx in 1:chainLength ) {
362      cellp[,,chainIdx] = ( lambda12Sample[,,chainIdx]
363                           / sum( lambda12Sample[,,chainIdx] ) )
364  }
365  # Display credible cell probabilities
366  windows((datalist$Nx1Lvl)*2.75,(datalist$Nx2Lvl)*2.25)
367  layoutMat = matrix( 1:(datalist$Nx2Lvl*datalist$Nx1Lvl) ,
368                     nrow=(datalist$Nx2Lvl) , ncol=(datalist$Nx1Lvl) , byrow=T )
369  layout( layoutMat )
370  par( mar=c(4,1.5,2.5,0.5) , mgp=c(2,0.7,0) )
371  maxp = max( cellp )
372  for ( x2idx in 1:datalist$Nx2Lvl ) {
373    for ( x1idx in 1:datalist$Nx1Lvl ) {
374      histinfo = plotPost( cellp[x1idx,x2idx,] ,
375                 breaks=seq(0,maxp,length=51) , xlim=c(0,maxp) ,
376                 xlab=bquote(probability[list(x1==.(x1idx),x2==.(x2idx))]) ,
377                 main=paste("x1:",x1names[x1idx],", x2:",x2names[x2idx]) ,
378                 HDItextPlace=0.95 )
379    }
380  }
381  dev.copy2eps(file=paste(fnroot,"CellP.eps",sep=""))
382
383
384  #===============================================================================
385  # Conduct NHST Pearson chi-square test of independence.
386
387  # Convert dataFrame to frequency table:
388  obsFreq = matrix( 0 , nrow=Nx1Lvl , ncol=Nx2Lvl )
389  for ( x1idx in 1:Nx1Lvl ) {
390    for ( x2idx in 1:Nx2Lvl ) {
391      obsFreq[x1idx,x2idx] = y[ dataFrame[,2]==x1names[x1idx]
392                               & dataFrame[,3]==x2names[x2idx] ]
393    }
```

```
394    }
395    obsFreq = t(obsFreq) # merely to match orientation of histogram display
396    chisqtest = chisq.test( obsFreq )
397    print( "obs :" )
398    print( chisqtest$observed )
399    print( "( obs - exp )^2 / exp :" )
400    print( ( chisqtest$observed - chisqtest$expected )^2 / chisqtest$expected )
401    print( chisqtest )
402
403    #============================================================================
```

22.5 EXERCISES

Exercise 22.1. [Purpose: To observe the influence of sample size on the precision of the estimate.] Consider the data regarding hair color and eye color in Table 22.1, which is available for use in the code of Section 22.4 (PoissonExponentialBrugs.R).

(A) Divide the original data frequencies by 2 and take the ceiling() of the result. Thus, the modified data are ceiling(c(68,...,16) / 2). This modification (nearly) preserves the relative proportions in each cell but halves the sample size. Run the program with these modified data, show the histograms, and compute the width of the 95% HDIs for the contrasts.

(B) Multiply the original data vector by 5 so that all the cell frequencies are five times larger than the original. This modification preserves the relative proportions in each cell but quintuples the sample size. Run the program again. Show the histograms. Compute the width of the 95% HDIs for the contrasts. How do the precisions of the contrasts differ for the reduced and the enlarged data?

Exercise 22.2. [Purpose: To explore main effects and interactions in a contingency table.] The data section of the program in Section 22.4 (PoissonExponentialBrugs.R) includes data regarding the type of crime committed by convicted criminals and whether or not the criminal is a regular drinker of alcoholic beverages (Snee, 1974).

(A) Run the program with these data selected. Is there a credible noninde-pendence of the attributes? Describe the interaction in terms of the actual levels of the attributes (i.e., type of crime and drinker or nondrinker).

(B) Is there a ("simple") main effect of fraud versus rape? That is, marginal-izing across drinking, is there a credible difference in the proportion of criminals who committed fraud or committed rape? Show the results of an appropriate contrast.

(C) Is the difference from the previous part the same among drinkers and nondrinkers? In other words, is the effect of criminal type the same at all levels of drinking? Show the results of an appropriate interaction contrast.

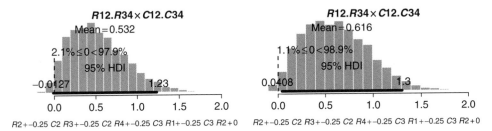

FIGURE 22.6

For Exercise 22.3. Left histogram shows posterior interaction contrast for the "toy" data when `dataMultiplier = 6`. A chi-square test (not shown) indicates that $p > 0.05$. Right histogram shows posterior interaction contrast for the "toy" data when `dataMultiplier = 7`. A chi-square test (not shown) indicates that $p < 0.05$. (The x-axis labels are truncated because of insufficient space.)

Exercise 22.3. [Purpose: To compare Bayesian analysis with chi-square test of independence.] This exercise assumes that you have some familiarity with traditional chi-square tests of independence. In the code of Section 22.4 (`PoissonExponentialBrugs.R`), the data section includes some simple fictitious ("toy") data, which we can manipulate and then compare with results of Bayesian and chi-square analyses.

(A) Set the `dataMultiplier` to 6, and run the program. Include the histograms of the β parameters and of the interaction contrast. Does the 95% HDI of the contrast include zero? The end of the program runs a chi-square test. What is the p value from the test? (Hint: See Figure 22.6.)

(B) Set the `dataMultiplier` to 7, and run the program. Include the histograms of the β parameters and of the interaction contrast. Does the 95% HDI of the contrast include zero? The end of the program runs a chi-square test. What is the p value from the test? (Hint: See Figure 22.6.)

(C) Typical chi-square tests rely on approximating the sampling distribution of *discrete* Pearson chi-square values (i.e., $\sum_r \sum_c (f_{rc} - \hat{f}_{rc})^2 / \hat{f}_{rc}$) with the *continuous* chi-square distribution (which derives from the sum of standardized normal samples). When the expected frequencies, $\hat{f}_{rc}$, are too small, then the approximation is not good and the estimated p value may be wrong. A usual heuristic for declaring the chi-square test to be suspect is when (at least 10% of the) expected frequencies, $\hat{f}_{rc}$, are less than 5. This is why computer packages issue warnings when expected values are too small. Bayesian analysis has no such problems. Set the `dataMultiplier` to 2, and run the program. Does the Bayesian analysis complain or do anything wrong? (No.) The end of the program runs a chi-square test. Is there a warning message? (Yes; report what it is.)

Tools in the Trunk

CONTENTS

She changes her hair, and he changes his style,
She paints on her face, and he wears a fake smile,
She shrink wraps her head, and he stretches the truth;
But they'll always be stuck with their done wasted youth.

This chapter includes a few important topics that can apply to many different models throughout the book. The first topic is how to report a Bayesian analysis in a scientific journal. The second topic is the details behind computing an HDI.

The third issue is something that has not come up explicitly very often in the book but lurks in the shadows all the time. This is the topic of reparameterization. For example, the beta distribution has shape parameters a and b, but

Doing Bayesian Data Analysis: A Tutorial with R and BUGS. DOI: 10.1016/B978-0-12-381485-2.00023-7
© 2011, Elsevier Inc. All rights reserved.

we regularly transformed them into parameters μ and κ. The complication of reparameterization is that a probability distribution on a parameter is not the same on the transformed parameter. In particular, if we specify a prior on a parameter but then transform the parameter, the implied prior on the transformed parameter is different. (The quatrain at the beginning of the chapter has reparameterization in mind.)

23.1 REPORTING A BAYESIAN ANALYSIS

Bayesian data analyses are not yet standard procedure in many fields of research, and no conventional format for reporting them has been established. Therefore, the researcher who reports a Bayesian analysis must be sensitive to the background knowledge of his or her specific audience and must frame the description accordingly. I once assigned an exercise to students in which they had to write up the results of a Bayesian analysis as it would appear in a research journal. Because I am a psychologist, a student then earnestly asked me, "Do we have to write it as if it were for a *psychology* journal or for a *science* journal?" After swallowing my feeling of injury at the implication that psychology is not science, I politely asked the student to clarify the question. The student said, "For a psychology journal you have to explain what you did, but for a science journal the reader has to puzzle it out."

23.1.1 Essential Points

When reporting a Bayesian analysis, the writer must address the following essential points:

- *Motivate the use of Bayesian (non-NHST) analysis.* Many audiences, including journal editors and reviewers, are used to NHST and unfamiliar with Bayesian methods. You may motivate your use of Bayesian data analysis on several grounds, depending in part on the particular application. For example, Bayesian models are designed to be appropriate to the data structure, without having to make approximation assumptions typical in NHST. The inferences from a Bayesian analysis are richer and more informative than NHST. And, of course, there is no reliance on p values.
- *Clearly describe the model and its parameters.* Because the posterior distribution is a distribution over parameter values, the parameters must be clearly defined. This task of describing the model can be arduous for complex hierarchical models, but it is necessary and crucial if your analysis is to mean anything to your audience. Because the model refers to the data and their structure, the model description requires that the structure of the data has already been explained.

■ *Clearly describe and justify the prior.* It is crucial to convince your audience that your prior is appropriate and does not predetermine the outcome. There is no escape from this requirement, even when using "objective" or "uninformed" priors, because there is no unique choice of such priors, and even they can have major consequences for model comparison. The prior should be amenable to a skeptical audience. The prior should be at least mildly informed to match the scale of the data being modeled. If there is copious previous research using similar methods, it should not be ignored. Optionally, as mentioned again later, it may be helpful to try different priors and report the robustness of the posterior.

■ *Mention the MCMC details, especially evidence that the chains were converged (plenty of burn-in and no orphaned chains) and not clumpy (low autocorrelation from sufficient thinning).* Also state how many points there are in the final MCMC sample. Usually this section of the report can be brief, if your audience believes that you know what you're doing and therefore took adequate caution to generate a trustworthy and representative sample from the posterior. See Section 23.2 for a reminder of how to do this.

■ *Interpret the posterior.* Many models have dozens or even hundreds of parameters; therefore it is impossible to summarize all of them. The choice of which parameters or contrasts to report is driven by domain-specific theory and by the results themselves. You want to report the parameters and contrasts that are theoretically meaningful and those that showed effects driven by data, whether expected or not. Reporting of HDIs can be done in text alone, to save space. (Histograms of posteriors are useful for explanation in a textbook but may be unnecessary in a concise report.) Include scatter plots of correlated important parameters if the scatter is not bivariate Gaussian. Be sure to describe effects of shrinkage if appropriate. If your model includes interactions of predictors, be careful how you interpret lower-order effects. Finally, if you are using a ROPE for posterior interpretation, justify its limits.

23.1.2 Optional Points

The following points are not necessarily crucial to address in every report, but they certainly should be considered. Whether to include these points depends on the particulars of the application domain, the points the reporter wants to make, and the audience to which the report is being addressed:

■ *Robustness of the posterior for different priors.* When there is real or imagined contention about the prior, it can be most convincing simply to conduct the analysis with different priors and demonstrate that the essential conclusions from the posterior do not change. Which priors should be used? Those that are amenable to your audience, such as reviewers and editors of the journal to which the report is submitted. This uncertainty in the choice of priors may seem unappealing, but in fact it accurately reflects the way science gets done, by incorporating previously conducted research and addressing currently active researchers.

- *Posterior predictive check.* By generating simulated data from the credible parameter values of the model and examining the qualities of the simulated data, the veracity of the model can be further bolstered, if the simulated data do resemble the actual data. On the other hand, if the simulated data are discrepant from the actual data in systematic and interpretable ways, then the posterior predictive check can inspire new research and new models.

- *Power analysis.* For example, if there is only a weak effect in your results, what sample size would be needed to achieve some desired precision in the estimate? If you found a credible difference, what was the retrospective power of your experiment and what is its replication power? This sort of information can be useful not only for the researcher's own planning, but it can also be useful to the audience of the report to anticipate potential follow-up research and to assess the robustness of the currently reported results.

23.1.3 Helpful Points

Finally, to help science be cumulative, make your results available on the web:

- *Post the raw data.* There are two benefits of posting the original data. One benefit is that subsequent researchers can analyze the data with different models. New insights can be gained by alternative modeling interpretations. The longevity of the original research is enhanced. A second benefit is that if an exact or near-exact replication is conducted, the original data set can be concatenated with the new data set to enhance sensitivity of the new data.

- *Post the MCMC sample of the posterior.* There are two benefits of making the posterior publicly available. One is that other researchers can explore the posterior for effects and comparisons that were not in the report. Complex models have many parameters, and no single report can cover every possible perspective on the posterior distribution. The longevity and impact of the research is thereby enhanced. A second benefit is that if subsequent researchers do follow-up research with a similar design and model, then the posted posterior can inform the prior of the subsequent analysis. Because the full posterior automatically incorporates all the covariations between all the parameters, the full posterior can be more useful than summaries of marginal distributions in a report.

23.2 MCMC BURN-IN AND THINNING

When using a Markov chain to generate a Monte Carlo sample from a distribution, we want to be sure that the resulting chain is a truly *representative* sample from the distribution. There are several ways in which the chain might fail to be representative.

One problem is that the chains might start far from the true mode(s) of the distribution, and therefore the early steps in the chain might be unrepresentative of the distribution. The main way to mitigate this problem is to start the chains intelligently, instead of randomly, if you can. If you know a point that is likely to be in the midst of the distribution, try starting the chains there. Many of the programs in this book use this method. For example, in the multiple linear regression program of Section 17.5.1 (`MultipleLinearRegressionBrugs.R`), the chains were started at the maximum likelihood estimate, because it was assumed that the posterior will be dominated by the data, not by the prior. Although a good starting point reduces the time it takes for the chains to find the bulk of the distribution, the early steps are still not necessarily representative because we still need to let the chains run a while to dilute the influence of the initial value; see Figure 7.2, p. 123. This initial set of steps is called the *burn-in* period.

Another problem is that the chains might not change value much from step to step and therefore take a long time to generate values from the entire breadth of the distribution. When consecutive steps have similar values, the chain is said to be highly *autocorrelated*. The problem is not merely taking a long time to generate the full range of the distribution. The main problem is that an autocorrelated chain is "clumpy": It over-represents some values and under-represents other values. The primary way to mitigate this problem is to *thin* the chain: Instead of using every step in the chain, we only use every m^{th} step, where m could be 50 or 500 or larger, depending on the model and data.

Autocorrelation is measured simply as the correlation of the chain values with the values L steps behind (or ahead), where L is called the *lag*. Expressed in terms of R code, if the chain is the vector v and it has W components, then the autocorrelation at lag L is ACF(L) = cor(v[1:(W-L)] , v[(L+1):W]). At $L = 0$, the ACF(0) is 1.0, of course. If consecutive steps in the chain are similar, then ACF(1) will be close to 1.0, but if consecutive steps in the chain are uncorrelated, then ACF(1) will be close to zero.

Another problem is that it is possible for chains to get stuck in unrepresentative regions of parameter space if they are initialized poorly or take a bad jump. As a check that the chains are actually exploring the bulk of the distribution and are not stuck in some unrepresentative zone, we run several chains at once and examine whether they are well mixed, without any outlying chain.

One measure of mixing considers the variability between chains relative to the variability within chains. If the chains are well mixed, then this ratio should be about 1 (i.e., there should be no more variability between chains than within chains). But if the chains are not well mixed and one or more chains is lingering far from the others, then the variance between chains will be larger than the variance within chains. This ratio is merely an F ratio for the chain values over a specified window of steps. In BUGS it is called the *bgr statistic*, after Brooks, Gelman, and Rubin (see Brooks & Gelman, 1998).

To illustrate these ideas, consider the σ_β parameter from oneway ANOVA, near the top of the hierarchical diagram of Figure 18.1, p. 494. This standard

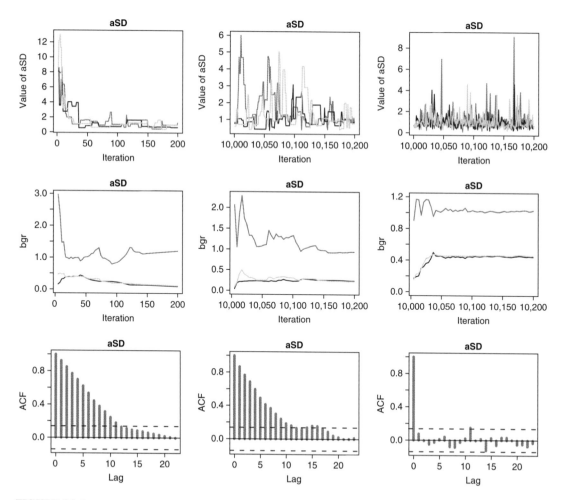

FIGURE 23.1
Effects of burn-in and thinning on MCMC chains. *Left column*: Zero burn-in steps and no thinning. *Middle column*: 10,000 burn-in steps and no thinning. *Right column*: 10,000 burn-in steps and thinning to 1 every 500 steps.

deviation expresses the estimated variation across groups. In the BRugs program of Section 18.4.1 (ANOVAonewayBRugs.R), this parameter is denoted aSD. The top-left panel of Figure 23.1 shows the initial steps of three chains when they are intentionally started far from truly representative values. You can see that it takes about 50 steps (labeled as "iterations" in the graph) for the chains to descend to representative values. The top-middle panel of Figure 23.1 shows the chains after 10,000 burn-in steps (which is overkill for this particular application). The chains appear to be converged, without systematically increasing or decreasing.

The bottom row of panels in Figure 23.1 shows the autocorrelation function, ACF. The left and middle columns show results when there is no thinning of the chains, and you can see that the ACF remains high for lags up to 15 or 20 steps. The autocorrelation is visible in the chains themselves as sustained plateaus during which the values from step to step barely change. The right column shows results when the chains were thinned, keeping only 1 step in every 500. You can see that after even one of the thinned steps, ACF($L = 1$) is nearly zero.

The middle row of Figure 23.1 shows the bgr statistic, which measures how well mixed the chains are, over a limited length of the chains. As mentioned earlier, the bgr should be around 1.0. The plots show the bgr as a curve (plotted as red by BUGS) hovering near a value of 1.0. The two other curves, lower in the graphs (and plotted as green and blue by BUGS), show the between-chain and within-chain variances, which should be nearly overlapping each other if their ratio is near 1.0. In this particular application, the chains converge quickly and well.

The graphs in Figure 23.1 were produced by the following R program, which can be called after running any BUGS model. The argument nodename is a string that is the name of any variable monitored by BUGS (plotChains.R):

```
1   plotChains = function( nodename , saveplots=F , filenameroot="DeleteMe" ) {
2       summarytable = samplesStats(nodename)
3       show( summarytable )
4       nCompon = NROW(summarytable)
5       nPlotPerRow = 5
6       nPlotRow = ceiling(nCompon/nPlotPerRow)
7       nPlotCol = ceiling(nCompon/nPlotRow)
8       windows(3.75*nPlotCol,3.5*nPlotRow)
9       par( mar=c(4,4,3,1) , mgp=c(2,0.7,0) )
10      samplesHistory( nodename , ask=F , mfrow=c(nPlotRow,nPlotCol) ,
11                      cex.lab=1.5 , cex.main=1.5 )
12      if ( saveplots ) {
13        dev.copy2eps( file=paste( filenameroot , toupper(nodename) ,
14                      "history.eps" , sep="" )) }
15      windows(3.75*nPlotCol,3.5*nPlotRow)
16      par( mar=c(4,4,3,1) , mgp=c(2,0.7,0) )
17      samplesAutoC( nodename , chain=1 , ask=F , mfrow=c(nPlotRow,nPlotCol) ,
18                    cex.lab=1.5 , cex.main=1.5 )
```

```
19        if ( saveplots ) {
20           dev.copy2eps( file=paste( filenameroot , toupper(nodename) ,
21                          "autocorr.eps" , sep="" )) }
22        windows(3.75*nPlotCol,3.5*nPlotRow)
23        par( mar=c(4,4,3,1) , mgp=c(2,0.7,0) )
24        samplesBgr( nodename , ask=F , mfrow=c(nPlotRow,nPlotCol) ,
25                       cex.lab=1.5 , cex.main=1.5 )
26        if ( saveplots ) {
27           dev.copy2eps( file=paste( filenameroot , toupper(nodename) ,
28                          "bgr.eps" , sep="" )) }
29        return( summarytable )
30     }
```

23.3 FUNCTIONS FOR APPROXIMATING HIGHEST DENSITY INTERVALS

HDIs have been used routinely throughout the book to describe distributions. This section provides details regarding how the HDIs are computed. The algorithm for computing an HDI on a grid approximation applies to any dimensionality and any shape distribution. The algorithms for computing an HDI of an MCMC sample or for a function apply only to single parameters with unimodal distributions.

23.3.1 R Code for Computing HDI of a Grid Approximation

This function was first used in the R code of Section 6.7.1 (BernGrid.R), and again in Section 8.8.1 (BernTwoGrid.R).

We can imagine the grid approximation of a distribution as a landscape of poles sticking up from each point on the parameter grid, with the height of each pole indicating the probability mass at that discrete point. We can imagine the highest density region by visualizing a rising tide: We gradually flood the landscape, monitoring the total mass of the poles that protrude above water, stopping the flood when 95% (say) of the mass remains protruding. The waterline at that moment defines the highest density region.

The program that follows finds the approximate highest density region in a somewhat analogous way. It uses one extra trick at the beginning, however. It first sorts all the poles in order of height, from tallest to shortest. The idea is to move down the sorted queue of poles until the cumulative probability has just barely exceeded 95% (or whatever). The resulting height is the "waterline" that defines all points inside the highest density. See the comments in the top of the code for details of how to use the function.

(HDIofGrid.R)

```
1    HDIofGrid = function( probMassVec , credMass=0.95 ) {
2        # Arguments:
3        #    probMassVec is a vector of probability masses at each grid point.
4        #    credMass is the desired mass of the HDI region.
5        # Return value:
6        #    A list with components:
7        #    indices is a vector of indices that are in the HDI
8        #    mass is the total mass of the included indices
9        #    height is the smallest component probability mass in the HDI
10       # Example of use: For determining HDI of a beta(30,12) distribution
11       #    approximated on a grid:
12       #    > probDensityVec = dbeta( seq(0,1,length=201) , 30 , 12 )
13       #    > probMassVec = probDensityVec / sum( probDensityVec )
14       #    > HDIinfo = HDIofGrid( probMassVec )
15       #    > show( HDIinfo )
16       sortedProbMass = sort( probMassVec , decreasing=T )
17       HDIheightIdx = min( which( cumsum( sortedProbMass ) >= credMass ) )
18       HDIheight = sortedProbMass[ HDIheightIdx ]
19       HDImass = sum( probMassVec[ probMassVec >= HDIheight ] )
20       return( list( indices = which( probMassVec >= HDIheight ) ,
21                     mass = HDImass , height = HDIheight ) )
22   }
```

23.3.2 R Code for Computing HDI of an MCMC Sample

The algorithms for computing the HDI for an MCMC sample or for a function rely on a crucial property: For a unimodal probability distribution on a single variable, the HDI of mass M is the *narrowest* possible interval of that mass. Figure 23.2 illustrates why this is true. Consider the 90% HDI as shown. We construct another interval of 90% mass by moving the limits of the HDI to the right, such that each limit is moved to a point that covers 4%, as marked in gray in Figure 23.2. The new interval does indeed cover 90%, because the 4% lost on the left is replaced by the 4% gained on the right.

Consider the blue regions in Figure 23.2. Their left edges have the same height, because the left edges are defined by the HDI. Their areas are the same, because, by definition, the areas are both 4%. Notice, however, that the left blue area is taller than the right blue area, because the left edge is at a point where the distribution is increasing in density, but the right edge is at a point where the distribution is decreasing in density. Therefore, because the areas are the same but the left is taller than the right, *the width of the left blue zone must be less than the width of the right blue zone*. Consequently, the distance between right edges of the two blue zones must be greater than the HDI width. The exact widths are marked in Figure 23.2.

This argument applies for any size of blue zone and for any mass HDI. The argument relies on unimodality, however. Given the argument and diagram in

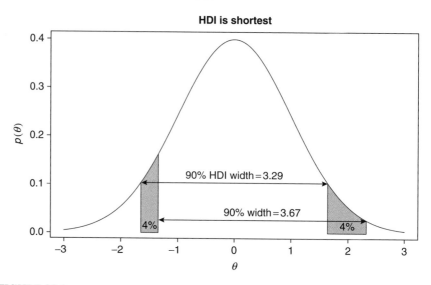

FIGURE 23.2

For a unimodal distribution, the HDI is the narrowest interval of that mass. This figure shows the 90% HDI and another interval that has 90% mass.

Figure 23.2, it is not too hard to believe the converse: For a unimodal distribution on one variable, for any mass M, the interval containing mass M that has the narrowest width is the HDI for that mass. The algorithms described as follows are based on this property of HDIs. The algorithms find the HDI by searching among candidate intervals of mass M. The shortest one found is declared to be the approximate HDI. It is an approximation, of course. (See Chen & Shao (1999) for more details, and Chen et al. (2003) for dealing with the unusual situation of multimodal distributions.)

Here is the code for finding the HDI of an MCMC sample. It is brief and self-explanatory after the preceding discussion.

(HDIofMCMC.R)

```
1   HDIofMCMC = function( sampleVec , credMass=0.95 ) {
2       # Computes highest density interval from a sample of representative values,
3       #   estimated as shortest credible interval.
4       # Arguments:
5       #   sampleVec
6       #     is a vector of representative values from a probability distribution.
7       #   credMass
8       #     is a scalar between 0 and 1, indicating the mass within the credible
9       #     interval that is to be estimated.
10      # Value:
11      #   HDIlim is a vector containing the limits of the HDI
12      sortedPts = sort( sampleVec )
13      ciIdxInc = floor( credMass * length( sortedPts ) )
```

```
14      nCIs = length( sortedPts ) - ciIdxInc
15      ciWidth = rep( 0 , nCIs )
16      for ( i in 1:nCIs ) {
17          ciWidth[ i ] = sortedPts[ i + ciIdxInc ] - sortedPts[ i ]
18      }
19      HDImin = sortedPts[ which.min( ciWidth ) ]
20      HDImax = sortedPts[ which.min( ciWidth ) + ciIdxInc ]
21      HDIlim = c( HDImin , HDImax )
22      return( HDIlim )
23  }
```

23.3.3 R Code for Computing HDI of a Function

This program finds the HDI of a probability density function that is specified mathematically in R. For example, it can find HDIs of normal densities or of beta densities or gamma densities, because those densities are specified as functions in R.

What the program accomplishes is just a search of HDIs, converging to the shortest one, but it does this by using some commands and R functions that have not been used much, or at all, elsewhere in the book. One function that the program uses is the *inverse cumulative density function* (ICDF) for whatever probability distribution is being targeted. We have seen one case of an ICDF previously, namely, the probit function, which is the inverse of the cumulative-density function for the normal distribution. In R, the ICDF of the normal is the qnorm(x) function, where the argument x is a value between zero and one. The program for finding the HDI takes, as one of its arguments, R's name for the ICDF of the function. For example, if we want to find an HDI of a normal density, we pass in ICDFname="qnorm".

The crucial function called by the program is R's optimize routine. The optimize routine searches for the minimum of a specified function over a specified domain. In the program that follows, we define a function called intervalWidth that returns the width of the interval that starts at lowTailPr and has 95% mass. This intervalWidth function is repeatedly called from the optimize routine until it converges to a minimum.

(HDIofICDF.R)
```
1   HDIofICDF = function( ICDFname , credMass=0.95 , tol=1e-8 , ... ) {
2       # Arguments:
3       #   ICDFname is R's name for the inverse cumulative density function
4       #     of the distribution.
5       #   credMass is the desired mass of the HDI region.
6       #   tol is passed to R's optimize function.
7       # Return value:
8       #   Highest density iterval (HDI) limits in a vector.
9       # Example of use: For determining HDI of a beta(30,12) distribution, type
10      #   HDIofICDF( qbeta , shape1 = 30 , shape2 = 12 )
```

```
11    #    Notice that the parameters of the ICDFname must be explicitly named;
12    #    e.g., HDIofICDF( qbeta , 30 , 12 ) does not work.
13    # Adapted and corrected from Greg Snow's TeachingDemos package.
14    incredMass =  1.0 - credMass
15    intervalWidth = function( lowTailPr , ICDFname , credMass , ... ) {
16        ICDFname( credMass + lowTailPr , ... ) - ICDFname( lowTailPr , ... )
17    }
18    optInfo = optimize( intervalWidth , c( 0 , incredMass ) , ICDFname=ICDFname ,
19                        credMass=credMass , tol=tol , ... )
20    HDIlowTailPr = optInfo$minimum
21    return( c( ICDFname( HDIlowTailPr , ... ) ,
22                ICDFname( credMass + HDIlowTailPr , ... ) ) )
23    }
```

23.4 REPARAMETERIZATION OF PROBABILITY DISTRIBUTIONS

There are situations in which it is natural to express a distribution on one scale but to parameterize it mathematically on a different scale. For example, we may think intuitively of the standard deviation of a normal distribution, but we have to parameterize it in terms of the precision (i.e., reciprocal of the variance). As another example, we may think intuitively of an underlying bias on an infinite scale, but we have to parameterize it on a zero-to-one interval. The question is, if we express a probability distribution on one scale, what is the equivalent distribution on a transformed scale?

The answer is not difficult to figure out, especially for single parameters. Let the "destination" parameter be denoted θ, and suppose that $\theta = f(\phi)$ for the "source" parameter ϕ, with a monotonic and differentiable function f. Let the probability distribution on ϕ be denoted $p(\phi)$. Then the corresponding probability distribution on θ is $p(\theta) = p\big(f^{-1}(\theta)\big)/\big|f'\big(f^{-1}(\theta)\big)\big|$, where $f'(\phi)$ is the derivative of f with respect to ϕ.

Here's why. Consider a small (actually infinitesimal) interval under the distribution $p(\phi)$, at a particular value ϕ. Call the width of the interval $d\phi$. The probability mass in that interval is the product of the density and the width, $p(\phi)d\phi$. We want to construct a density on θ, which we denote $p(\theta) = p\big(f(\phi)\big)$, that has the same probability mass in the corresponding interval at $\theta = f(\phi)$. The width of the corresponding interval on θ is, by definition of derivative, $d\theta = d\phi|f'(\phi)|$. So the probability mass in that interval is $p(\theta)d\theta = p\big(f(\phi)\big)d\phi|f'(\phi)|$. Therefore, to equate the probability masses in the corresponding intervals, we require that $p\big(f(\phi)\big)d\phi|f'(\phi)| = p(\phi)d\phi$, which, when rearranged, yields $p\big(f(\phi)\big) = p(\phi)/|f'(\phi)|$.

23.4.1 Examples

We can apply the general formula to the case in which $\theta = f(\phi) = \text{sig}(\phi) = 1/[1 + \exp(-\phi)]$, and the distribution on ϕ is $p(\phi) = (f(\phi))^a (1 - f(\phi))^b / B(a, b) = \theta^a (1 - \theta)^b / B(a, b)$. Notice that the derivative of f is $f'(\phi) = \exp(-\phi)/[1 + \exp(-\phi)]^2 = \text{sig}(\phi)(1 - \text{sig}(\phi)) = \theta(1 - \theta)$. Therefore, the equivalent probability density at $\theta = f(\phi)$ is $p(\theta) = p(\phi)/f'(\phi) = \theta^a (1 - \theta)^b / [\theta(1 - \theta)B(\theta; a, b)] = \theta^{(a-1)}(1 - \theta)^{(b-1)}/B(a, b) = \text{beta}(\theta; a, b)$. The upper row of Figure 23.3 shows this situation when $a = b = 1$. An intuitive way to think of this situation is that the probability on ϕ is dense near $\phi = 0$, but that is exactly where the sigmoidal transformation stretches the distribution. On the other hand, the probability on ϕ is sparse at large positive or large negative values, but that is exactly where the sigmoidal transformation compresses the distribution.

As another example, consider a case in which $\theta = f(\psi) = 1 - \exp(-\psi)$, with the probability density $p(\psi) = \exp(-\psi)$. Notice that the derivative of the

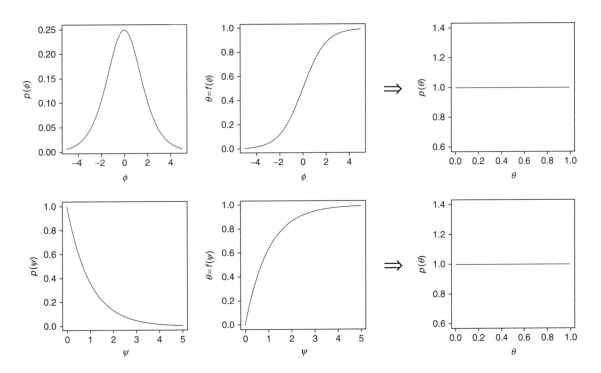

FIGURE 23.3
Top row shows a reparameterization that maps $[-\infty, +\infty]$ to $[0, 1]$. Bottom row shows a reparameterization that maps $[0, +\infty]$ to $[0, 1]$.

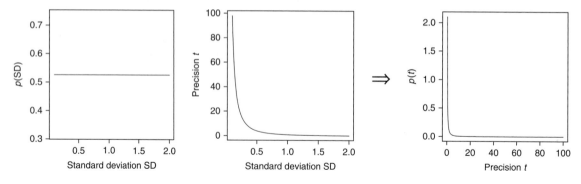

FIGURE 23.4
A uniform distribution on standard deviation, mapped to the corresponding distribution on precision.

transformation is $f'(\psi) = \exp(-\psi)$, and therefore the equivalent density at $\theta = f(\phi)$ is $p\left(f(\psi)\right) = p(\psi)/f'(\psi) = 1$. In other words, the equivalent density on θ is the uniform density, as shown in the lower panel of Figure 23.3.

As a final example, Figure 23.4 shows a uniform distribution on standard deviation transformed to the corresponding distribution on precision. By definition, precision is the reciprocal of squared standard deviation.

23.4.2 Reparameterization of Two Parameters

When more than one parameter is being transformed, the calculus becomes a little more involved. Suppose we have a probability density on a two-parameter space, $p(\alpha_1, \alpha_2)$. Let $\beta_1 = f_1(\alpha_1, \alpha_2)$ and $\beta_2 = f_2(\alpha_1, \alpha_2)$. Our goal is to find the probability density $p(\beta_1, \beta_2)$ that corresponds to $p(\alpha_1, \alpha_2)$. We do this by considering infinitesimal corresponding regions. Consider a point α_1, α_2. The probability *mass* of a small region near that point is the density at that point times the area of the small region: $p(\alpha_1, \alpha_2)d\alpha_1\, d\alpha_2$. The corresponding region in the transformed parameters should have the same mass. That mass is the density at the transformed point times the area of the region that is mapped to from the originating region. In vector calculus textbooks, you can find discussions demonstrating that the area of the mapped-to region is $|\det(J)|\, d\alpha_1\, d\alpha_2$ where J is the Jacobian matrix: $J_{rc} = df_r(\alpha_1, \alpha_2)/d\alpha_c$ and $\det(J)$ is the determinant of the Jacobian matrix. Setting the two masses equal and rearranging yields $p(\beta_1, \beta_2) = p(\alpha_1, \alpha_2)/|\det(J)|$.

As we have had no occasions to apply this transformation, no examples will be provided here. But the method has been mentioned for those intrepid souls who may wish to venture into the wilderness of multivariate probability distributions with nothing more than pen and paper.

References

Adcock, C. J. (1997). Sample size determination: a review. *The Statistician, 46,* 261–283.

Agresti, A., & Hitchcock, D. B. (2005). Bayesian inference for categorical data analysis. *Statistical Methods & Applications, 14*(3), 297–330.

Albert, J. H., & Rossman, A. J. (2001). *Workshop statistics: Discovery with data, a Bayesian approach.* Emeryville, CA: Key College Publishing.

Berger, J. O. (1985). *Statistical decision theory and Bayesian analysis, 2nd edition.* New York: Springer.

Berger, J. O., & Berry, D. A. (1988). Statistical analysis and the illusion of objectivity. *American Scientist, 76*(2), 159–165.

Berger, R. L., Boos, D. D., & Guess, F. M. (1988). Tests and confidence sets for comparing two mean residual life functions. *Biometrics, 44*(1), 103–115.

Berry, D. A. (1996). *Statistics: A Bayesian perspective.* Belmont, CA: Duxbury Press/Wadsworth.

Berry, D. A., & Hochberg, Y. (1999). Bayesian perspectives on multiple comparisons. *Journal of Statistical Planning and Inference, 82*(1–2), 215–227.

Bishop, C. M. (2006). *Pattern recognition and machine learning.* New York: Springer.

Bliss, C. I. (1934). The method of probits. *Science, 79*(2037), 38–39.

Bolstad, W. M. (2007). *Introduction to Bayesian statistics* (2nd ed.). Hoboken, NJ: Wiley.

Brambor, T., Clark, W. R., & Golder, M. (2006). Understanding interaction models: Improving empirical analyses. *Political Analysis, 14,* 63–82.

Braumoeller, B. F. (2004). Hypothesis testing and multiplicative interaction terms. *International Organization, 58*(04), 807–820.

Brehmer, B. (1974). Hypotheses about relations between scaled variables in the learning of probabilistic inference tasks. *Organizational Behavior and Human Performance, 11,* 1–27.

Brooks, S. P., & Gelman, A. (1998). Alternative methods for monitoring convergence of iterative simulations. *Journal of Computational and Graphical Statistics, 7,* 434–455.

Carlin, B. P., & Chib, S. (1995). Bayesian model choice via Markov chain Monte Carlo methods. *Journal of the Royal Statistical Society, B, 57*(3), 473–484.

Carlin, B. P., & Louis, T. A. (2009). *Bayesian methods for data analysis* (3rd ed.). Boca Raton, FL: CRC Press.

Casella, G., & Moreno, E. (2006). Objective Bayesian variable selection. *Journal of the American Statistical Association, 101*(473), 157–167.

Chaloner, K., & Verdinelli, I. (1995). Bayesian experimental design: A review. *Statistical Science, 10*(3), 273–304.

Chen, M.-H., He, X., Shao, Q.-M., & Xu, H. (2003). A Monte Carlo gap test in computing HPD regions. In H. Zhang and J. Huang (Eds.), *Development of Modern Statistics and Related Topics: In Celebration of Professor Yaoting Zhang's 70th Birthday*, 38–52. Singapore: World Scientific.

Chen, M. H., & Shao, Q. M. (1999). Monte Carlo estimation of Bayesian credible and HPD intervals. *Journal of Computational and Graphical Statistics, 8,* 69–92.

Clyde, M., & George, E. I. (2004). Model uncertainty. *Statistical Science,* 81–94.

Congdon, P. (2005). *Bayesian models for categorical data.* West Sussex, England: Wiley.

Damgaard, L. H. (2007). Technical note: How to use WinBUGS to draw inferences in animal models. *Journal of Animal Science, 85,* 1363–1368.

Dawes, J. (2008). Do data characteristics change according to the number of scale points used? an experiment using 5-point, 7-point and 10-point scales. *International Journal of Market Research, 50*(1), 61–77.

de Saint-Exupery, A. (1943). *The little prince.* San Diego, CA: Harcourt.

De Santis, F. (2004). Statistical evidence and sample size determination for Bayesian hypothesis testing. *Journal of Statistical Planning and Inference, 124,* 121–144.

De Santis, F. (2007). Using historical data for Bayesian sample size determination. *Journal of the Royal Statistical Society: Series A, 170,* 95–113.

DeGroot, M. H. (2004). *Optimal statistical decisions.* New York: Wiley Interscience.

Edwards, W., Lindman, H., & Savage, L. J. (1963). Bayesian statistical inference for psychological research. *Psychological Review, 70,* 193–242.

Feldman, H. A. (1988). Families of lines: random effects in linear regression analysis. *Journal of Applied Physiology, 64*(4), 1721–1732.

Freedman, L. S., Lowe, D., & Macaskill, P. (1984). Stopping rules for clinical trials incorporating clinical opinion. *Biometrics, 40,* 575–586.

Gallistel, C. R. (2009). The importance of proving the null. *Psychological Review, 116*(2), 439–453.

Gelfand, A. E., & Dey, D. K. (1994). Bayesian model choice: asymptotics and exact calculations. *Journal of the Royal Statistical Society, Series B, 56,* 501–514.

Gelman, A. (2005). Analysis of variance—why it is more important than ever. *The Annals of Statistics, 33*(1), 1–53.

Gelman, A. (2006). Prior distributions for variance parameters in hierarchical models. *Bayesian Analysis, 1*(3), 515–533.

Gelman, A., Carlin, J. B., Stern, H. S., & Rubin, D. B. (2004). *Bayesian data analysis* (2nd ed.). Boca Raton, FL: CRC Press.

Gelman, A., & Hill, J. (2007). *Data analysis using regression and multilievel/hierarchical models.* New York: Cambridge University Press.

Gelman, A., Hill, J., & Yajima, M. (2009). *Why we (usually) don't have to worry about multiple comparisons.* Available from www.stat.columbia.edu/~gelman/research/unpublished/multiple2.pdf.

Geman, S., & Geman, D. (1984). Stochastic relaxation, Gibbs distributions, and the Bayesian restoration of images. *IEEE Transactions on Pattern Analysis and Machine Intelligence, 6,* 721–741.

George, D. N., & Pearce, J. M. (1999). Acquired distinctiveness is controlled by stimulus relevance not correlation with reward. *Journal of Experimental Psychology: Animal Behavior Processes, 25*(3), 363–373.

George, E. I. (2000). The variable selection problem. *Journal of the American Statistical Association, 95*(452).

Gigerenzer, G., & Hoffrage, U. (1995). How to improve Bayesian reasoning without instruction: Frequency formats. *Psychological Review, 102,* 684–704.

Gigerenzer, G., Krauss, S., & Vitouch, O. (2004). The null ritual: What you always wanted to know about significance testing but were afraid to ask. In D. Kaplan (Ed.), *The Sage handbook of quantitative methodology for the social sciences* (pp. 391–408). Thousand Oaks, CA: Sage.

Gilks, W. R., Thomas, A., & Spiegelhalter, D. J. (1994). A language and program for complex Bayesian modelling. *The Statistician, 43*(1), 169–177.

Gill, J. (2002). *Bayesian methods for the social and behavioral sciences.* Boca Raton, FL: CRC Press.

Gilovich, T., Vallone, R., & Tversky, A. (1985). The hot hand in basketball: On the misperception of random sequences. *Cognitive Psychology, 17*(3), 295–314.

Gopalan, R., & Berry, D. (1998). Bayesian multiple comparisons using Dirichlet process priors. *Journal of the American Statistical Association,* 1130–1139.

Gosset, W. S. (1908). The probable error of a mean. *Biometrika, 6,* 1–25.

Greenland, S. (2008). Invited commentary: Variable selection versus shrinkage in the control of multiple confounders. *American Journal of Epidemiology, 167*(5), 523–529.

Guber, D. L. (1999). Getting what you pay for: The debate over equity in public school expenditures. *Journal of Statistics Education, 7*(2). Available from www.amstat.org/publications/JSE/secure/v7n2/datasets.guber.cfm.

Hahn, U., Chater, N., & Richardson, L. B. (2003). Similarity as transformation. *Cognition, 87*(1), 1–32.

Han, C., & Carlin, B. P. (2001). Markov chain Monte Carlo methods for computing Bayes factors: A comparative review. *Journal of the American Statistical Association, 96*(455), 1122–1132.

Hand, D. J., Daly, F., Lunn, A. D., McConway, K. J., & Ostrowski, E. (1994). *A handbook of small data sets.* London: Chaprman & Hall.

Hobbs, B. P., & Carlin, B. P. (2008, January). Practical Bayesian design and analysis for drug and device clinical trials. *Journal of Biopharmaceutical Statistics, 18*(1), 54–80.

Hoffman, P. J., Earle, T. C., & Slovic, P. (1981). Multidimensional functional learning (MFL) and some new conceptions of feedback. *Organizational Behavior and Human Performance, 27*(1), 75–102.

Holcomb, J., & Spalsbury, A. (2005). Teaching students to use summary statistics and graphics to clean and analyze data. *Journal of Statistics Education, 13*(3). Available from www.amstat.org/publications/jse/v13n3/datasets.holcomb.html.

Hyndman, R. J. (1996). Computing and graphing highest density regions. *The American statistician, 50*(2), 120–126.

Jackman, S. (2009). *Bayesian analysis for the social sciences.* New York: Wiley.

Joseph, L., Wolfson, D. B., & du Berger, R. (1995a). Sample size calculations for binomial proportions via highest posterior density intervals. *The Statistician, 44,* 143–154.

Joseph, L., Wolfson, D. B., & du Berger, R. (1995b). Some comments on Bayesian sample size determination. *The Statistician, 44,* 167–171.

Kalish, M. L., Griffiths, T. L., & Lewandowsky, S. (2007). Iterated learning: Intergenerational knowledge transmission reveals inductive biases. *Psychonomic Bulletin & Review, 14*(2), 288.

Kass, R. E., & Raftery, A. E. (1995). Bayes factors. *Journal of the American Statistical Association, 90,* 773–795.

Keith, T. (2005). *Multiple regression and beyond.* Columbus, OH: Allyn & Bacon.

Kolmogorov, A. N. (1956). *Foundations of the theory of probability.* New York: Chelsea.

Krauss, S., Martignon, L., & Hoffrage, U. (1999). Simplifying Bayesian inference: The general case. In L. Magnani, N. J. Nersessian, & P. Thagard (Eds.), *Model-based reasoning in scientific discovery* (pp. 165–180). New York: Springer.

Kruschke, J. K. (1993). Human category learning: Implications for backpropagation models. *Connection Science, 5,* 3–36.

Kruschke, J. K. (1996). Dimensional relevance shifts in category learning. *Connection Science, 8*, 201–223.

Kruschke, J. K. (2008). Bayesian approaches to associative learning: From passive to active learning. *Learning & Behavior, 36*(3), 210–226.

Kruschke, J. K. (2009). Highlighting: A canonical experiment. In B. Ross (Ed.), *The psychology of learning and motivation* (Vol. 51, pp. 153–185). Academic Press.

Kruschke, J. K. (2010a). Bayesian data analysis. *Wiley Interdisciplinary Reviews: Cognitive Science, 1*, 658–676.

Kruschke, J. K. (2010b). What to believe: Bayesian methods for data analysis. *Trends in Cognitive Sciences, 14*, 293–300.

Learner, E. E. (1978). *Specification searches.* New York: Wiley.

Lee, M. D., & Webb, M. R. (2005). Modeling individual differences in cognition. *Psychonomic Bulletin & Review, 12*(4), 605–621.

Liang, F., Paulo, R., Molina, G., Clyde, M. A., & Berger, J. O. (2008). Mixtures of *g* priors for Bayesian variable selection. *Journal of the American Statistical Association, 103*, 410–423.

Likert, R. (1932). A technique for the measurement of attitudes. *Archives of Psychology, 140*, 1–55.

Lindley, D. V. (1997). The choice of sample size. *The Statistician, 46*, 129–138.

Lindley, D. V., & Phillips, L. D. (1976). Inference for a Bernoulli processs (a Bayesian view). *The American Statistician, 30*(3), 112–119.

Lindquist, M. A., & Gelman, A. (2009). Correlations and multiple comparisons in functional imaging – a statistical perspective. *Perspectives in Psychological Science, 4*(3), 310–313.

Liu, C. C., & Aitkin, M. (2008). Bayes factors: Prior sensitivity and model generalizability. *Journal of Mathematical Psychology, 52*, 362–375.

Lynch, S. M. (2007). *Introduction to applied Bayesian statistics and estimation for social scientists.* New York: Springer.

MacKay, D. J. C. (2003). *Information theory, inference & learning algorithms.* Cambridge, UK: Cambridge University Press.

Marin, J.-M., & Robert, C. P. (2007). *Bayesian core: A practical approach to computational Bayesian statistics.* New York: Springer.

Maxwell, S. E., & Delaney, H. D. (2004). *Designing experiments and analyzing data: a model comparison perspective* (2nd ed.). Mahwah, NJ: Erlbaum.

McCullagh, P., & Nelder, J. (1989). *Generalized linear models, 2nd ed.* Boca Raton, FL: Chapman and Hall/CRC.

McDonald, J. H. (2009). *Handbook of biological statistics (2nd ed.).* Baltimore, MD: Sparky House Publishing.

McDonald, J. H., Seed, R., & Koehn, R. K. (1991). Allozymes and morphometric characters of three species of Mytilus in the Northern and Southern Hemispheres. *Marine Biology, 111*(3), 323–333.

McIntyre, L. (1994). Using cigarette data for an introduction to multiple regression. *Journal of Statistics Education, 2*(1). Available from `www.amstat.org/publications/jse/v2n1/datasets.mcintyre.html`.

Meng, C. Y. K., & Dempster, A. P. (1987). A Bayesian approach to the multiplicity problem for significance testing with binomial data. *Biometrics, 43*(2), 301–311.

Metropolis, N., Rosenbluth, A. W., Rosenbluth, M. N., Teller, A. H., & Teller, E. (1953). Equations of state calulations by fast computing machines. *Journal of Chemical Physics, 21*, 1087–1091.

Meyer, R., & Yu, J. (2000). BUGS for a Bayesian analysis of stochastic volatility models. *Econometrics Journal, 3*(2), 198–215.

Miller, J. (2009). What is the probability of replicating a statistically significant effect? *Psychonomic Bulletin & Review, 16*(4), 617–640.

Moore, T. L. (2006). Paradoxes in film ratings. *Journal of Statistics Education, 14*(1). Available from www.amstat.org/publications/jse/v14n1/datasets.moore.html.

Mueller, P., Parmigiani, G., & Rice, K. (2007). FDR and Bayesian multiple comparisons rules. In J. M. Bernardo et al. (Eds.), *Bayesian statistics 8*. Oxford, UK: Oxford University Press.

Navarro, D. J., Griffiths, T. L., Steyvers, M., & Lee, M. D. (2006). Modeling individual differences using Dirichlet processes. *Journal of Mathematical Psychology, 50*, 101–122.

Nelder, J. A., & Wedderburn, R. W. M. (1972). Generalized linear models. *Journal of the Royal Statistical Society. Series A (General), 135*(3), 370–384.

Nietzsche, F. (1967). *The will to power*. New York: Random House. (Translated by W. Kaufmann and R. J. Hollingdale.)

Ntzoufras, I. (2009). *Bayesian modeling using WinBUGS*. Hoboken, NJ: Wiley.

Oswald, C. J. P., Yee, B. K., Rawlins, J. N. P., Bannerman, D. B., Good, M., & Honey, R. C. (2001). Involvement of the entorhinal cortex in a process of attentional modulation: Evidence from a novel variant of an IDS/EDS procedure. *Behavioral neuroscience, 115*(4), 841–849.

Pham-Gia, T., & Turkkan, N. (1992). Sample size determination in Bayesian analysis. *The Statistician, 41*, 389–392.

Poldrack, R. A. (2006). Can cognitive processes be inferred from neuroimaging data? *Trends in Cognitive Sciences, 10*(2), 59–63.

Proschan, F. (1963). Theoretical explanation of observed decreasing failure rate. *Technometrics*, 375–383.

Qian, S. S., & Shen, Z. (2007). Ecological applications of multilevel analysis of variance. *Ecology, 88*(10), 2489–2495.

Robert, C. P., & Casella, G. (2004). *Monte Carlo statistical methods* (2nd ed.). New York: Springer.

Rosa, L., Rosa, E., Sarner, L., & Barrett, S. (1998). A close look at therapeutic touch. *Journal of the American Medical Association, 279*(13), 1005–1010.

Rouder, J. N., & Lu, J. (2005). An introduction to Bayesian hierarchical models with an application in the theory of signal detection. *Psychonomic Bulletin & Review, 12*(4), 573–604.

Rouder, J. N., Lu, J., Speckman, P., Sun, D., & Jiang, Y. (2005). A hierarchical model for estimating response time distributions. *Psychonomic Bulletin & Review, 12*(2), 195–223.

Rouder, J. N., Speckman, P. L., Sun, D., Morey, R. D., & Iverson, G. (2009). Bayesian t-tests for accepting and rejecting the null hypothesis. *Psychonomic Bulletin & Review, 16*, 225–237.

Roy, A., Ghosal, S., & Rosenberger, W. F. (2009). Convergence properties of sequential Bayesian D-optimal designs. *Journal of Statistical Planning and Inference, 139*, 425–440.

Sadiku, M. N. O., & Tofighi, M. R. (1999). A tutorial on simulation of queueing models. *International Journal of Electrical Engineering Education, 36*, 102–120.

Scott, J. G., & Berger, J. O. (2006). An exploration of aspects of Bayesian multiple testing. *Journal of statistical planning and inference, 136*(7), 2144–2162.

Snee, R. D. (1974). Graphical display of two-way contingency tables. *The American Statistician, 28*(1), 9–12.

Solari, F., Liseo, B., & Sun, D. (2008). Some remarks on Bayesian inference for one-way ANOVA models. *Annals of the Institute of Statistical Mathematics, 60*, 483–498.

Spiegelhalter, D. J., Freedman, L. S., & Parmar, M. K. B. (1994). Bayesian approaches to randomized trials. *Journal of the Royal Statistical Society. Series A, 157*, 357–416.

Stevens, S. S. (1946). On the theory of scales of measurement. *Science, 103*(2684), 677–680.

Thomas, A. (2004). *BRugs user manual (the R interface to BUGS)*. Available from `http://mathstat.helsinki.fi/openbugs/data/Docu/BRugs%20Manual.html`.

Thomas, A., O'Hara, B., Ligges, U., & Sturtz, S. (2006, March). Making BUGS open. *R News, 6*(1), 12–17.

Tsionas, E. G. (2002). Bayesian inference in the noncentral Student-t model. *Journal of Computational and Graphical Statistics, 11*(1), 208–221.

Wagenmakers, E. J. (2007). A practical solution to the pervasive problems of *p* values. *Psychonomic Bulletin & Review, 14*(5), 779–804.

Walker, L. J., Gustafson, P., & Frimer, J. A. (2007). The application of Bayesian analysis to issues in developmental research. *International Journal of Behavioral Development, 31*(4), 366.

Wang, F., & Gelfand, A. E. (2002). A simulation-based approach to Bayesian sample size determination for performance under a given model and for separating models. *Statistical Science, 17*, 193–208.

Weiss, R. (1997). Bayesian sample size calculations for hypothesis testing. *The Statistician, 46*, 185–191.

Werner, M., Stabenau, J. R., & Pollin, W. (1970). Thematic apperception test method for the differentiation of families of schizophrenics, delinquents, and "normals." *Journal of Abnormal Psychology, 75*(2), 139–145.

Western, B., & Jackman, S. (1994). Bayesian inference for comparative research. *The American Political Science Review, 88*(2), 412–423.

Winer, B. J., Brown, D. R., & Michels, K. M. (1991). *Statistical principles in experimental design, 3rd ed.* New York: McGraw-Hill.

Index

Page numbers followed by "f" indicates figures and "t" indicates tables.

A